ENVIRONMENTAL SCIENCE, ENGINEERING AND TECHNOLOGY

SYNGAS

PRODUCTION, APPLICATIONS AND ENVIRONMENTAL IMPACT

ENVIRONMENTAL SCIENCE, ENGINEERING AND TECHNOLOGY

Additional books in this series can be found on Nova's website under the Series tab.

Additional e-books in this series can be found on Nova's website under the e-book tab.

ENVIRONMENTAL SCIENCE, ENGINEERING AND TECHNOLOGY

SYNGAS

PRODUCTION, APPLICATIONS AND ENVIRONMENTAL IMPACT

ANTONIUS INDARTO
AND
JELLIARKO PALGUANDI
EDITORS

nova publishers
New York

For permission to use material from this book please contact us:
Telephone 631-231-7269; Fax 631-231-8175
Web Site: http://www.novapublishers.com

Library of Congress Cataloging-in-Publication Data

Syngas : production, applications, and environmental impact / editors, Antonius Indarto, Jelliarko Palguandi.
p. cm.
Includes bibliographical references and index.
ISBN 978-1-62100-870-5 (hardcover)
1. Synthesis gas. 2. Biomass energy. 3. Methane. I. Indarto, Antonius. II. Palguandi, Jelliarko.
TP360.S946 2011
665.7'7--dc23
2011039656

Published by Nova Science Publishers, Inc. † New York

CONTENTS

Foreword **vii**

Introduction: Syngas Era and Energy Issue **ix**

Chapter 1 Synthesis Gas Production by Reforming Methane in a Chemical Compression Reactor **1**
V. Shmelev

Chapter 2 New Syngas Production Method Based on Noncatalytic Methane Reaction with Metal Oxides **29**
H. Ale Ebrahim

Chapter 3 Coke Oven Gas to Syngas Technology **67**
Yongfa Zhang and Guojie Zhang

Chapter 4 Thermodynamic Analysis of Synthesis Gas and Higher Hydrocarbons Production from Methane **99**
I. Istadi

Chapter 5 Syngas Production by CO_2 Reforming of CH_4 under Microwave Heating – Challenges and Opportunities **121**
B. Fidalgo and J. A. Menéndez

Chapter 6 Exploitation of BioSyngas: Hydrogen and Synthetic Diesel Production **151**
A. R. de la Osa, A. De Lucas, A. Romero, J. L. Valverde and P. Sánchez

Chapter 7 Plasma Gasification as a Promising Way of Wood Energy Use **179**
Ph. G. Rutberg, A. N. Bratsev, V. A. Kuznetsov, V. E. Popov and A. A. Ufimtsev

Chapter 8 Role of Electrolysis in Regenerative Syngas and Synfuel Production **209**
Qingxi Fu

Chapter 9 Decarbonized Energy via Syngas Routes **241**
Wojciech M. Budzianowski

Chapter 10 Syngas Production and Applications for Greenhouse Gas Mitigation **269**
NorAishah Saidina Amin and Muhammad Tahir

Chapter 11 Lean Premixed Syngas Combustion for Gas Turbines Applications **297**
Salvatore Daniele

Conclusion **349**

Glossary **351**

Index **355**

FOREWORD

It is a pleasure to write the preface for this valuable book titled Syngas: Production, Applications and Environmental impact.

Syngas is a classical and at the same time a topic of lively debate in chemical engineering. Syngas production is a robust and flexible technology with respect to reactants and products. The raw material can be any hydrocarbon: coal, fossil oil, natural gas, biomass. This large variety in raw materials forms a challenge process in development and operation but it also explains why syngas technology is so wide spread: syngas is produced and used all over the world and it is not expected that this will change.

The products generated depend on the conditions and the catalyst used: hydrogen, carbon monoxide, ammonia, methanol, parafins, olefins and, of course, electric power. Syngas production is one of the basis technologies for the petrochemical industry and the production of transport fuels. The impact to society is enormous. Not surprisingly, environmental aspects suggest themselves. Rightly, the editors decided to extend the focus of the topic to the environmental impact.

The book starts with a description of an original reactor concept for methane reforming: essentially an internal combustion engine. The main raw materials for syngas production are coal and methane. However, syngas technology is also an avenue for utilization of "difficult" hydrocarbon resources such as gas as by-product in oil production platforms. In the book several good ideas are presented for instance, the utilization of coke oven gas and specific examples of the large variety of biomass resources. In a sense to the same category belongs using syngas as an efficient way to store sustainable energy (solar, wind) in high density products such as transport fuels. The basis for this application is using the harvested electricity to produce syngas. This area is treated well in the book. It is shown that syngas routes in power generation allow low-carbon power cycles. Such cycles can contribute directly to a reduction of the Green House Effect.

Syngas is the common name for gas mixtures of several compositions. For an observer the field might seem chaotic but in most cases the composition is easily described by thermodynamics knowing feed composition and operating conditions. In the book a lucid thermodynamic analysis is given. It is for instance shown that thermodynamics is a tool for a quick process selection. In syngas related R and D many chemical routes might be considered. It is in such cases always advisable to perform a thermodynamic study.

The reforming of CH_4 with CO_2 is an appealing reaction and often seen as an ideal reaction for CO_2 utilization, usually referred to as "Dry reforming". In several chapters various aspects of this reaction are treated. In practice catalyst stability is an issue because of potentially excessive carbon deposition. Thermodynamics allows pinpointing the window of operation where coke deposition can occur. Cogeneration of syngas and hydrocarbons has often been proposed as an efficient route. However, from thermodynamic analysis the conclusion is drawn that syngas production in dry reforming is favorable but the cogeneration of C2 hydrocarbons is thermodynamically less favorable. This conclusion is a good reminder that in process development thermodynamics should not be forgotten.

Alternative technologies are discussed for dry reforming, viz, plasma and photo technologies. Very promising conversions and efficiencies are reported. It might be the right time that chemical engineers are brave enough to embark on new technologies taking into consideration plasmas and photons!

The book covers concepts in production and in applications. An important application is Integrated Gasification Combined Cycle (IGCC) technology, extensively treated in the last chapter. In combustor design fuel flexibility is highly desired. In this useful chapter the basic phenomena are described important in changing the feedstock from natural gas into syngas. An eye opener for the average chemical engineer!

Since the beginning of the previous century (one might even defend since the end of the 18th century) syngas is an important topic in chemical engineering and catalysis. Thus, is it a mature topic that only needs to be part of teaching courses and encyclopedia? I am convinced that the opposite is true. Syngas is crucial in realizing a sustainable technology and new processes are badly needed.

This book is issued at a right time and it will be helpful and inspiring for the chemists and chemical engineers (and combustion engineers) involved in production and processing syngas. The editors Dr Indarto and Dr Palgunadi did do an excellent job.

Jacob A. Moulijn
Delft University of Technology, Delft, Netherlands

INTRODUCTION: SYNGAS ERA AND ENERGY ISSUE

The discovery of a combustible gas mixture containing CO and H_2 (water gas) generated from the reaction of water steam with carbon at temperature over 500°C by an Italian physicist, Felice Fontana in 1780 became the industrial milestone of synthesis gas (syngas) [1]. The dawn era of fuels and chemicals synthesis from syngas, a mixture containing varying amounts of mostly CO and H_2 with appreciable amount of CO_2, N_2, and CH_4 was initiated in the beginning of the 20th century. The first synthesis of synthetic methane was carried out in 1902 by Sabatier and Sanderens by passing CO and H_2 over Ni, Fe, and Co catalysts. At about the same time, the first commercial hydrogen from syngas produced from steam methane reforming was launched to market. Following this success, in 1910, Haber and Bosch discovered the synthesis of ammonia from H_2 and N_2 and the first industrial ammonia synthesis plant was commissioned in 1913. In 1921, an effort of methanol synthesis from H_2 and CO catalyzed by ZnO was reported for the first time. The production of liquid hydrocarbons and oxygenates from syngas conversion over iron catalysts was discovered in 1923 by Fischer and Tropsch [1,2]. In the following years, variations on the Fischer-Tropsch synthesis pathway have been utilized for the selective production of methanol, mixed alcohols, oxygenates, and isosynthesis products.

The production and purification technology of syngas has received a huge attention from researchers and industries for the last few years because its development represents one of major efforts toward more efficient, sustainable, and environmentally benign use of the fossil hydrocarbon resources. Decline of global fossil fuels – petroleum, natural gas, and coal supplies and the need for clean and alternative energy have become major motives of research worldwide for sustainable energy development. The energy supply of the world today still heavily relies on combustion of fossil fuels for stationary systems, domestic use, and transportation vehicles. Alternative fuels, are needed to fill the supply gap that will continue to grow. Today, there are at least three major challenges in fuel issue [3]:

a) To supply alternative fuels to meet the increasing demands for liquid and gaseous fuels.
b) To increase the efficiency of energy utilization for fuels and electricity production.

c) To eliminate the pollutants and control the "greenhouse" gas emissions via the reduction of CO_2, CH_4, and N_2O in power plants, manufactures, and in transportation systems.

Therefore, the new processes and new energy systems should be developed to be much more efficient and also environmentally benign. Technology development on syngas production may have great potential in addressing those matters in the near future.

Traditionally, syngas has been utilized as building block material for the production of bulk chemical intermediates, such as ammonia and methanol. Nowadays, besides considered as an attractive feedstock for chemical synthesis, a great interest on syngas research and development is driven by the growing need on using clean synthetic fuels. The importance of syngas production also gets its momentum by the high growth in demand of hydrogen supply. Conversion of syngas to liquid fuels is increasingly lucrative because syngas can be produced from any carbon-based feedstock such as, natural gas, coal, petroleum coke, biomass, and municipal waste.

Synthetic fuel can be defined as the gasoline, kerosene and diesel fuels obtained from thermochemical processes of carbonaceous feedstock. For this purpose, the appropriate feedstock can be converted to syngas via reforming or gasification process.

Hydrocarbons including synthetic petroleum for fuels can be manufactured from syngas using either, GTL (Gas-To-Liquids), CTL (Coal-To-Liquids), or BTL (Biomass-To-Liquids) technology where the Fischer–Tropsch synthesis route has been the key process for these technologies.

Syngas Generation Technologies

Emerging technologies in the syngas production open greater opportunity for the syngas utilizations and change its economic value globally. There are several developed technologies available for syngas production including reforming and gasification. Air Blown Catalytic Reforming by Syntroleum, Air Products's Heat Exchange Reforming and Johnson Matthey's Gas Heated Reforming (now Davy's compact reforming) are some well known examples of the reforming technology.

In the recent decades reformation of natural gas is one of the most important methods for syngas manufacturing. Natural gas consisting primarily of methane (CH_4) is abundantly available and its composition is variable depending on the geological structure of the source. Conventional production methods via steam reforming of natural gas or liquid hydrocarbons have been the dominant and industrialized techniques for the hydrogen and syngas generation. Other techniques, such as catalytic CO_2 (dry) reforming and combined processes have been the subjects of scrutiny and limitedly implemented.

Each reforming process inherently generates syngas mixtures having appreciably different H_2/CO compositions. Steam reforming of natural gas for example, in which methane and steam are catalytically and endothermically converted to H_2 and CO [4] has been extensively employed to supply over 95% of hydrogen demand because it can generate syngas with a much higher H_2/CO ratio [5]. This characteristic accounts for its dominant role in the hydrogen production used for ammonia and fertilizer synthesis.

Meanwhile, partial oxidation of methane with oxygen under an exothermic reaction in the presence/absence of a catalyst and in addition, autothermal reforming generating lower H_2/CO ratio is of most interest for gasoline production [4,5].

In principle, any carbonaceous materials are potential feedstocks for the syngas generation. Materials include natural gas, naphtha, residual oil, petroleum coke, and coal have been used by large-scale syngas producers. The syngas production method depends on the raw materials input. The common conversion route to syngas from natural gas, liquid, or solid feedstock are briefly described as follow [4,6].

Steam Reforming

The first patent on supported nickel catalysts was used by Mittasch and coworkers in 1912 for steam reforming. The endothermic reaction of steam reforming is carried out under 750–850°C and 20–30 atm assisted by a catalyst where water-gas shift reaction also takes place simultaneously.

$$CH_4 + H_2O \rightarrow CO + 3H_2 \tag{I.1}$$

$$CO + H_2O \leftrightarrows CO_2 + H_2 \tag{I.2}$$

Under this circumstance, catalytic process of steam reforming method can generate syngas with high H_2 to CO ratio ($H_2/CO \geq 5$).

The major drawback of the steam reforming method is associated with the deactivation of the supported nickel catalysts due to the formation of carbonaceous materials on the catalyst surfaces at such reaction conditions.

Partial Oxidation

The relatively exothermic reaction of partial oxidation of methane can generate syngas with the ratio of $H_2/CO = 2$ which is desirable for methanol production plants or Fischer-Tropsch processes to produce synthetic fuels.

$$CH_4 + \tfrac{1}{2}O_2 \rightarrow CO + 2H_2 \tag{I.3}$$

Ni-based catalysts are widely used for this process, though deactivation due to coke deposition is still the major problem. Catalytic partial oxidation has several advantages over the steam reforming process because it includes autothermal operation, requiring no external heat input from furnaces or heat exchangers. Partial oxidation can be carried out in the absence of catalyst under more severe reaction condition.

CO_2 (Dry) Reforming

Release of large quantities of CO_2 in the atmosphere causes a global warming/greenhouse effect. Hence, research activities on the activation of carbon dioxide for its conversion to useful products are gaining more and more importance. It has been well established that the presence of CO_2 in a conventional steam methane reformer feed reduces the H_2/CO syngas ratio. In this attractive yet challenging method, highly endothermic reaction of CH_4 and CO_2 proceeds assisted by a catalyst, such as nickel-based material. If a more effective catalytic system against deactivation due to the coke formation can be discovered, dry reforming is a very promising route for mitigation of two major contributors among greenhouse gases [7].

$$CH_4 + CO_2 \rightarrow 2CO + 2H_2 \quad \text{(I.4)}$$

There is reverse water gas shift reaction accompanied the above reaction to give higher CO mole fraction that H_2 at the outlet. The syngas composition produced from this route is desirable for the production of the liquid hydrocarbons or petrochemical products.

Autothermal Reforming

Autothermal reforming uses oxygen and CO_2 or steam in a reaction with CH_4 to form syngas. In principle, autothermal reforming is the steam reforming with oxygen input [8]. Therefore, the required heat for endothermic steam reforming reaction is supplied by the exothermic partial oxidation. The process also requires the availability of a low-cost oxygen supply and produces a large quantity of excess steam.

The reactions can be described in the following equations, using CO_2:

$$2\,CH_4 + O_2 + CO_2 \rightarrow 3\,H_2 + 3\,CO + H_2O \quad \text{(I.5)}$$

and using steam:

$$4\,CH_4 + O_2 + 2\,H_2O \rightarrow 10\,H_2 + 4\,CO \quad \text{(I.6)}$$

Syngas with variable H_2/CO composition generated from the autothermal reforming method assisted by a catalyst is desirable for the GTL applications. Oxygen-blown autothermal reforming has been mentioned as the preferred technology for large-scale Fischer-Tropsch plants. Haldor Topsøe for example, developed autothermal reforming technology for production of syngas for Fischer–Tropsch synthesis using cobalt-based catalysts at a low steam to carbon ratio.

Overall, the choice of reformer technology will have an influence on the thermal efficiency of the plant and on the capital costs of the reformer and oxygen plant (where applicable). One of the biggest challenges is to optimize the energy integration between the syngas generation and syngas conversion sections. For illustration, the greatest impact on improving the fuels synthesis economy is associated with syngas syngas manufacture because it is responsible for nearly 60% from the total investments for GTL technology [8].

For all catalytic processes, development of a more effective catalyst material which can have good lifetime stability, sustain against carbon deposition, sour gases and sulphur containing compounds, and maintain the high productivity is the major challenge. For example, supported noble metal, such as rhodium or platinum which is more resistant to deactivation have been used on the catalytic partial oxidation process [9]. Otherwise, modifications to the process, alternative, or combined methods should be employed in order to achieve high and steady conversions.

It has been shown that the simultaneous steam and dry reforming of methane allows some limited control of the H_2/CO syngas ratio to a desired value while avoiding carbon deposition on the catalyst surface [10]. Three-reforming process consists of a combination of dry reforming, steam reforming, and partial oxidation has also been proposed [11]. Such combination offers several advantages because it is possible to adjust the final H_2/CO ratio in syngas by varying the $CO_2/H_2O/O_2$ ratio in the feed, or to balance the heat supply and heat consumption. Moreover, the presence of oxidant agent, O_2, inhibits the process of carbon deposition and the energy requirement of the process is reduced when O_2 is introduced due to the exothermic reaction of partial oxidation [12].

Novel techniques for syngas production relevant with the greenhouse gas emission control such as, plasma [13] and photocatalysis [14] have been introduced for CO_2 reforming of methane. The plasma technology allows higher conversion and selectivity with lower input energy while photocatalysis can utilize sunlight to generate energy for reforming with the aid of semiconductor materials. Microwave-assisted CO_2 reforming over carbon catalyst has also been tested to give higher conversions of CH_4 and CO_2 [15].

Gasification

Syngas production via a well established mature technology of gasification has very important role and bright future regarding the sustainable development and fuels supplies. A wide range of carbonaceous materials like coal, petroleum coke (pet coke), and heavy petroleum residues/fuel oil has been used as feedstock for gasification. Coal gasification processes were used in the past to manufacture town gas for gas lighting, cooking, and heating before electric power and the natural gas infrastructure became widely available. The vast interest on the gasification technology is illustrated in the Worldwide Gasification Database [16]. The "World Gasification Operating Capacity – by Feedstock" in 2010 indicates that coal now dominates as the feedstock in 51% or 36,315 MWth of syngas capacity, representing 53 plants. Petroleum (including fuel oil, refinery residue, and naphtha) is the second leading feedstock, with 17,938 MWth, or 25% of total gasification capacity, covering 56 plants. Natural gas provides 22%, with petcoke about 1% and biomass/waste accounting for 0.5%. Raw syngas obtained from coal, pet coke, petroleum residues, etc. typically contains (v/v) 25–30% H_2, 30–60% CO, 5–15% CO_2, and 2–3% H_2O [17]. The 2010 Worldwide Gasification Database indicates that chemicals and liquid transportation fuels represent the leading products with 45% and 38%, respectively, of the world gasification capacity. Chemicals are generated at 112 plants and liquid transportation fuels at five plants. Other products are power (11%) and gaseous fuels (6%) [16]. in 2010, the existing world gasification capacity has grown significantly to 70,817 MWth of syngas output from 144 operating plants and 412 gasifiers. This represents a 57% increase compared to the 2004 database level of 45,001 MWth.

Prior to gasification, coal is converted to coke by pyrolysis (destructive distillation). Subsequently, alternating blasts of steam and air are forced to contact with the coke bed. In principle, major reactions involved in the coke gasification process are combustion (reaction with O_2), Boudouard reaction (reaction with CO_2) and steam gasification (reaction with steam) [17].

$$C + \frac{1}{2} O_2 \rightarrow CO \text{ (Gasification with oxygen); } \Delta H^\circ_{298} = -110.5 \text{ kJ mol}^{-1} \quad (I.7)$$

$$C + O_2 \rightarrow CO_2 \text{ (Combustion with oxygen) } \Delta H^\circ_{298} = -393.0 \text{ kJ mol}^{-1} \quad (I.8)$$

$$C + CO_2 \rightarrow 2\,CO \text{ (Gasification with carbon dioxide); } \Delta H^\circ_{298} = +172.0 \text{ kJ mol}^{-1} \quad (I.9)$$

$$C + H_2O \rightarrow CO + H_2 \text{ (Gasification with steam): } \Delta H^\circ_{298} = +131.4 \text{ kJ mol}^{-1} \quad (I.10)$$

Gasification of coal for syngas accounts for the largest amount in industry because of the low price of coal. Gasification technology also is now under intensive investigation and looks to be very compatible with biomass and municipal solid waste [18]. Importantly, the driving force of using municipal waste as the feedstock is because the syngas produced in large waste-to-energy gasification facilities can be utilized to generate electricity. The presence of much lower sulfur content in biomass (also in coal) favors these feedstocks for the production of cleaner syngas. However, the presence of high ash content in coal (>41–46% in some cases) may cause operational problems in gasifier. Meanwhile, municipal solid wastes contain considerable amount (~50%) of easily combustible materials and thus can be gasified at lower temperature [17]. Therefore, co-gasification of biomass with coal may offers advantages related with the low ash and sulphur content of biomass.

The concept of integrated multigeneration of power and chemicals has been a hot topic of research as it improves the plant economy and reduces the emissions. To implement this idea, modern gasification units can be constructed based on integrated gasification combined cycle (IGCC) to generate power along with hydrogen, chemicals, methanol/Fischer–Tropsch liquid fuels, synthetic natural gas or any combinations of these [19]. ELCOGAS power plant, located in Puertollano (Spain), is an example of this technology, participating in two projects for generation, capture and storage of CO_2 and H_2 via WGSR and for obtaining diesel fuels via Fischer–Tropsch.

Antonius Indarto
Universita di Torino, Turin, Italy
Jelliarko Palgunadi
University of Kyunghee, Seoul, South Korea

REFERENCES

[1] H.- W. Häring (ed.), Industrial gases processing, Wiley-VCH Verlag GmbH and Co., Weinheim (2008).

[2] P.L. Spath and D.C. Dayton, *Preliminary Screening–Technical and Economic Assessment of Synthesis Gas to Fuels and Chemicals with Emphasis on the Potential for Biomass-Derived Syngas*, Technical Report, NREL/TP-510-34929, December 2003.

[3] K. Liu, C. Song, and V. Subramani (Eds.), *Hydrogen and Syngas Production and Purification Technologies*, John Wiley and Sons, Inc., Hoboken, New Jersey (2010).

[4] D. J. Wilhelm, D. R. Simbeck, A. D. Karp, R. L. Dickenson, *Fuel Proces. Technol.* 71, 139 (2001).

[5] N. Patel, K. Ludwig, and P. Morris, *Insert flexibility into your hydrogen network—Part 1*, Special Report, Hydrocarbon Processing, Air Products and Chemicals, Inc., Allentown, Pennsylvania, September 75 (2005).

[6] Gas Production in Ullmann's Encyclopedia of Industrial Chemistry, On-line Edition, Wiley-VCH Verlag GmbH and Co., Weinheim (2011).

[7] J. R. Rostrup-Nielsen, J. Sehested and J. K. Norskov, *Adv. Catal.* 47, 65 (2002).

[8] J. R. Rostrup-Nielsen, Syngas in perspective, *Catal. Today* 71, 243 (2002).

[9] B. J. Dreyer, *Synthesis Gas and Olefins from the Catalytic Autothermal Reforming of Volatile and Non-volatile Liquids*, A Ph.D Dissertation, University of Minnesota, September 2007.

[10] C. Li, Y. Fu, G. Bian, Y. Xie, T. Hu, J. Zhang, *Kinetics Catal.* 45, 679 (2004).

[11] V. R. Choudhary and A. S. Mamman, *J. Chem. Technol. Biotechnol.* 73, 345 (1998).

[12] S. Wang, G. Q. Lu and G. J. Millar, *Energy Fuels* 10, 896 (1996).

[13] N. Seyed-Matin, A. H. Jalili, M. H. Jenab, S. M. Zekordi, A. Afzali, C. Rasouli and A. Zamaniyan, *Plasma Chem. Plasma Proc.* 30, 333 (2010).

[14] L. Yuliati, H. Itoh and H. Yoshida, *Chem. Phys. Let.* 452, 178 (2008).

[15] B. Fidalgo, A. Domínguez, J.J. Pis and J.A. Menéndez, *Int.l J. Hydrogen Energy.* 33, 4337 (2008).

[16] World Gasification Database, http://www.netl.doe.gov/technologies/index.html.

[17] P. Mondal, G.S. Dang, M.O. Garg, *Fuel Proc. Technol.* 92 1395 (2011).

[18] K. Göransson, U. Söderlind, J. He, and W. Zhang, *Renewable Sustainable Energy Rev.* 15, 482 (2011).

[19] R. W. Breault, *Energies*, 3, 216 (2010).

In: Syngas
Editors: Antonius Indarto and Jelliarko Palguandi
ISBN: 978-1-62100-870-5

Chapter 1

Synthesis Gas Production by Reforming Methane in a Chemical Compression Reactor

V. Shmelev*

Combustion Lab, Institution of Russian Academy of Sciences Semenov Institute of Chemical Physics RAS, Russia

Abstract

The various technologies are studied to convert methane, a main ingredient of natural gas, into hydrogen. These investigations are part of the total efforts to develop new low pollution energies and reduce greenhouse gases. The discussion, different from existing methods of reformation, suggests several steps in approach to a noncatalytic partial oxidation methane in the methane-air mixture with usage of the internal combustion engine, which can be modified to a chemical compression reactor with the heat recuperation.

The use of a chemical compression reactor (CCR) is quite promising for syngas and hydrogen production. There is wide industrial base for manufacturing cheap CCRs on the base of internal combustion engine. However, analysis shows that it is required to use additional techniques for ensuring ignition and the occurrence of the chemical conversion of rich hydrocarbon-air mixtures. Such effective method is preheating of the mixture prior to supply to a reaction vessel. It is considered several methods. The mixture can be heated directly outside the CCR, for example, in an electrical heater or by heat regeneration, and internal heat recovery processes can be used. The outside preheating is the simplest method and can be considered as a first step for investigation and development of this technology.

This chapter includes the theoretical analysis for CCR with external electrical heating mixture and experimental results with the methane-air mixtures; a method for noncatalytic partial oxidation of methane in a CCR with heat regeneration is proposed and the appropriate theory is developed. The heat regenerator is located in the channel of a combined exhaust and intake manifold, through which reaction products and reactants alternately pass; a method for the partial oxidation of methane in a CCR with internal heat recovery is proposed and examined. The reactor cylinder divided by a partition with

* E-mail: shmelev@chph.ras.ru.

a cross flow orifice into two chambers is considering. Internal heat recovery is implemented using a thermal activator installed in the smaller (auxiliary) chambe.

NOMENCLATURE

α	air– methane equivalence ratio
ε	burning completeness coefficient
θ	dimensionless temperature
τ	dimensionless time
χ	fraction of reacted methane
ρ	gas density, kg·m^{-3}
ω	parameter of the heat exchange
η	volume of syngas produced per unit energy, m^{3}·kWh^{-1}
β	volume ratio for the working and auxiliary chambers
λ	volumetric compression ratio
ξ	volumetric concentrations of the components
δ	width of slot channels, m
α_a	coefficient of heat transfer, J·sec^{-1}·m^{-2}·K^{-1}
φ	volume of syngas produced per unit time per unit volume,m^{3}·h^{-1}·L^{-1}
A	empirical parameter, (mol·m^{-3})$^{1\text{-A-B-C}}$·sec^{-1}·K^{-d}
C	specific heat, J·kg^{-1}·K^{-1}
CCR	Chemical Compression Reactor
d	orifice diameter, mm
E	specific internal energy, J·kg^{-1}
E_a	effective activation energy, J·mol^{-1}
H	specific enthalpy, J·kg^{-1}
HRE	Heat-Retaining Element
ICE	Internal Combustion Engine
K	rate constants of the reaction, m^{3}·mol^{-1}·sec^{-1}
m_a	mass of thermal activator, kg
N	crankshaft speed, rpm
N_i	concentrations of the components, %
Nu	Nusselt number
P	pressure, Pa
Pe	Peclet number
q	overall mixture consumption rate, kg·sec^{-1}
Q_f	total flow, L·min^{-1}
R	universal gas constant, J·mol^{-1}·K^{-1}
S_a	heat transfer surface area, m^{2}
T	temperature, K
t	time, sec
TA	Thermal Activator
U	piston velocity, m^{3}·sec^{-1}

V	volume, m^3
W	power per unit volume and per cycle, $kW{\cdot}L^{-1}$
W_i	heat release rate or heat transfer, $J{\cdot}sec^{-1}$
W_m	overall hydrocarbon oxidation rate, $mol{\cdot}m^{-3}{\cdot}sec^{-1}$
x_i	molar coefficients

INTRODUCTION

The widespread use of natural gas and hydrogen in the power industry, transport, and as raw materials for the chemical industry is one of the primary contemporary technological tendencies. An overwhelming share of hydrogen is synthesized in the processes of natural gas and coal conversion into synthesis gas, and also in reforming and other oil refining and petrochemical processes. From an ecological point of view, it is attractive to use hydrogen in motor transport for the power supply of onboard fuel cell–based engines. However, there are some serious problems in the way of its practical realization; one of them is creation an efficient infrastructure for hydrogen production and distribution. Because of their technical complexity and high-energy consumption, the contemporary industrial hydrogen synthesis methods, which are based on steam or steam–oxygen natural gas conversion, will hardly be able to claim a significant role in a scattered production of hydrogen.

It is necessary to note, that the specified method is a catalytic conversion. The use of catalysts considerably complicates the process of hydrocarbon oxidation. Therefore, in some cases, it is expedient to aim for the noncatalytic and cheaper process of conversion. Such a promising method for producing synthesis gas is the partial oxidation of methane in exothermal reactions. Thereby, one of the important problems of chemical engineering is the development of new economically and ecologically efficient methods for processing natural reactants into a final target product and the development of apparatuses for their implementation.

A rather attractive process of noncatalytic partial oxidation of methane is the combustion of hydrocarbon-rich mixtures in a chemical compression reactor (CCR) that uses the operating principle of an internal combustion engine (ICE) [1-5]. Such a CCR is a heat engine of cyclic operation in which adiabatic (ideally) a piston accomplishes compression of reactants up to the temperature necessary for rapid occurrence of chemical conversion. In the subsequent expansion stroke, the reaction products are cooled and hardened, and the mechanical energy supplied to the gas during its compression is returned during the expansion as useful work. Under conditions of a sufficient heat effect of the chemical conversion, the reactor can produce excess mechanical energy. This makes it possible to organize a high-efficiency system for recovering the heat of chemical conversion during the operation of the reactor.

Thus, a CCR combines in itself a heater; a reactor; a hardening device; and, probably, an energy generator and is a unique apparatus in chemical engineering that combines in itself so many necessary qualities. However, there is a principal limitation on the wide application of these reactors for carrying out many reactions of practical importance [1,6], in particular, for partial oxidation of hydrocarbons. It is hardly possible to heat a rich hydrocarbon–air mixture without adding oxygen with a low adiabatic exponent by compression to the necessary ignition temperature or reach the reaction temperature (~1200 - 1500 K) of very rich hydrocarbon–air mixture at a limited compression pressure (~100 atm). To achieve this temperature, too high a pressure of compression of about 1000 atm is necessary, which is intolerable for state-of-the-art ICEs. It is required to use additional techniques for ensuring ignition and the occurrence of the chemical conversion of rich hydrocarbon–air mixtures.

Studies of "explosive" methane reforming in internal combustion engines were begun in Russian and German [2,3] in the last century. The relatively slow speed spark ignition-type engine was used [2-4]. It was shown that syngas could be produced in an engine with spark ignition when the mixture is enriched to an air–fuel equivalence ratio $\alpha = 0.38$. However, to sustain the process the air should be enriched with oxygen to 80%. Nevertheless, it was stated that in view of the additional oxygen cost the production of syngas in an internal combustion engine can be profitable and attractive and that a large-scale production could be easily implemented. In addition to syngas, the process could concurrently produce power at a high efficiency. The use of a high-speed compression ignition engine for synthesis gas production was very attractive. As a very powerful ignition system was needed to ignite this mixture, a more suitable option was an engine with compression ignition. The pilot liquid fuel in [5] provided ignition sources distributed over a large gas volume. The engine with a compression ratio of 14.2 operated by a two-component fuel mixture consisting of methane and 10% diesel fuel was used to achieve a syngas content up to 80% in the exhaust gases at $\alpha = 0.42$ under of 1000 rpm.

The partial oxidation of methane was also accomplished by the method of pulsed mixture compression in free-piston ballistic compressors [1]. A high conversion of methane was reached when a mixture with a $CH_4/O_2 \approx 2$ ratio was compressed to a temperature higher than 1500 K. However, to reach this temperature, it was necessary to dilute the mixture with argon to its concentration in the mixture exceeding 80%. This resulted in an increase in the mixture adiabatic index, which made it possible to reach the required temperature at a limited compression ratio of about 30.

The authors of [7] proposed a method for the partial oxidation of methane with air in a CCR constructed on the basis of an ICE. In this apparatus, a rich mixture, with an air excess ratio of ~ 0.5, is inflamed by a torch of combustion products of a stoichiometric mixture. The authors of [8] proposed idea of preliminarily heating the reaction mixture to 200 - 450°C before it is introduced into the reactor. In [6,10,11], a method for gas heating in the mode of "superadiabatic" compression in the CCR cylinder is proposed that makes it possible to attain higher temperatures in comparison to adiabatic compression to the same maximum pressure or compression ratio. This process can be organized in multistage (in particular, two-stage) compression of gas during its flow from one vessel to another (with an increase in entropy), for example, in a divided cylinder of an ICE. An increase in entropy occurs when gas flows through a hole in a partition between the vessels.

The effect of a rise in temperature is explained by the possibility of delivering larger work on the gas (as a result of one or more intermediate stages of compression to a maximum pressure). In [6], it is proposed to heat the gas in two sequential strokes with superadiabatic compression in order to considerably increase the temperature. Further development of the considered technology is connected with improvement of the combustion process in ICE reactors and improvement of the gasdynamic and thermodynamic processes in pulse compressive reactors with the free piston of the single and two stages compression [12-14]. In this chapter, the partial oxidation of methane in a CCR with a outside preheater or hypothetical regenerators that are set outside or inside the cylinder space is theoretically studied, and the capabilities of these methods are evaluated. The outside preheating is the simplest method and can be considered as a first step for investigation and development of this direction.

REACTOR WITH INTERNAL HEAT RECOVERY

A method for heating a gas in the superadiabatic regime in an ICE cylinder was developed as mention in [11]. Among other things, it involves the internal recovery of heat, a technique that makes it possible to use low exothermic mixtures. It is implemented as follows: an ICE cylinder is divided into two chambers (working and auxiliary) by a partition with a crossflow orifice. The auxiliary chamber accommodates a heat recuperator with a developed surface (also known as a thermal activator (TA) [11]), for example, a set of metal plates or a porous metallic or ceramic body (Figure 1.1). The TA absorbs part of the heat ΔQ stored by the reaction products during the expansion stroke and releases it into the initial mixture during the compression stroke, thereby increasing the temperature and entropy of the mixture. During the intake and exhaust strokes, the TA only slightly interacts with the mixture and reaction products, since it is separated from the working chamber by the partition. Thus, a closed cycle of heat transfer is realized immediately in the ICE. The superadiabatic regime of compression of the reaction mixture to a higher temperature occurs without cycle-average energy expenditures on the heating of the TA. The use of a TA makes it possible to extend the region of chemical transformations to fuel-richer mixtures.

A similar idea, the use of a porous medium in the cylinders of a diesel engine for enhancing fuel evaporation, was put forward and realized in [15]. The authors of [11] experimentally demonstrated that this idea can be implemented in a simple way in an ICE.

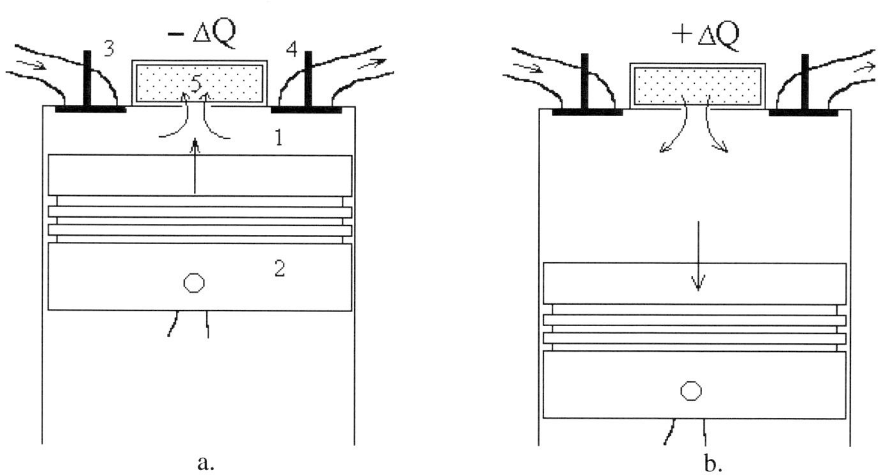

Note: 1. Cylinder, 2. Piston, 3. Intake valve, 4. Exhaust valve, and 5. Thermal activator.

Figure 1.1. Schematic of a CCR with internal heat recovery: (a) Compression stroke; (b) Expansion stroke.

The aim of this paragraph is to theoretically analyze whether the partial oxidation of methane in rich methane–air mixtures can be accomplished using an CCR on the basis of an ICE with internal heat recovery by means of a TA [16].

Calculation Model

Consider the partial oxidation of methane in rich methane–air mixtures containing ξ_{m0} = 0.17–0.3 mole fractions of methane. At the two limiting methane concentrations, this process can be approximately described by the equations

$$CH_4 + O_2 = CO + H_2 + H_2O \qquad \xi_{m0} = 0.173,\ \Delta H = -277.5 \text{ kJ/mol} \qquad (1.1)$$

$$CH_4 + \tfrac{1}{2} O_2 = CO + 2\,H_2 \qquad \xi_{m0} = 0.3,\ \Delta H = -35.6 \text{ kJ/mol} \qquad (1.2)$$

where ΔH is heat of the reaction.

Part of the thermal energy released by the first reaction can be converted into useful work. Note, however, that the conversion of methane according to the second equation releases only a small amount of heat, and, therefore, it needs an external energy input to occur.

We calculated the characteristics of the partial oxidation of methane in an ICE-based CCR with a two chamber cylinder (separated by a partition with a crossflow orifice shown in Figure 1.1). Let the volumes of the working and auxiliary chambers be V_1 and V_2, respectively, with $\beta = V_2 / V_0 \ll 1$, where $V_0 = V_1 + V_2$ is the volume of the cylinder and let the auxiliary chamber accommodate a TA. As the piston moves toward the partition, it compresses the gas mixture in the working chamber, with the final degree of compression being $\lambda \leq 1/\beta$ (the maximum degree of compression $\lambda = 1/\beta$ is attained when all the gas overflows into the auxiliary chamber). It was assumed that, at the initial instant of time, the piston is positioned at the bottom dead center, with the mixture being uniformly distributed over the cylinder at pressure P_0 and temperature T_0. As the piston moves to the top dead center, it forces the cold mixture to flow into the auxiliary chamber, where it receives heat from the TA heated in the previous cycle to a temperature T_a. The superadiabatic heating of the mixture occurs due to the joint action of three factors: compression, inflow with entropy increase [9], and heat transfer from the TA. We numerically simulated only two stages, the compression and expansion strokes, assuming that the intake and exhaust strokes are not accompanied by changes in the thermal parameters of the TA.

The chemical conversion of methane was described using global modeling [17]. According to this method, a detailed kinetic scheme can be reduced to a limited number of kinetic equations. In principle, the consumption of a hydrocarbon can be described with acceptable accuracy by one equation. In this case, the reaction orders with respect to the reactants and the effective activation energy of the process are determined from experiment. The overall hydrocarbon oxidation rate W_m is given by the global kinetic equation

$$W_m = \frac{dN_m}{dt} = -A\, N_m^a\, N_{ox}^b\, N_n^c\, T^d \exp(-E_a / RT), \qquad (1.3)$$

where *A, a, b, c, d* are empirical parameters; E_a is the effective activation energy; N_m, N_{ox}, and N_n are, respectively, the molar concentrations of hydrocarbon, oxygen, and nitrogen in the hydrocarbon–air mixture; *t* is the time; *T* is the current temperature; and *R* is the universal gas constant. The empirical coefficients were chosen based on the results of experiments with rich methane-air mixtures [18-20].

At high temperatures, it took into account the dissociation of water vapor, with the composition of the reaction products being calculated with consideration given to the processes involving water gas. An analysis demonstrated that the extent of conversion of methane can be estimated rather accurately without performing detailed kinetic calculations of the composition of the reaction products. The dissociation of water is described by the equation:

$$\frac{dN_{H_2O}}{dt} = K_1 N_{H_2O} N_g - K_2 N_r^2 N_g \qquad (1.4)$$

where *Ng, Nr* are the molar concentrations of the gaseous components and dissociation products (R_0 = H or OH) and K_1 and K_2 are the rate constants of the direct and reverse reactions, respectively [21,22].

The reaction of water gas can be described as:

$$\frac{dN_{CO_2}}{dt} = K_3 N_{CO} N_{H_2O} - K_4 N_{CO_2} N_{H_2} \qquad (1.5)$$

where N_{CO2}, N_{CO}, N_{H2O}, and N_{H2} are the molar concentrations of carbon dioxide, carbon monoxide, water vapor, and hydrogen, respectively, and K_3 and K_4 are the rate constant of the direct and reverse reactions [22].

In some calculations, the kinetics of soot formation was described [22]. It turned out, however, that the process was too short to produce an appreciable amount of soot; therefore, we report here no data on soot formation. We solved a system of equations describing the changes in the internal energy and mass of the methane-air reaction mixture in the working and auxiliary chambers of the reactor during compression, crossflow, and expansion [6,10,24]:

$$\begin{aligned}
&\frac{d(E_i m_i)}{dt} = P_i \frac{dV_{gi}}{dt} + H_k q_k + W_{ci} - W_{ti} + W_{ai} \\
&\frac{dm_i}{dt} = q_k \\
&\frac{d\xi_{ji}}{dt} = w_{ji} + \tilde{q}_{jk} \\
&\frac{dV_{gi}}{dt} = U(t) \\
&\frac{d(c_a m_a T_a)}{dt} = -\alpha_a S_a (T_a - T_2) = -W_{a2}
\end{aligned} \qquad (1.6)$$

Equation set of (1.6) was supplemented with the Clapeyron equation.

The last equation in (1.6) describes the temperature of the TA, which is controlled by heat transfer between it and the gas mixture in the auxiliary chamber. Here, H and E are the specific enthalpy and internal energy of the gas mixture; P is the pressure; V_{gi} is the volume of the compressed gas ($V_{g2} = V_2$); i and k (i, k = 1, 2) are the indices that denote the chambers of the reactor, with $k = i$ if the gas outflows from the ith chamber; W_c is the heat release rate; W_t is the heat transfer (or heat loss) rate, which was determined using the Woshni model [25]; W_{a2} is the rate of heat transfer from the TA to the gas mixture and vice versa ($W_{a1} = 0$, $W_{ai} = -W_{a2}$); ξ_j and w_j are, respectively, the relative concentrations of the initial components in the gas mixture with mass m_i or those of the reaction products (CH_4, CO, CO_2, H_2, H_2O, and R_0) that are labeled by j and the dimensionless rate of their transformations in the reaction; q and $\tilde{q}_j$ are the overall mixture consumption rate and the relative cross-flow rates of the components, respectively [24]; U is the piston velocity [25]; c_a is the specific heat of the material of the TA; m_a, S_a, and T_a are the mass, heat transfer surface area, and temperature of the TA; and α_a is the coefficient of heat transfer between the TA and gas mixture. The temperature dependences of the heat capacities of air, methane, and the reaction product components were borrowed from [26].

The concentrations of the reaction products were found from the material balance equation in view of kinetic equations (1.3) and (1.4):

$$x_m CH_4 + O_2 \rightarrow (1-\chi)(x_m CH_4 + O_2) + \chi(x_1 CO + x_2 H_2 + x_3 CO_2 + x_4 H_2O) \tag{1.7}$$

where x_m is the input parameter of the problem, and the unknown molar coefficients x_1, x_2, and x_4 were calculated from the carbon, hydrogen, and oxygen balance; the coefficient x_3 was calculated from kinetic equation (1.5); and the fraction of reacted methane χ was calculated from kinetic equation (1.3). The process of water dissociation, equation (1.4) was not included in the consideration of the material balance due to the insignificance of its contribution.

The calculations were performed at the following values of the input parameters: the cylinder diameter, 10 cm; working chamber length, 10 cm; piston stroke length, 10 cm; crossflow orifice diameter, d = 5 - 15 mm; β = 0.01 - 0.1; crankshaft rotational speed, N = 1000 rpm; and working cylinder wall temperature, T_0 = 360 K. The maximum pressure during compression and chemical reaction did not exceed 100 - 200 atm, whereas the final degree of compression was λ = 10 - 20, with the geometric degree of compression in the working chamber being:

$$\lambda_1 = \lambda \frac{1-\beta}{1-\lambda\beta} \tag{1.8}$$

The TA was modeled by a 0.1-mm-thick ribbon wound into a spiral with an intercoil distance of 1 mm, a heat transfer surface area of S_a = 790 - 1710 cm^2, and a mass of m_a = 30 - 70 g.

CALCULATION RESULTS

The calculation results show that, for the conventional one-stage compression of a mixture with a methane concentration of $\xi_{m0} = 0.167$ to a degree of $\lambda = 20$ (typical of diesels), no methane conversion is observed. At a crossflow orifice diameter of $d = 5$ mm, $N = 1000$ rpm, and $\beta = 0.048$, the maximum temperature in the working chamber T_{f1} attains a temperature of nearly 900 K at a pressure of $P_{f1} = 65$ atm; in the auxiliary chamber, $T_{f2} \approx 1000$ K at a peak pressure of $P_{f2} = 55$ atm. These conditions proved to be too mild to initiate the reaction.

When the thermal activator (Figure 1.1) was installed in the auxiliary chamber, the picture changed drastically, even for mixtures with $\xi_{m0} > 0.167$.

At the very beginning of the compression stroke, the mixture is additionally heated by the TA (heated to $T_a = 1140$ K in the previous cycle) and explodes at a temperature just above 1000K, with the degree of methane conversion attaining $\chi = 0.87$. Initially, the temperature of the TA decreases by 3.7 K (curve 3 of Figure 1.2b), but, as the process of compression continues and the chemical reaction occurs, the temperature of the mixture grows and eventually exceeds that of the thermal activator. The ensuing heat transfer from the mixture to the activator raises its temperature by about $\Delta T_a = 1$ K compared to the steady-state temperature T_a. Once the steady state of the periodic process is established, the TA temperature at the end of the expansion stroke attains its initial value, $T_a = 1140$ K; i.e., the temperature at the end of each cycle remains unchanged (curve 3 of Figure 1.2b), a behavior indicative of internal heat recovery. The peak temperatures and pressures in the working and auxiliary chambers were found to be $T_{f1} = 1410$ K, $P_{f1} = 79.2$ atm and $T_{f2} = 1710$ K, $P_{f2} = 53.8$ atm, respectively. The composition of the conversion products for a number of calculation variants (at $\lambda = 15$) and other parameters of the cycle are given in the table. Here W is the power per unit volume of the setup (1 litter) per cycle, that is, per two crankshaft revolutions (a positive value corresponds to energy production, and a negative value to a process powered by an external drive); η is the energy cost of 1 m^3 of synthesis gas; φ is the volume yield of synthesis gas per hour per 1 liter of the setup; and ξ_{mk}, ξ_{H2}, ξ_{CO}, ξ_{H2O}, and ξ_{CO2} are the final mole fractions of CH_4, H_2, CO, H_2O, and CO_2 in reaction products.

Note that, for a given mixture and fixed values of the input parameters of compression, the condition of the cycle being closed in T_a (the equality of the temperatures at the beginning and end of the thermodynamic cycle) can be met at various values of T_a, in particular, $T_a = 1140$ K and $T_a = 1090$ K (the second and third row of the table). At $T_a = 1090$ K, the peak pressures and temperatures in the chambers are lower than those at $T_a = 1140$ K; as a result, the extent of conversion is lower ($\chi = 0.7$), as are the concentrations of the partial oxidation products (H_2 and CO).

Figure 1.3 shows the regions of existence of a recovery cycle. In the region to the left of the curves, the regime of closed recovery is realized, for which the temperatures at the beginning and end of the cycle are equal to each other. To the right of the curves, the condition of existence of a closed cycle for the TA temperature is not satisfied. As the concentration ξ_{m0} increases (even at a higher degree of compression λ), the region of existence of a recovery cycle shifts to smaller crossflow orifice diameters d. This is associated with the fact that the compression temperature needed to ignite the mixture increases with the fuel-to-air ratio.

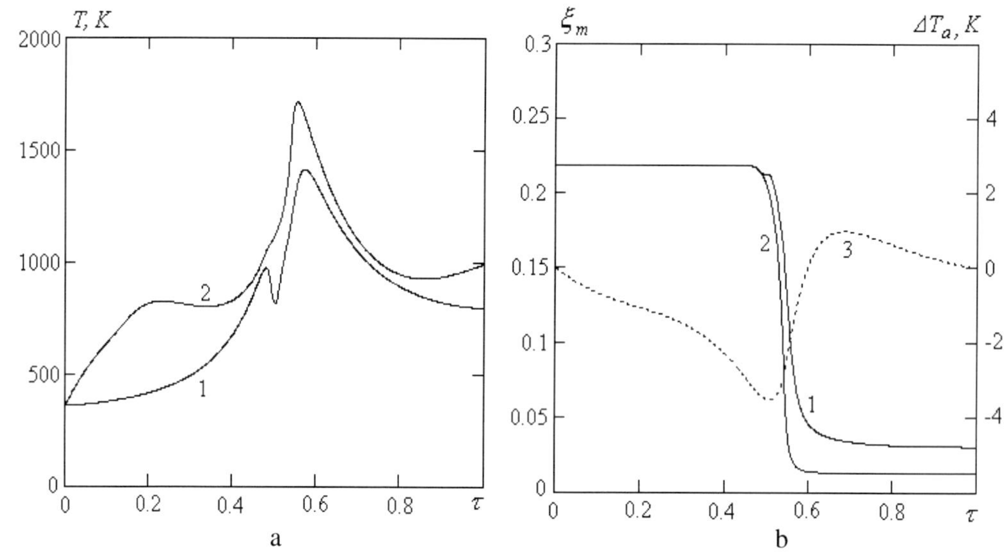

Note: τ is the dimensionless time (current time in units of crankshaft rotational period).

Figure 1.2. Time evolution of the (a) temperature and (b) methane concentration in (1) working and (2) auxiliary chambers and (3) the temperature of the TA in the auxiliary chamber at $\xi_{m0} = 0.22$, $\lambda = 15$, $\beta = 0.064$, $d = 0.47$ cm, and $T_a = 1140$ K.

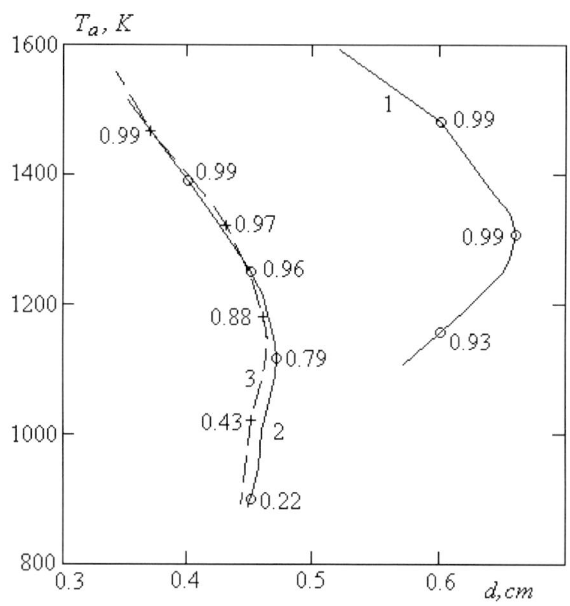

Note: The numbers at the curves denote extent of methane conversion χ.

Figure 1.3. The temperature dependence of the TA on the crossflow orifice diameter (boundaries of the region of existence of a heat recovery cycle) at (1) $\xi_{m0} = 0.167$, $\lambda = 12$, and $\beta = 0.081$; (2) $\xi_{m0} = 0.22$, $\lambda = 15$, and $\beta = 0.064$; (3) $\xi_{m0} = 0.22$, $\lambda = 12$, and $\beta = 0.081$.

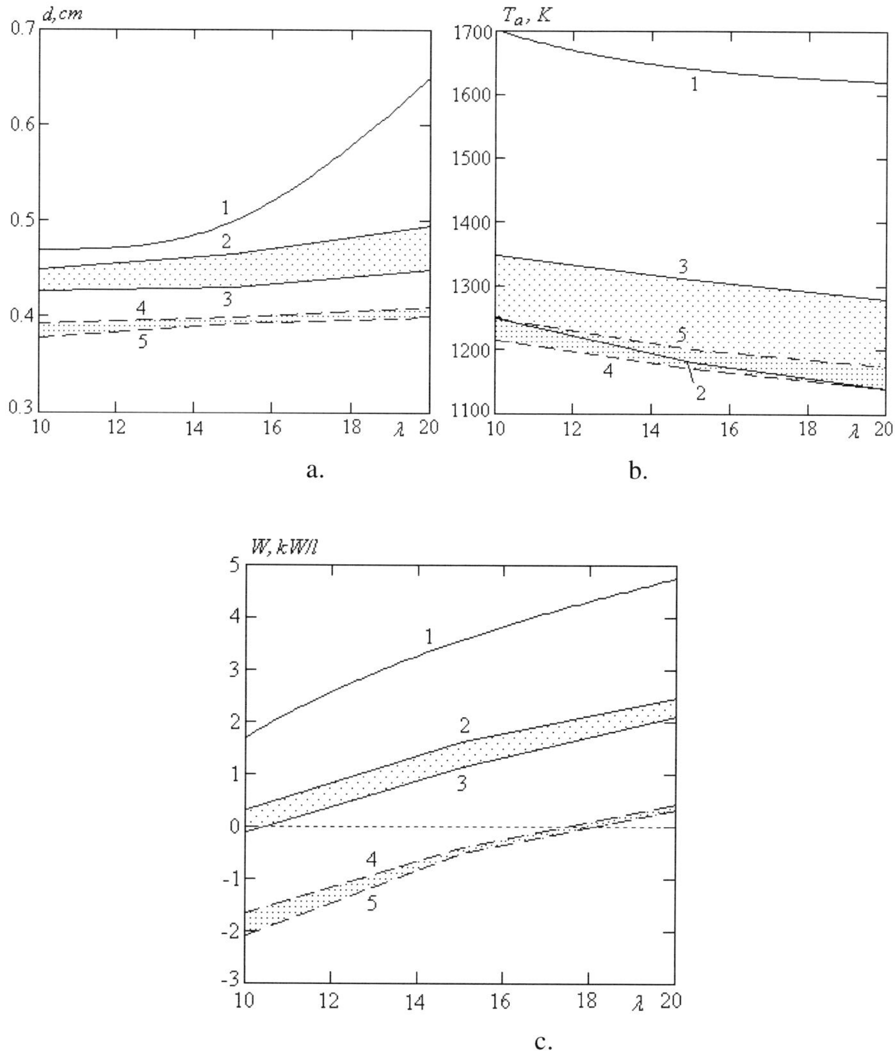

Note: The regions of intermediate χ values are hatched.

Figure 1.4. (a) Crossflow orifice diameter, (b) TA temperature, and (c) power released (absorbed) per unit volume of the reactor as functions of the degree of compression at (1) $\xi_{m0} = 0.167, \chi = 0.99$; (2) $\xi_{m0} = 0.22, \chi = 0.93$; (3) $\xi_{m0} = 0.22, \chi = 0.97$; (4) $\xi_{m0} = 0.242, \chi = 0.9$; and (5) $\xi_{m0} = 0.242, \chi = 0.93$.

At a fixed ξ_{m0} (curves 2 and 3), a decrease in λ does not results in an appreciable change in the crossflow orifice diameter, a behavior that can be explained by the fact that the parameter d produces a stronger effect on the compression temperature than λ does.

Figure 1.4 shows how the crossflow orifice diameter d, TA temperature T_a, and the output (input) energy W depend on the degree of compression λ for methane–air mixtures with various fuel-to-air ratios at various extents of conversion χ. As can be seen from Figure 1.4a, the diameter d increases with the degree of compression λ at all ξ_{m0} values, since the decrease in the compression temperature in the working chamber V_1 caused by an increase in d is compensated for by its increase associated with an increase in λ.

Table 1.1. Composition of the conversion product and other parameters of the cycle

ξ_{m0}	d, cm	T_a, K	P_{f1}, atm	P_{f2}, atm	T_{f1}, K	T_{f2}, K	W, kW/l	η, m^3/(kWh)	φ, m^3/hl	χ	ξ_{mk}	ξ_{H2}	ξ_{CO}	ξ_{H2O}	ξ_{CO2}
0.167	0.95	1280	81.0	81.1	2160	2280	+5.22	+1.37	6.02	0.99	0.002	0.18	0.103	0.103	0.039
0.219	0.47	1140	79.2	53.8	1410	1710	+1.4	+6.89	8.1	0.87	0.03	0.246	0.132	0.048	0.015
0.219	0.47	1090	78.9	45.8	1210	1560	+0.76	+10.24	6.56	0.70	0.068	0.195	0.108	0.041	0.01
0.242	0.41	1130	125.1	54.8	1245	1560	-0.4	-26.18	8.77	0.83	0.043	0.27	0.142	0.029	0.008

Note: W is the power per unit volume and per cycle (two rotations of the crankshaft) (the positive and negative signs correspond to energy generation and consumption, respectively); η is the volume of syngas produced per unit energy; φ is the volume of syngas produced per unit time per unit volume of the reactor; and ξ_{mk}, ξ_{H2}, ξ_{CO}, ξ_{H2O} and ξ_{CO2} are the final mole fractions of CH_4, H_2, CO, H_2O, and CO_2 in the reaction products.

At all ξ_{m0} values, T_a monotonically decreases with increasing λ. In this case, the decrease in the temperature of the mixture due to the decrease in T_a is compensated for by the appropriate increase in λ. As can be seen from Figure 1.4c, when operating with mixtures with ξ_{m0} = 0.167 and ξ_{m0} = 0.22, the CCR produces concurrently syngas and useful work (except for a small area near λ = 10 at ξ_{m0} = 0.22). At ξ_{m0} = 0.24, a small amount of energy can be generated only at $\lambda > 18$.

We simulated the composition of the products of the oxidation of methane in rich methane–air mixtures in an ICE cylinder measured in [7, 8]. Unfortunately, the authors of [8] failed to report the degree of compression, whereas the authors of [7] provide no information on either the degree of compression or the crankshaft rotational speed. Nevertheless, using typical values of these parameters, λ = 10 - 15 and N = 350 - 1000 rpm, we satisfactorily predicted the extent of methane conversion, oxidation temperature, and concentrations of the reaction products. The calculated and measured concentrations of hydrogen, carbon monoxide, carbon dioxide, and water vapor were found to range within ξ_{H2} = 0.18 - 0.22, ξ_{CO} = 0.10–0.12, ξ_{CO2} = 0.02–0.04, and ξ_{H2O} = 0.08–0.11, respectively.

Reactor with Outside Heat Regeneration

Consider another variant of heat recuperation [27]. It is proposed to place the regenerator, i.e. a heat-retaining element (HRE) outside the CCR cylinder so that reaction products and reactants alternately pass through it (Figure 1.5). In the exhaust stroke, when the piston moves to the top dead center (Figure 1.5a), highly heated reaction products, passing through the regenerator, give a portion of the heat ΔQ to the regenerator, which results in the heating of the HRE.

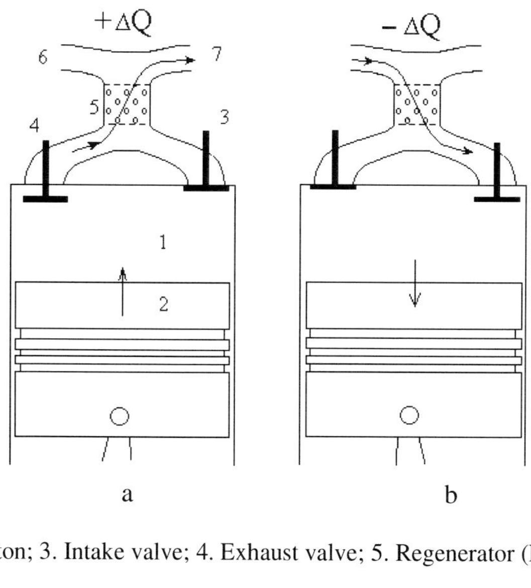

Note: 1. Cylinder; 2. Piston; 3. Intake valve; 4. Exhaust valve; 5. Regenerator (HRE); 6. Inlet pipe; 7. Outlet pipe.

Figure 1.5. Scheme of the CCR based on the ICE with a regenerator: (a) exhaust stroke, (b) intake stroke.

In the intake stroke, when the piston moves to the bottom dead center (Figure 1.5b), the reactants take a portion of the heat ΔQ, thus cooling the HRE and heating up to the necessary temperature. Thus, it becomes possible to use a CCR for the chemical conversion of rich hydrocarbon–air mixtures with retention of all of its known advantages due to the preheating of the original mixture in a regenerative cycle.

Different designs of the regenerator are possible. For example, the motion of the reaction products and reactants in the regenerator can occur in separated flows through adjoining and alternating channels. In the other variant, the motion of the reaction products and reactants can occur alternately through all the channels in the total flow. In this case, the regenerator space will be filled with residual reaction products or reactants upon completion of the exhaust and intake strokes, respectively. This leads to undesirable dilution of the fresh mixture and products with the opposite component.

In this paragraph, the partial oxidation of methane in a CCR with a hypothetical regenerator that is set outside the cylinder space is theoretically studied, and the capabilities of this method are evaluated.

Calculation Model

A mathematical model of the process is described in detail above. However, there are some changes in equations of 1.6 because a single chamber cylinder, i = 1 only and all crossflow rates are zero. Beside we have to write additional equation reflecting preheating gas mixture:

$$\frac{d(Em)}{dt} = \alpha_a S_a (T_a - T) \tag{1.9}$$

where $m \equiv m_1$ and $T \equiv T_1$ are the gas mass and temperature respectively. The initial conditions for the intake stroke of a fresh mixture of reactants in the regenerator are as follows: $T = T_0$, $\xi_j = 0$, $\xi_m = \xi_{m0}$, $T_a = T_{a0}$; in cylinder: $T = T_{r0}$, $\xi_j = \xi_{rj}$, $\xi_m = \xi_{m0}$, $V = V_0/\lambda$. Here, T_0 is the initial temperature of the mixture; ξ_{rj} is the concentration of components in the residual gases of the cylinder; T_{r0} is the temperature of the residual gas of the cylinder; V_0 is the volume of the cylinder.

An important issue of the problem is the choice of the regenerator design and parameters. The regenerator operates under non-steady-state conditions according to the following scheme: cooling of the reaction products and heating of the regenerator mass → cooling of the regenerator mass and heating of the reactants that enter the reactor → the expectation stage. Note that, instead of the stage of expectation, active operation of the regenerator in the intake and exhaust strokes of the second cylinder of the reactor is possible. To ensure the efficient operation of the regenerator, it is required that the characteristic time of the heating of the reactants in the intake stroke of a working mixture in the cylinder should be comparable to the characteristic time of the piston motion t_0 = 60/4N, where N is the crankshaft speed. Let us write the equation (1.9) that describes the heat exchange of the gas mixture with the HRE in the dimensionless form:

$$\frac{d\theta}{d\tau} = \omega(\theta_a - \theta) \tag{1.10}$$

here, $\theta = T/T_0$; $\theta_a = T_a/T_0$, is the parameter of the heat exchange, which is about unity; c is the specific heat of gas; and $\tau = t/t_c$, where $t_c = 4t_0$ is the characteristic time of a complete cycle.

Let the regenerator, which is constructed in cylindrical or rectangular geometry, consists of thin metal plates that are separated by slot channels with width δ. We express the mass of the gas and the reach of the heat exchange through the volume of the cylinder V_0 and the volume of the HRE V_a, respectively. Since $m = \rho \cdot V_0$ (ρ is the gas density) and the reach of the heat exchange at a large number of plates and their small thickness as compared to δ is $S_a \approx V_a/\delta$, we derive a relationship between the volumes:

$$\frac{V_a}{V_0} = \frac{\rho c \delta}{\alpha_a t_0 \omega} \tag{1.11}$$

The Nusselt number Nu for a flat slot at laminar flow and the parameter almost do not depend on the rate of the gas flow and $Nu = 7.5$ [28], where Pe is the Peclet number and l_a is the characteristic length of the volume of the heat exchange. The value of the heat transfer coefficient α_a between the gas and the plates of the regenerator can be estimated to be $\alpha_a = 50$ - 100 W/(m^2 K).

If the motion of the reaction products and the reactants in the regenerator occurs in separated flows through adjoining and alternating channels, the estimates by formula (1.9) yield the ratio $V_a/V_0 \approx 1 - 2$ at $\omega = 1$.

If the motion of the reaction products and the reactants in the regenerator occurs alternately through all the channels in the total flow, the estimates by formula (1.9) yield $V_a/V_0 \approx 0.5 - 1$, since the area of the heat exchange in this case is doubled (gas flows past the inner plates of the HRE from both sides). Introduction of additional cross partitions into the design of the regenerator will probably make it possible to lower the ratio V_a/V_0 down to the limiting value $V_a/V_0 \approx 0.3 - 0.5$.

In the regenerator of the given design, upon completion of the exhaust and intake strokes, there is a relatively large quantity of reaction products or a fresh mixture of reactants, respectively. Residual reaction products are returned to the cylinder of the reactor in the intake stroke, which should lead to an abrupt decrease in the production rate of the reactor and even the impossibility of its operation. The reaction products that leave the reactor are diluted with the reactants of the fresh mixture, which considerably decreases the quality of the final product.

This situation can be improved by different methods, in particular, by pumping residual reaction products and a residual fresh mixture from the regenerator to the corresponding phases of the cycles.

CALCULATION RESULTS

Calculations were carried out at the following input parameters of the problem: the cylinder diameter is 0.1 m, the piston stroke is 0.1 m, and the compression ratio is 14 - 30. The initial concentration of methane in the methane–air mixture is 0.138 - 0.242, and the crankshaft speed is 500 - 3000 rpm. The initial temperature of the mixture of reactants was chosen to be 300 K, and the temperature of the cylinder walls was taken to be 360 K. The final pressure of the compression was limited to a value of approximately 100 atm at a compression ratio of 30.

The parameter of the heat exchange of the regenerator ω varied in the range of 0.2 - 1 depending on the crankshaft speed. The regenerator with the given values of the parameter ω can be represented, for example, as a set of concentric ring layers made of metal foil with a thickness of 0.1 mm with a gap between the layers of 1 mm, which are located in a heat-insulated vessel.

The diameter of the HRE is 0.066 m, its length is 0.3 m, the total area of the heat exchange is 2 m^2, and mass is 0.8 kg. Thus, the volume of the regenerator was comparable to the volume of the cylinder.

To ensure the operation of this regenerator in the mode of the flows of the reaction products and reactants that are alternating and not separated in channels, the removal of the residual reaction products from the volume of the regenerator prior to the stage of the intake of a fresh mixture of reactants was supposed in the computational model.

Figure 1.6 shows the variation of the pressure and temperatures in the CCR cylinder and in the HRE as a function of the dimensionless time $\tau = t/t_c$ in the working cycle of the CCR. The initial and final temperature of the regenerator T_{a0} was taken to be 620 K, the initial concentration of methane in the mixture ξ_{m0} was 0.231, the compression ratio λ was 18, and the speed N was 1000 rpm.

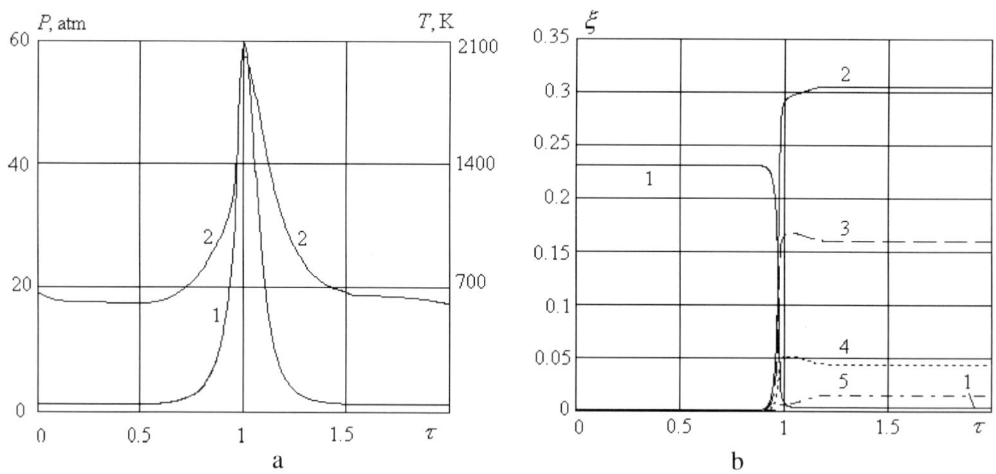

Figure 1.6. (a) Variation of (1) pressure and (2) temperature in the cylinder of the CCR; (b) Variation of the concentration of (1) methane, (2) hydrogen, (3) carbon monoxide, (4) steam, and (5) carbon dioxide in the cylinder of the CCR at $\xi_{m0} = 0.231$, $\lambda = 18$, $N = 1000$ rpm, and $T_{a0} = 620$ K.

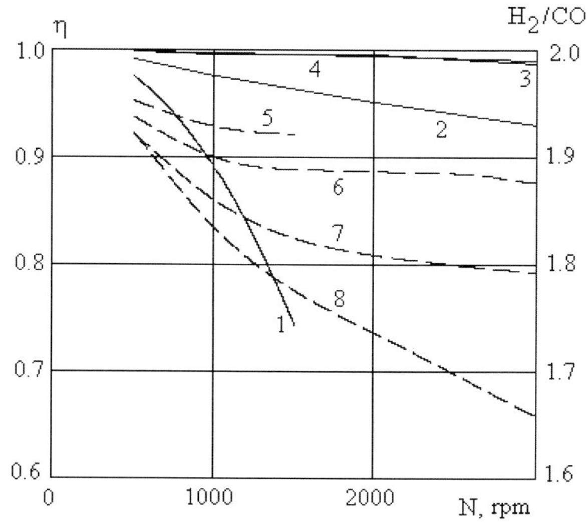

Figure 1.7. Degree of methane conversion (solid lines 1 - 4) and the H_2/CO ratio in the conversion products (dotted lines 5 - 8) as a function of the speed of the CCR crankshaft at $\lambda = 18$ and $\xi_{m0} = 0.242$ (line 1 and 5), 0.231 (line 2 and 6), 0.193 (line 3 and 7), and 0.138 (line 4 and 8).

It is seen from Figure 1.6a that the maximum pressure in the CCR cylinder does not exceed 60 atm, and the maximum temperature of the gas in the cylinder reaches 2010 K. The temperature of the HRE changes little (the maximum change in temperature ΔT_a was less than one degree) due to the higher heat capacity of the material of the regenerator as compared to the heat capacity of the gas mixture. In launching the CCR with a cold HRE, the system reaches a quasi-steady-state mode in ~ $T_{a0}/\Delta T_a$ ~ 1000 - 3000 cycles; i.e., the HRE heating up to the maximum temperature of the regenerative cycle T_{a0} occurs for 1 - 3 min of the operation of the CCR. In the quasi-steady-state mode of the reactor operation, a high degree of methane conversion to products $\eta = 99\%$ is attained (Figure 1.6b, curve 1). In this case, the concentrations of carbon monoxide $\xi_{CO} = 0.16$ and hydrogen $\xi_{H2} = 0.305$ in the reaction products are 0.16 and 0.305, respectively (the ratio of H_2/CO is 1.91). The concentrations of the remaining products (carbon dioxide and water vapor) are as follows: $\xi_{CO2} = 0.015$ and $\xi_{H2O} = 0.044$ ($CO/CO_2 = 10.7$, $H_2/H_2O = 6.9$); i.e., the amount of the products of complete oxidation does not exceed 6%. In this case, in the complete duty cycle, the reactor can produce energy that is 1.15 kW/(L·s) from the unity volume of the setup, and the energy price of 1 m^3 of the synthesis gas is 0.07 kW h at the volumetric yield of the synthesis gas of 13.5 m^3/h from 1 L of the setup. A decrease in the reactor crankshaft speed leads to an increase in the degree of methane conversion at any composition of the original mixture $\xi_{m0} = 0.138$ - 0.242, since the mixture residence time at the maximum temperature increases (Figure 1.7). At $\xi_{m0} = 0.242$, the degree of methane conversion increases from $\eta = 74.2$ to 97.4% with a decrease in N from 1500 to 500 rpm (curve 1). In the case of less rich mixtures, the degree of methane conversion is not less than 90% at $N \leq 3000$ rpm (curves 2 - 4) and is not less than 98% at $N \leq 1000$ rpm. This means that the factor of the time of the reaction occurrence is very substantial. For a rich mixture with $\xi_{m0} = 0.242$ (curve 1) at $N > 1500$ rpm, it is not possible to satisfy the condition of the closure of the regeneration “ring”, i.e., the equality of the temperature T_{a0} at the beginning of the compression stroke and at the end of the expansion stroke, due to the low calorific value of the mixture. The ratio of H_2/CO in the synthesis gas increases (Figure 1.7) with the enrichment of the mixture with methane and a decrease in N. The ratio of H_2/CO reaches 1.9 - 1.95 at $N = 500$ rpm for all the studied compositions of the original mixture. With the enrichment of the mixture, the ratio of unoxidized products or the products of incomplete oxidation to the corresponding products of complete oxidation, i.e., the ratio of H_2/CO and CO/CO_2, increases, whereas their dependence on the reactor crankshaft speed is insignificant. For example, the ratio of CO/CO_2 increases approximately from 1 to 15 and the ratio of H_2/H_2O increases from 1 to 10 at $N = 1000$ rpm and $\lambda = 18$ with an increase in the concentration of methane in the original mixture from $\xi_{m0} = 0.138$ to 0.242. At $N = 3000$ rpm (a decrease in the time of the reaction), the ratio of CO/CO_2 increases approximately from 1 to 13 and the ratio of H_2/H_2O increases from 1 to 7 with an increase in the concentration of methane in the original mixture from $\xi_{m0} = 0.138$ to 0.231.

The temperature of the HRE T_{a0} rises with an increase in the calorific value of the mixture when approaching the stoichiometric composition, but the dependence of T_{a0} on the reactor crankshaft speed is insignificant. For example, T_{a0} increases approximately from 600 to 1050 K at $N = 1000$ rpm and a pressure of compression of 60 atm with a decrease in the concentration of methane in the original mixture from $\xi_{m0} = 0.242$ to 0.138. At $N = 3000$ rpm, T_{a0} increases approximately from 700 to 1020 K with a decrease in the concentration of methane in the original mixture from $\xi_{m0} = 0.231$ to 0.138. An increase in the compression ratio λ leads to an increase in the degree of methane conversion η at any composition of the original mixture (Figure 1.8), since the maximum temperature of the mixture in the cylinder of the reactor rises. For $\xi_{m0} = 0.242$, the degree of methane conversion increases from $\eta =$ 88.9 to 95.7% with an increase in λ from 18 to 30 (curve 1). At $\lambda < 18$, it is not possible to satisfy the condition of the closure of the regeneration ring due to the low calorific value of the mixture. In the case of less rich mixtures, the degree of methane conversion exceeds 98.6% at all the values of $\lambda \geq 14$ (curves 2 and 3).

With an increase in the compression ratio, the value of the H_2/CO in the reaction products decreases (Figure 1.8); the value of H_2/CO is lower the closer the composition of the mixture is to the stoichiometrical composition. This can be associated with a decrease in the maximum temperature of the regenerative cycle with an increase in λ and with the depletion of the mixture in methane.

With the enrichment of the mixture, the ratio of unoxidized products or the products of incomplete oxidation to the corresponding products of complete oxidation, i.e., the ratio of H_2/CO and CO/CO_2, increases, whereas their dependence on the compression ratio is insignificant. The ratio of CO/CO_2 increases approximately from 4 to 15 and the ratio of H_2/H_2O increases from 3 to 10 at $\lambda = 18$ and $N = 1000$ rpm with an increase in the concentration of methane in the original mixture from $\xi_{m0} = 0.193$ to 0.242.

The temperature of the HRE T_{a0} decreases with an increase in the compression ratio at any composition of the original mixture, since a decrease in T_{a0} is offset by a corresponding increase in λ. With an increase in the calorific value of the mixture, i.e., with a decrease in the content of methane in the mixture and approaching the stoichiometric composition, the value of T_{a0} increases. For example, T_{a0} increases approximately from 580 to 830 K at $\lambda = 18$ and $N = 1000$ rpm with a decrease in the concentration of methane in the original mixture from $\xi_{m0} = 0.242$ to 0.193. At $\lambda = 30$ T_{a0} increases approximately from 490 to 760 K with a decrease in the concentration of methane in the same interval.

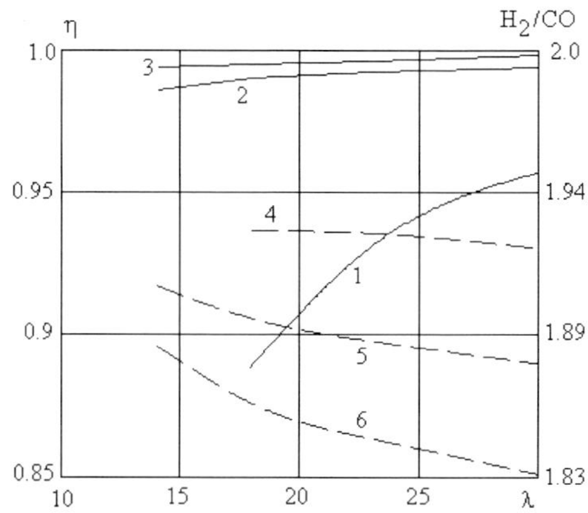

Figure 1.8. Degree of methane conversion (solid lines 1–3) in the methane–air mixture and the H_2/CO ratio in the conversion products (dotted lines 4–6) as a function of the compression ratio in the CCR at $N = 1000$ rpm and $\xi_{m0} = 0.242$ (line 1 and 4), 0.219 (line 2 and 5), and 0.193 (line 3 and 6).

Reactor with Outside Heater

Consider the simplest method of outside preheating mixture, which can be considered as a first step for investigation of reforming methane in a chemical compression reactor. It is proposed to place the electrical heater outside the CCR cylinder so preheating of the mixture prior to supply to a reaction vessel (Figure 1.9).

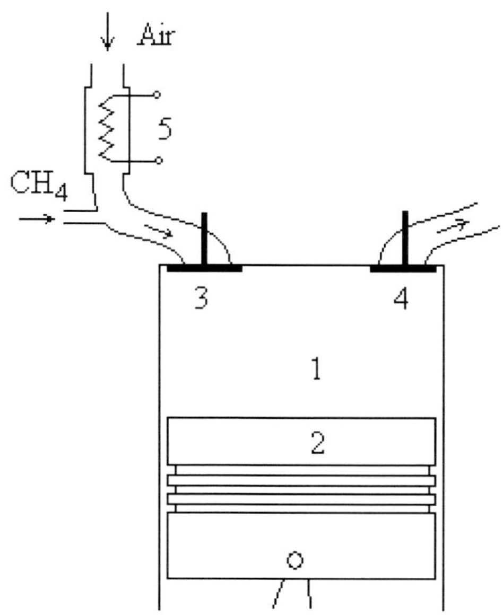

Note: 1. Cylinder; 2. Piston; 3. Intake valve; 4. Exhaust valve; 5. Heater.

Figure 1.9. Scheme of the CCR based on the ICE with outside electrical heater.

In the intake stroke, when the piston moves to the bottom dead center (Figure 1.9), the reactants take a portion of the heat ΔQ , thus heating up to the necessary temperature. Thus, it becomes possible to use a CCR for the chemical conversion of rich hydrocarbon–air mixtures with retention of all of its known advantages due to the outside preheating of the original mixture. In principle, the electrical heating can be changed by thermo heating.

Calculation Model

A mathematical model of the process is described in detail above with reference to a CCR with a outside heat regenerator. The calculations were carried out at input values of the problem applicable to design of the diesel engine, which was used in the experiment. The temperature of the engine walls was chosen 360 K, the total flow rate Q_f (cold air + methane) was changed from 40 up to 150 L/min. Because not all intake mixture could take part in the chemical reaction, the burning completeness coefficient ε was used, which was varied in the interval of $0.7 \leq \varepsilon \leq 1$.

Experimental Method

Experiments were carried out in Prof. Y. N. Chun laboratory of Chusun University, Republic of Korea. A commercial diesel engine ND130DIE was used as a CCR and its technical data are following: cylinder bore of 94.5mm, the stroke 90.5 mm, replacement volume 670cm^3. The compression ratio λ_s was 18, the crankshaft speed N changed from 1000 up to 2000 rpm.

The gas supply line was composed of the methane supply line and air supply line. Methane was supplied through the Compressed Natural Gas (CNG) cylinder filled up with a high pressure of 220 atm and was injected through the mixer via a regulator, a flowmeter (Dwyer, RMB-5, USA) and a surge tank (7.5 L) for mixture with air. The air supply line consist of an orifice flow meter (KFE, OLA-N, Korea), a surge tank (19 L), a diaphragm with a 10 mm orifice diameter, a safety valve and 6 kW electric heater. A mixer from an LPG motor vehicle was used for preparing the methane-air mixture. The spark ignition of the mixture was used to avoid hard operation of the engine.

The measuring system is composed of the devices for measurement of the temperature and the engine's rotational frequency. For measurement of the temperature, a thermocouple (K-type, outside diameter of 6 mm) was installed onto the intake manifold and exhaust manifold and a data logger (FLUKE, Hydra Data Logger 2625A, USA) was used for the temperature monitoring. A magnetic-type temperature control device was also made and used to control the temperature of the heater. For measurement of the rotational frequency, a tachometer (HIOKI, 3404 TACHO HiTESTER, Japan) was installed. As for the exhaust gas analysis system, a sampling probe was inserted into the exhaust line at the point of 390 mm away from the exhaust valve. Exhaust gas was inhaled through a vacuum pump (Gast Manufacturing Inc., 15D1150-102-1021, USA) and then passed through the impinger so that soot and moisture could be got rid of. After that, the waste gas was analyzed by the gas chromatograph (SHIMADZU, 14B, Japan). A thermal conductive detector (TCD) detector was used for the analysis, with Molecular Sieve 5A (80/100 mesh) for hydrogen, Molecular Sieve 13X (80/100 mesh) for carbon monoxide and HayeSep R (100/120 mesh) for carbon dioxide and C_mH_n.

The engine was first test-run with gasoline during 15 minutes and, only after stabilization of the temperature inside the engine, the refining fuels of methane and air were injected. Then, experiments were carried out in accordance with each of the variables, including an oxygen/methane ratio, a total flow rate under fixed intake preheating temperature.

RESULTS AND DISCUSSION

The intake temperature T_I of the gas mixture was kept of 600K in the all presented experiments. The respective calculations predicts that the hydrogen concentration have to reach the great value ~ 30 % at the equivalence ratio $\alpha \approx 0.35$ and the total intake flow Q_f of 106.5 L/min if all methane - air mixture has time for conversion. However, experiment shows the maximum value was near 20 % (Figure 1.10). As the equivalence ratio decreased, so the hydrogen concentration fell. If the equivalence ratio increased at $\alpha > 0.35$ the hydrogen concentration fell again.

The curve 4 reflects the measured crankshaft speed of the engine. The theory provides a satisfied fit if to suppose that only some part of the mixture has time for burning. It means that the crankshaft speed in the experiments was too high so required time τ_b for complete burning all mixture exceeds the characteristic time of the piston displacement τ_d. Calculation shows of 20-30% unburned mixture, i.e. the burning completeness coefficient $\varepsilon = 0.7 - 0{,}8$ (Figure 1.10).

Decreasing hydrogen yield for $\alpha < 0.35$ is because of the combustibility of the rich mixture is worsened due to lack of oxygen, which untenable to promote quick partial oxidation, i.e. $\tau_b > \tau_d$ in spite of decreasing the crankshaft speed. On the contrary, the hydrogen concentration decreases after it has reached the maximum point, despite the continual increase in the equivalence ratio. It is because of the excessive crankshaft speed, which does not provide full burning of the mixture, i.e. $\tau_b > \tau_d$ is conserved in this region of α. Similar dependences were found out for the carbon monoxide (Figure 1.11).

The results of uncompleted burning are displayed in Figure 1.12. The carbon dioxide and unconverted methane concentrations in the combustion products increase with decreasing equivalence ratio while the oxygen concentration grows.

It can be noted when the equivalence ratio was below 0.3, the excess of fuel and the discharge losses caused misfire. In the area where the ratio got above 0.7, a normal engine operation became difficult due to the pre-ignition and knocking.

The measurement of the H_2 and CO concentrations in the dry combustion products of reformed gas and the engine operation speed are displayed in Figure 1.13 and 1.14, where the total flow is changed between 40 and 130 L/min. The equivalence ratio was fixed at 0.35 and the intake preheating temperature was $T_I = 600$K.

When the total gas flow stood at 110 - 120 L/min, the concentration of hydrogen and carbon monoxide reached the maximum of ~ 20 and 15%, respectively. As the total flow decreases from $Q_f = 120$ L/min, amount of hydrogen and carbon monoxide in the synthesis gas also decreased, due to reducing the chemical reaction rate under the dropping pressure in spite of increasing τ_d.

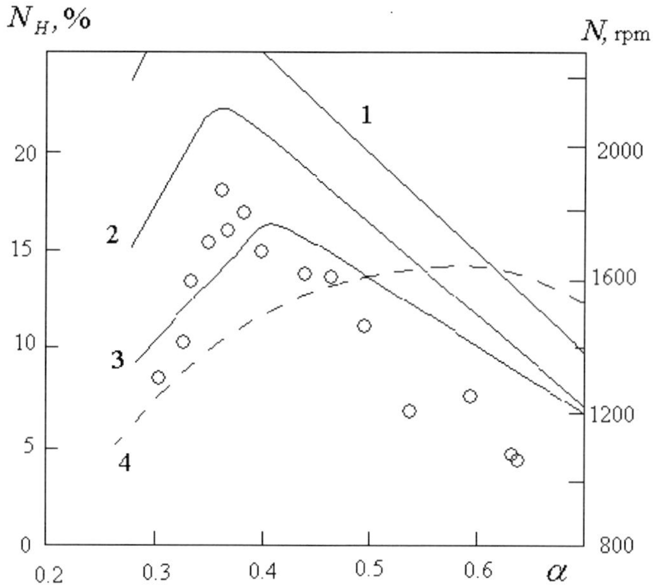

Figure 1.10. Comparison of experiment (point) and model prediction (curves 1 - 3) for the hydrogen concentration in the dry combustion products at $\varepsilon = 1$ (line 1), 0.8 (line 2), 0.7 (line 3) and the experimental engine speed (line 4).

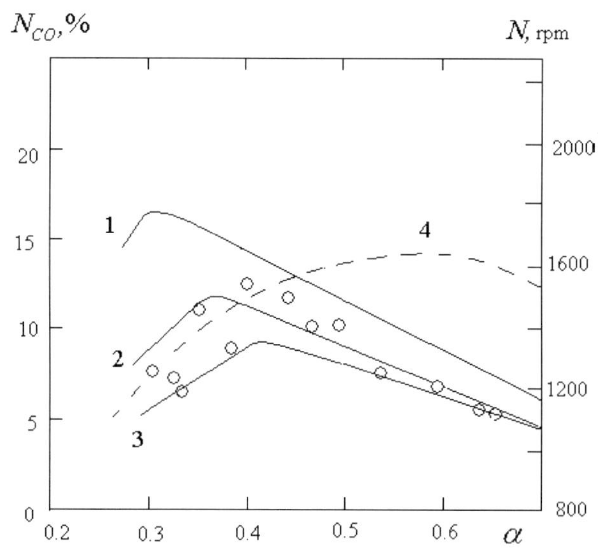

Figure 1.11. Comparison of experiment (point) and model prediction (curves 1 - 3) for the carbon monoxide concentration in the dry combustion products at $\varepsilon = 1$ (line 1), 0.8 (line 2), 0.7 (line 3) and the experimental engine speed (line 4).

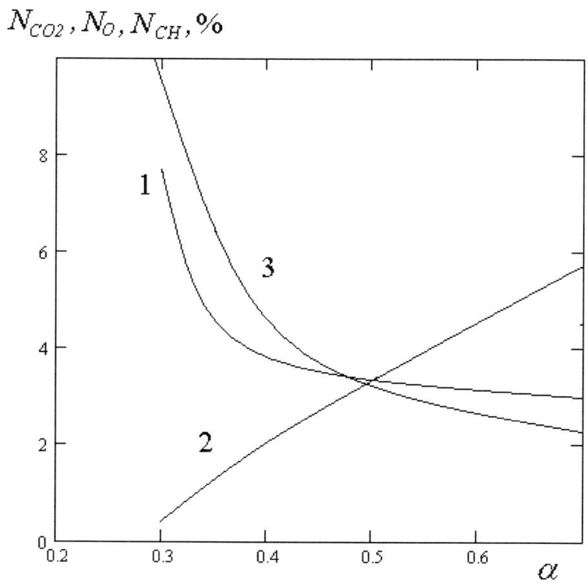

Figure 1.12. Dependence on the carbon dioxide (1), oxygen (2) and unconverted methane (3) concentrations in the dry combustion products from equivalence ratio for CH_4/Air mixture at $\varepsilon = 0.8$.

For $Q_f > 120$ L/min the hydrogen and carbon monoxide concentration fall down after reaching the peak. It is because the increased amount of mixed gas causes reducing the coefficient ε in region of the great crankshaft speed. The concentrations of methane and oxygen in the reformed gas are displayed in Figure 1.15 and 1.16. Here the crankshaft speed dependence is same as for Figure 1.13 and 1.14.

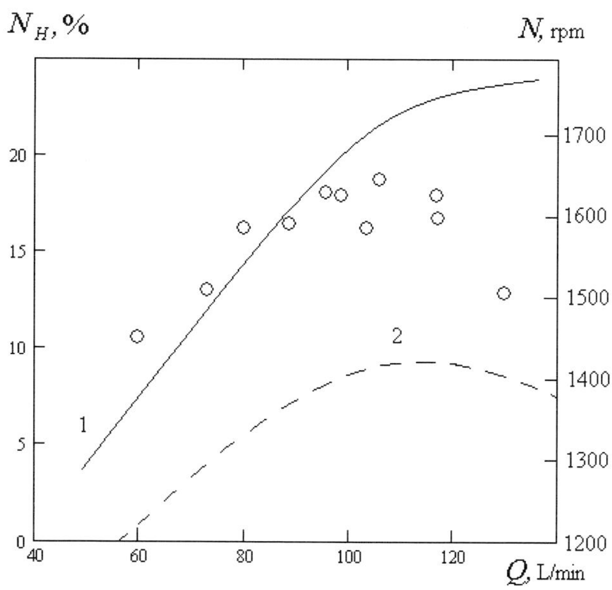

Figure 1.13. The hydrogen yield (1) and crankshaft speed (2) for variable gas consumption at $\varepsilon = 0.8$.

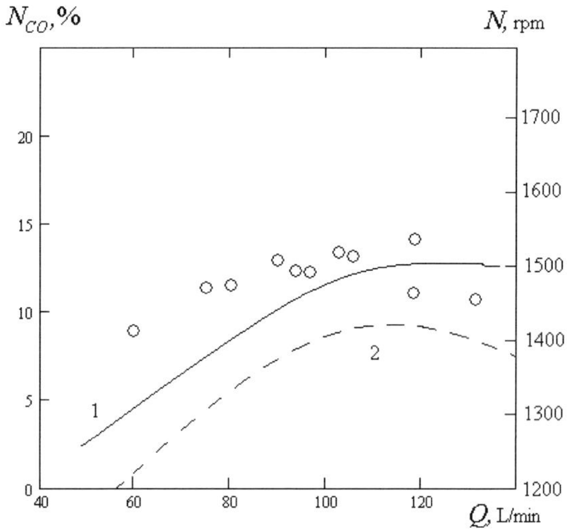

Figure 1.14. The carbon monoxide concentration (1) and crankshaft speed (2) for variable gas consumption at $\varepsilon = 0.8$.

The theory satisfactorily describes the experiment in the region $80 > Q_f > 120$ L/min. For $Q_f > 120$ L/min the experimental oxygen and rest methane concentrations in the combustion products increase unlike the theoretical values. The explanation of this effect is like as for hydrogen. Increased amount of the mixture causes reducing the coefficient ε in region of the great crankshaft speed. Besides the value ε can be function from Q_f, $\varepsilon = f(Q_f)$ reducing if Q_f grows.

It was difficult to expect the good describing experimental data by the theoretical simulation because of the simplified 0D theoretical model has been used. There are several particularities, which used model does not take into account.

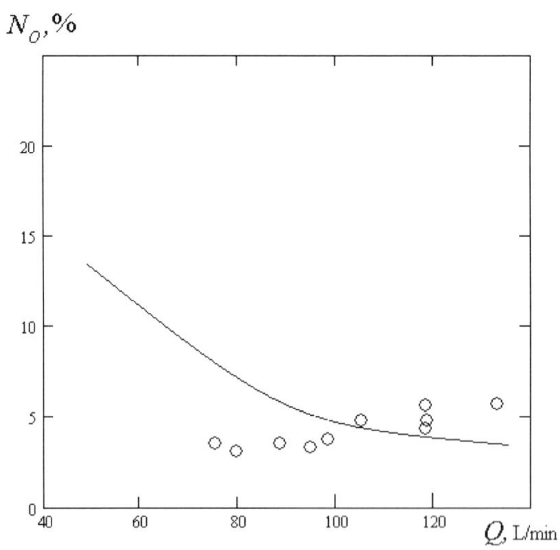

Figure 1.15. Dependence on the oxygen concentration in the combustion products from the gas consumption at $\varepsilon = 0.8$.

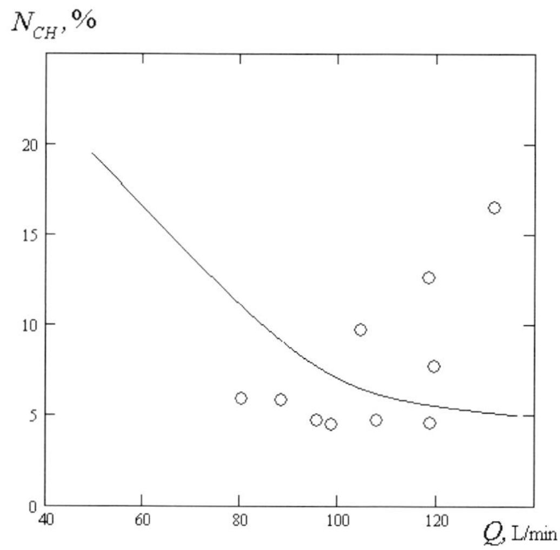

Figure 1.16. Dependence on the unconverted methane concentration in the combustion from the gas consumption at $\varepsilon = 0.8$.

In first, the theoretical model makes assumption on space ignition of the mixture by compression while the distributed burning occurs from spark ignition in the real experiment. In second, the model proposes the temperature homogeneity of the mixture in the cylinder. The real picture has 3D natures. It is possible strong cooling the heated mixture in the boundary layers in the intake pipe and cylinder when filling the cylinder by fresh heated mixture.

The mixture in the cylinder will not be space uniform in the temperature. Because ignition time delay is sensitive to the gas temperature that a chemical conversion is possible only in the volumes, where the mixture temperature exceeds the critical value. The small characteristic time of the piston displacement τ_d (great speed of the crankshaft) aggravate this situation.

Effect of the partial combustion of the mixture is confirmed that circumstance that the engine power falls (speed of the crankshaft decreases) and experimental points of the hydrogen and carbon monoxide concentrations (Figure 1.13 and 1.14) lies noticeably below calculations at increasing the gas consumption in the high magnitudes Q_f region. Simultaneously, the concentration of unburned methane is found above the theoretical curve. So using the fitting parameter ε, it takes into account the share of the mixture taking part in the chemical reaction, wholly justified. If we presume that the mixture is temperature uniform in the volume, that is possible to enter a correction in the heat exchange coefficient α_a or in the initial gas temperature T_1. However, the calculations show, that increase of α_a or reduction of T_1 cause a shift of the maximum on the hydrogen and carbon monoxide yields in the area of high α comparatively to the experimental data. Herewith, if there is a satisfactory description of the experiment $N_H(\alpha)$ that the dependency $N_H(Q_f)$ impossible to describe.

However the comparison of the experimental data with calculations with using simple theoretical model is useful because allows us to conclude about strong effect of the crankshaft speed on the syngas yield.

Conclusion

In this chapter, a method for noncatalytic partial oxidation of methane in methane–air mixtures in a chemical compression reactor is proposed and the appropriate theory is developed. Several reactors was considered that uses the operating principle of an internal combustion engine with an internal heat recovery, heat regenerator located in the channel of a combined exhaust and intake manifold, through which reaction products and reactants alternately pass, and with external electrical heating mixture.

It was demonstrated that internal heat recovery in a CCR equipped with a thermal activator makes it possible to conduct the conversion of methane in methane-air mixtures with a hydrocarbon content of up to 24% and degrees of compression of 10 to 20, and a peak pressure of less than 200 atm. A region (determined by the crossflow orifice diameter, degree of compression, and initial mixture composition) within which the condition of existence of a heat recovery cycle is met was identified. It was demonstrated that, for mixtures containing 22% methane or less, the extent of conversion exceeds 93%, a value that makes it possible to generate useful energy at degrees of compression above 10.

Reactor with heat regenerator makes it possible to implement noncatalytic conversion of methane in methane-air mixtures with the content of a hydrocarbon of up to 24% at ratios of compression of 14 - 30; the degree of methane conversion can reach above 97%, and, in this case, useful power is produced.

The theoretical analysis for CCR with external electrical heating mixture was done and the experiments were carried out with the methane-air mixtures. Results showed that the concentration of hydrogen and carbon monoxide could reach almost 20% and 15%, respectively, under the optimal standard conditions with equivalence ratio of 0.35, total flow rate of 106.5 L/min and intake preheating temperature of 600 K. It was shown the considerable role of the crankshaft speed on syngas yield.

The crucial issues of the possibility of practical implementation of these methods are associated with further detailed calculations and design study of the reactors.

Acknowledgments

I thank Dr. V. Nikolaev for help in calculations and students of Chosun University Mr. Mun Sup Lim, and Mr. Myung Seok Hong for participation in the experiments.

Author's Biography

Vladimir Shmelev was born in Russia on June 30, 1940. He received the B.S. degree from Moscow Engineering Physical Institute, Moscow, Russia, in 1964 and the Ph.D. and D.Sc. degrees in physics from the Institute of Chemical Physics, Russian Academy of Science, Moscow, in 1968 and 1982, respectively. Since 1964, he has been with the Institute of Chemical Physics, Russian Academy of Science, where he is currently the Head of the Combustion Laboratory. His current research interests include the fundamentals of the combustion theory, gas lasers on combustion, plasma chemistry. He has developed some technologies in the combustion field, which were marked by golden medals on the international exhibitions "Arhimed".

REFERENCES

[1] Yu. A. Kolbanovskii, *et al.*, *Pulse Compression of Gases in Chemistry and Technology*, Nauka, Moscow (1982), [in Russian].

[2] N. I. Kobozev, *et al.*, *The Explosive Conversion of Methane*, Trudy Vses. Sov. Po Kom. Khim. Per. Neft., Report in Three parts, Acad. Sci. USSR, Moscow (1956), [in Russian].

[3] Von I. Szesich, *Riv. Combust.* 3, 204 (1956).

[4] T. Yamamoto, *et al.*, "Production of Synthesis Gas by an Internal Combustion Engine", 6th World Petroleum Congress, Sec. IV-Paper 42-PD, 9, 1963 (Frankfurt am Main, Germ.) pp429-437.

[5] G. A. Karim, *British Chemical Engineering*, 8, 392 (1963).

[6] V. M. Shmelev, and V. M. Nikolaev, *Theor. Found. Chem. Eng.* 42, 19 (2008).

[7] Vladimir Genkin, *et al.*, *Russian Federation Patent* No. 2120913, Febr. 24, 1998.

[8] Vladimir Grunval'd, *et al.*, *Russian Federation Patent* No. 2096313, Nov. 20, 1997.

[9] V. M. Shmelev, and V. M. Nikolaev, *Problemy Mashinostroeniya Avtomatizatsii,* 4, 57 (2002), [in Russian].

[10] V. M Nikolaev, and V. M. Shmelev, "Combustin of the Lean Mixtures in an ICE with Superadiabatic Compression", Proc. X Int. Symp. on Fundametal and Applied Problems of Improvement of Reciprocating Engines, Vladimirsk. Gos. Univ., CD paper no. 05-01-043, 2005 (Vladimir, Russia), [in Russian].

[11] V. M Shmelev, and *et al.*, "The fuel gasification in an ICE cylinder", Proceeding of X International Symposium on Fundametal and Applied Problems of Improvement of Reciprocating Engines, Vladimirsk. Gos. Univ., CD paper no. 05-01-044, 2005 (Vladimir, Russia), [in Russian].

[12] M. Glouchenkov, and A. Kronberg, "Pulsed Compression Technology", A Breakthrough in the Production of Hydrogen, 16th World Hydrogen Energy Conference (WHEC), 7, 2006. (Lyon, France) pp1-7.

[13] V. M. Shmelev, and *et al.*, *High Temperature,* 46, 77 (2008).

[14] V. M. Shmelev., and *et al.*, *Combustion and Explosive*, Edited by S. Frolov, Torus Press, Moscow, 2, 12 (2009), [in Russian].

[15] F. Durst, and M. Weclas, *Proc. Proc. Inst. Mech. Eng., Part D,* 215, 63 (2001).

[16] V. M. Shmelev, and V. M. Nikolaev, *Russian Journal of Physical Chemistry B*, Pleiades Publishing, Ltd., 2, 431 (2008).

[17] V. M. Zamanskii, , and A. A. Borisov, *Itogi Nauki Tekhn., Ser. Kinet. Katal.*, 19, 156 (1989), [in Russian].

[18] G. I. Kozlov, *Proc. 3rd All-Union Meeting on the Theory of Combustion,* Akad. Nauk SSSR, Moscow, 1, 30 (1960), [in Russian].

[19] A. Nemeth, and R. Sawyer, *J. Phys. Chem.,* 73, 2421 (1969).

[20] E. L. Petersen, , and *et al.*, *Combust. Flame,* 117, 272 (1999).

[21] V. Ya. Basevich, and *et al.*, *Khim. Fiz.,* 17, 112 (1998), [in Russian].

[22] N. A. Dvornikov, *Combustion, Explosion and Shock Waves,* 35, 230 (1999).
[23] Z. A. Mansurov, *Combustion, Explosion and Shock Waves,* 41, 727 (2005).
[24] V. M. Shmelev, and *et al., Khim. Fiz.,* 25, 42 (2006), [in Russian].
[25] J. I. Ramos, *Internal Combustion Engine Modelling,* HPB, New York (1989).
[26] R. C. Reid, and *et al., Properties of Gases and Liquids,* McGraw-Hill, New York (1977).
[27] V. M Shmelev, and V.M. Nikolaev, *Theor. Found. Chem. Eng.,* Pleiades Publishing, Ltd., 43, 50 (2009).
[28] V. N. Lukanin, *et al. Teplotekhnika (Heat Engineering),*Vysshaya shkola, Moscow (2002), [in Russian].

In: Syngas
Editors: Antonius Indarto and Jelliarko Palguandi
ISBN: 978-1-62100-870-5

Chapter 2

NEW SYNGAS PRODUCTION METHOD BASED ON NONCATALYTIC METHANE REACTION WITH METAL OXIDES

H. Ale Ebrahim*
Chemical Engineering Department, Amirkabir University, Iran

ABSTRACT

Syngas is the most important intermedia
te for the natural gas conversion to petrochemical products. The usual methods for syngas production are catalytic steam reforming, partial oxidation, autothermal reforming, and recently dry reforming. The coke deposition and catalyst deactivation are the major problems for these processes.

In this chapter a new method for syngas production based on noncatalytic methane reaction with metal oxides is presented. Two suitable metal oxides for this new method are ZnO and NiO. For CH_4+ZnO reaction, the products are metallic zinc (after condensation) and syngas with $H_2/CO = 2$. Zinc can be used as an energy carrier for natural gas transportation, and it will produce hydrogen (by water-splitting) or electricity (in fuel-cells) in destination.

The syngas with H_2/CO ratio about two is also an ideal feed for methanol or gas to liquids (GTL) plants. The results of CH_4+ZnO reaction are described in a laboratory scale (thermogravimetery with an online mass spectrometer), and then in a small pilot plant reactor.

The mathematical modeling for these systems is also presented in details. The CH_4+NiO reaction is now under investigation, and some of its preliminary results are presented. This reaction can produce the syngas at lower temperatures compared to the reaction with ZnO.

---

* E-mail address: alebrm@aut.ac.ir. H. Ale Ebrahim is associate professor in the chemical engineering department of Amirkabir University (Tehran, Iran). The expertise fields of this author are mathematical modeling and experimental investigation of non-catalytic gas-solid reactions in the chemical and extractive metallurgical processes, simulation of the crude oil chemical refining units, natural gas conversions, and environmental gaseous pollution control reactions.

eliminate the furnace (and its greenhouse gas) of the steam reforming, or the costly oxygen separation unit from air of the partial oxidation.

NOMENCLATURE

ξ	dimensionless distance from the bed inlet, $= x/R_p$
ζ	dimensionless gas concentration within the pellet, $= c'/c_0$
β	dimensionless parameter in gas balance equation, $= 3\cdot(1-e)\ k_m\cdot R_p/e\cdot D_L$
ρ	dimensionless radial distance within pellet, $= R/R_p$
τ	dimensionless time for the grain model, $= k\cdot c_0\cdot M_B\cdot t/\rho_B\cdot r_{g0}$
Λ	dimensionless total length of the bed, $= L/R_p$
ε	pellet porosity
ϕ	Thiele modulus for reaction in the pellet, $= R_p\sqrt{F_g k(1-\varepsilon_0)/D_{e0} r_{g0}}$
ρ_B	true molar density of solid reactant, gmole·cm^{-3}
ρ_D	true molar density of solid product, gmole·cm^{-3}
σ_g	reaction modulus for the grains, $= \sqrt{k r_{g0}/2D_p F_g}$
Bi	Biot number, $= k_m\cdot R_p/D_e$
BL_i	the baseline of component i in the mass spectrometer (MS), mbar
c'	gas concentration within the solid pellet, gmole·cm^{-3}
C	gas concentration in the bulk phase, gmole·cm^{-3}
CP_i	cracking pattern parameter of component i
D_e	effective diffusivity within the pellet, cm^2·min^{-1}
D_L	axial dispersion coefficient within the bed, cm^2·min^{-1}
D_P	effective diffusivity of the gas in the product layer, cm^2.min^{-1}
e	bed void fraction
$f(C_{Ab})$	concentration dependency of reaction rate, gmole·cm^{-3}
F_{Ai}	volumetric flow rate of inlet methane, cm^3·min^{-1}
F_e	volumetric flow rate of exit gases, cm^3·min^{-1}
F_g	shape factor of grains 1, 2, and 3 for slab, cylinder, and sphere respectively
F_{Ii}	volumetric flow rate of inlet dilutant inert, cm^3·min^{-1}
IS_i	ionization sensitivity of component i
k	surface rate constant, cm·min^{-1}
k_m	mass transfer coefficient, cm·min^{-1}
L	total length of the bed, cm
M	molecular weight, g·gmole^{-1}
P_0	pressure of the gases in the thermogravimetry (TG), atm
Pe	Peclet number, $= u\cdot R_p/D_L$
PH_i	peak height of component i in the MS, mbar
P_i	pressure of the inlet gas, atm
PP_i	partial pressure of component i in the MS, mbar

R	radial distance in pellet for the grain model, cm
r	radius of the spherical nonporous pellet at each time, cm
r^{*}	dimensionless unreacted radius in the grain, = r_g/r_{g0}
r^{**}	dimensionless outer grain radius at each time, = $\left(Z+(1-Z)r^{*3}\right)^{1/3}$
Rec.	percentage of the zinc recovery in the liquid form
R_g	gas constant, cal.gmole^{-1}·K^{-1}
r_g	radius of unreacted core in the grain, cm
R_p	pellet radius for the grain model, cm
t	time, min
T	temperature, K
T_0	temperature of the gases in the TG, K
t_1	the initial time of a peak on MS, min
t_2	the final time of a peak on MS, min
T_i	temperature of inlet gas stream, K
TP	total pressure of gases in the MS, mbar
u	interfacial velocity, cm·min^{-1}
$W(t)$	weight of the pellet at each time, mg
x	axial distance from front of the bed, cm
$X(t)$	conversion of the ZnO at each time
y	dimensionless bulk gas concentration, = c/c_0
$y(t)$	gaseous mole fraction as function of time
Z	molar volume ratio of solid product to the solid reactant, = $\rho_B \cdot M_D/\rho_D \cdot M_B$

Subscripts and Superscripts

0	initial condition
CH_4	index for methane
B	index for solid reactant
D	index for solid product
G	index for gaseous product
i	index for component *i*
O	index for oxygen
Zn	index for zinc
ZnO	index for zinc oxide

INTRODUCTION

Introduction of this chapter is divided to three sub-sections: 1. Natural gas preferences; 2. Syngas importance; and 3. Common syngas production methods.

Natural Gas Preferences

Natural gas is one of the most important energy sources, due to the great available reservoirs and also environmental preferences. The world natural gas reserves are estimated at 3000 billion barrel oil equivalent (each barrel of oil is equivalent to about 170 m^3 of natural gas), compared with the 2000 billion barrel of known crude oil reservoirs [1].

Environmental preferences of natural gas over crude oil or coal are evident. Natural gas sweetening by amine and alkali solutions is a simple absorption process leading to complete H_2S removal [2], whereas the crude oil fractions hydrotreating is a tedious process. The latter needs special catalysts for the resistant hydrodesulphurization heavy compounds such as 4,6 dimethyldibenzothiophene to reach the level of 15 ppm remaining sulfur in diesel fuel, for the new standards [3-4]. Finally, the coal desulphurization is almost impossible and therefore, the costly flue gas desulphurization is needed [5-6]. For example, the SO2 emission from natural gas based power plants can be completely eliminated compared to the coal based power plants with about 15 kg of SO_2/MWh [7]. The main flue gas desulphurization for the coal based power plants is the lime-throwaway method. However, this reaction shows pore-mouth closure and incomplete CaO conversion due to high expansion of solid product [8]. Therefore, the efficiency of $CaO+SO_2$ reaction is relatively low.

The second environmental advantage of the natural gas over heavy fuels is lower greenhouse gas (CO_2) production due to minimum C/H ratio of the methane. For example, the greenhouse gas emissions from combined-cycle natural gas based power plants (with 53% efficiency) is about 400 kg of CO_2/MWh and is less than the 900 kg of CO_2/MWh of steam turbine coal based power plants (38% efficiency) [9]. One of the most important environmental problems in the recent decades is global warming due to greenhouse gas effect. Greenhouse gases (mainly CO_2) are accumulating in the atmosphere as a result of human activity and various industries [10-11]. There are extensive new papers about the methods for CO_2 concentration from the flue gas and further sequestration or conversion of CO_2 in the literature [12-18]. Since the above mentioned methods are very expensive, another interesting approach is to propose new production methods with inherently reduced greenhouse gas emission. For example, in the direct reduction of iron ore (by reducing gases from natural gas reforming), the CO_2 emission can be decreased considerably with respect to the usual coke-based blast furnace method [19].

The main problem against extensive consumption of natural gas is its transportation difficulties [20]. The transport of natural gas by pipelines is only economic for short distances and high flow rates. In addition, converting of natural gas to electricity and its transportation is normally designed for short distances. On the other hand, liquefied natural gas (LNG) may be transported to far distances. However, the cryogenic cost and price of special LNG ships are very high. Therefore, an interesting alternative is chemical conversion of natural gas to the transportable liquid hydrocarbons (such as mid-distillates) via gas to liquids or GTL process, and petrochemical products (such as methanol). In this field, the main intermediate material for natural gas conversion into final products is syngas (CO and H_2 mixtures).

Syngas Importance

Methane, the main component of natural gas, is a stable molecule. Therefore, direct methane conversion into final products shows low yields. Consequently, natural gas conversion to the petrochemical products is accomplished by indirect processes via syngas (mixtures of CO and H_2) as an intermediate.

The main syngas applications in the petrochemical industries are as follows [21-22]:

1. Production of ammonia, which needs hydrogen-rich syngas.
2. Methanol production via syngas with a H_2/CO ratio about two.
3. Hydrogen, acetaldehyde, acetic acid, ethylene glycol, dimethyl ether, and vinyl acetate can be produced from syngas directly.
4. On the other hand, ammonia is used for the production of nitric acid, ammonium nitrate, and urea.
5. Methanol is an important material for the production of formaldehyde, acetic acid, methyl acetate, methyl tertiary butyl ether, and acetic anhydride. Moreover, methanol can be used as an energy carrier for natural gas transportation.

In addition, heavy hydrocarbons can be produced from syngas by GTL process. The GTL process is divided into direct and indirect methods. The direct methane conversion to liquid hydrocarbons is usually a low yield process [23]. The indirect methods may be based on either methanol or syngas as intermediates. The methanol-based GTL process consists of larger number of steps, and often produces gasoline via methanol to olefins and olefin oligomerization [24]. The syngas-based GTL process steps are syngas production, Fischer-Tropsch reaction, and wax hydrocracking for conversion of natural gas to the high quality diesel fuel as the main product [1]. The suitable H_2/CO ratio in syngas for the Fischer-Tropsch reaction is about 2.1 [1].

The catalysts for the Fischer-Tropsch reaction are usually based on either iron or cobalt. The process with iron catalyst is often a high temperature reaction conducted in fluidized bed reactors with gasoline as the predominant product. The cobalt-based process on the other hand, is a low temperature reaction taking place in slurry reactors with diesel and wax as the main products [24]. After final hydrocracking, about 80% of this wax is also converted to diesel by cracking and hydrogenation [1].

The quality of diesel fuel from the low temperature Fischer-Tropsch reaction is excellent. At this condition, the diesel is essentially linear (with low aromatics) and has a cetane number of about 75, and its sulfur content is zero [1]. A comparison between a GTL diesel fuel, and a conventional diesel fuel exhaust emissions verified the environmental advantage of GTL diesel [25].

This comparison demonstrated that GTL diesel reduced unburned hydrocarbons (8%), carbon monoxide (19%), carbon dioxide greenhouse gas (3%), NO_x (10%), and completely eliminated SO_2 acid gas from the exhaust.

Finally, syngas is known as the reducing gas in the metallurgical industries. This means that it is possible to reduce iron oxides by syngas instead of coke. The direct reduction process is commercially used for the production of sponge iron by reducing gases from steam and dry reforming of natural gas. Direct reduction is accomplished in a moving packed bed reactor at the moderate temperatures [26].

Common Syngas Production Methods

The conventional methods for syngas production from natural gas are divided to the following groups:

Steam Reforming

The catalytic steam reforming method produces syngas with high H_2/CO ratio that is suitable for ammonia production [27]. The steam reforming is an endothermic reaction as follows:

$$CH_4 + H_2O \rightarrow CO + 3H_2 \tag{2.1}$$

This reaction is performed in the catalytic packed-tubes between 750-850°C and 20-30 atm, which are heated indirectly in a furnace [27]. The furnace flares and tubes are designed as top-fired or side-fired configurations [28]. In addition, simultaneous water-gas shift reaction takes place as:

$$CO + H_2O = CO_2 + H_2 \tag{2.2}$$

Therefore, the H_2/CO ratio of the final syngas in this method is above five [29]. Moreover, a part of carbon monoxide is converted to carbon dioxide. For example, SRI reformer outlet composition is as 45.3% H_2, 8.1% CO, 5.6% CO_2, 1.0% CH_4, 40.0% H_2O [30].

The basic problem of the steam reforming method is deactivation of the Ni-based catalysts due to filamentous carbon production [31]. This phenomenon can even lead to blockage of the reactor tubes in severe conditions [32]. Therefore, precise control of the process and the excess steam/methane ratio are applied [29]. The complete removal of traces SH_2 from natural gas is also needed because of the sensitivity for Ni-based catalysts [27].

The kinetics of the steam reforming method have been studied extensively [33]. The mathematical modeling for the steam reforming tubes must be two-dimensional, due to external heating and axial and specially radial temperature gradients. The heterogeneous and two-dimensional steam reformer model equations have been presented in the literature [34-35].

The steam reforming method has a thermodynamic constraint in the low temperatures (desirable for low deactivation) and high pressures (favorable for high production rates). For example, the equilibrium methane conversion at 800°C and 20 atm is about 70% for steam to carbon ratio of 2.5 [36]. The palladium membrane reactor has been proposed for selective hydrogen removal and consequently solving the above mentioned problem [37].

Partial Oxidation

The partial oxidation method offers preferences with respect to the steam reforming method. The partial oxidation of methane is a relatively exothermic reaction as follows [38]:

$$CH_4 + \tfrac{1}{2}O_2 \rightarrow CO + 2H_2 \tag{2.3}$$

The ratio of $H_2/CO = 2$ in the syngas is desirable for Fischer-Tropsch units or methanol production plants [39]. However, oxygen separation cost from the air by distillation is relatively high. There is an alternative of high oxygen permeability ceramic membrane reactor for in-situ oxygen separation [40]. In the partial oxidation method, the Ni-group catalysts are widely used. But the deactivation problem for these catalysts is serious due to coke deposition [39]. Therefore, special methods for producing high dispersion Ni catalysts, such as solid phase crystallization have been checked in recent works [39-41]. The noble metal catalysts for partial oxidation are reactive and selective, but are very expensive [39]. However, the mechanism of partial oxidation of methane on noble catalysts has been studied [42,43]. The exothermic reaction in the partial oxidation method causes hot points and heat waves in fixed bed reactors [44]. Therefore, simulation of these reactors is highly recommended because of the safety considerations. Fluidized bed reactors have also been proposed for solving the hot point problem [45]. The partial oxidation method has a thermodynamic constraint in the high pressures and high temperatures [46]. Selective removal of a reaction product can overcome this problem [47]. Therefore, palladium membrane reactors have been proposed for the complete reaction in high pressure conditions [48-49]. The partial oxidation method has been proposed for even sour natural gas conversion into syngas, by a non-catalytic oxidation method [50].

Dry Reforming

The CO_2 or dry reforming has been suggested as the following reaction in the recent years for decreasing the greenhouse gas emission [51]:

$$CH_4 + CO_2 \rightarrow 2CO + 2H_2 \tag{2.4}$$

The above equation is a highly endothermic reaction with nickel based catalysts. In this method, there is a preference of greenhouse gas consumption for production of the liquid hydrocarbons or petrochemical products. The complete kinetic study of the carbon dioxide reforming of natural gas has been presented in the literature [52]. The catalysts of this reaction are also deactivated by the coke deposition even faster than steam reforming method [53]. The simulation of packed bed reactors for the dry reforming is based on two-dimensional pseudo-homogeneous model [54]. However, there is reverse water gas shift reaction simultaneous with equation (2.4). Consequently, outlet CO mole fraction is greater than H_2 mole fraction.

Autothermal Reforming

Autothermal reforming is the steam reforming with oxygen input [28]. Therefore, the required heat for endothermic steam reforming reaction is supplied by the exothermic partial oxidation. The autothermal reforming method can be used for syngas production for the GTL applications [55]. Syngas production and oxygen plant comprise about half of the total capital cost of a GTL plant. Recent developments in autothermal reforming for syngas production in GTL plants have been presented in the literature [56].

The simulation for autothermal reforming has been done by a one-dimensional heterogeneous model [57]. In this system there is a maximum temperature for the catalyst surface in the initial reactor lengths. Then, the catalyst surface and bulk gas temperatures remain approximately constant [57].

Combined Methods

It is possible to use dry (CO_2) reforming, accompanying with the steam reforming to decrease the high H_2/CO ratio of steam reforming to a desired value [58-59]. However, both the above reactions are highly endothermic. In addition, tri-reforming process consists of a combination of dry reforming, steam reforming, and partial oxidation has been proposed [60]. In such a system, it is possible to adjust the final H_2/CO ratio in syngas [61], or to balance the heat supply and heat consumption [62].

METHANE AS REDUCING AGENT FOR METALLURGICAL REACTIONS

Using natural gas as the reducing agent in chemical and metallurgical reactions has been a research subject of Chemical Engineering Department of Amirkabir University from about 1990. The results of this research with the related publications are now described in two experimental and modeling sections briefly.

Reduction of ZnO and NiO by CH_4

The kinetics of zinc oxide reduction by methane was determined at 840-930°C by thermogravimetry and applying the sharp interface model [63]. These operating temperatures are lower than of about 1200°C for the coke-based pyrometallurgical zinc production furnaces. Effects of external mass transfer and bulk flow on the above reaction were also considered [64]. The gaseous products of $ZnO+CH_4$ reaction were analyzed by an online mass spectrometer. This test showed that syngas with H_2/CO ratio of about two was produced which can be used as feed of a petrochemical plant [65]. Therefore, the greenhouse gas emission from such a metallurgical unit can be eliminated completely by combination with a petrochemical unit. Finally, the results of $ZnO+CH_4$ reaction for syngas production at a small packed bed reactor were presented in a recent paper [66]. Complete description of the above new method for syngas production is presented in next section of this chapter.

The kinetics of nickel oxide reduction by methane were determined at 600-725°C by thermogravimetry and applying the simple grain model [67]. The outlet gaseous analysis of this reaction and applying a more real mathematical model are now under consideration. Some of the preliminary results of this reaction for syngas production are also presented in next section of this chapter.

Reduction of ZnO-PbO, $BaSO_4$, and $SrSO_4$ by CH_4

In the ZnO-PbO mixture system, the molten lead can be trapped the zinc vapor. Therefore, the lead splash condensers of the zinc production furnaces may be omitted. The kinetics of ZnO-PbO reduction by CH_4 were also determined by thermogravimetry [68].

The kinetics of barite reduction by methane to barium sulfide were determined at 850-950°C by thermogravimetry and applying the sharp interface model with product layer resistance [69]. In addition, the catalytic effect of ZnO on the above reaction was investigated successfully [70]. The kinetics of celestite reduction by methane to strontium sulfide were determined at 850-1000 °C by thermogravimetry and applying the grain model with product layer resistance [71]. In such a new process, the greenhouse gas emission can be decreased up to 50% with respect to the usual coke-based method for strontium sulfide and strontium carbonate production.

Proposing a New Method for Natural Gas Energy Transport

A new method for natural gas energy transport by metallic zinc, and hydrogen production in destination was proposed in a patent [72]. A schematic diagram of this method is presented in Figure 2.1. As this figure shows, the $ZnO+CH_4$ reaction converts natural gas into syngas and zinc. Syngas can be used as the feed of petrochemical plants or GTL units. On the other hand, zinc is transported by conventional ships simply. In destination, zinc can be used for hydrogen production by water-splitting reaction, or in zinc/air fuel cells. Finally, the produced ZnO is recycled to the system again.

Hydrogen may be the clean fuel of the automobiles. The problem of fossil fuels is producing gaseous pollutants such as NO_x, CO, and even SO_2 (from incomplete-hydrotreated fuels) which need catalytic converters, and greenhouse gas (CO_2) emission from exhaust with its drastic effect on global warming. Hydrogen combustion in engines will eliminate CO, CO_2, and SO_2 but with remaining NO_x. Finally, hydrogen usage in a fuel cell in the automobiles will omit this NO_x too.

Some of the related investigations for solar energy storage in the metals (for example ZnO dissociation by solar energy) and hydrogen production from metals by water-splitting (for example $Zn+H_2O$ reaction) were also presented in the literature [73-79].

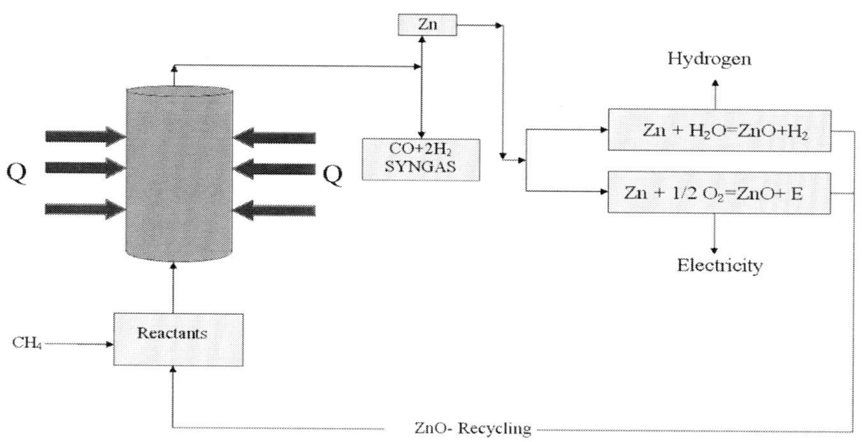

Figure 2.1. Proposed diagram for natural gas energy transport by zinc [66].

Modeling Achievements

Fluid-solid reactions are very important in many chemical and metallurgical industries. Some examples of such reactions are reduction of metallic oxides [80-81], roasting of metal sulfides [82], adsorption of acid gases [83-84], catalyst regeneration [85], phosphoric acid production [86], and active carbon preparation [87]. Chemical reaction engineering of these processes is based on a series of mathematical models [88]. These models for a solid porous pellet consist of volume reaction model [89-90], the grain model [91-93], nucleation model [94], and random pore model [95-97].

The theoretical results of our research about gas-solid reaction modeling and applying solution techniques for their differential equations are as follows:

1. A new solution technique for half-order volume reaction model was proposed [98].
2. A new method for solution of simple grain model equations was also suggested [99].
3. A similar solution technique for the nucleation model was applied successfully [100].
4. For the modified grain model with structural changes, a new solution technique was proposed [101].
5. For liquid-solid reactions with accumulation term in the fluid balance equation, a similar solution method was applied [102].
6. All of gas-solid reaction model equations were solved by the orthogonal collocation method [103].
7. The finite element method was applied to the gas-solid reaction equations with moving boundary [104].
8. Finite element method was also used for the gas-solid models with structural changes [105].
9. The random pore model was applied to the SO_2 removal system by lime successfully [106].
10. Direct iron oxide reduction method was simulated based on the grain model with product layer resistance [107].

ZnO+CH_4 System for Syngas Production

The new proposed ZnO+CH_4 system for syngas production is now described in details. First of all, the kinetic study and online gaseous analysis of the above reaction are explained in a laboratory scale. Then, the possibility of syngas production in the small pilot plant reactors is described.

This research shows that it is possible to reduce zinc oxide by methane to metallic zinc successfully [63]. Another valuable product of this reaction is syngas ($CO+2H_2$) from interaction of methane with oxygen of zinc oxide. The H_2/CO ratio of this syngas is about 2, which is very suitable for methanol production or GTL plants [63]. Therefore, this cheap source of the syngas can be very interesting with respect to the usual catalytic steam reforming or partial oxidation units. Other possible application of these gaseous products is as the reducing agent for the direct reduction steel plants.

In addition, this method can be used as a clean technology for the zinc production. Since, the common pyrometallurgical zinc plants produce a large amount of the greenhouse gas

(CO_2) which discharges directly to the atmosphere [108]. On the other hand, the gaseous products of the $ZnO+CH_4$ reaction may be the raw material for preparing petrochemical products. Therefore, it seems that the catalytic steam reformer units of such petrochemical plants and also CO_2 emission from the usual coke-based zinc production furnaces can be eliminated by proposing of a combined unit for the related metallurgical and petrochemical plants [65].

In present section, the nature of gaseous products of the $ZnO+CH_4$ reaction is considered. Experimental data were based on the thermogravimetry (TG) and simultaneous and continuous gas analysis by a mass spectrometer (MS). From a quantitative viewpoint of MS results, the mole fractions of hydrogen and carbon monoxide (gaseous products) are plotted versus time. Then the conversion-time profiles from two TG and MS methods are compared. Finally, the kinetic parameters of this reaction for design of the industrial plants are presented.

Thermodynamic Study

In the common pyrometallurgical methods for zinc production, zinc oxide is reduced by coke at high temperatures as follows [109]:

$$ZnO + C = Zn_{(g)} + CO \qquad (2.5)$$

Although the CO continues its reduction and produces CO_2, which, in contact with excess coke, converts to CO again [109].

Methane is a strong reducing agent, and reacts at moderate temperatures (below boiling point of zinc) simply [63]:

$$ZnO + CH_4 = Zn_{(l)} + CO + 2H_2 \qquad (2.6)$$

The equilibrium constants for ZnO reduction by coke and methane are presented in Table 2.1. Table 2.1 shows that the thermodynamic reactivity of methane with ZnO is very much greater than that of coke. Therefore, the operating temperature can be decreased from about 1200°C (industrial furnaces with coke) to about 900°C with methane. Moreover, the gaseous products of methane reduction ($CO+2H_2$) can be used as syngas for petrochemical industries. The environmental preference of such combined plants is omitting the greenhouse gas emission from the metallurgical unit to the atmosphere.

The thermodynamic reactivity of other reducing agents, such as CO and H_2 at moderate temperatures is very low. For example, the equilibrium constants for ZnO reduction with CO and H_2 at 850°C are 0.0050 and 0.0054 respectively. Therefore, CO and H_2 can not reduce zinc oxide at moderate temperatures in the industrial scale.

Table 2.1. Equilibrium constants for ZnO reduction by coke and CH4

Temperature	Coke	Methane
850 °C	0.079	2.7
950 °C	0.951	72.6

Raw Materials, Equipment, and Design

Kinetic study on unknown reactions must be done with pure materials. Therefore, the most pure raw materials were prepared. The zinc oxide is a guaranteed reagent from Merck (Art. No. 8849). The methane was purchased from Air Products with 99.95% purity. The inert gases (argon and nitrogen) are also high purity (99.999%) grades.

In noncatalytic gas-solid reactions, the weight of the solid pellet and concentration of the gaseous products vary with the progress of the reaction. Therefore, these reactions need special equipments for their kinetic studies. The usual system for kinetic study of noncatalytic gas-solid reactions is a single pellet reactor with an excess gas stream [110-111].

In this work, a thermogravimeter (TG) and a mass spectrometer (MS) from the Rheometric Scientific and Leda Mass companies were used. This TG can work under a reducing gas stream and has an isothermal facility with 1 °C deviation. The MS is an online gas analyzer and can plot the apparent partial pressures of up to 12 gaseous species (with sub ppm sensitivity) versus time. Because of the high vacuum in the MS and narrow capillary between the TG and MS, the time lag between the TG and MS is very small (less than 0.1 s). Therefore, side reactions in the gas phase (after TG) can not be affected the MS results.

It is conventional to use nonporous pellets for kinetic studies of unknown reactions [110]. Because modeling of nonporous reactions has an analytical solution and leads to explicit conversion-time relations that are suitable for estimation of the kinetic parameters [88]. The cylindrical pellets with height equal to diameter (5 mm), which is similar to spherical pellets, were built by high pressure (2000 bar) hydraulic press. The porosity of these pellets was about 50% (from the apparent density method). Therefore, such pellets were sintered at 1000 °C for 5 h, in order to eliminate their porosity. The porosity of the sintered pellets was below 1% (from the mercury porosimetry method).

The flow diagram of the system is presented in Figure 2.2. The solid pellet was put on a tungsten wire basket cell in the TG. The system is heated to the desired temperature under an argon gas stream (gas 1). Then, the isothermal period begins, and after temperature stabilization, the reducing gas (a mixture of CH_4/Ar through gas 2) is introduced from the bottom of the TG.

The outlet gas stream leaves the system from the top through a bubbler (for slight positive pressure control). The MS suction is also from the outlet gas stream and through a narrow heated capillary. At this isothermal temperature, the weight of the pellet and apparent partial pressures of the outlet gases are plotted versus time.

One of the most important continuous methods for analysis of multi-component gas mixtures is mass spectrometry [112]. But a predefined calibration recipe has been proposed for quantitative gas analysis by MS [113].

Only a method for semi-quantitative (or comparative) multi-gas analysis by mass spectrometry exists in the literature [114]. This semi-quantitative method was also used for interpretation of FTIR results [115]. In the present section, a basic method for quantitative gas analysis by MS is introduced, and its accuracy is verified by decomposition gases from a standard material.

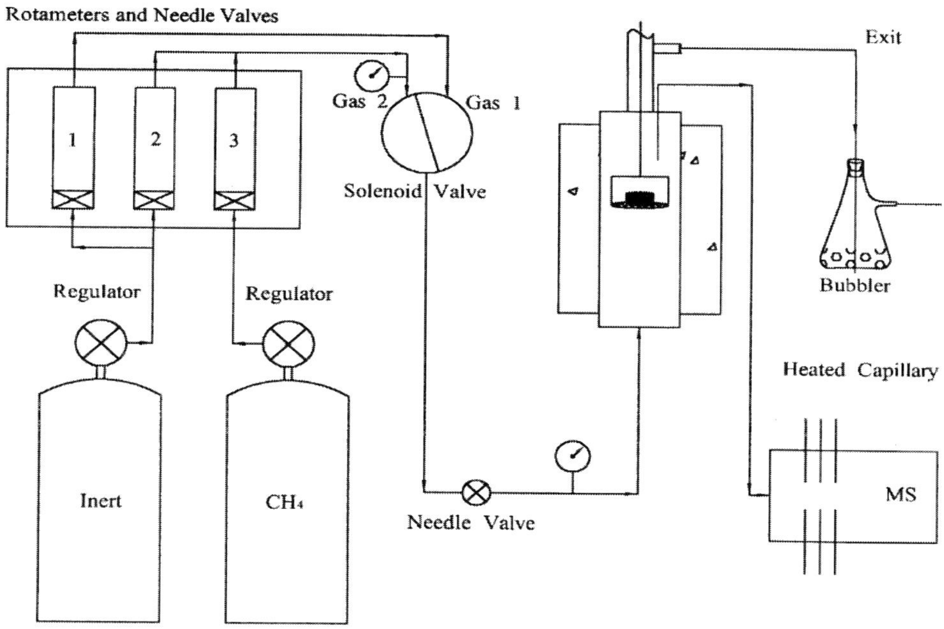

Figure 2.2. Flow diagram of system for the kinetic study of $ZnO+CH_4$ reaction.

The MS plots the apparent partial pressures of the gaseous spices versus time. However, these apparent partial pressures are not true pressures, and we named them peak heights (PH). In the MS, each gas cracks to a series of ionized fragments by electron bombardment. For example, argon is cracked to ionized argon (40 amu, and base peak) and double ionized argon (20 amu) with its height about 13% of the base peak. Therefore, the cracking pattern parameter (CP) of 1.13 must be considered for argon. Moreover, the ionization sensitivities of different gases are different. For example ionization sensitivity (IS) of argon is 1.2, while the IS of hydrogen is 0.44 (more stable than argon). Finally, the baseline (BL) of the peaks in the blank test must be subtracted from the reduction tests. These baselines are due to thermal cracking of methane, MS filament emission, trapped humidity in the MS, and the small leak of air to the MS. Now, the true partial pressure (PP) of each gas can be computed from the following equation [65]:

$$PP_i = \frac{PH_i . CP_i}{IS_i} - BL_i \tag{2.7}$$

The total pressure (*TP*) is the summation of the partial pressures (*PP*). But in this system with an excess methane stream, it is approximately equal to the summation of the methane and argon partial pressures and remains relatively constant:

$$TP = \sum PP_i \cong PP_{CH4} + PP_{Ar} \tag{2.8}$$

Now, by dividing each gaseous partial pressure to the total pressure, the mole fraction profile of each gas can be computed:

$$y_i(t)=\frac{PP_i(t)}{TP} \tag{2.9}$$

The base peak (*BP*), cracking pattern parameter (*CP*), and ionization sensitivity (*IS*) for some common gases are presented in Table 2.2 [116].

Table 2.2. BP, CP, and IS for some gases

Gas	*BP* (amu)	CP	IS
CH_4	16	2.11	1.6
Ar	40	1.13	1.2
H_2	2	1.03	0.44
CO	28	1.08	1.06
H_2O	18	1.22	1.0
CO_2	44	1.24	1.4

Because of high volatility of molten zinc above 800 °C, most of the liquid zinc in equation (2.6) evaporates, and the TG results must be corrected. This correction is based on the true liquid zinc recovery and can be expressed as follows [63]:

$$X(t)=\frac{W_0-W(t)}{W_0\dfrac{M_O+(1-\mathrm{Re}c.).M_{Zn}}{M_{ZnO}}} \tag{2.10}$$

The method of estimating liquid zinc recovery on the TG is based on final heating period under an inert atmosphere. Therefore, the remaining liquid zinc on the pellet surface evaporates and it can be determined from the final weight loss. For example, liquid zinc recoveries were 22% and 17% at 850°C and 900°C respectively [63]. This correction gives very good agreement with respect to the MS results, which is discussed later. In addition, by mixing ZnO with PbO it is possible to enhance the above liquid zinc recoveries. For example, the liquid zinc recovery was improved from 22% to 54% at 850 °C for a mixed pellet with Zn/Pb=0.17 [68].

This work used two simultaneous TG and MS methods successfully, which is an important advantage. Kinetic studies of simple noncatalytic gas-solid reactions are usually based on only one method. For example, Guger and Manning used the TG method for a kinetic study of the ZnO+CO reaction at high temperatures [117], and Gioia et al. used water vapor analysis for the ZnO+H_2 reaction [118]. However, for complicated reactions such as ZnO+CH_4 at moderate temperatures with partial zinc evaporation, two simultaneous methods must be used for getting accurate results.

From the MS results, the mole fraction profiles of the gaseous products of the reaction can be plotted versus time. For the ZnO+CH_4 reaction, the conversion of ZnO can be expressed from the peak area of such mole fraction profiles as follows [65]:

$$X(t) = \frac{M_{ZnO}}{W_0} \frac{P_i}{R_g T_i} (F_{Ai} + F_{Ii}) \int_0^t y_{CO}(t)\, dt \tag{2.11}$$

Preliminary Tests

Compatibility tests between the MS and flowmeters results on the CH_4/inert mixtures are indicated in Table 2.3. The other important preliminary test is compatibility of the TG and MS results. This test is based on the thermal decomposition of calcium oxalate monohydrate. Calcium oxalate monohydrate decomposes in three stages. Figure 2.3 shows the TG curve for this multiple decomposition under an argon stream. The MS results for this test are also presented in Figure 2.4 with the peaks of H_2O, CO, and CO_2 respectively.

Table 2.3. Percent of methane in the CH_4/inert mixtures

Gas mixture	CH_4/N_2	CH_4/Ar
Percent of methane from flowmeters	29.9	48.8
Percent of methane from MS	32.0	49.2

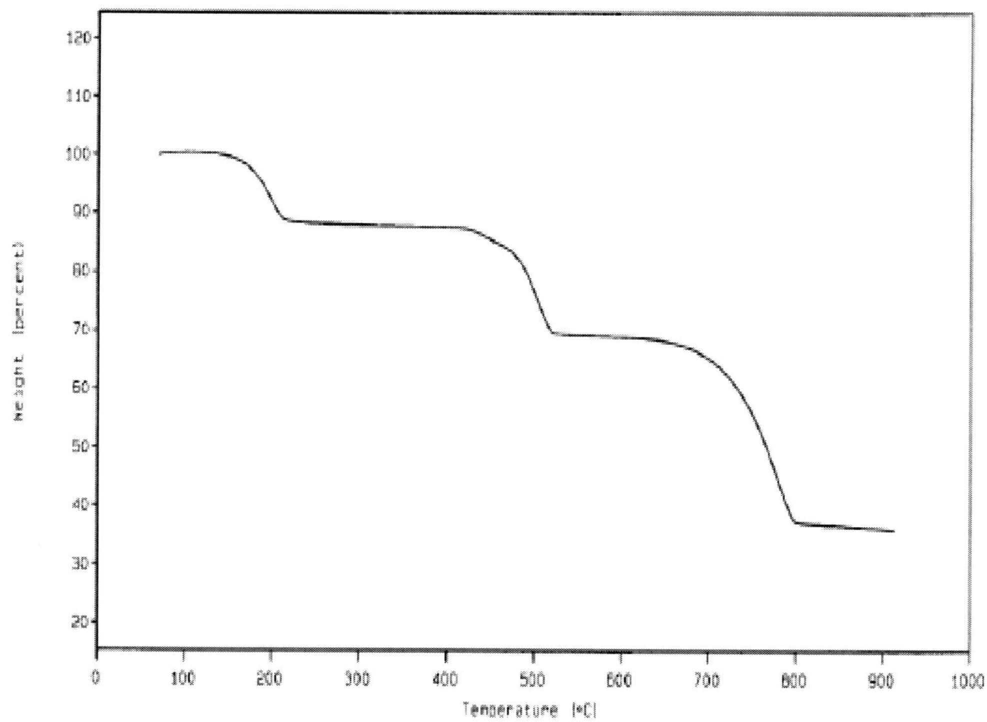

Figure 2.3. TG results of calcium oxalate decomposition.

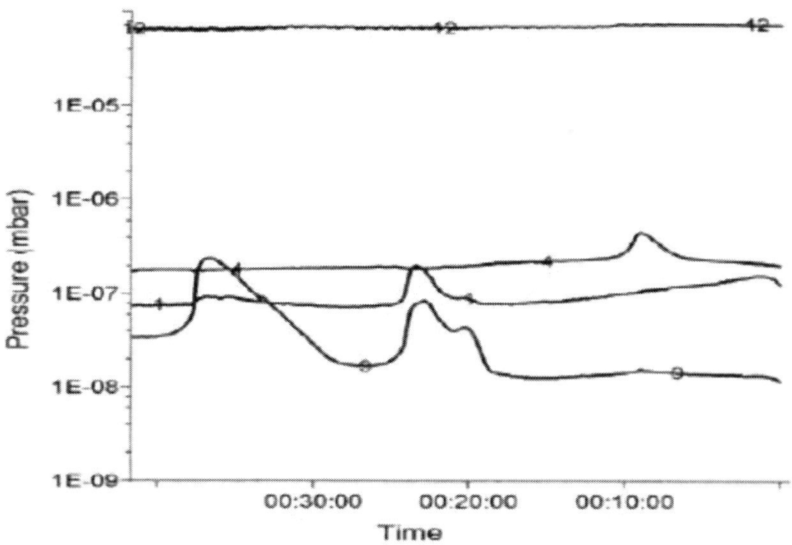

Figure 2.4. MS results of calcium oxalate decomposition under argon stream.

The weights of these gaseous products from the TG curve (Figure 2.3) can be determined simply.

Moreover, the weights of the gaseous products can be computed from the peak area of the MS curves as follows [65]:

$$W_g = M_g \frac{P_i}{R_g T_i} (F_{Ai} + F_{Ii}) \int_{t_1}^{t_2} y_g(t)\, dt \tag{2.12}$$

The computed weights of the gaseous products from these two methods are in good agreement. This comparison is presented in Table 2.4.

Finally, temperature calibration of the TG was accomplished by magnetic method (Curie point of trafoperm).

Table 2.4. Decomposition products of calcium oxalate

Gaseous products	H_2O	CO	CO_2
Weight from TG (mg)	0.76	1.18	2.06
Weight from MS (mg)	0.68	1.30	2.28

Final Tests

Reduction of semi-spherical sintered ZnO pellets in the TG was studied between 840 - 930°C and 20 - 60% CH_4/Ar. Some of the TG curves for ZnO reduction by methane with final liquid zinc evaporation period are presented in Figures 2.5 and 2.6. Figure 2.5 is for 860°C test with 40% CH_4/Ar, and Figure 2.6 shows the results of reduction at 900°C with 30% CH_4/Ar mixture.

Now the gaseous products of the reaction are considered. Some of the MS curves for the ZnO+CH_4 reaction are presented in Figures 2.7 – 2.8. Figure 2.7 is for unsintered ZnO pellet reduction at 880°C with 40% CH_4/Ar. Figure 2.8 shows the results for a sintered ZnO pellet reduction at 900°C by 60% CH_4/Ar.

Also in Figure 2.9, the continuous decrease of the peak heights of CO and H_2 by progress of the reaction is presented more clearly. Therefore, the gaseous products of the reaction are carbon monoxide and hydrogen. By equation (2.7) – (2.9), the mole fraction profiles of H_2 and CO can be obtained from the MS results. These mole fraction profiles are presented in Figures 2.10 and 2.11 for the reduction tests at 880°C (unsintered ZnO) and 900°C (sintered ZnO), respectively.

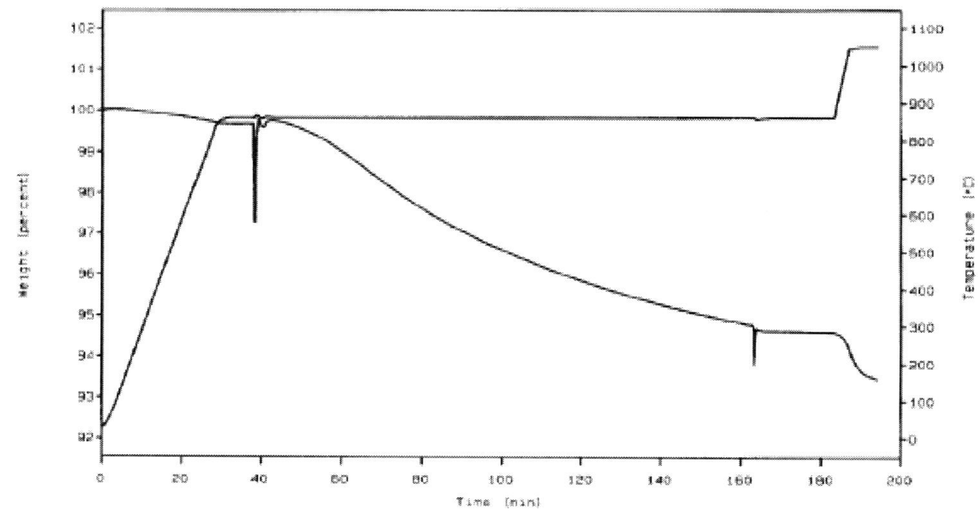

Figure 2.5. TG results for ZnO reduction at 860°C and CH_4/Ar ratio of 40%.

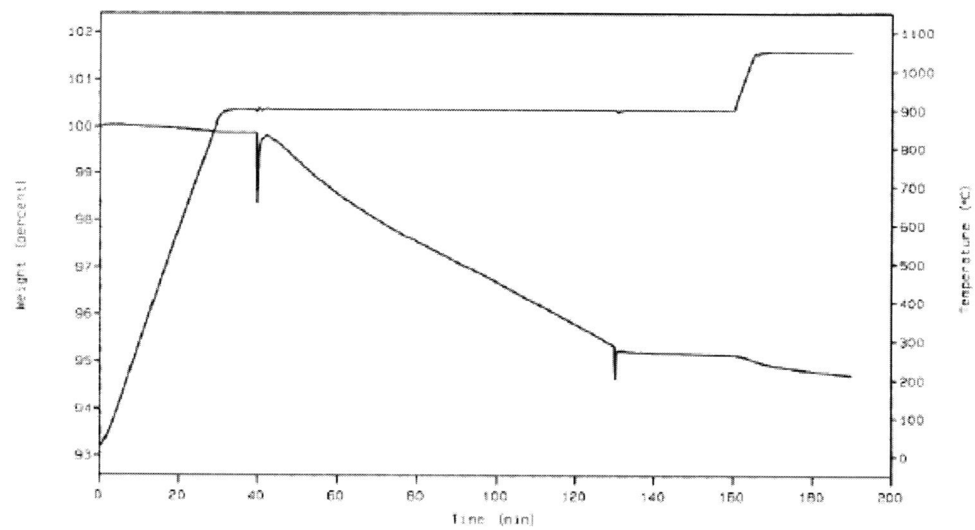

Figure 2.6. TG results for ZnO reduction at 900°C and CH_4/Ar ratio of 30%.

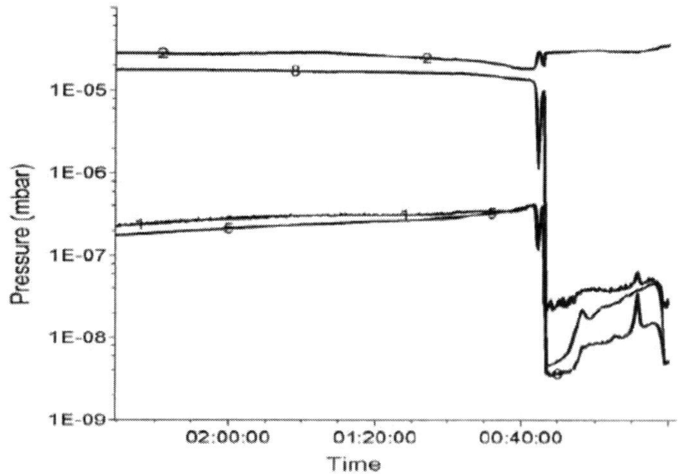

Figure 2.7. MS results for unsintered ZnO reduction at 880°C and CH_4/Ar ratio of 40%.

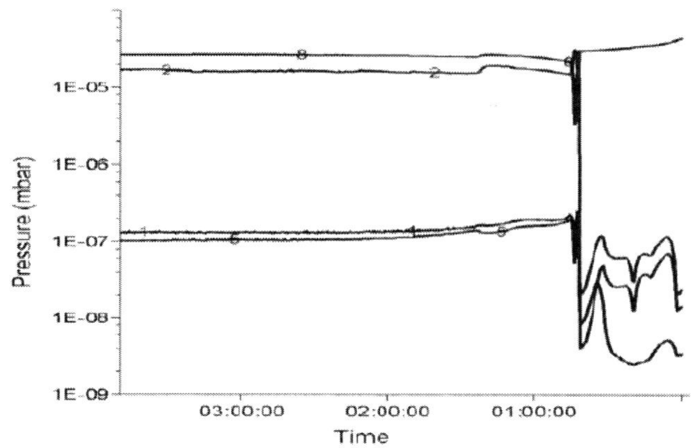

Figure 2.8. MS results for ZnO reduction at 900°C and CH_4/Ar ratio of 60%.

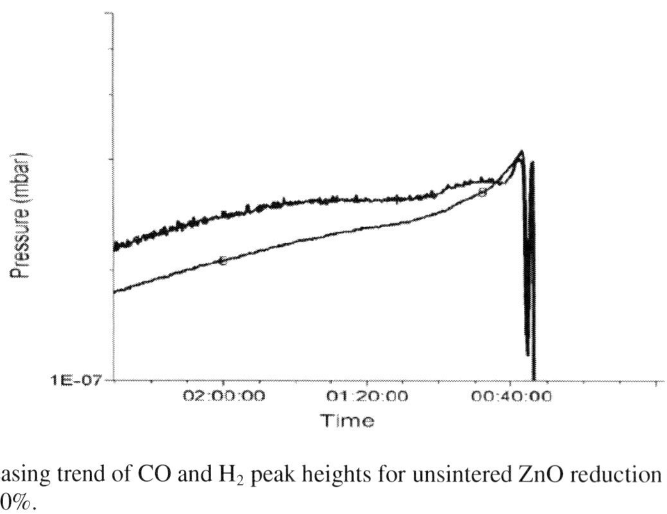

Figure 2.9. Decreasing trend of CO and H_2 peak heights for unsintered ZnO reduction at 880°C and CH_4/Ar ratio of 40%.

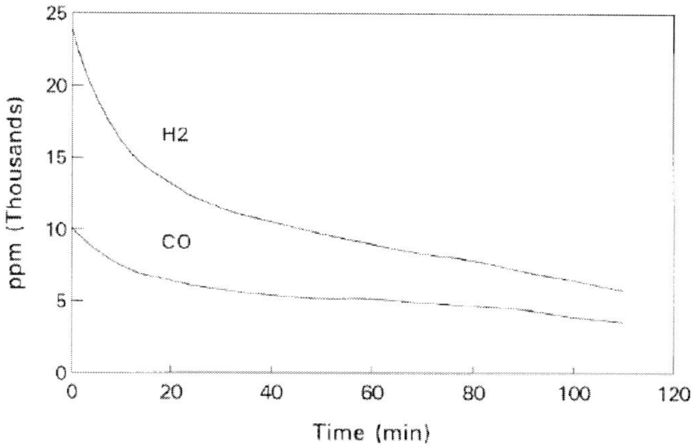

Figure 2.10. Mole fraction profile of gaseous products for unsintered ZnO reduction at 880°C and CH_4/Ar ratio of 40.

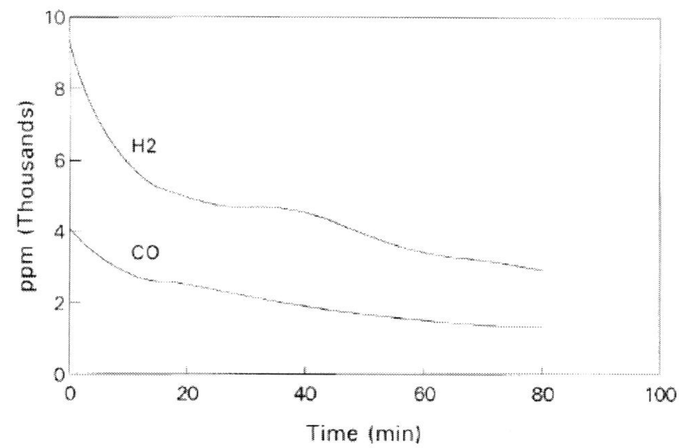

Figure 2.11. Mole fraction profile of gaseous products for ZnO reduction at 900°C and CH_4/Ar ratio of 60%.

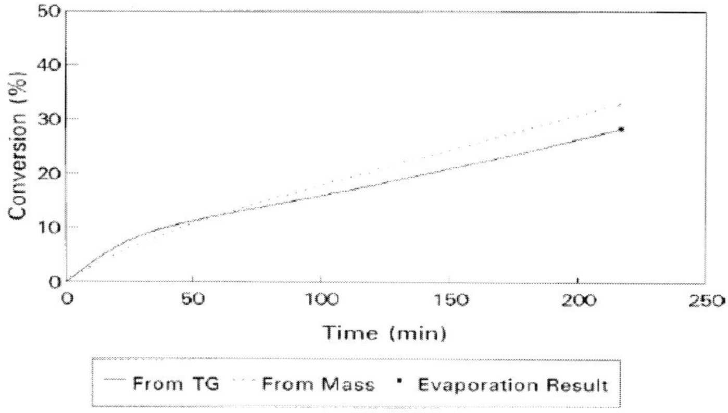

Note: TG = thermalgravimetry; Mass = mass spectrometer.

Figure 2.12. Conversion-time profile for ZnO reduction at 900°C and CH_4/Ar ratio of 60%.

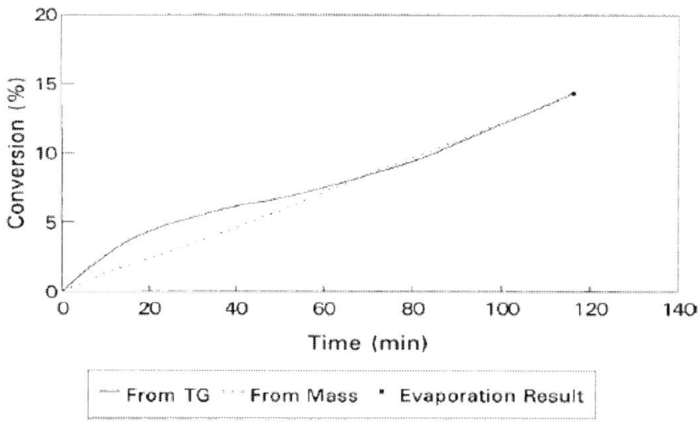

Note: TG = thermalgravimetry; Mass = mass spectrometer.

Figure 2.13. Conversion-time profile for ZnO reduction at 910°C and CH_4/Ar ratio of 40%.

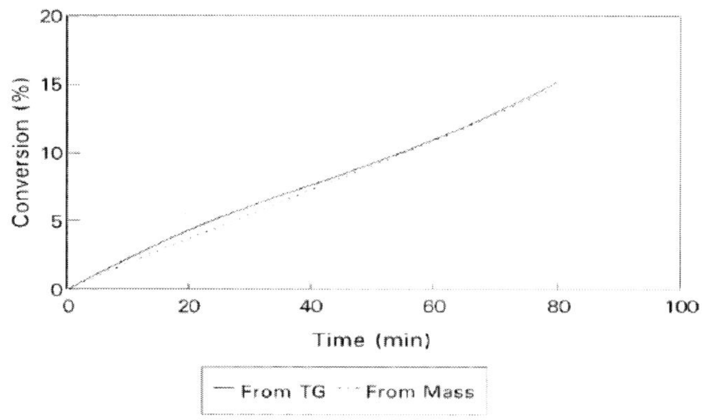

Note: TG = thermalgravimetry; Mass = mass spectrometer.

Figure 2.14. Conversion-time profile for ZnO reduction at 920°C and CH_4/Ar ratio of 40%.

The ratios of H_2 to CO in Figures 2.10 and 2.11 are about two and equal to its stoichiometric reaction ratio of equation (2.6).

Now by computing the peak area of CO, equation (2.11), the conversion-time profiles can be obtained from the MS results. Figures 2.12 – 2.14 show the conversion-time profiles for reductions at 900, 910, and 920°C, respectively. In these figures, a good agreement between the corrected TG results, equation (2.10), and the MS results, equation (2.11), was achieved, which is one of the important findings of this work [65].

Kinetic Parameters Estimation

Mathematical modeling of experimental data is very important for determination of the kinetic parameters. Moreover, by using a suitable model, it is possible to design an industrial reactor. Some of the important gas-solid reaction models have been reviewed by Ramachandran and Doraiswamy [88].

In this section, nonporous sintered ZnO pellets were used, and therefore, the sharp interface model was applied. The complete kinetic study of the $ZnO+CH_4$ reaction at the chemical control regime on a TG scale has been published [63]. The conversion function profile of the spherical pellets at the chemical control regime can be expressed as follows [63]:

$$\frac{k\,f(C_{Ab})t}{\rho_B\, r_0}=1-[1-X(t)]^{1/3} \tag{2.13}$$

One of the conversion function profiles is indicated in Figure 2.15.

The slopes of conversion function profiles at 900°C and various methane concentrations were used for determination of concentration dependency of the reaction. The concentration dependency of the $ZnO+CH_4$ reaction between 20 - 60% CH_4/Ar was expressed by the Langmuir-Hinshelwood relation [63]:

$$f\left(C_{Ab}\right)=\frac{C_{Ab}}{1+46136C_{Ab}} \tag{2.14}$$

The above equation was obtained from Figure 2.16 at 900°C and CH_4/Ar range of 20 - 60%.

Then from slopes of the conversion function profiles at various temperatures, the rate constants were determined. The temperature dependency of the $ZnO+CH_4$ reaction rate constants between 840 - 930°C is as follows [63]:

$$k=4.95\times10^{12}\exp\left(-\frac{67090}{R_gT}\right) \tag{2.15}$$

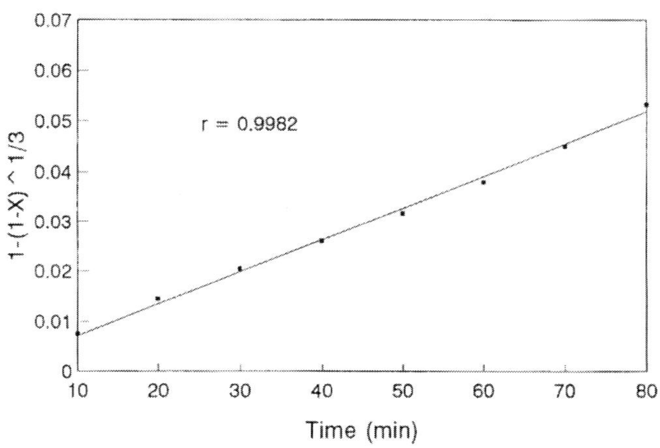

Figure 2.15. Conversion function for ZnO reduction at 920°C and CH_4/Ar ratio of 40%.

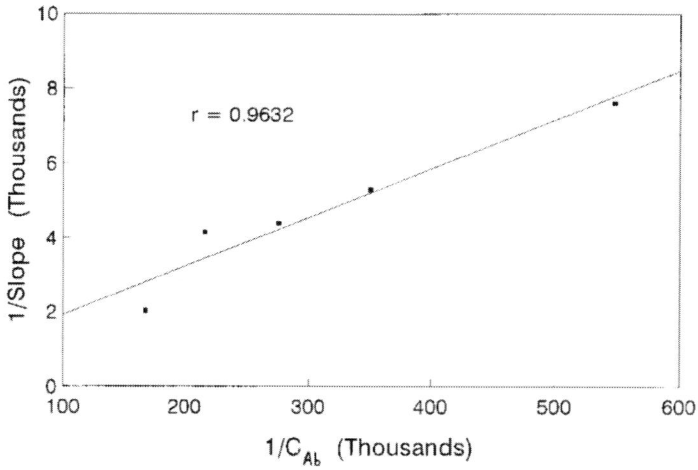

Figure 2.16. Concentration dependency of ZnO reaction at 900°C and CH_4/Ar ratio of 20-60%.

The above equation was obtained from Arrhenius plot, which is presented in Figure 2.17.

The effects of external mass transfer and bulk flow will be important in industrial $ZnO+CH_4$ systems with large pellets. Because the mass transfer coefficient is nearly proportional to the inverse of the pellet radius. A mixed control mathematical model for the mass transfer and bulk flow effects and surface reaction has been presented elsewhere [64]. The following implicit conversion-time equations were proposed for the mixed control regime:

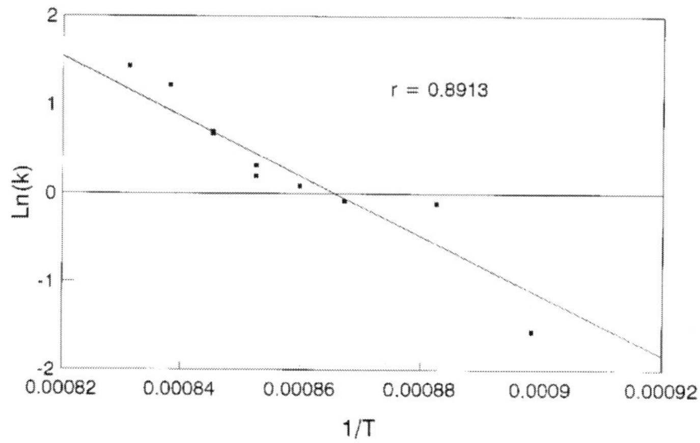

Figure 2.17. Temperature dependency of $ZnO+CH_4$ reaction at 840 - 930°C.

$$r = r_0 (1-X)^{1/3} \tag{2.16}$$

$$k_m = \frac{a_1}{r} + \frac{a_2}{\sqrt{r}} \tag{2.17}$$

$$t = \frac{2\rho_B(1+3y_{Ab})}{C_{Ab} a_2^4}[a_1^3 \ln(\frac{a_1 + a_2\sqrt{r}}{a_1 + a_2\sqrt{r_0}}) + \frac{a_1 a_2^2}{2}(r-r_0) - a_1^2 a_2(\sqrt{r}-\sqrt{r_0}) - \frac{a_2^3}{3}(r\sqrt{r} - r_0\sqrt{r_0})] - \frac{\rho_B(r-r_0)}{kC_{Ab}} \tag{2.18}$$

The rate constants of the $ZnO+CH_4$ reaction at high temperatures (1000 - 1025°C) were also presented in reference [64]. In addition, a good agreement between the mixed control model and experimental data was achieved at high temperatures such as in Figure 2.18.

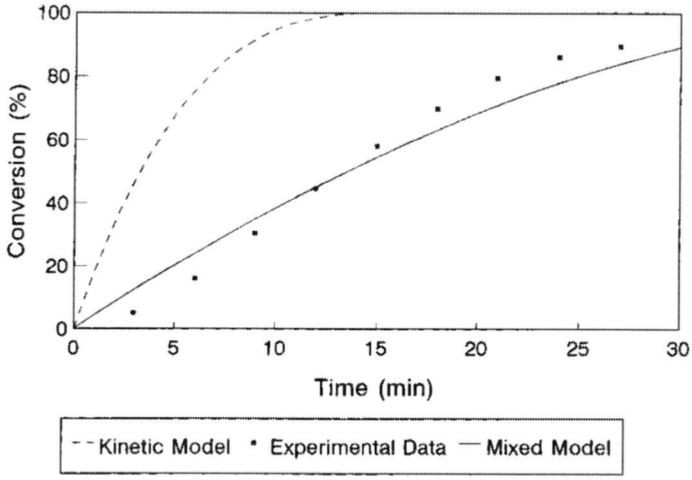

Note: The data was obtained at T of 1025°C, CH_4/Ar ratio of 26%, flow of 125 cm^3/min, and pellet radius of 0.24 cm.

Figure 2.18. Conversion-time profiles prediction from models and comparison with experimental data for $ZnO+CH_4$ reaction.

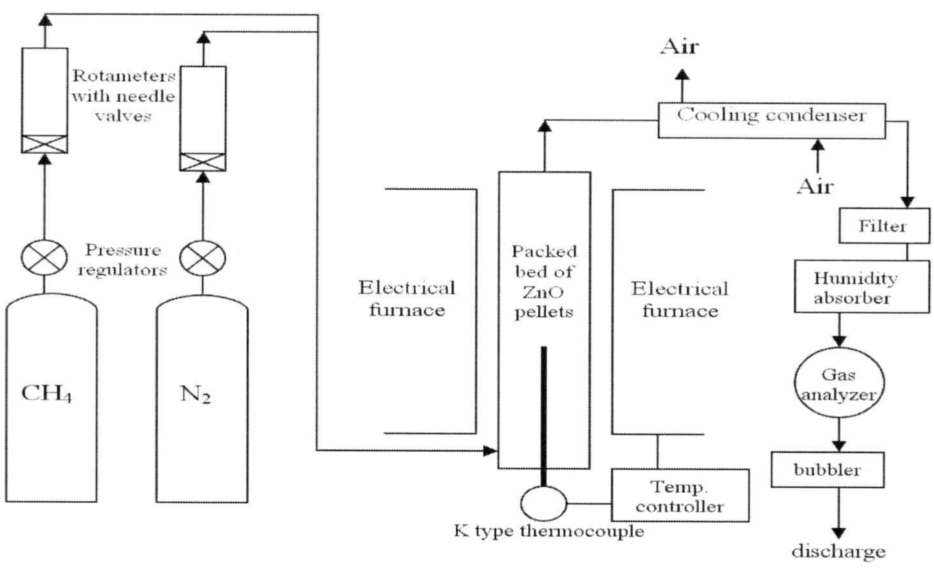

Figure 2.19. Flow diagram of the packed bed experimental set-up.

All of the necessary mathematical equations and kinetic parameters for the design of industrial plants based on this new method for syngas and zinc production from the $ZnO+CH_4$ reaction were presented in equation (2.13) – (2.18).

In addition, by combination of a metallurgical and a petrochemical unit, an economical plant can be expected. In such an integrated system, the greenhouse gas emission of a conventional metallurgical unit will be eliminated. Moreover, the catalytic steam reformers of the petrochemical unit will be omitted. This method seems to be competitive with respect to other greenhouse gas elimination proposals [17,119].

Pilot Plant Scale Tests

Since the syngas mole fraction in the previous laboratory scale experiments with a single ZnO pellet was low (for example about 3% in Figure 2.10), small pilot plant tests were designed for reaching to a nearly pure syngas stream.

In this step of research, a small pilot plant packed bed reactor has been employed for syngas production from the $ZnO+CH_4$ reaction. The simple prototype reactor has been constructed and tested at different temperatures, and various bed heights. The outlet gaseous analysis showed high quality syngas production without any effect of methane cracking or undesired side reactions. Finally, the modified grain model has been applied to this system successfully [66].

The resulting zinc product can be employed in the zinc /air fuel cells, or it can be used in the water splitting reaction to produce hydrogen at destination and then zinc returns to zinc oxide again. Thus zinc takes the role of an energy carrier in the proposed cycle (Figure 2.1), or a solid fuel [66].

Figure 2.19 shows a schematic flow diagram of the experimental set up for the packed bed reactor. The reactor is a cylindrical heat resistance steel alloy with an inner diameter of 5 cm and a length of 120 cm. The reactor was mounted vertically on the inside of an electrical furnace. The heating zone of the furnace is 75 cm. The bottom of the reactor was filled with inert packing ceramic material in order to preheat methane before its entry into the reaction zone. The height of the ceramic material in the heating zone was 25 cm. The flows of inlet and reducing gases were adjusted by two rotameters. The spherical pellets of ZnO were placed over the packed ceramics. The packed bed temperature was monitored and controlled during the reaction with a type-K thermocouple of length 50 cm inserted through the bottom of the reactor. Therefore, the thermocouple sensed the temperature of the ZnO packed bed. The reaction products exited the reactor and flowed through a simple horizontal steel pipe which was cooled with air. In this air cooled condenser, the zinc vapor was condensed and then solidified. Some portion of the fine zinc particles which escape from the condenser was collected downstream by a filter. The composition of the product gases were quantitatively analyzed online by FTIR. Samples of the solid products collected from the condenser were analyzed by both SEM and XRD tests.

During the experiments, the reactor was heated to the desired temperature under the inert gas stream. The nitrogen flow during the start-up was 300 $ml \cdot min^{-1}$. After establishing the isothermal temperature, the flow of nitrogen was substituted with methane flow. The reaction begins by contacting the methane with hot spherical ZnO pellets. All of the experimental tests in this work have been carried out with a pure methane flow at atmospheric pressure. The inlet methane flow rate was in the range of 150-750 $ml.min^{-1}$. The gaseous products including zinc vapor left the reactor and then entered a horizontal condenser. Then the solid zinc was separated from the syngas in this condenser. Finally, the outlet gases were conducted to a FTIR system for on line gas analysis.

The ZnO powder was 99.7% purity with a mean particle size of about 0.5 micron, bulk density of 0.8 g/cm^3, and specific surface area of 4.6 m^2.g^{-1}. This powder was mixed with water to produce a clay form. Then this clay was used for producing of the semi-spherical pellets. These pellets were dried at room temperature for 48 h, and then the dried pellets were heated at 500°C for 1 h in a furnace. Finally, the pellets showed a satisfactory mechanical strength. The porosity of the dried spherical pellets was about 62% from apparent density method. The results of nitrogen adsorption (BET method) showed a surface area of 3.5 m^2.g^{-1} for the final pellets.

Experimental runs were carried out with three packed bed heights at different temperatures. In a typical run with a fixed packed bed height, the methane flow rate was manually controlled so that the unreacted methane mole fraction of the outlet gases becomes negligible at the operating temperature. After such test, the methane flow was substituted with inert flow of nitrogen and then the reactor was heated to the next desired higher temperature. Three different temperatures, i.e., 860, 940, and 995 °C were considered in each run. The experimental results of the packed bed system are presented in Table 2.5. This table clearly shows the effect of operating temperature and height of the reactant bed on the maximum syngas production capacity of the system. These syngas flowrates were determined in a condition for approximately complete methane reforming. The outlet experimental gaseous mole fractions are also presented in Table 2.5, which is discussed later.

A packed bed drier ($CaCl_2$) was installed just prior to the gas analyzer in order to adsorb the humidity content of the outlet gaseous stream. Therefore, the outlet gas analyses by FTIR are dry basis mole fractions.

Table 2.5. Operating conditions and experimental gaseous mole fractions for the packed bed system with pellet diameter of 3 cm

bed height (cm)	bed temp. (°C)	Inlet CH_4 flow rate (ml/min)	Outlet CH_4	Outlet H_2	Outlet CO	Outlet CO_2	Outlet H_2O	H_2/CO ratio
40	860	330	0.045	0.600	0.314	0.033	0.004	1.91
	940	583	0.029	0.614	0.313	0.034	0.008	1.96
	995	744	0.009	0.590	0.296	0.035	0.077	1.99
30	860	240	0.077	0.588	0.305	0.024	0.006	1.92
	940	417	0.010	0.627	0.316	0.029	0.008	1.98
	995	583	0.010	0.610	0.302	0.030	0.048	2.01
20	860	149	0.040	0.620	0.305	0.028	0.007	2.03
	940	285	0.009	0.601	0.300	0.030	0.055	2.00
	995	501	0.009	0.590	0.297	0.032	0.069	1.99

Very low amounts of CO_2 have been detected by the gas analyzer in the temperature range of the experimental tests (see Table 2.5). Figure 2.20 is a FTIR spectrum of the outlet gas analysis for one of the experimental tests (at 995°C). It should be noted that non-polar symmetric molecules such as H_2 can not be detected by FTIR. Thus, the hydrogen mole fraction was obtained by difference and also it was verified by mass balance equations.

There are five components in the product gas stream in this system, i.e., CH_4, CO, H_2, CO_2, and H_2O. Table 2.5 includes the outlet gaseous mole fractions for these components at various operating conditions.

These mole fractions were obtained by dry basis FTIR percentages for CH_4, CO, and CO_2 and mass balance conservations for carbon and hydrogen.

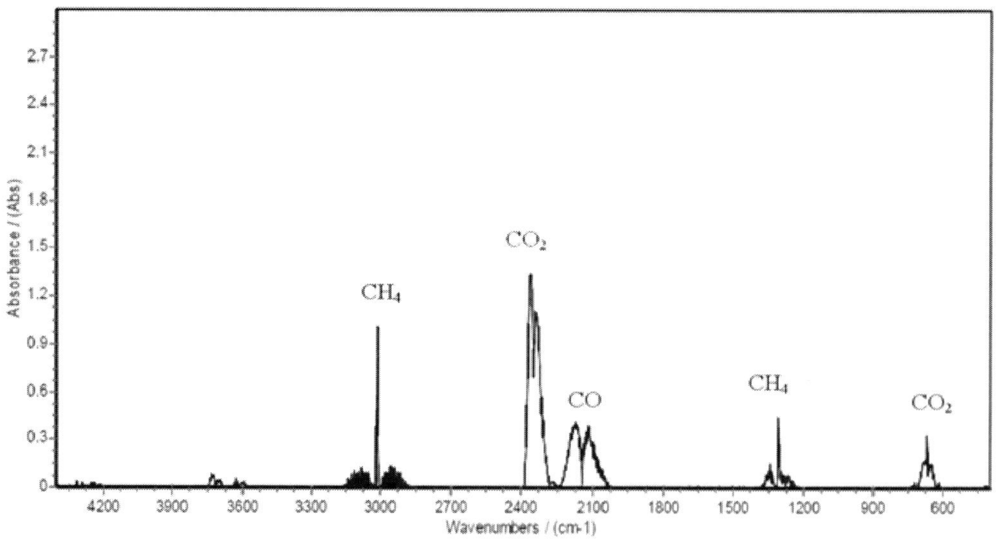

Figure 2.20. FTIR spectrum of the outlet gases for bed height 40 cm, and 995°C.

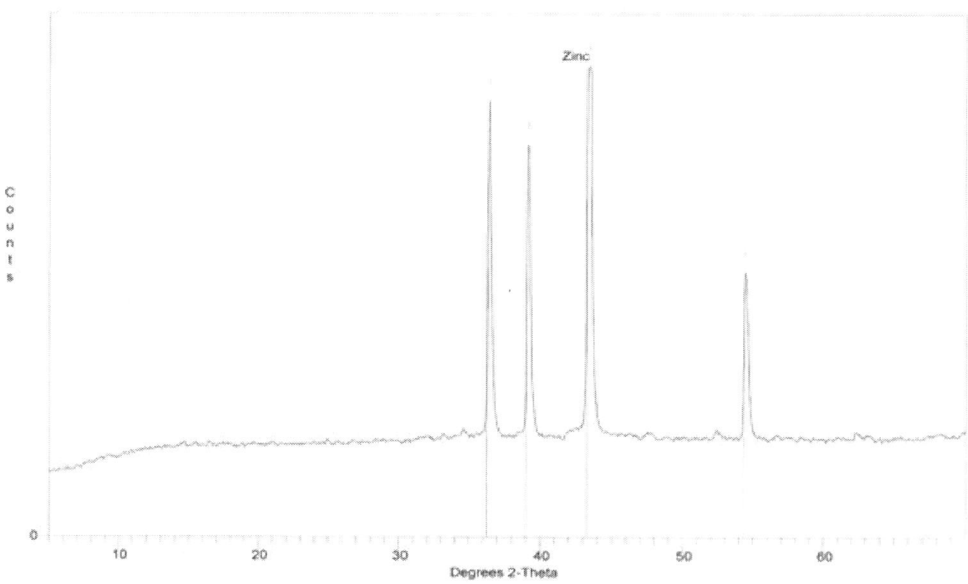

Figure 2.21. XRD pattern obtained from the solid product powders.

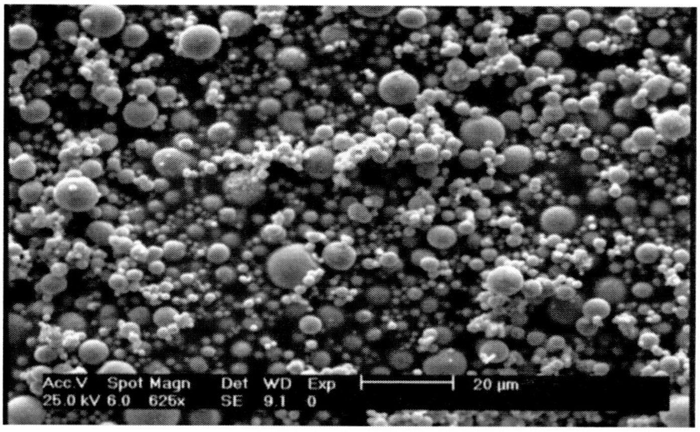

Figure 2.22. SEM result of the collected solid zinc powders from the condenser.

The reproducibility of the gaseous mole fraction analysis by this method was very good with an error of about 1%. This method was also validated by gas sample injection and analysis by mass spectrometer.

As Table 2.5 indicates, high quality syngas with negligible CH_4, CO_2, and H_2O was obtained. Moreover, the molar ratios of H_2/CO are very close to 2. This ratio is specially suitable for methanol synthesis or GTL plants.

The solid powders collected from the condenser were analyzed by XRD test. SEM images were also obtained from these collected solid powders. Figure 2.21 is the XRD pattern of the solid powders collected from the condenser. As this figure indicates, the XRD analyzer detects the presence of zinc in the sample. Figure 2.22 is the SEM of the collected solid powders, which shows spherical zinc particles after condensation and solidification of the zinc vapor. Another method for determination of the amount of zinc in the solid products is based on its reaction with hydrochloric acid and comparing of the evolved hydrogen with its stoichiometric value. This test showed that all samples are approximately pure zinc.

The objective of this section is to apply a suitable model for co-production of syngas and zinc in a noncatalytic packed-bed reactor. These reactors have been simulated using a one-dimensional heterogeneous model [120]. Both external and intera-particle concentration gradients are taken into account. The variation of the solid reactant by the progress of reaction is also considered. The governing dimensionless equations of the modified grain model in a packed bed reactor with the initial and boundary conditions are as follows [8,120]:

$$\frac{\partial^2 y}{\partial \xi^2} - Pe\frac{\partial y}{\partial \xi} = \beta\left(y - \zeta\,|_{\rho=1}\right) \tag{2.19}$$

$$\xi = 0:\quad \frac{\partial y}{\partial \xi} = Pe\left(y - 1\right) \tag{2.20}$$

$$\xi = \Lambda:\quad \frac{\partial y}{\partial \xi} = 0 \tag{2.21}$$

$$\frac{1}{\rho^2}\frac{\partial}{\partial\rho}\left(\delta\,\rho^2\frac{\partial\zeta}{\partial\rho}\right)=\phi^2\frac{\zeta\,r^{*2}}{1+6\sigma_g^2\left(r^*-\left(r^{*2}/r^{**}\right)\right)} \quad (2.22)$$

$$\rho=0:\ \frac{\partial\zeta}{\partial\rho}=0 \quad (2.23)$$

$$\rho=1:\ \frac{\partial\zeta}{\partial\rho}=Bi\left(y-\zeta\right) \quad (2.24)$$

$$\frac{\partial r^*}{\partial\tau}=-\frac{\zeta}{1+6\sigma_g^2\left(r^*-\left(r^{*2}/r^{**}\right)\right)} \quad (2.25)$$

$$\tau=0:\ r^*=1 \quad (2.26)$$

The dimensionless parameters are defined in the nomenclature. Equation (2.19) is a mass balance for the gaseous reactant in the bulk along the reactor. Equation (2.22) is the gas conservation in the solid pellet with diffusion and reaction. Finally, equation (2.25) shows the variation of the dimensionless unreacted core in the grains of the solid pellets with progress of the reaction. Equation (2.19) and (2.22) are related to each other by the term $\zeta|_{\rho=1}$. This term is the gas concentration at the surface of the solid pellet along the bed.

The following assumptions have been considered in the above modeling:

- The gas phase accumulation terms in the bed and in the pellets have been neglected.
- The reaction is irreversible and first order with respect to the gaseous reactant.
- The system is isothermal.

The porosity of the pellets at each time can be expressed as follows [121]:

$$\frac{\varepsilon}{\varepsilon_0}=1-\frac{(Z-1)(1-\varepsilon_0)(1-r^{*3})}{\varepsilon_0} \quad (2.27)$$

In order to relate diffusion to the pellet porosity, the following equation was used [122]:

$$\delta=\frac{D_e}{D_{e0}}=\left(\frac{\varepsilon}{\varepsilon_0}\right)^2=\left[1-\frac{(Z-1)(1-\varepsilon_0)(1-r^{*3})}{\varepsilon_0}\right]^2 \quad (2.28)$$

In this model, the solid conversion for a spherical pellet at each point of the column can be calculated by the following equation:

$$X(\theta)=1-3\int_0^1 \rho^2 r^{*3}(\tau,\rho)\,d\rho \tag{2.29}$$

The solution procedure of equation (2.19) - (2.26) is based on Rayleigh-Ritz finite element method [104,105]. The reaction rate constant which appears in the reaction Thiele modulus (ϕ) is obtained from references [63,64] for the operating temperatures.

The parameters and operational conditions for 860°C test are considered for the simulation. Figure 2.23 shows the development of the methane mole fraction profile along the reactor at various dimensionless times.

Figure 2.24 is the predicted gaseous reactant and product mole fractions versus the axial coordinate of the reactor at a specific dimensionless time. The production of syngas with the molar ratio of $H_2/CO = 2$ is observed in this figure. Considering this figure, the methane is still unconverted until approximatly $\xi=2$. This means that the solid bed up to about was already reacted at the previous times, and after this point the reaction between methane and unreacted solid phase begins. Therefore, syngas appears in the reaction zone. In addition, a comparison of the simulated results with experimental data is performed in Figure 2.24 successfully. The simulation results in this figure show that all of methane is converted to the syngas at the outlet of the reactor.

Figure 2.25 is the solid phase conversion calculated by equation (2.29) at each point of the bed, along the reactor at dimensionless time. At this dimensionless time, about 50% of the packed bed has been converted. Figure 2.26 is the breakthrough curve for methane in the packed bed. This figure shows the appearance of methane at the outlet of the reactor by progress of time.

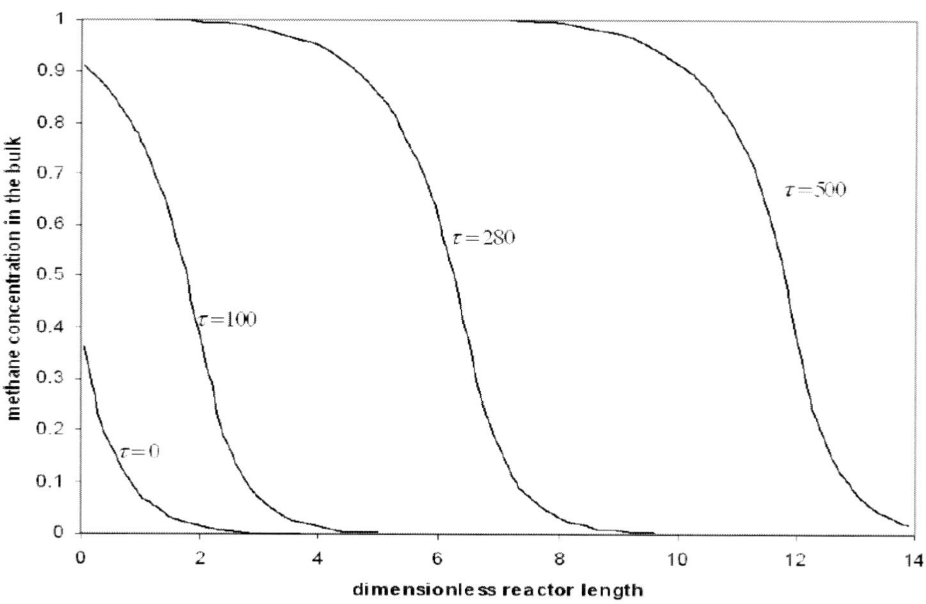

Note: the data were obtained at $\phi = 5.96$, Pe = 1, Bi = 2.39, $\beta = 6.31$, Z = 0, $\varepsilon_0 = 0.62$, T = 860°C, $\sigma_g = 0$, and $R_p = 1.5$ cm.

Figure 2.23. Simulation results for methane mole fraction profile along the packed bed.

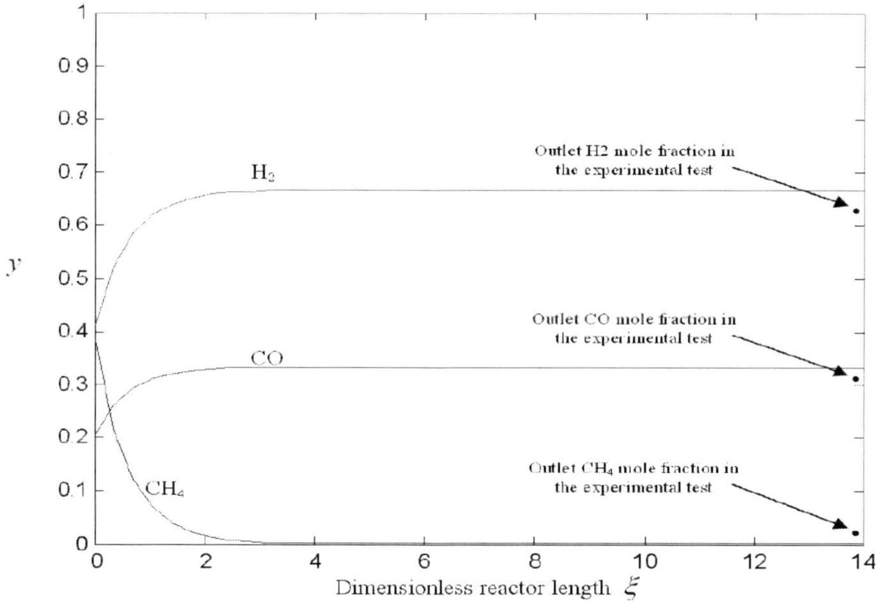

Note: the data were obtained at $\phi = 5.96$, Pe = 1, Bi = 2.39, $\beta = 6.31$, Z = 0, $\varepsilon_0 = 0.62$, T = 860°C, $\sigma_g = 0$, $R_p = 1.5$ cm, and $\tau = 0.2$.

Figure 2.24. Prediction of gaseous mole fractions versus axial coordinate.

For example at $\tau = 550$, about 20% unreacted methane has been appeared in the outlet gas stream. In the other words, the capacity of the solid bed for methane reforming will be finished at the times greater than this breakthrough point.

The following endothermic reactions describe the reduction of ZnO by hydrogen and carbon monoxide as the possible side reactions, respectively.

$$H_2 + ZnO \rightarrow H_2O + Zn \qquad \Delta G_{1300K} = 27.9\ kJ{\cdot}mol^{-1} \tag{2.30}$$

$$CO + ZnO \rightarrow CO_2 + Zn \qquad \Delta G_{1300K} = 33.9\ kJ{\cdot}mol^{-1} \tag{2.31}$$

A significant influence of the above side reactions on the syngas composition begins at elevated temperatures. In addition, the water-gas shift reaction can be considered as another possible exothermic side reaction as follows:

$$H_2O + CO \rightarrow H_2 + CO_2 \qquad \Delta G_{1300K} = 5.9\ kJ{\cdot}mol^{-1} \tag{2.32}$$

By considering the data from Table 2.5, it is clear that the effect of the above undesired side reactions is almost negligible.

No carbon deposition was observed in the solid products or on the remaining reactant packed bed of this new noncatalytic method after each run. While the coke formation from methane cracking or Boudward reaction is the major problem in the conventional catalytic steam reformers which leads to deactivation of the catalyst.

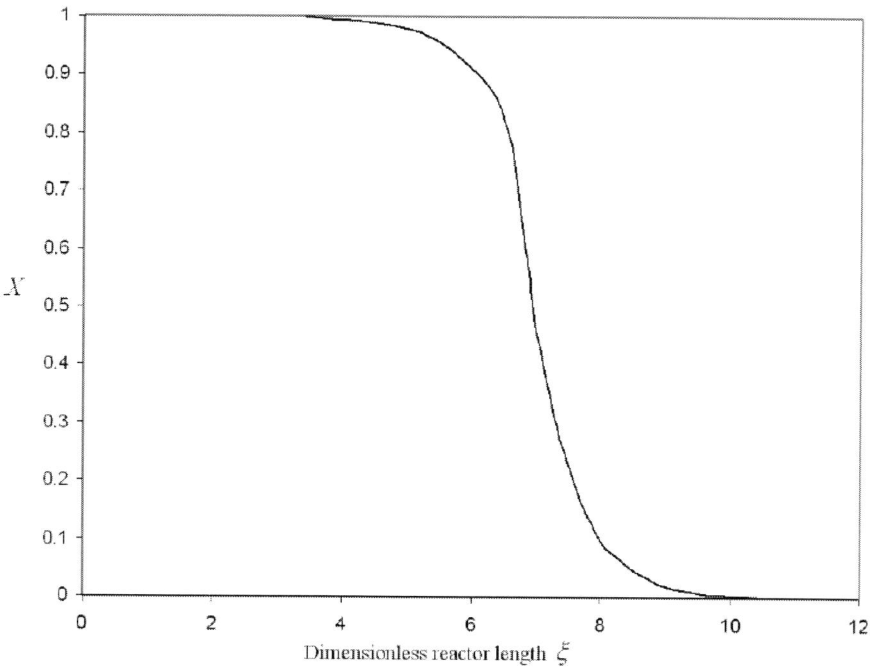

Note: the data were obtained at $\phi = 5.96$, Pe = 1, Bi = 2.39, $\beta = 6.31$, Z = 0, $\varepsilon_0 = 0.62$, T = 860°C, $\sigma_g = 0$, $R_p = 1.5$ cm, and $\tau = 280$.

Figure 2.25. Simulation results of solid phase conversion versus axial coordinate.

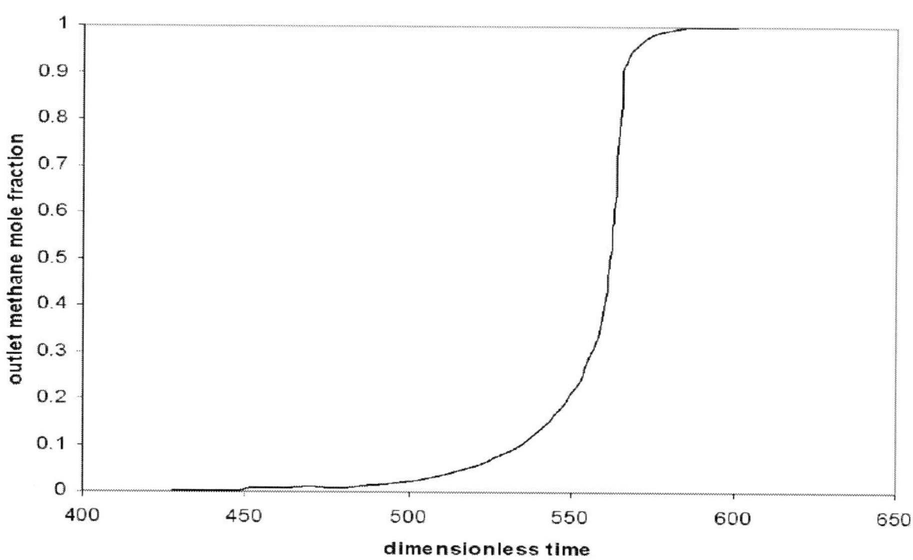

Note: the data were obtained at $\phi = 5.96$, Pe = 1, Bi = 2.39, $\beta = 6.31$, Z = 0, $\varepsilon_0 = 0.62$, T = 860°C, $\sigma_g = 0$, $R_p = 1.5$ cm, and $\xi = 14$.

Figure 2.26. Prediction of the breakthrough curve for methane.

A continuous solid powder feeding to a lifting reactor for syngas production by $ZnO+CH_4$ reaction is now under investigation. The preliminary results of this system are very interesting. For example, the syngas production capacity in the lifting pilot plant can be 14 times greater than of a similar packed bed system at the same temperature. This is because of using the fine ZnO powder, and thus existence of a very high reaction surface per unit mass. Moreover, ZnO powder was continuously fed to the reactor by a special screw feeder, and then this fine powder was lifted by methane stream in a vertical tube for the reaction. The comprehensive mathematical modeling of this lifting reactor and comparison of its predictions with experimental data for syngas composition at the various temperatures and flow-rates are now under consideration.

The combined reforming of methane and reduction of ZnO in a noncatalytic packed bed reactor was presented in this section. The outlet gaseous analysis showed high quality syngas production with the H_2/CO molar ratios of about two. The solid products analysis verified the presence of pure metallic zinc powder in the condenser. The partial differential equations for modeling of a packed bed reactor were solved and the results of the simulation were successfully compared with the experimental data. This system can be considered as a new method for production of high quality syngas without the use of catalyst. In addition, the solid zinc product can be used as an energy carrier for in-situ hydrogen production at far destinations, or a green fuel for the zinc/air fuel cells.

$NiO+CH_4$ System for Syngas Production

A new suggested system for syngas production by metal oxides reduction with methane is $NiO+CH_4$ reaction. Such a system with cyclic streams has been also proposed as the chemical looping reforming (CLR) process for syngas production in recent years [123-126].

The main goals of this research are as:

1. Kinetic investigation of $NiO+CH_4$ reaction in TG/MS system
2. Considering gaseous compositions of $NiO+CH_4$ reaction in TG/MS system
3. Gaseous analysis of $NiO+CH_4$ reaction in a packed bed reactor
4. Consideration of temperature and flow effects on H_2/CO ratio, unreacted CH_4, and CO_2 content
5. Applying a suitable gas-solid reaction model with structural changes
6. Studying the effect of reaction rate, bulk flow, mass transfer, and product layer diffusion

The preferences of this research are predicted as:

1. A new syngas production method without catalyst, or oxygen separation unit
2. Syngas production with $H_2/CO = 2$ which is suitable for methanol and GTL plants
3. Produced Ni is a clean fuel and can supply the heat for endothermic $NiO+CH_4$ reaction
4. Decreasing of operating temperature from 1000°C (for ZnO) to about 700°C

Thermodynamics of $NiO+CH_4$ Reaction

The chemical reaction for this method is as:

$$NiO + CH_4 = Ni_{(s)} + CO + 2H_2 \quad (2.33)$$

The equilibrium constants for the above reaction at 700 and 800°C are about 2,382 and 27,464 respectively. Therefore, this reaction is completely irreversible at such operating temperatures.

The Gibbs free energy minimization was applied for predicting the composition of ingredients for this reaction. The results of such computation are presented in Table 2.6 at various temperatures. As Table 2.6 shows, there are $y_{CH4} = 0.03$, $y_{H2}/y_{CO} = 2.016$, and $y_{CO2} = 0.012$ in syngas at 800°C. Therefore, production of a high quality syngas is predicted from $NiO+CH_4$ reaction by thermodynamics.

Table 2.6. Equilibrium compositions for $NiO+CH_4$ reaction

T (°K)	800	900	1000	1100	1200
y_{CH4}	0.3197	0.1696	0.0618	0.0208	0.0068
y_{H2}	0.2923	0.4665	0.5899	0.6389	0.6571
y_{CO}	0.0683	0.1942	0.2865	0.3194	0.3294
y_{H2O}	0.1612	0.0871	0.0356	0.0139	0.0051
y_{CO2}	0.1585	0.0826	0.0262	0.0069	0.0017
y_{H2}/y_{CO}	4.2796	2.4022	2.0590	2.0003	1.9948
$y_{H2}+y_{CO}$	0.3606	0.6607	0.8764	0.9583	0.9865

The equilibrium conversions of such reactions at high pressure are important for their future industrial applications. For example, the equilibrium conversion of $NiO+CH_4$ reaction at 700°C and 30 atm is about 0.87. While, for $ZnO+CH_4$ reaction at 950°C and 30 atm, the equilibrium conversion is calculated as 0.22. Therefore, $NiO+CH_4$ reaction shows a thermodynamic advantage over $ZnO+CH_4$ reaction at high pressures.

Some Preliminary Tests of $NiO+CH_4$ Reaction

Some of the experimental conversion-time profiles for $NiO+CH_4$ reaction are presented in Figure 2.27 at various operating temperatures. As this figure shows, the slope of conversion-time profiles is increased at higher temperatures.

One of the preliminary gaseous products tests of $NiO+CH_4$ reaction by MS showed initial CO_2 and H_2O peaks. Then these peaks decreased, and high CO and H_2 peaks as the main gaseous products appeared. The similar mechanism for partial oxidation of methane by oxygen has been reported in the literature [127]. This means that partial oxidation is accomplished via total oxidation and subsequent reforming of the remaining methane by CO_2 and H_2O.

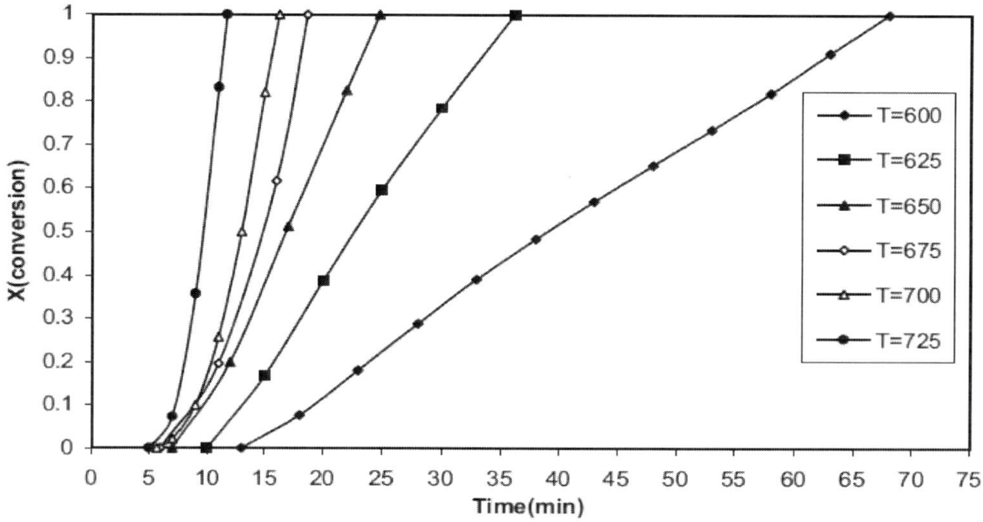

Figure 2.27. Conversion-time profiles for NiO+CH_4 reaction.

CONCLUSION

A new noncatalytic method for syngas production by metal oxides reduction with methane was introduced in this chapter. The operating temperature for the ZnO+CH_4 reaction is about 1000°C, and the zinc byproduct can be used as an energy carrier and hydrogen storage material. On the other hand, NiO+CH_4 reaction was accomplished at about 700°C, and produced nickel can supply the required heat by its clean combustion. The H_2/CO ratio in syngas for this new method was about two, which is very suitable for the methanol or GTL plants. Finally, the rate constants and mathematical model equations for the ZnO+CH_4 reaction system were presented.

REFERENCES

[1] M. E. Dry, *J. Chem. Technol. Biotechnol.* 77, 43 (2001).

[2] S. Mokhatab, W. A. Poe and J. G. Speight, Handbook of Natural Gas Transmission and Processing, Gulf Publishing, New York (2006).

[3] P. D. Costa, C. Potvin, J. M. Manoli, J. L. Lemberton, G. Perot and G. D. Mariadassou, *J. Mol. Catal. A: Chem.* 184, 323 (2002).

[4] M. Bhaskar, G. Valvarasu, B. Sariam, K. S. Balaraman and K. Balu, *Ind. Eng. Chem. Res.* 43, 6654 (2004).

[5] T. Shimizu, M. Peglow, K. Yamagiwa, M. Tamaka, S. Sakuno, N. Misawa, N. Suzuki, H. Udea, H. Sasatsu and H. Gotou, *Chem. Eng. Sci.* 57, 4117 (2002).

[6] S. Zarkanitis and S. V. Sotirchos, *AIChE J.* 35, 821 (1989).

[7] Kirk Othmer, Encyclopedia of Chemical Technology, John-Wiley and Sons, New York, Vol. 12, pp 318-340 (1991).

[8] C. Georgakis, C. W. Chang and J. Szekely, *Chem. Eng. Sci.* 34, 1072 (1979).

[9] R. A. Dawe, Modern Petroleum Technology, John Wiley and Sons, New York, Vol. 1, pp 337-382 (2000).

[10] S. Chakraborty, A. V. Tiedemann and P. S. Teng, *Environ. Pollut.* 108, 317 (2000).

[11] D. Leaf, H. J. H. Verolme and W. F. Hunt, *Environ. Int.* 29, 303 (2003).

[12] D. Bonenfant, M. Mimeault and R. Hausler, *Ind. Eng. Chem. Res.* 42, 3179 (2003).

[13] H. Gupta and L. S. Fan, *Ind. Eng. Chem. Res.* 41, 4035 (2002).

[14] B. K. Na, H. Lee, K. K. Koo and H. K. Song, *Ind. Eng. Chem. Res.* 41, 5498 (2002).

[15] O. R. West, C. Tsouris, S. Lee, S. D. McCallum and L. Liang, *AIChE J.* 49, 283 (2003).

[16] T. Kodama, Y. Kondoh, T. Tamagawa, A. Funatoh, K. I. Shimizu and Y. Kitayama, *Energy Fuels* 16, 1264 (2002).

[17] J. H. Edwards, K. T. Do, A. M. Maitra, S. Schuck, W. Fok and W. Stein, *Energy Convers. Manage.* 37, 1339 (1996).

[18] S. Wang, G. O. Lu and G. J. Millar, *Energy Fuels* 10, 896 (1996).

[19] A. Morrison, S. Hietkamp and D. S. van Vuuren, *Ironmaking Steelmaking* 31, 285 (2004).

[20] S. Thomas and R. A. Dawe, *Energy* 28, 1461 (2003).

[21] A. Chauvel and G. Lefebvre, *Petrochemical Processes*, Editions Techn., Paris (1989).

[22] H. L. List, *Petrochemical Technology*, Prentice-Hall, New Jersey (1986).

[23] G. V. Echevsky, E. G. Kodenev, O. V. Kikhtyanin and V. N. Parmon, *Appl. Catal. A: General* 258, 159 (2004).

[24] R. Krishna, J. Ellenberger and S.T. Sie, *Chem. Eng. Sci.* 51, 2041 (1996).

[25] C. Knottenbelt, *Catalysis Today* 71, 437 (2002).

[26] D. R. Parisi and M. A. Laborde, *Chem. Eng. J.* 104, 35 (2004).

[27] Kirk-Othmer, Encyclopedia of Chemical Technology, John Wiley and Sons, New York, Vol. 2, pp 638-691 (1995).

[28] J. Dybkjaer, *Fuel Process. Technol.* 42, 85 (1995).

[29] Kirk-Othmer, Encyclopedia of Chemical Technology, John Wiley and Sons, New York, Vol. 16, pp 537-556 (1995).

[30] SRI Report, *Synthesis Gas by Catalytic Steam Reforming of Natural Gas*, (1976).

[31] J. W. Snoeck, G. F. Froment and M. Fowles, *Ind. Eng. Chem. Res.* 41, 3548 (2002).

[32] J. W. Snoeck, G. F. Froment and M. Fowles, *Ind. Eng. Chem. Res.* 41, 4252 (2002).

[33] K. Hou and R. Hughes, *Chem. Eng. J.* 82, 311 (2001).

[34] M. N. Pedernera, J. Pina, D. O. Borio and V. Bucala, *Chem. Eng. J.* 94, 29 (2003).

[35] S. Grevskott, T. Rusten, M. Hillestad, E. Edwin and O. Olsvik, *Chem. Eng. Sci.* 56, 597 (2001).

[36] K. Liu, C. Song and V. Subramani, Hydrogen and Syngas Production and Purification Technologies, John Wiley and Sons, New Jersey (2010).

[37] I. A. Abba, J. R. Grace and H. T. Bi, *Ind. Eng. Chem. Res.* 42, 2736 (2003).

[38] O. Deutschmann and L. D. Schmidt, *AIChE J.* 44, 2465 (1998).

[39] H. Morioka, Y. Shimizu, M. Sukenobu, K. Ito, E. Tanabe, T. Shishido and K. Takehira, *Appl. Catalysis A: General* 215, 11 (2001).

[40] J. T. Ritchie, J. T. Richardson and D. Luss, *AIChE J.* 47, 2092 (2001).

[41] T. Zhu and M. F. Stephanopoulos, *Appl. Catalysis A: General* 208, 403 (2001).

[42] K. Nakagawa, N. Ikenaga, T. Kobayashi and T. Suzuki, *Catalysis Today* 64, 31 (2001).

[43] F. Basile, G. Fornasari, F. Trifiro and A. Vaccari, *Catalysis Today* 64, 21 (2001).

[44] K. Gosiewski, *Chem. Eng. Sci.* 56, 1501 (2001).
[45] Y. Ji, W. Li, H. Xu and Y. Chen, *Appl. Catalysis A: General* 213, 25 (2001).
[46] L. Mleczko, T. Ostrowski and T. Wurzel, *Chem. Eng. Sci.* 51, 3187 (1996).
[47] L. Paturzo and A. Basile, *Ind. Eng. Chem. Res.* 41, 1703 (2002).
[48] R. S. Souleimanova, A. S. Mukasyan and A. Varma, *AIChE J.* 48, 262 (2002).
[49] T. M. Raybold and M. C. Huff, *AIChE J.* 48, 1051 (2002).
[50] H. K. Abdel-Aal, M. A. Shalabi, D. K. Al-Harbi and T. *Hakeem Ind. Eng. Chem. Res.* 38, 1069 (1999).
[51] M. M. V. M. Souza, D. A. G. Aranda and M. Schmal, *Ind. Eng. Chem. Res.* 41, 4681 (2002).
[52] G. F. Froment, *J. Molecular Catalysis A: Chem.* 163, 147 (2000).
[53] L. Ji, S. Tang, H. C. Zeng, J. Lin and K. L. Tan, *Appl. Catalysis A: General* 207, 247 (2001).
[54] E. Akpan, Y. Sun, P. Kumar, H. Ibrahim, A. Aboudheir and R. Idem, *Chem. Eng. Sci.* 62, 4012 (2007).
[55] D. J. Wilhelm, D. R. Simbeck, A. D. Karp and R. L. Dickenson, *Fuel Process. Technol.* 71, 139 (2001).
[56] K. A. Petersen, T. S. Christensen, C. S. Nielsen and I. Dybkjaer, *Fuel Process. Technol.* 83, 253 (2003).
[57] C. R. H. Smet, M. H. J. M. Croon, R. J. Berger, G. B. Marin and J. C. Schouten, *Chem. Eng. Sci.* 56, 4849 (2001).
[58] V. R. Choudhary and A. S. Mamman, *Appl. Energy* 66, 161 (2000).
[59] V. A. Tsipouriari and X. E. Verykios, *Catalysis Today* 64, 83 (2001).
[60] V. R. Choudhary and A. S. Mamman, *J. Chem. Technol. Biotechnol.* 73, 345 (1998).
[61] C. Song and W. Pan, *Catalysis Today* 98, 463 (2004).
[62] M. Halmann and A. Steinfeld, *Catalysis Today* 115, 170 (2006).
[63] H. Ale Ebrahim and E. Jamshidi, *Chem. Eng. Res. Design* 79, 62 (2001).
[64] H. Ale Ebrahim and E. Jamshidi, *Ind. Eng. Chem. Res.* 41, 2630 (2002).
[65] H. Ale Ebrahim and E. Jamshidi, *Energy Conv. Manag.* 45, 345 (2004).
[66] A. Afshar Ebrahimi, H. Ale Ebrahim, E. Jamshidi and A. H. Faramarzi, *Chem. Eng. Technol. 33*, 1989 (2010).
[67] R. Alizadeh, E. Jamshidi and H. Ale Ebrahim, *Chem. Eng. Technol.* 30, 1123 (2007).
[68] H. Ale Ebrahim and E. Jamshidi, *Ind. Eng. Chem. Res.* 44, 495 (2005).
[69] E. Jamshidi and H. Ale Ebrahim, *Chem. Eng. Process.* 47, 1567 (2008).
[70] R. Alizadeh, E. Jamshidi and H. Ale Ebrahim, *Thermochimica Acta* 460, 44 (2007).
[71] H. Ale Ebrahim and E. Jamshidi, Mineral Process. *Extr. Metall.* 118, 194 (2009).
[72] E. Jamshidi and H. Ale Ebrahim, Iranian Patent No.31321, 2005
[73] R. Muller, P. Haeberling and R. D. Palumbo, *Solar Energy* 80, 500 (2006).
[74] S. Kraupl and A. Steinfeld, Trans. *ASME* 127, 102 (2005).
[75] M. Werder and A. Steinfeld, *Energy* 25, 395 (2000).
[76] S. Kraupl and A. Steinfeld, Trans. *ASME* 125, 124 (2003).
[77] A. Weidenkaff, A. W. Reller, A. Wokaun and A. Steinfeld, *Thermochimica Acta* 359, 69 (2000).
[78] A. Berman and M. Epstein, *Int. J. Hydrogen Energy* 25, 957 (2000).
[79] K. Wegner, H. C. Ly, R. J. Weiss, S. E. Pratsinis and A. Steinfeld, *Int. J. Hydrogen Energy* 31, 55 (2006).

[80] J. Szekely and C. I. Lin, *Metall. Mater. Trans. B* 7, 493 (1976).
[81] K. L. Breg and S. E. Olsen, *Metall. Mater. Trans. B* 31, 477 (2000).
[82] S. Kimura, AIChE J. 35, 339 (1989).
[83] S. Wu, M. A. Uddin, C. Su, S. Nagamine and E. Sasaoka, *Ind. Eng. Chem. Res.* 41, 5455 (2002).
[84] J.B. Gibson and D.P. Harrison, *Ind. Eng. Proc. Des. Dev.* 19, 231 (1980).
[85] P. A. Ramachandran, M. H. Rashid and R. Hughes, *Chem. Eng. Sci.* 30, 1391 (1975).
[86] S. I. Abu Eishan and N. M. Abu Jabal, *Chem. Eng. J.* 81, 231 (2001).
[87] G. Q. Lu and D. D. Do, *Carbon* 30, 21 (1992).
[88] P. A. Ramachandran and L. K. Doraiswamy, *AIChE J.* 28, 881 (1982).
[89] M. P. Dudukovic and H. S. Lamba, *Chem. Eng. Sci.* 33, 303 (1978).
[90] C. Y. Wen and M. Ishida, *AIChE. J.* 14, 311 (1968).
[91] J. Szekely and J. W. Evans, *Metall. Trans.* 2, 1691 (1971).
[92] H. Y. Sohn and J. Szekely, *Chem. Eng . Sci.* 29, 630 (1974).
[93] Q. Garza-Garze and M. P. Dudukovic, *Chem. Eng. J.* 24, 35 (1982).
[94] H. Y. Sohn, *Metall. Trans. B* 9, 89 (1978).
[95] S. K. Bhatia and D. D. Perlmutter, *AIChE J.* 26, 379 (1980).
[96] S. K. Bhatia and D. D. Perlmutter, *AIChE J.* 27, 247 (1981).
[97] S. K. Bhatia and D. D. Perlmutter, *AIChE J.* 27, 226 (1981).
[98] E. Jamshidi and H. Ale Ebrahim, *Chem. Eng. J.* 63, 79 (1996).
[99] E. Jamshidi and H. Ale Ebrahim, *Chem. Eng. Sci.* 51, 4253 (1996).
[100] E. Jamshidi and H. Ale Ebrahim, *Chem. Eng. J.* 68, 1 (1997).
[101] E. Jamshidi and H. Ale Ebrahim, *Chem. Eng. Sci.* 54, 859 (1999).
[102] A. Shiravani, E. Jamshidi and H. Ale Ebrahim, *Chem. Eng. J.* 140, 264 (2008).
[103] A. Afshar Ebrahimi, H. Ale Ebrahim and E. Jamshidi, *Comput. Chem. Eng.* 32, 1746 (2008).
[104] A. Afshar Ebrahimi, H. Ale Ebrahim, M. Hatam and E. Jamshidi, *Chem. Eng. J.* 144, 110 (2008).
[105] A. Afshar Ebrahimi, H. Ale Ebrahim, M. Hatam and E. Jamshidi, *Chem. Eng. J.* 148, 533 (2009).
[106] H. Ale Ebrahim, *Ind. Eng. Chem. Res.* 49, 117 (2010).
[107] S. M. M. Nouri, H. Ale Ebrahim and E. Jamshidi, *Chem. Eng. J.* 166, 704 (2011).
[108] Kirk-Othmer, Encyclopedia of Chemical Technology, John Wiley and Sons, New York, Vol. 24, pp 807-862 (1984).
[109] J. J. Moore, Chemical Metallurgy, Butterworths, London (1981).
[110] J. Szekely, J. W. Evans and H. Y. Sohn, Gas-Solid Reactions, Academic Press, New York (1976).
[111] J. M. Smith, Chemical Engineering Kinetics, McGraw-Hill, New York (1981).
[112] T. Hatakeyama and Z. Liu, Handbook of Thermal Analysis, John Wiley and Sons, New York (1998).
[113] Mini-Lab Catalogue, On-Line Gas Analyzer by Quadrupole Mass Spectrometer, MKS Spectra Products.
[114] R. Garcia, A. Arenillas, J. L. Crespo, J. J. Pis and S. R. Moinelo, *Energy Fuels* 16, 935 (2002).
[115] R. V. Siriwardane and S. Woodruff, *Ind. Eng. Chem. Res.* 36, 5277 (1997).
[116] Residual Gas Analyzers for Windows Manual, Spectra Instruments Inc.

[117] C. E. Guger and F. S. *Manning, Metall. Trans.* 2, 3083 (1971).
[118] F. Gioia, G. Mura and A. Viola, *Chem. Eng. Sci.* 32, 1401 (1977).
[119] T. Weimer, K. Schaber, M. Specht and A. Bandi, *Energy Conv. Manag.* 37, 1351 (1996).
[120] N. Fernandes and G.R. Gavalas, *AIChE J.* 41, 2549 (1995).
[121] E.A. Efthimiadis and S.V. Sotirchos, *Chem. Eng. Sci.* 48, 829 (1993).
[122] N. Wakao and J. M. Smith, *Chem. Eng. Sci.* 17, 825 (1962).
[123] M. Ryden, A. Lyngfelt and T. Mattisson, *Energy Fuels*, 22, 2585 (2008).
[124] M. Ryden, A. Lyngfelt and T. *Mattisson, Fuel* 85, 1631 (2006).
[125] L. F. de Diego, M. Ortiz, F. G. Labiano, J. Adanez, A. Abad and P. Gayan, *Energy Procedia* 1, 3 (2009).
[126] L. F. de Diego, M. Ortiz, J. Adanez, F. G. Labiano, A. Abad and P. *Gayan, Chem. Eng. J.* 144, 289 (2008).
[127] H. Nishimoto, K. Nakagawa, N. Ikenaga and T. Suzuki, *Catalysis Letters* 82, 161 (2002).

In: Syngas
Editors: Antonius Indarto and Jelliarko Palguandi ISBN: 978-1-62100-870-5

Chapter 3

COKE OVEN GAS TO SYNGAS TECHNOLOGY

Yongfa Zhang and Guojie Zhang*
Key Laboratory of Coal Science and Technology,
Taiyuan University of Technology, Shanxi, China

ABSTRACT

World's annual output of coke is more than 500 million tons since 2006 years. Coke oven gas has become a large tonnage of energy resources. The investigation shows that there are more than forty-five billion cubic meters coke oven gas without recovery and utilization every year, this amount more than 2 times that designed by the "West-East Gas Transmission of China". The synthesis gas, H_2 and CO as the main components, has extremely wide range uses in the chemical industry. Typical applications include liquid phase synthesis of methanol, oxo synthesis of acetic acid and synthesis of many oxygenated compounds, such as dimethyl ether and dimethyl carbonate etc. In addition, the synthesis gas can be used as raw materials of basic chemical products, ammonia and so on. Coke oven gas is very precious energy and resource, contains 25% methane, 58% of the hydrogen.

By using several methods, we can convert waste from coke oven gas to useful material, accordingly, effective recover coke oven gas reduce the coke oven gas that direct combustion emitting into the air. Coke oven gas converted into synthesis gas (Hydrogen and carbon monoxide) and then produced to form chemical products is the reasonable and scientific use of an effective way, while protecting the environment against pollution.

At present, the technology of synthesis gas in the world is followed: catalytic steam reforming, catalytic partial oxidation, combined reforming of catalytic steam reforming and Catalytic Partial Oxidation, autothermal reforming, non-catalytic partial oxidation and catalytic carbon dioxide reforming. In this chapter, the theory, industrial applications and research development of the above-mentioned technology of synthesis gas are mainly analyzed and discussed.

* E-mail address: yongfaz@tyut.edu.cn (Yongfa Zhang), zhgjdoc@126.com (Guojie Zhang).

NOMENCLATURE

AGHR	Advanced Gas Heated Reformer
COG	Coke Oven Gas
CPOX	Catalytic Partial Oxidation
D	diameter, cm
GHR	Gas Heated Reformer
HTER	Heat Exchange Reforming
K	equilibrium constant
L	length, cm
LSOX	Local Strong Oxidation
M	radical compound
POM	Partial Oxidation of Methane
POX	Partial Oxidation
R	axial direction, cm
R	radial direction, cm
ΔH	reaction enthalpy, kJ mol^{-1}

Superscripe and Subscribe

298	index for tempearture 298K

INTRODUCTION

World's annual output of coke is more than 500 million tons since 2006 year [1-4]. In 2010, the coke production was about 630 million tons in the world. While coke oven gas (COG) can be considered a byproduct of coking plants, consist mainly of H_2 (55 - 60%), CH_4 (23 - 27%), CO (5 - 8%) and N_2 (3 - 5%), along with other hydrocarbons, H_2S and NH_3 in small proportions. Most of this gas is used as fuel in the coke ovens and other processes of the steel plant, but very often the excess of COG cannot be used in this way and so it is burnt in torches. The investigation showed that about forty billion cubic meters COG was discharged into the air annually, this amount more than two times that designed by the "West-East Gas Transmission of China" [5,6]. These discharges not only pollute the environment but also waste precious fossil fuel resources heavily. Coke oven gas has become a large tonnage of energy resources. Therefore, reasonable utilization of oven gas is of significance both for the economy and for the environment.

An alternative option for the excess COG is for it to be valorized by means of hydrogen separation or syngas production through partial oxidation, steam reforming or dry reforming. The synthesis gas, H_2 and CO as the main components, has extremely wide range uses in the chemical industry. Typical applications include liquid phase synthesis of methanol, oxo synthesis of acetic acid and synthesis of many oxygenated compounds, such as dimethyl ether and dimethyl carbonate etc. In addition, the synthesis gas can be used as raw materials of basic chemical products, such as: ammonia, methanol, etc. Coke oven gas is very precious energy and resource, contains 25% methane, 58% of the hydrogen. By using several methods, we can convert waste from coke oven gas to useful material, accordingly, effective recover coke oven gas reduce the coke oven gas that direct combustion emitting into the air. Coke oven gas converted into synthesis gas (hydrogen and carbon monoxide) and then produced to form chemical products is the reasonable and scientific use of an effective way, while protecting the environment against pollution. At present, the technology of synthesis gas in the world is followed [7-13]: catalytic steam reforming (CSR), catalytic partial oxidation (CPOX), combined reforming of catalytic steam reforming and catalytic partial oxidation, autothermal reforming (ATR), non-catalytic partial oxidation and catalytic carbon dioxide reforming. In this chapter, the theory, industrial applications and research development of the above-mentioned technology of synthesis gas are mainly analyzed and discussed.

TECHNICAL FEATURES OF CH_4 TO SYNGAS

Catalytic Steam Reforming

Catalytic steam reforming is used to produce hydrogen by splitting water. Steam reacts with methane to produce hydrogen and carbon monoxide, when heated to very high temperatures and is in the presence of a metal based catalyst. Catalytic steam reforming is a different process to catalytic reforming, a process used to produce high octane fuel and hydrogen [14-17]. The main chemical reactions in the course of a catalytic steam reforming are:

$$CH_4 + H_2O = CO + 3H_2 \qquad \Delta H_{298} = 206.4 \text{ kJ mol}^{-1} \qquad (3.1)$$

$$CH_4 + 2H_2O = CO2 + 4H_2 \qquad \Delta H_{298} = 165.4 \text{ kJ mol}^{-1} \qquad (3.2)$$

$$CH_4 + CO_2 = 2CO + 2H_2 \qquad \Delta H_{298} = 247.4 \text{ kJ mol}^{-1} \qquad (3.3)$$

$$CO + H_2O = CO_2 + H_2 \qquad \Delta H_{298} = -41.0 \text{ kJ mol}^{-1} \qquad (3.4)$$

In the above reactions, reactions (9.1) - (9.3) are the main reactions in the catalytic steam reforming, and reactions (9.1) - (9.3) are endothermic; reaction (9.4) acts as the control step that dominates the reformation result of the process.

The process of natural gas intermittent conversion to syngas is a catalytic steam reforming, which has some advantages, such as low investment cost, less equipments, low operation cost. The technology is used by small ammonia plant using natural gas as raw material in 1970s.

However, it has disadvantages, such as high energy consumption, low production capacity, short catalyst life. So, the intermittent conversion process is gradually phased out. The process is not suitable for coke oven gas conversion to syngas and it has not been reported, to our knowledge.

Catalytic Partial Oxidation

Partial oxidation (POX) is a type of chemical reaction. It occurs when a substoichiometric fuel-air mixture is partially combusted in a reformer, creating a hydrogen-rich syngas which can then be put to further use, for example, in a fuel cell. A distinction is made between thermal partial oxidation (TPOX) and catalytic partial oxidation [18].

Catalytic partial oxidation is partially combusted in a chamber where the methane is partially oxidized. The reaction is exothermic due to the oxidation. The gas mixture is now passed into a reformer which converts most of the remaining methane into CO and H_2 over a catalyst of nickel oxide. When natural gas is used as a raw material to produce syngas, the main chemical reactions in the course of oxidative combustion are:

$$CH_4 + 2O_2 = CO_2 + 2H_2O \quad \Delta H_{298} = -802.6 \text{ kJ mol}^{-1} \tag{3.5}$$

$$CH_4 + O_2 = CO + H_2 + H_2O \quad \Delta H_{298} = -278.0 \text{ kJ mol}^{-1} \tag{3.6}$$

$$2CH_4 + O_2 = 2CO + 4H_2 \quad \Delta H_{298} = -71.2 \text{ kJ mol}^{-1} \tag{3.7}$$

$$H_2 + 1/2O_2 = H_2O \quad \Delta H_{298} = -241.0 \text{ kJ mol}^{-1} \tag{3.8}$$

When COG or the primary reaction products is used as the only raw material, with its large number of H_2, the main oxidative combustion reaction is (3.8). Because hydrogen has a flame speed faster than other, such as three times faster than butane, five times faster than propane and six times faster than methane. Hydrogen burns more rapidly than hydrocarbon fuels because hydrogen molecules are smaller and enter combustion reaction chambers at a higher velocity than other gases, which results in more molecular collisions than those that occur with heavier molecules. That is, at the top of the combustion chamber, the combustion of hydrogen and O_2 is the main reaction, generating H_2O and a great amount of heat. In the combustion process, a small proportion of CH_4, and CO combustion (3.4) - (3.7) may be accompanied, a small amount of water vapor reaction (3.1) and (3.4) may also occur. If the water vapor content is low or the reaction temperature is too high, decomposition reactions will occur for methane ($CH_4 = C + 2H_2$; $\Delta H_{298} = -77.9$ kJ mol^{-1}) leading to deposit formation, the deposit gasification reaction with water vapor will occur ($C + H_2O = CO + H_2$; $\Delta H_{298} = -127.7$ kJ mol^{-1}). The above reactions are exothermic, the temperature of reaction system increases. When gas mixture, after burning, reaches the catalyst layer of the reactor, almost all the oxygen is consumed (oxygen conversion >99.0%). The main reactions in the catalysts layer are (3.1) - (3.4). The total reactions in the catalyst layer are as follows:

$$CH_4 + ½O_2 = CO + 2H_2 \quad \Delta H_{298} = -36 \text{ kJ mol}^{-1} \tag{3.9}$$

$$CH_4 + 3/2O_2 = CO + 2H_2O \quad \Delta H_{298} = -519 \text{ kJ mol}^{-1} \tag{3.10}$$

In fact, the catalyst used in this process is still catalytic steam reforming catalyst (high-temperature resistant catalyst). The essence of catalytic partial oxidation belongs to the catalytic steam reforming. It was believed that this process is called autothermal partial oxidation or steam reforming, to be *more precise.*

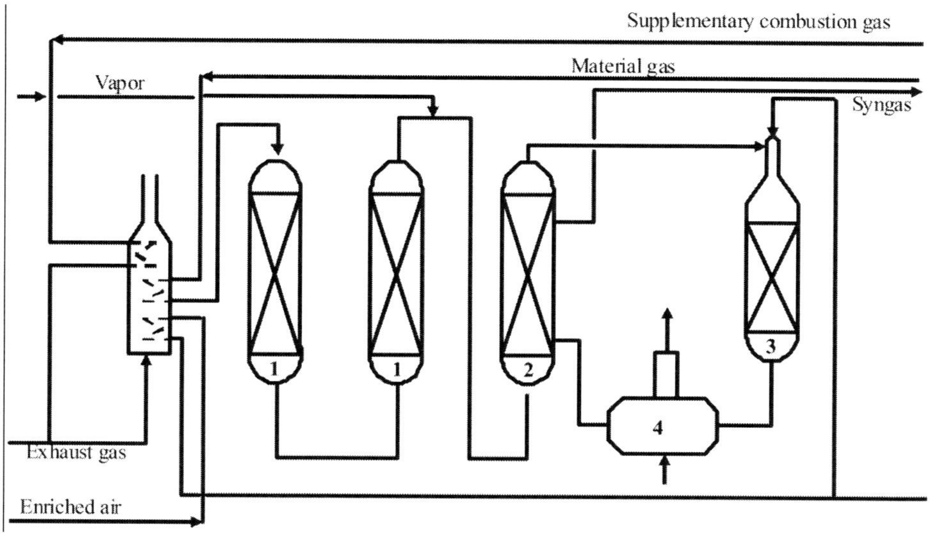

Note: 1. Devulcanizer; 2. Heat exchanger; 3. Converter; 4. Waste heat boiler; 5. Heating furnace.

Figure 3.1. The process flow diagram of catalytic partial oxidation.

The process flow diagram of catalytic partial oxidation is shown in Figure 3.1. Autothermal catalytic partial oxidation conversion technology is a partial oxidation process that provides heat for the reaction system. As the nickel-containing catalysts are sensitive to poisons, any sulfur compounds present in the hydrocarbon feedstock have to be removed by hydrodesulfurization, generally with a combination of cobalt-molybdenum and zinc oxide catalysts. The mixture of coke oven gas and vapor, after preliminary heating and desulfurizing, enters heat exchanger (2, in Figure 3.1). Further heating up to 500°C, the mixture enters converter (3). Then the mixture and preheating oxygen-enriched air (oxygen) mix and reform. When the synthesis gas is used as raw material for ammonia, adding nitrogen content in the air should meet the requirements of ammonia. The synthesis gas is used as raw material for alcohol ether, using pure oxygen rather than air. Hydrogen or CH_4 partial oxidation in converter releases a lot of reaction heat. The heat released from reactions will be supplied to the conversion reaction system. The outlet gas temperature is above 900°C. The residual methane is less than 0.5%. The advantage of this technology is mild exothermic and can produce syngas with a molar ratio of H_2 to CO of 2/1, which can be directly used as feed for methanol synthesis or the Fischer-Tropsch synthesis. Moreover, the partial oxidation of methane (POM) process can greatly speed up the production of syngas since it can be operated at very high space velocities. In CPOX (catalytic partial oxidation) the use of a catalyst reduces the required temperature to around 800°C - 900°C. Compressed natural gas (CNG) conversion industry has been using this technology in1960s. However, the choice of reforming technique depends on the sulfur content of the fuel being used. CPOX can be employed if the sulfur content is below 50 ppm. A higher sulfur content would be poison the catalyst. The proportions of (H_2-CO_2) to (CO+CO_2) at the reformer exit are 2.6 - 2.7, which are away from the optimal condition, $(H_2\text{-}CO_2) / (CO+CO_2) = 2.05$, required by methanol synthesis with no shift-reaction process. Furthermore, there are a lot of carbon black productions forming because of excess H_2. CPOX has been generally replaced by combined reforming. However, for heavier feedstock it remains the only feasible method. CPOX is a technology in common use today, with a lower efficiency than SR but a wider range of fuel which can be used for processing. High investment costs are due to the cryogenic air separation to provide the oxygen. For large scale de-carbonization applications in pre-combustion, however, hydrogen used as fuel will have to be diluted with either N_2 or steam. Hydrogen in high concentrations causes high flame temperatures in the combustion process in the gas turbine and must be reduced through dilution. In this case, air instead of oxygen as the oxidant may be the more efficient solution. CH_4 content of coke oven gas is lower than that of natural gas, COG can be converted into syngas by CPOX. China's patent publication No. CN1385501A discloses the invention, heat exchange COG pressure catalytic partial oxidation to synthesis gas technology, which is belonged to CPOX. In the reaction process, part of the hydrogen in the coke oven gas is burned to provide heat for conversion reaction system. At high temperatures (950 – 1150°C), pressure (2.07 MPa), steam (H_2O) reacts with methane (CH_4) to yield syngas using nickel compounds as catalysts. It has some advantages, such as conserving raw materials and energy, low lost, safety and protection of the catalyst [19].

Combined Reforming

In order to solve the problem described above, combined reforming has been practiced in industrial units since the 1980s [19-23], the combination of both catalytic steam reforming and CPOX. According to the different primary reformer heating, combined reforming is divided into two kinds: Tubular reforming and heat exchange reforming.

Tubular Reforming

In this process, the primary reactor is a tube-type furnace which is catalytic steam reforming reactor. The secondary reactor is a partial oxidation reactor. The production gas of the primary reactor, after being mixed with oxygen-enriched air (oxygen) being heated to 450 - 550°C in the gas preheater, enter the secondary reactor upper. First, hydrogen of the production gas from the primary reaction reacts with O_2 evolution of considerable heat, which let the secondary reactor temperature up to 1000°C. Then H_2O and CH_4 are transformed into CO and H_2 under the carbon catalyst in the secondary reactor. The outlet syngas of the secondary reactor is about 900°C, the residual methane is less than 0.5%. Then the syngas enters fine desulfurization, to make the sulfur content meet the requirements of chemical synthesis. The process flow diagram of heat exchange reforming is shown in Figure 3.2. In this process, the primary reactor is a recuperative reactor which is catalytic steam reforming reactor. Source gas is fed to a desulfurization reactor, 2 in Figure 3.2. At the same time, steam is also supplied to the desulfurization reactor. The source gas that has undergone the desulfurization process, is mixed with steam, and is fed to the heat exchanger, 3. A certain proportion of the carbon and water, after being mixed and being heated to 500-550°C in heat exchanger 3, the source gas is fed to the first reformer, 4. The first reformer includes a catalyst bed there within, the catalyst bed being filled with a catalyst having a metal, such as nickel and/or ruthenium, supported on a carrier, e.g., aluminum oxide. In the reactor 4, 30% CH_4 in the source gas react with the steam by the action of the catalyst and are converted to H_2 and CO. Operation conditions in the first recuperative reactor, 4, is as follows: the inlet temperature is set between 500°C and 550°C; the outlet temperature is set between 600°C and 650°C; and the added amount of the steam is set to above 2.5 mol, with respect to 1 mol of carbon in the source gas. The produced gas from the primary reforming reactor, 4, is fed to the two-stage converter, upper side of 5. Oxygen-enriched or oxygen, heated to about 530°C in heat furnace 1, which is supplied separately via a pipe to the two-stage converter 5 in which the oxygen is mixed with the mixed gas of the produced gas described above and the steam. Then hydrogen of the production gas reacts with O_2 evolution of considerable heat.

The heat released from reactions will be supplied to catalytic CH_4-steam reforming. Then H_2O and CH_4 are transformed into CO and H_2 under the carbon catalyst in the secondary reactor. The outlet syngas of the secondary reactor is about 950-1000°C, the residual methane is less than 0.35%.

Heat Exchange Reforming

The principle of heat exchange reforming is not new. It has been practiced in industrial units since the 1980s, albeit at conditions much less severe than those prevailing in GTL plants [24-26]. Different arrangements are possible, as illustrated in Figure 3.3. In the series' arrangement, all gas passes through the steam reforming unit and then through the ATR. This will mean that the steam reforming catalyst may set the lower limit for the steam to the carbon ratio.

In the parallel arrangement the two reformers are fed independently, giving freedom to optimize the steam to the carbon ratio individually. However, the heat exchange reformer must operate at a higher temperature than in series arrangement in order to obtain a low methane concentration in the synthesis gas [27].

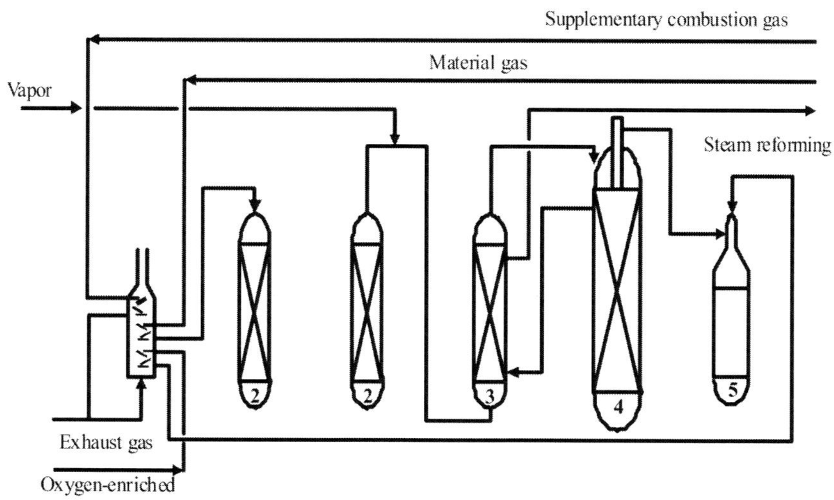

Note: 1. Heating furnace; 2. Devulcanizer; 3. Heat exchanger; 4. Primary reformer; 5. Secondary reformer.

Figure 3.2. The process flow diagram of heat exchange reforming.

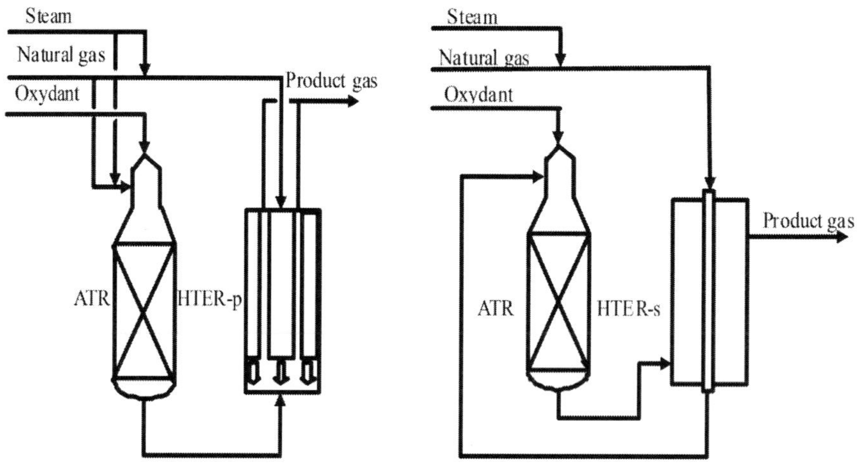

Figure 3.3. The combination between ATR and heat exchange reforming (HTER).

The combined reforming process has the following characteristics:

1. The secondary reactor has high operating temperature. CH_4 conversion is high, outlet gas CH_4 concentration low; Outlet H to CO molar ratios can easily be controlled by adjusting the oxygen concentration in the inlet gas mixture.
2. The primary steam reforming furnace is operating at low loads, transition temperature also decreases, and thus to a lower requirement for equipment and material. As results, good effect and low cost can be obtained.
3. The secondary reformer outlet gases supply the primary reforming heat reduces energy use is a lowering of CO_2 emissions.
4. Costs are reduced because a compressor is not required.
5. There are two reactors and construction costs are high.

An undesired side reaction in the combustion zone is the formation of carbon or soot, which leads to solid carbon deposition on the catalysts answer subsequent coking and deactivate. Gas phase carbon forms soot on the downstream surface thereby causing equipment damage and heat transfer problems. Excessive local temperatures also lead to burner damage. In the combined transformation process, the composition of the product gas from the primary reforming reactor is similar to COG. Therefore, the secondary reactor of the combined transformation is a partial oxidation reactor, which can transform COG to synthesis gas.

Other Reforming Technologies [28-32]

A recent development which avoids a fired primary reformer is the exchanger reformer, which with some simplification may be viewed as tubular heater exchanger with the catalyst inside the tubes, which are heated by the hot secondary reformer effluent flowing on the shell-side. In some designs the tubes may be open at the lower end, in which case the gas flow on the shell side consists of a mixture and from the reformer tubes. Commercially operating designs are the gas heated reformer (GHR) of ICI (Imperial Chemical Industries) and the

KRES (Kellogg exchanger reformer) system of M.W. Kellogg. Similar concepts are offered by other licensors and contracts e.g., Braun and Root or Topsoe.

ICI has came out with a modified design, the AGHR (Advanced gas heated reformer), with "A" standing for "advanced". As shown in Figure 3.4, the bayonet tubes are replaced by normal tubes attached to a bottom tubesheet using a special seal to allow some expansion [28,29]. In this way the delicate double tubesheet of the GHR is avoided. The seal which prevents leakage of methane-rich gas to the secondary reformer effluent following on the shell side has a unique design which is subject to patent application of ICI. The AGHR will allow a single-line concept for worldscale plants whereas with the GHR several parallel units for large plants would be necessary.

In the Kellogg reformer exchanger system, abbreviated KRES, Figure 3.5, the gas flow pattern is different [28, 29]. The tubes are open at the lower end and the reformed gas mixes with the hotter effluent of the secondary reformer. The mixed gas stream flow up-ward on the shell side to heat the reformer tubes. Thus primary reforming and secondary reforming reaction proceed in parallel in contrast to the ICI concept where the two reactions proceed in series. The Kellogg process uses enriched air. The complete elimination of the fired tubular furnace leads to a drastic reduction of NO_x emission, because there is only flue gas from much smaller fired heaters required for feed and process air preheat.

An even more progressive exchanger reformer presently operating in a demoplant is Uhde's CAR (Combined Autothermal Reformer) which not only replaces the catalytic secondary reforming step by a non catalytic partial oxydation step but also combines this with the exchanger reformer in one single vessel [29-31]. This combined autothermal reformer design, shown in Figure 3.6 was operated in a demonstration unit producing 13000 m^3/h of synthesis gas. Preliminary industry tests showed that the combined autothermal reformer investment is low (30%) and reduces energy consumption 27%.

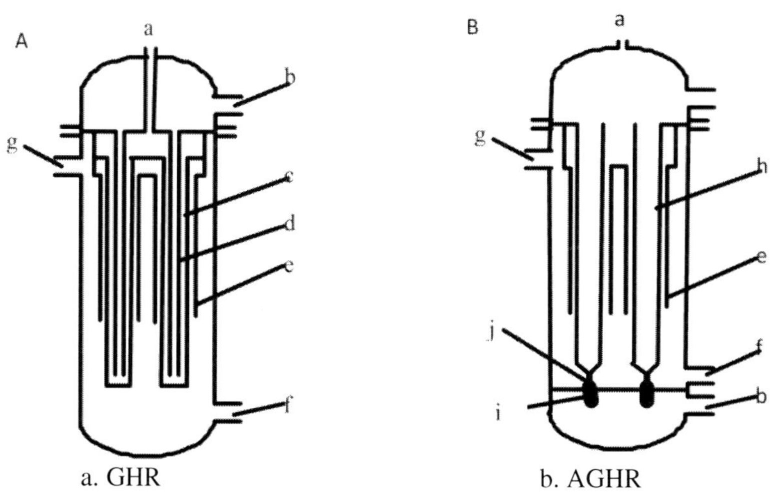

Note: a. Tubeside inlet; b. Tubeside outlet; c. Scabbard tube; d. Bayonet tube; e. Sheath tube; f. shellside inlet; g. Shellside outlet; h. Catalyst tube; i. Tail pipe; k. Catalyst; l. Refractory linling.

Figure 3.4. ICI gas heated reformer.

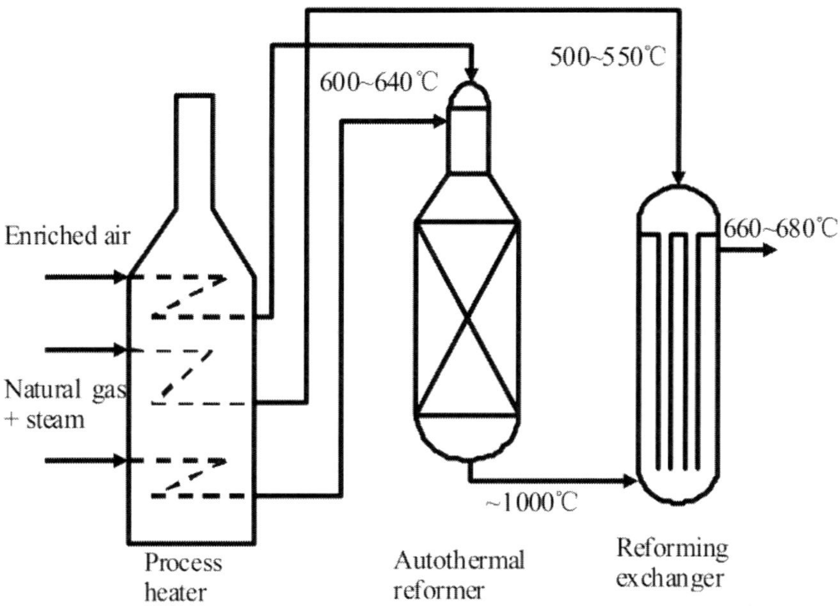

Figure 3.5. Kellogg reforming exchanger system (KRES).

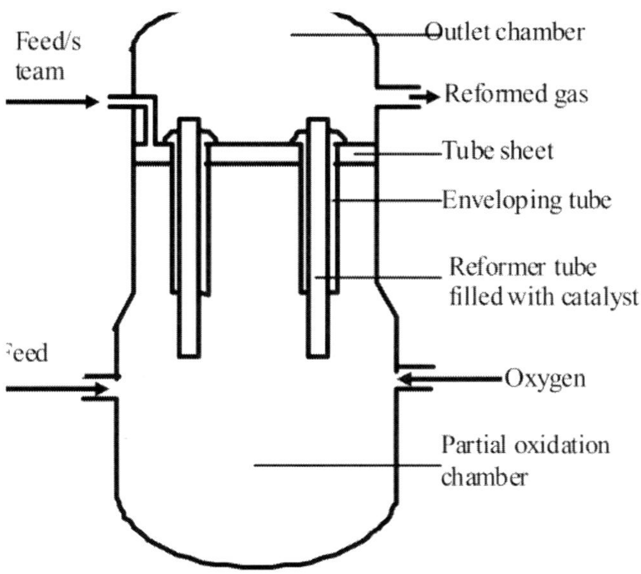

Figure 3.6. Uhde combined auothermal reformer.

The Topsoe autothermal reforming uses oxygen and carbon dioxide or steam in a reaction with methane to form syngas [32]. The feedstock is mixed with oxygen and steam in a mixer/burner. The reaction takes place in a single chamber where the methane is partially oxidized.

The reaction is exothermic due to the oxidation. The advantage of ATR is that the H_2:CO can be varied and this is particularly useful for producing certain second generation biofuels such as DME which requires a 1:1 of H_2 to CO ratio. When the ATR uses carbon dioxide, the produced H_2 to CO ratio is 1:1 and when the ATR uses steam, the produced H_2 to CO ratio is 2.5:1. The outlet temperature of the syngas is between 950-1100°C and outlet pressure can be as high as 100 bar.

Among the drawbacks, on the other hand, is the fact that large volumes of CO_2 are produced and the top of the combustion compartment reaches high temperatures close to 2000°C.

In order to avoid soot formation within the high temperature region, steam to natural gas ratios must be employed relatively high in existing commercial ATRs, greater than about 0.6. Additionally, ignition means, such as burner nozzles and related mechanical equipments in existing commercial ATRs are complex and have limited operating life due to the stresses associated with high temperature operations.

In actuality these technologies are points along a spectrum that spans combustion and partial oxidation through steam reforming (SR). Autothermal reforming, combines partial oxidation (POX) and SR, in a single process. POX reaction is exothermic or produces heat, while SR reaction is endothermic and heat must be generated external to the reformer process. Other exothermic reactions that may simultaneously occur in ATR include water gas shift (WGS) and methanation reactions. Typically, ATR reactions are considered to be thermally self-sustaining, and therefore, do not produce or consume external thermal energy. Catalysts are commonly used to enhance the reaction rate of the reforming processes at lower temperatures.

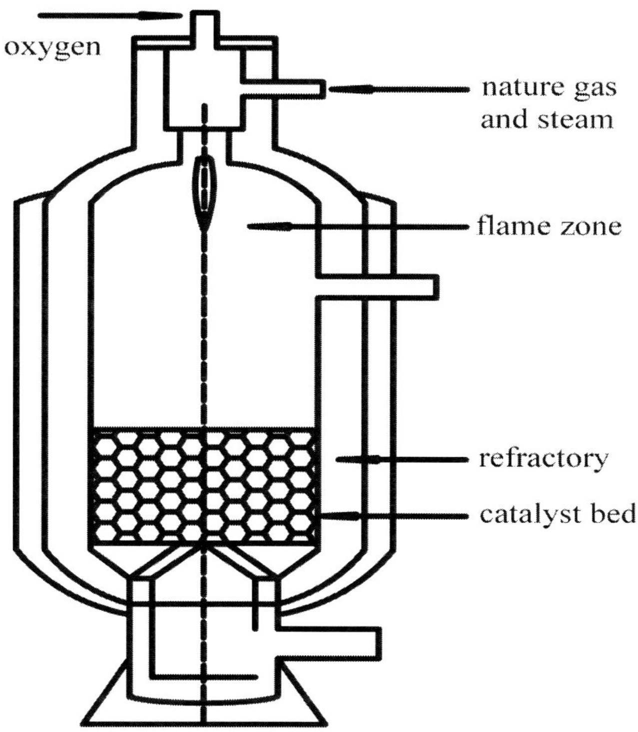

Figure 3.7. ATR reactor construction.

Non-Catalytic Partial Oxidation [33,34]

The key problem of non-catalytic partial oxidation is burner and flow-field. Either catalytic or non catalytic reformer, coke oven gas and oxygen would be mixed in the burner. The burner not only promotes mixed COG with oxygen, but also matches with the furnace to form the appropriate flow field, and then forms the appropriate temperature distribution.

For non-catalytic partial oxidation COG, the burner has two primary functions: making sure the fuel (COG) and oxygen mix in the correct proportions and figuring the direction and the shape of the flame. The burner is the key element for the oxygen-fired reformer. The burner mixes the hydrocarbon and oxygen feedstocks. Careful design of the burner nozzles ensures a flow pattern with efficient mixing that protects the refractory and burner from the hot flame core. In non-catalytic partial oxidation reforming the reactor's high aspect ratio (L/D) requires more than 3:1 in order to ensure proper mixing and residence time.

The Flow Property of Non-Catalytic Partial Oxidation and Temperature

The main chemical reactions in the course of a non-catalytic POX of oven gas are:

$$CH_4 + 2O_2 = CO_2 + 2H_2O \qquad \Delta H_{298} = -802.60 \text{ kJ mol}^{-1} \tag{3.11}$$

$$H_2 + 1/2\, O_2 = H_2O \qquad \Delta H_{298} = -241.84 \text{ kJ mol}^{-1} \tag{3.12}$$

$$CO + 1/2O_2 = CO_2 \qquad \Delta H_{298} = -282.96 \text{ kJ mol}^{-1} \tag{3.13}$$

$$CH_4 + 1/2O_2 = CO + 2H_2 \qquad \Delta H_{298} = -27.32 \text{ kJ mol}^{-1} \tag{3.14}$$

$$CH_4 + H_2O = CO + 3H_2 \qquad \Delta H_{298} = 206.15 \text{ kJ mol}^{-1} \tag{3.15}$$

$$CH_4 + CO_2 = 2\, CO + 2H_2 \qquad \Delta H_{298} = 247.27 \text{ kJ mol}^{-1} \tag{3.16}$$

$$CO_2 + H_2 = CO + H_2O \qquad \Delta H_{298} = 115.98 \text{ kJ mol}^{-1} \tag{3.17}$$

$$CH_4 = C + 2H_2 \qquad \Delta H_{298} = 71.56 \text{ kJ mol}^{-1} \tag{3.18}$$

In the non-catalytic POX of oven gas, the main reactions are reactions (3.11) - (3.17). Reactions (3.11) - (3.14) refer to the combustion of oven gas with a reaction rate as fast as a millisecond, called primary reactions while reactions (3.15) - (3.17) act as the control steps that dominate the reformation result of the process, would be complete in 1-2 seconds, called secondary reactions. In order to ensure the completion of the secondary reactions, the reactor must have a proper height-to-diameter (H/D) ratio. Reaction (3.18) is less important in the non-catalytic POX of oven gas.

Reactions (3.11) - (3.14) are exothermic, whereas reactions (3.15) - (3.17) are endothermic.

The heat released from combustion reactions (3.11) - (3.14) will be supplied to reforming reactions (3.15) - (3.17), i.e., exothermic reactions (3.11) - (3.14) are coupled with endothermic reactions (3.15) - (3.17) [34].

Since reactions (3.11) - (3.14) have very large equilibrium constants, and thus the reactions (3.11) - (3.14) may be regarded as irreversible reactions. In endothermic reactions (3.15) - (3.17), increase in temperature increases the K value. In other words, raising the temperature is advantageous to the conversion of COG from the thermodynamic aspect. Of course, a higher temperature but n a poor fuel mixture will lead to the occurrence of methane decomposition reaction (3.18), and thus to a higher requirement for equipment and material. Overall, in the non-catalytic POX, the reaction rate is fast and the control step for oven gas reformation is the mixing process.

Analysis of Reformer Region

The non-catalytic POX of COG is normally completed in an entrained-flow bed. According to hydrodynamic characteristics, the reformer can be divided into three regions [35]: a jet-flow region, a recirculation-flow region and a tube-flow region. A schematic diagram of typical entrained-flow bed reactor sections equipped with a WHB is show in Figure 3.1.

Corresponding to the three flow regions, the reformer can be also divided into three zones according to different chemical reaction characteristics, that is, a primary reaction zone (combustion zone), a secondary reaction zone, and a coexistence zone for primary and secondary reactions.

The reactions in these zones can be further subdivided into two types: one is a combustion reaction of COG (called the primary reaction), CO and H_2 in recirculation gas, and the other is a COG reformation reaction and the product of the primary reaction, called the secondary reaction, with different characteristics.

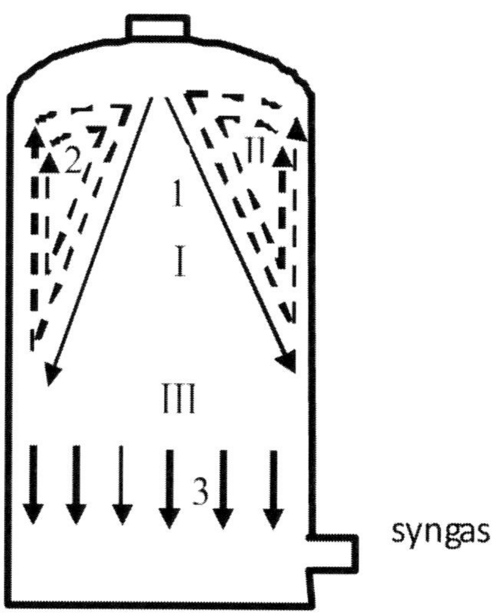

Note: I. Jet flow region; II. Recirculation-flow region; III. Tube flow region; 1. Primary reactions zone; 2. Primary reactions and secondary reactions co-existing zone; 3. Secondary reactions zone.

Figure 3.8. Schematic diagram of zones in oven gas reforming furnace.

Undoubtedly, combustion mainly occurs in the jet-flow region corresponding to the primary reaction zone (combustion zone). Generally, the reaction in the tube-flow region corresponds to the secondary reaction zone, while the combustion zone may extend to the tube-flow region depending on the mixing effectiveness. The recirculation flow region corresponds to a coexistence zone for the primary and secondary reactions, in which both combustion and secondary reaction happen at the same time due to the turbulent diffusion of oxygen.

Experimental studies had shown that [36-38], for a confined jet impingement, a recirculation flow would form at up to 10 times the jet diameter downstream the jet centerline. This flow recirculation would bring CO and H_2 in the downstream to the vicinity of the jet-flow region.

In the considered jet plate-to-impingement plate configuration, the jet after impingement was restricted to flow only in two opposite directions parallel to the x-axis. An increase of the jet plate width means an increase of the heated-surface area in the direction (y-direction) perpendicular to the main flow. Whether the main reaction in the combustion zone is the combustion of COG or the combustion of recirculation gas depends on the size of the macro-mixing time scale and the combustion time scale.

Reaction features in other zones is mainly affected by two factors: one is the size of the mixing (micro-mixing or macro-mixing) time scale, the other is the reaction time scale.

Since the micro-mixing time scale is close to the macro-mixing time scale in the reformer, on the condition that both the macro-mixing control zone and the micro-mixing control zone are developing simultaneously, the reactants, in partially-separated form, are dispersed in the reformer, and the mixing process of the reactants happens both in the bulk and in the local at the same time.

1. Primary reaction zone

Since the time scale of reactant macro-mixing is much bigger than that of oven gas combustion (in milliseconds), the main reaction in the primary reaction zone is the combustion of oven gas, whereas the combustion of CO and H_2 in the recirculation gas entrained by jet flow is not considered.

2. Secondary reaction zone

The gas product of primary reactions, CO_2, CH_4, H_2O, CO, H_2, and a very small amount of soot formed by cracking CH_4 in COG, arrives secondary zone which precedes the secondary reactions:

$$C + CO_2 = 2\,CO \tag{3.19}$$

$$C + H_2O = CO + H_2 \tag{3.20}$$

$$CH_4 + H_2O = CO + 3H_2 \tag{3.21}$$

$$CH_4 + CO_2 = 2\,H_2 + 2CO \tag{3.22}$$

$$CO_2 + H_2 = CO + H_2O \quad (3.23)$$

At the reformation temperature, the reactions (3.21) - (3.23) occur **very** quickly which are all homogeneous-phase reactions.

The Process of Non-Catalytic Partial Oxidation

The process of non-catalytic partial oxidation COG is shown in Figure 3.9. The key problem in technology is burner and flow-field.

In the non-catalytic POX section of an autothermal reformer, for the case of methane and oxygen or air as the feed, oxygen is mixed with the methane and combusted in a turbulent diffusion flame [39-41].

The exothermic combustion reactions are very fast so that all oxygen is consumed by the reaction with methane.

Methane combustion reaction occurred through many radical reactions [42]. Non-catalytic partial oxidation COG has not been used industrially, but there are the following several advantages [43-54]:

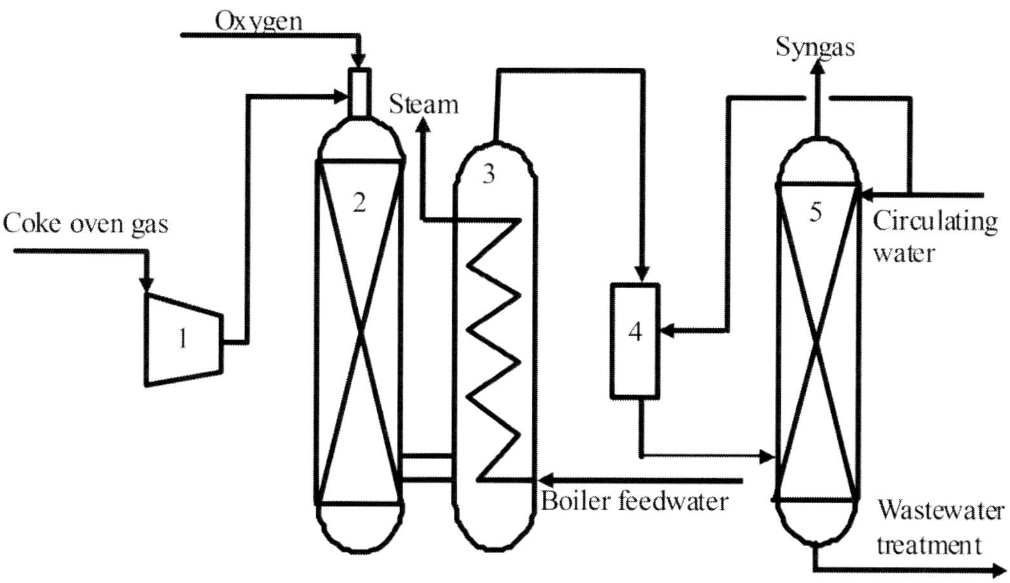

Note: 1. Compressor; 2. Converter; 3. Waste heat boiler; 4. Scrubber; 5. Spray column.

Figure 3.9. The process flow diagram of COG non-catalytic partial oxidation.

1. Non-catalytic partial oxidation with pure oxygen is an ideal process for H_2/CO ratio smaller than 2.
2. Exothermic reaction, no supplementary heat is required.
3. Being used pure oxygen partial oxidation, the inert gas content of loop gas is low in methanol synthesis, conducive to energy saving.

4. Non-catalytic partial oxidation does not use catalyst and the catalyst poisoning problem would be resolved.

Before gas is reformed, the removal of sulphur compounds (desulfurizer) is not necessary.

In the high temperature (1200°C), the organic sulfur compounds (thiophene, sulfide, thiols) are transformed of H_2S and COS, which can then be more easily removal. And it can greatly simplify the purification process, increase the desulfurization precision, reduce the cost of cleaning and reduce secondary pollution to the environment. It was believed that non-catalytic partial oxidation is the future direction of coke oven gas purifying and reforming development.

Non-Catalytic Local Strong Oxidation

Basic Principles of Non-Catalytic Local Strong Oxidation for COG

Under certain temperature conditions, CO_2 or $H_2O_{(g)}$ reforming of CH_4 occurs to produce H_2 and CO ($CH_4 + H_2O = CO + 3H_2 + 206$ kJ mol^{-1}; $CH_4 + CO_2 = 2CO + 2H_2 + 247$ kJ mol^{-1}). Due to the high exothermic character of the methanation reactions, the temperature has important implications in chemical reaction rates [15]. In catalytic partial oxidation the use of a catalyst reduces the required temperature to around 800-1050°C [16-18]. Temperatures typically in the range of 1250 - 1400°C occur due to the exothermic reaction in non-catalytic POX [19,20]. Thus, either non-catalytic POX or CPOX, appropriately increasing reaction temperature become the key factor increasing CH_4 conversion and enlarging reactor.

Raising the temperature has relatively high oxygen consumption. Of course, a higher temperature will lead to a higher requirement for equipment and material [21-23]. While not producing any increase in oxygen consumption, non-catalytic local strong oxidation would make CH_4 high-efficiency conversion by forming local high temperature in the oxidation zone.

Temperature Distribution

Table 3.1 shows the reactor chamber operating temperature of pyrolysis and CPOX, non-catalytic POX and non-catalytic LSOX (non-catalytic local strong oxidation). As can be seen from the table1, CPOX requires temperature lower (800-1150°C). The conversion temperature of non-catalytic partial oxidation is higher, about 1310-1430°C. Many factors will affect transition temperature, such as the types and quantities of reactants, oxygen consumption, Preheat temperature, and so forth. Adding a small amount of steam (0.5 - 1.0 m^3 m^{-3}) into the feed gas not only can significantly improve the catalytic stability by inhibiting coke formation, but also can lower the oxidation zone temperature during the reaction. Simulation shows that the operating temperature of CPOX non-catalytic POX drops about 110°C and 260°C as adding water vapor, respectively. Because CH_4 transformation is a thermodynamic equilibrium process (endothermic reaction), the increment of temperature would significantly improve the equilibrium conversion. In other words, raising the temperature (e.g. maintaining the operating temperature at 1200 - 1300°C) is advantageous to the conversion of methane in oven gas from the thermodynamic aspect. So conventional non-catalytic partial oxidation has a high demand for oxygen and has a relatively high oxygen consumption.

Although it does not add water vapor to the reaction in non-catalytic LSOX, H_2O is formed during combustion because coke oven gas contains a notable amount of H_2. The reaction pathway of CH_4 in the coke oven gas in the membrane reactor is as follows: firstly, H_2 in the coke oven gas reacts with the O_2, and generates H_2O, then CH_4 reforms with the H_2O to produce H_2 and CO. Steam reforming CH_4 happens when oxygen is the reactants of. The results show that non-catalytic LSOX and CPOX can be completed at the same temperature. The combustion zone (oxidation zone) average temperature of non-catalytic LSOX is around 1250°C, which is lower 170°C than the average temperature of non-catalytic POX, and which is lower 60°C than the average temperature reported in the literature [15,17]. Non-catalytic LSOX can obviously lower average oxidation temperature, reduce oxygen consumption. Compared with CPOX, non-catalytic LSOX has obvious advantages in reforming: no catalyst for reforms, no pre-desulfurization units and auxiliary units.

Experimental and simulation results show that the temperature distribution of non-catalytic LSOX and non-catalytic POX has a significantly different in the reformer. The temperature distribution of non-catalytic LSOX and non-catalytic POX is shown in Figure 3.10. Figure 3.10 shows a typical temperature distribution of non-catalytic LSOX and non-catalytic POX. As can been from Figure 3.10, non-catalytic LSOX temperature distributes concentrates in a localized zone instead of throughout a large region. The local temperature is above 1700°C. The temperature distribution non-catalytic POX is relatively uniform. In the conventional non-catalytic partial oxidation process, feed gas and oxygen are pre- mixture before oxidation reaction, by the time jet mixture reaches the oxidation zone, molecular mixing is fairly complete. When COG is used as raw material, with its large number of H_2, the main oxidative combustion reaction is ($H_2 + ½ O_2 = H_2O$; $\Delta H_{298} = -241.0$ kJ mol^{-1}). Because hydrogen has a flame speed faster than CH_4 and CO. Hydrogen burns more rapidly than hydrocarbon fuels because hydrogen molecules are smaller and enter combustion reaction chambers at a higher velocity than other gases. The main oxidation process is hydrogen and O_2 in the non-catalytic partial oxidation process. Because H_2, CH_4, and O_2, have been mixed, the overall temperature is uniform, up to 1300°C.

Table 3.1. The reactor operating temperature of pyrolysis and CPOX, non-catalytic POX and non-catalytic LSOX [6-8]

CPOX (°C)	Non-catalytic POX (°C)	Non-catalytic LSOX, (°C)		
		Top	Midst	Outlet
1050 - 1150	1310 ± 10	1167.0	1072.0	982.0
850 - 1050	1410 - 1430	1193.0	1078.0	971.6
—	—	1218.0	1081.0	956.7

Note: CPOX = Catalytic Partial Oxidation; POX = Partial Oxidation; LSOX = Local Strong Oxidation.

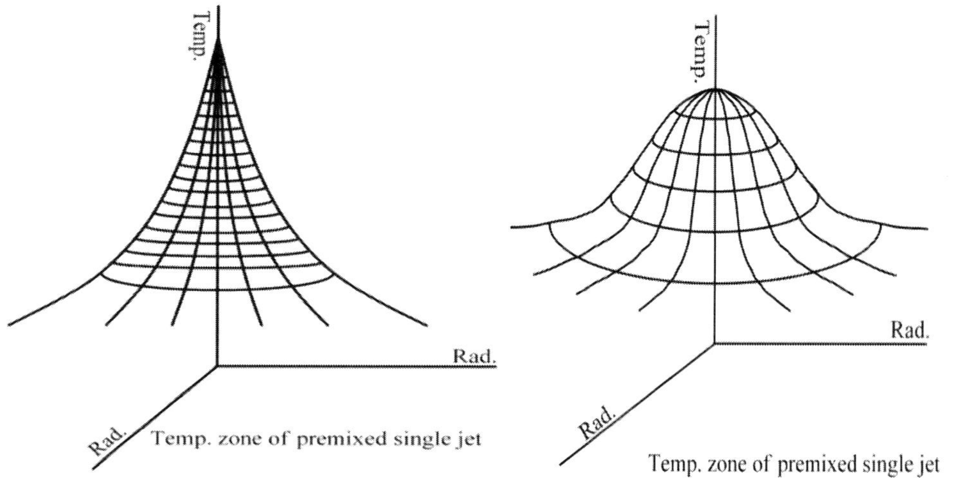

a. Temperature distribution of non-CLSOX b. Temperature distribution of non-CPOX

Figure 3.10. Temperature distribution of reforming process.

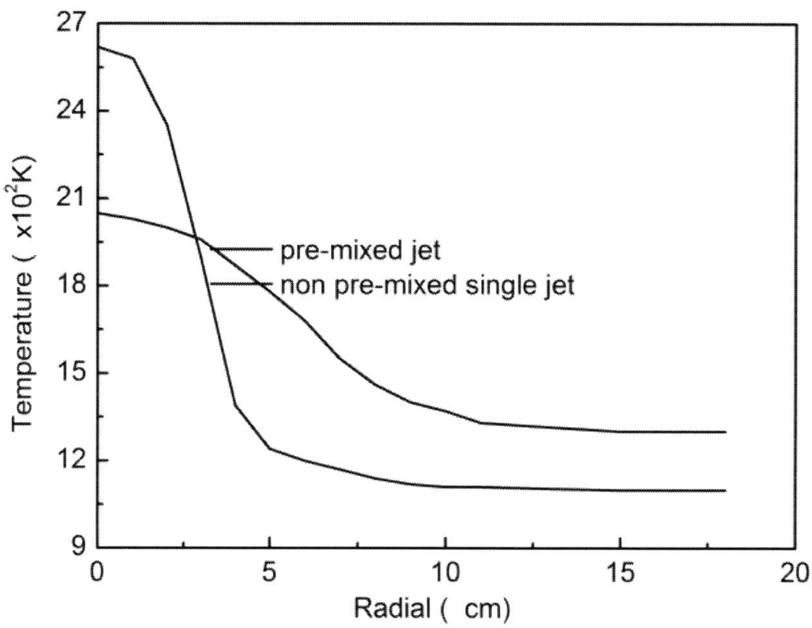

Figure 3.11. The radial temperature distribution curve of reformer.

In non-catalytic LSOX process, feed gas and oxygen are non-premixed before oxidation reaction, by the time jet mixture reaches the oxidation zone, the diffusion process cannot be quickly completed. A certain proportion of CH4 and CO combustion may be accompanied. CH_4 and CO combustion has two obvious advantages: 1. CH_4 reforming concentration Reduces, fewer energy requirements; 2. As CH_4 and CO oxidation reaction heat is significantly higher than H_2 oxidation (802 kJ mol^{-1}, 282 kJ mol^{-1}, and 241 kJ mol^{-1}, respectively for each oxidation reaction), heat release increases forming local temperature, which is beneficial HO-, O-, and other active species generations, accelerates CH_4 reforming.

CH_4 reforming is significantly divided into two parts: one section is oxidized in the oxidation zone; another part is restored in the reduction zone. From the Figures 3.11 and 3.12, it can be seen, local strong oxidation shortens oxidation zone, forming local high temperature which is beneficial HO-, O-, and other active species generations. The active species can accelerate CH_4 transformation. Calculated temperature profile indicates that oxygen react strongly with coke oven gas in the region near the oxygen nozzle and the temperature in this region increase sharply. In the reduction (secondary reactions) zone, the residual CH_4 continue to be transformed where the system temperature is gradually reduced. The upper oxidized zone shortens. The lower reduced zone lengthens. CH_4 is a fully transformation.

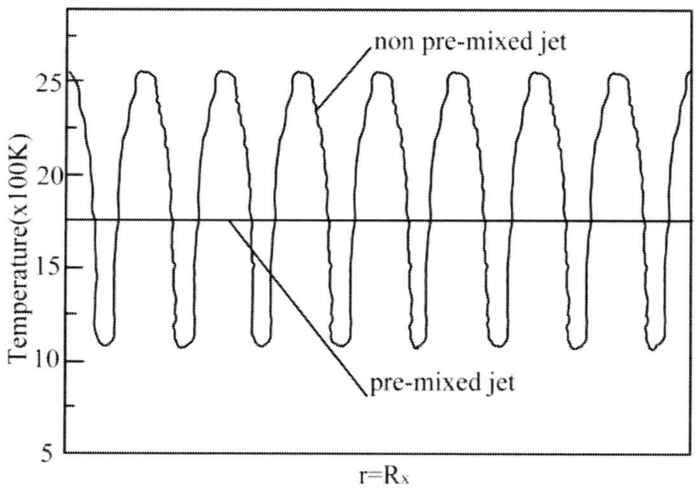

Note: Wavy curve: interface temperature of non-catalytic LSOX; Straight line: interface temperature of non-catalytic CPOX.

Figure 3.12. The axial temperature distribution curve of reformer (r = R_X).

Analysis of Reaction Mechanism

The primary reaction is the oxidation process. The main reaction is an ionic chain reaction during non-catalytic LSOX, H_2, Co and CH_4 oxidized. In the local high temperature zone, the oxidation reaction is significantly faster than the normal oxidation reaction, and generates a large number of HO-, O-, and other active species. H_2, CO, and CH_4 combustion theory supports this view [23].

H_2 branched chain reaction speed against the time graph for the non-catalytic LSOX is shown in Figure 3.13. During the reaction, a large number of HO-, O-and other active species are generated. At the initial stage of reaction, the reaction rate is low because the active species' concentration is low. This transient time to the steady state of reaction is called the induction period. The reaction is similar to the general combustion reaction. Further reaction, the active species increases because of branched chains, the reaction speed rapidly increases and then reaches the maximum value.

The main chain reactions during a non-catalytic LSOX of oven gas are:

1. Chain of production:

$$H_2 + O_2 = 2OH \tag{3.24}$$

$$H_2 + M = 2H + M \tag{3.25}$$

$$O_2 + O_2 = O_3 + O \tag{3.26}$$

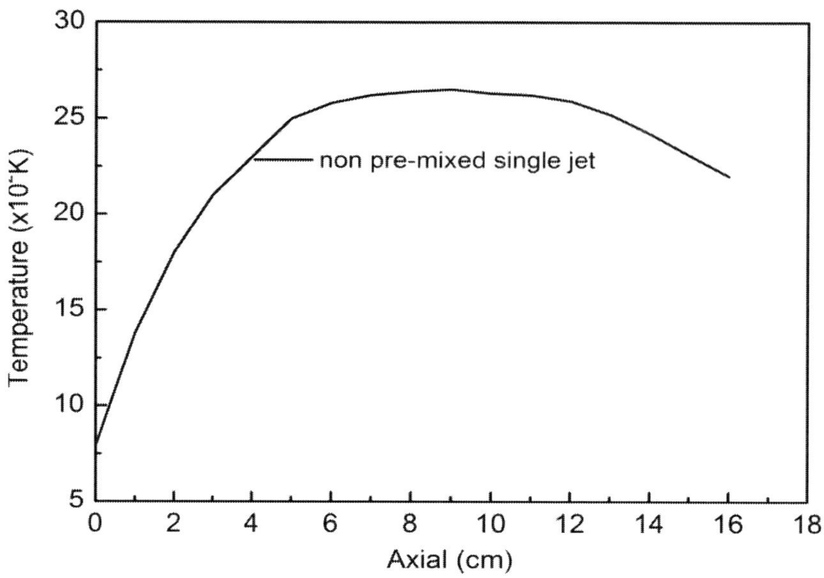

Figure 3.13. The graph of H_2 branched chain reaction speed against time.

2. Branched chains (continuation):

$$H + O_2 = OH + O \tag{3.27}$$

$$OH + H_2 = H_2O \tag{3.28}$$

$$O + H_2 = OH + H \tag{3.29}$$

3. Chain scission at the wall:

$$H + wall = ½ H_2 \tag{3.30}$$

$$OH + wall = ½ H_2O_2 \tag{3.31}$$

$$O + wall = ½ O_2 \tag{3.32}$$

4. Space broken chain:

$$H + O_2 + M = HO_2 + M^* \tag{3.33}$$

$$O + O_2 + M = O_2 + M^* \tag{3.34}$$

$$O + H_2 + M = H_2O + M^* \quad (3.35)$$

The characteristics of carbon monoxide oxidation are same to hydrogen oxidation. However, CO oxidation will only occur when there is enough H_2O existence. The reaction mechanism is as followed:

1. Chain of production:

$$H_2O + CO = H_2 + CO_2 \quad (3.36)$$

$$H_2 + O_2 = 2OH \quad (3.37)$$

2. Chains continuing:

$$OH + CO = CO_2 + H \quad (3.38)$$

3. Branched chains:

$$H + O_2 = OH + O \quad (3.39)$$

$$O + H_2 = OH + H \quad (3.40)$$

4. Broken chain:

$$H + Wall = \frac{1}{2} H_2 \quad (3.41)$$

$$CO + O = CO_2 \quad (3.42)$$

Because it does not add water vapor to the reaction in non-catalytic LSOX, the CO content in the coke oven gas is relatively low (about the mole fraction 6 - 7%). So the CO conversion mainly occurs at the lower part of the reactor. There has a little effect for high temperature zone.

The temperature of non-catalytic LSOX is above 1700°C. The methane does not only react in oxidation chain reaction, but also react in decomposition reactions under high temperatures. The main chemical reactions include: the partial oxidation of methane and the combustion of formaldehyde.

$$CH_4 + O_2 = HCHO + H_2O \quad (3.43)$$

$$HCHO + O_2 = H_2O + CO_2 \quad (3.44)$$

The reaction mechanism is as followed:

1. The primary reaction:

$$CH_4 = CH_3 + H \quad (3.45)$$

$$CH_3 + O_2 = HCHO + OH \quad (3.46)$$

$$H + O_2 = OH + O \quad (3.47)$$

$$CH_4 + OH = CH_3 + H_2O \quad (3.48)$$

$$CH_4 + O = CH_3 + OH \quad (3.49)$$

2. The secondary reaction is as followed:

$$HCHO = HCO + H \quad (3.50)$$

$$H + O_2 = OH + O \quad (3.51)$$

$$HCO + O_2 = CO + HO_2 = CO + OH + O \quad (3.52)$$

$$HCHO + OH = HCO + H_2O \quad (3.53)$$

$$HCHO + O = HCO + OH \quad (3.54)$$

3. HCO broken into CO:

$$HCO = CO + H \quad (3.55)$$

$$HCO + O = CO + OH \quad (3.56)$$

$$HCO + OH = CO + H_2O \quad (3.57)$$

4. CO oxidation:

$$CO + OH = CO_2 + H \quad (3.58)$$

$$CO + O = CO_2 \quad (3.59)$$

As it was shown from the above description, CH_4 oxidation mechanism is more complex than H_2 and CO. CH_4 begins to oxide at a lower temperature. The methane does not only react in oxidation chain reaction, but also in decomposition under high temperatures. The oxidation reaction mechanism is degenerated branched chain reaction. Not only the induction period became long, but also the reaction speed is slower than the branched chain reaction.

In non-catalytic CPOX process, feed gas and oxygen are premixed before oxidation reaction, by the time jet mixture reaches the oxidation zone, the diffusion process can be quickly completed.

The main chemical reaction of the oxidation zone is H_2 combustion reaction. CH_4 transformation is mainly in the reduction zone: $CH_4 + H_2O = CO + H_2$ or $CH_4 + CO_2 = CO + H_2$. In non-catalytic LSOX process, a certain proportion of CH_4 and CO combustion may be accompanied because the diffusion process cannot be quickly completed. CH_4 and CO oxidized in the oxidation zone, a large of heat released forms local temperature, which is beneficial HO-, O-, and other active species generations, accelerates CH_4 reforming. A reformer is obtained that has an outlet temperature lower, and thus has a lower oxygen consumption.

Non-Catalytic LSOX Process

The process flow diagram of heat exchange reforming is shown in Figure 3.14. This process is the atmospheric pressure reaction system. COG (25 - 35°C) from Coking Plant is fed to a saturated humidifier (25 - 35°C). COG is fed to heat exchanger, being heated to about 300°C.

The preheated COG is then fed to tube furnace and heated to 550°C - 600°C in tube furnace. COG is also injected to the converter with preheated oxygen is also supplied to the converter at the same time. According to the mechanism, the reactor is divided into two parts: upper local oxidation zone and lower reduction zone (1150 - 1350°C).

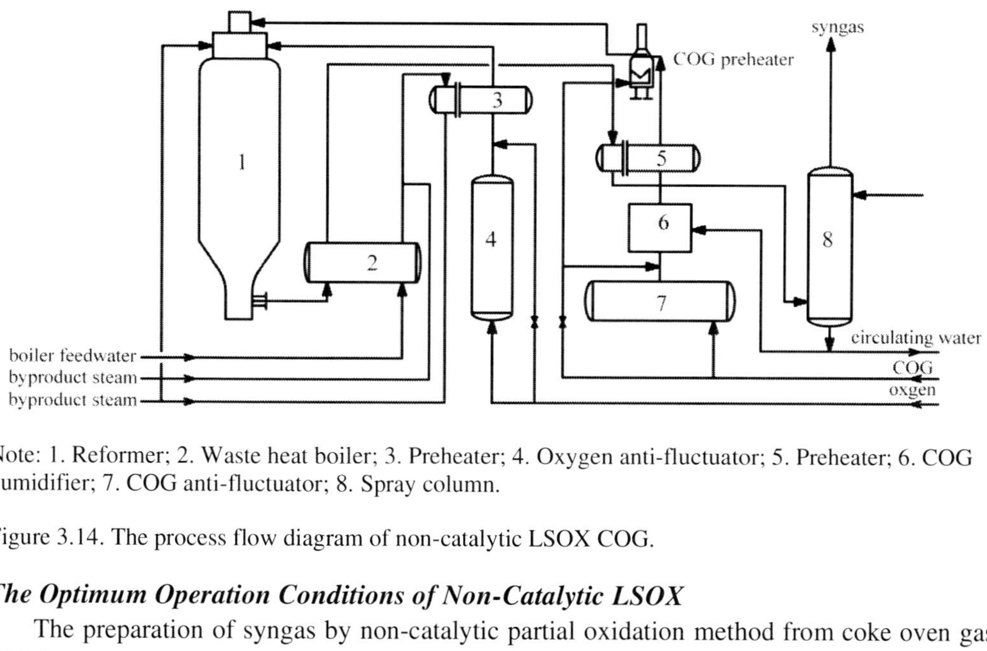

Note: 1. Reformer; 2. Waste heat boiler; 3. Preheater; 4. Oxygen anti-fluctuator; 5. Preheater; 6. COG humidifier; 7. COG anti-fluctuator; 8. Spray column.

Figure 3.14. The process flow diagram of non-catalytic LSOX COG.

The Optimum Operation Conditions of Non-Catalytic LSOX

The preparation of syngas by non-catalytic partial oxidation method from coke oven gas (COG) was carried out on a non-catalytic LSOX COG reactor. Influences of two kinds of burners (volute and partial oxidation) on the distribution of temperature and methane conversion were compared and investigated. It was found that the partial oxidation burner showed a desirable distribution in the temperature and flow field compared with the volute one under the pure oxygen conditions, reducing methane from 17% to less than 1% in gas product.

As a result, a higher conversion rate of methane and a lower oxygen consumption were obtained. The optimum operation conditions were established: oxygen flow rate 9 m^3/h , O_2 / COG = 0.22 - 0.24, the top temperature of the rector 1200 ~ 1250°C, outlet temperature 960 - 1000°C.

$COG\text{-}CO_2$ Reforming over Carbonaceous Catalyst ("Dual-Gas Resources" Catalyzed by C-Catalyst)

Although most authors have concentrated their attention on the steam reforming of COG, in the last few years the dry reforming of COG has also been investigated [55-66], due to the numerous advantages that it offers compared to steam reforming, such as the saving of energy or CO_2 consumption.

Another important advantage of the dry reforming of COG is the possibility of obtaining a syngas with a H_2/CO ratio of about 2, which is the ideal proportion for methanol synthesis, in only one step provided the process is carried out under stoichiometric conditions of CH_4 and CO_2. The process can be considered as a way of "partial recycling" of CO_2 since it consumes, at least theoretically, half of the CO_2 produced when methanol is burnt. The prospects for this technology are far-reaching, since the demand for methanol for vehicle fuel, as a source of hydrogen for fuel cells or biodiesel production is rapidly increasing.

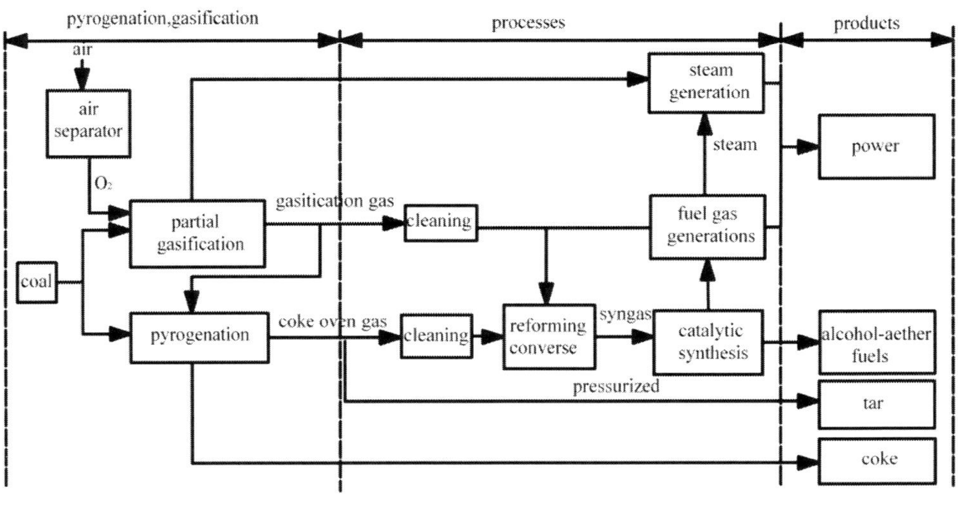

Figure 3.15. A schematic diagram of the new polygeneration system.

A large scale polygeneration system based on the coal is recognized as a clean coal technology at present, which is a perfect combination of coal power and chemical industry [67-69]. The different polygeneration system approaches have been represented in the world in recent years. In 1998, the United States proposed "Vision 21" polygeneration system which is based on gasification, power generation, hydrogen and liquid fuel. Shell Company of Netherlands proposed a polygeneration system which was close to the concept of the ecological park. Japan's new energy development organization, also in 1998, put forward EAGLE polygeneration plan on clean gasification gas, gas-fired power generation, fuel cell power generation and synthetic liquid fuels as the main productions. In order to achieve the ultimate goal of "zero discharge" in the whole process, the system also plans CO_2 recovery. Based on taking full account of the facts that china produces mass coke and about 20 billion cubic meters coke oven gas has not been used rationally per year, a polygeneration system called as gasification gas and pyrolysis gas reforming into syngas ($CO + H_2$) system, or called as dual-gas resources polygeneration system, shown in Figure 3.15 was proposed by Taiyuan

University of technology in 2005. China pushed forward, in its "973" program, "dual fuel" methanol-power poly-generation by utilizing catalytic CO_2 and CH_4 reforming without shift. CO and H_2 can be produced by catalytic reforming of CO_2 and CH_4. The ratio of carbon to hydrogen, which is required by the follow-up chemical synthesis, can be adjusted by different combinations of CO_2 rich gasification gas and CH_4 rich coke oven gas. In this way, the gasification gas can avoid the water-gas shift reaction to adjust the ratio of carbon to hydrogen and catalytically reform with the coke oven gas, which effectively increases the amount of effective gas. This scheme makes full use of coke oven gas, reduces CO_2 emission and system energy loss [69].

The Basic Idea of Dual-Gas Resources Polygeneration System

The main components in the raw gas from the ash agglomerating fluidized bed gasification are CO (~35%), H_2 (~40%), CO_2 (~20%). If the raw gas used to produce alcohol-ether flues, the H_2/CO ratio of raw gas should be adjusted to 2/1 through water-gas shift reaction ($CO + H_2O = CO_2 + H_2$) and the decarburization treatment is needed. The coke oven gas composes mainly of CO (5% - 8%), H_2 (55% - 60%) and CH_4 (23% - 27%). It is a kind of H_2-rich resource. It is used to produce syngas using the method of steam reforming. However, the H_2 content in the product gas from steam reforming is higher (H_2:CO = 3:1) than that required by methanol and ether production. If the coke oven gas and gasification gas are mixed in a certain percentage and the CH_4 in the coke oven gas and CO_2 in the gasification gas can be reformed ($CH_4 + CO_2 = 2\ CO + 2\ H_2$; $\Delta H_{298} = 247.4$ kJ mol^{-1}) into syngas (CO and H_2), the goals of both adjustments of H_2 without water-gas shift reaction and CO_2 emission reduction will be achieved. It was found that CH_4 in coke oven gas and CO_2 in gasification gas can be reformed into syngas over the carbonaceous catalyst in the high-temperature system. According to the idea and findings, the syngas production process of "dual-gas resources" catalyzed by c-catalyst was formed. The ratio of hydrogen and carbon is adjusted through CH_4 - CO_2 reforming reaction. The main reaction is: $CH_4 + CO_2 = 2\ CO + 2\ H_2$.

Since this reaction is a strong endothermic process, it could adopt the autothermal partial oxidation process. Under catalysis of carbon, CH_4 and CO_2 are converted into CO and H_2 through CH_4 and CO_2 reforming reaction (about 3000 kPa, 1000°C) [70].

The Dual-Gas Resources CH_4–CO_2 Reforming Process [70]

The mixture of coke oven gas and gasification gas, after oxidation, enters the catalyst layer to carry through CH_4 and CO_2 reforming process under catalysis. The reactor is composed of a combustion chamber and a reaction chamber. After mixing in an appropriate proportion and partially combusting in the combustion chamber, gasification gas, coke oven gas and O_2, enter the reaction chamber to further react into synthesis gas. However, if coke oven gas is used as the only raw material, with its large number of H_2, the main combustion reaction is ($H_2 + ½O_2 = H_2O$, $\Delta H_{298} = -241.0$ kJ mol^{-1}). The reason for this is that the combustion rate of H_2 is 1000 to 100 million times of the rate of the other reactions, far greater than that of CH_4 and CO. That is, at the top of the combustion chamber, the combustion of hydrogen and O_2 is the main reaction, generating H_2O and a great amount of heat. In the combustion process, a small proportion of CH_4, and CO combustion may be accompanied. When gas mixture, after burning, reaches the catalyst layer of the reactor, almost all the oxygen is consumed.

The characteristics of "dual fuel" auto-thermal reforming process are [71-75]:

1. Making use of both coke oven gas and CO_2;
2. Compared with steam conversion, it saves a lot of water;
3. Based on carbon catalyst, lowering the requirements of feed gas; coke oven gas needing no deep desulphurization and purification; without catalyst sulfur poisoning issues;
4. No carbon deposition poisoning problem.

TECHNOLOGY COMPARISON OF COG REFORMING TO SYNGAS

The main process of COG reforming can be divided into two categories: non-catalytic POX and CPOX. The Former Soviet Union has developed a non-catalytic POX coke oven gas to synthesis gas technology since 1940s, but no industrial applications reported. In recent years, Sedin engineering Co, Ltd (Former second Design Institute of Chemical Industry, China) has made significant breakthroughs in this field.

Table 3.2. Technology comparison of COG reforming to syngas

	Country	Maturity	Steam	Oxide zone T (°C)	Component content, %			
					H_2	CO	CO_2	CH_4
Non- POX	Japan	Mature	~10%	1,230	—	—	—	—
Non - POX	-	not industrialized yet	Yes	1,300 - 1,400	—	—	—	—
CPOX-SR	China	Mature	Yes	1,050 - 1,150	71.2	18.6	7.5	0.7
CPOX	China	Mature	Yes	1,310	55-59	15-20	6.5-8	2 - 4
Non - POX	FSU	not industrialized	—	—	53.5	35.8	2.2	0.3
Non- LSOX	China	pilot plant	—	1,150 – 1,250	66.6	24.1	4.0	0.7

Note: T = temperature; POX = Partial Oxidation; CPOX = Catalytic Partial Oxidation; SR = Steam Reforming; LSOX = Local Strong Oxidation; FSU = Former Soviet Union.

The technology has been applied for the Industrial Production. Japan becomes also very active in developing COG reforming, and made great progress. In 2005, Chengdu Huaxi Chemical Research Institute has been applied non-catalytic POX COG for the Industrial Production. However, the technology is only to produce synthesis gas for ammonia, using oxygen-enriched as oxidative. Up to now, it is not found yet that coke oven gas as raw material has been used to produce syngas with non-catalytic technology, i.e. for methanol synthesis. Technology and technical/economic evaluation of COG reforming to syngas are shown in Table 3.2 and 3.3 below [76-79].

Table 3.3. Technical/economic evaluation of COG reforming to syngas

	Maturity	COG consumption ($m^3 \cdot h^{-1}$)	O_2 consumption ($m^3 \cdot h^{-1}$)	steam consumption ($t \cdot h^{-1}$)	Cost in Yuan (1000 m^3syngas)$^{-1}$
Non- POX	not industrialized yet	38,300	8,360	3.8[c]	302.68
CPOX	mature	29,700	5,700	-4.3[d]	184.06
CPOX-SR	mature	29,500	5,400	-5[d]	220.30
Non-LSOX	pilot plant	2,950	5,920[a]; 5,244[b]	-5	-

Note: A single set of 10 million t/a methanol unit; a. Non-preheat; b. Preheat at 600°C; c. Condition at 2.7 Mpa, 250°C; d. 3.82 Mpa, 450°C.

ACKNOWLEDGMENTS

Project supported by the National Basic Research Program of China (Grant No.2005CB221202), Innovation Team Funds Scheme, the Natural Science Foundation of China (Grant No. 21006066), and Shanxi Provincial Nature Science Foundation (2010011014-1).

AUTHORS' BIOGRAPHIES

Yongfa Zhang, Eng.D. is Professor and Vice Director of Key Laboratory of Coal Science and Technology of Ministry of Education, Taiyuan University of Technology. He received his BS in Engineering from the East China University of Science and Technology and Doctor of Engineering from Taiyuan University of Technology. He is the Secretary General of the Shanxi Energy and Conservation Association and an associate editor of Journal of the China Coal Society, the Coal Process and Application and the China Journal of Energy and Conservation. Dr. Yongfa Zhang has published over 90 refereed papers in academic and professional journals, mainly related to coal, energy, various issues in clean coal technology, and coal education. His papers have appeared in Fuel, Carbon, Fuel Science and Technology Int'l, Catalysis Today, Chemical Engineering Journal, Journal of the China Coal Society, Chemical Engineering (China), Modern Chemical Industry, Chemical Reaction Engineering and Technology, The Chinese Journal of Process Engineering and Frontiers of Chemical Engineering in China. Prof. Zhang is a world-famous scholar who has published the most research works internationally among clean coal technology. His major teaching and research areas are clean coal and coal gasification. He is a consultant expert of the Organization Department of the CPC Central Committee.

Guojie Zhang, Eng.D. is teacher of Key Laboratory of Coal Science and Technology of Ministry of Education, Taiyuan University of Technology and project-planning engineer at Shanxi Urban and Rural Planning Design Institute, Taiyuan, Shanxi. Dr. Zhang received both his B.S. in Engineering from the Taiyuan University of Technology. He is a Fellow of the Shanxi Energy and Conservation Association. He teaches Process analysis and optimization,

Coal Chemistry and Coking technology. Dr. Zhang has numerous publications in academic and professional journals, mainly related to issues in energy and environment. Current research interests include clean coal technology, CH_4/CO_2 reforming.

REFERENCES

[1] Y. Zhang, P. Wang, K. Xie, *Shanxi Energy Conserv.* 37, 13 (2005).

[2] Z. Liu, "Industrial policy direction of China's energy and coke industry", *The First China's International Forum on Coal, Coke and Chemicals Industry,* 15, May, 2005 (Beijing, China) pp26-35.

[3] W. Zheng, Z. Yu, "Technology of coke production in China", *The First China's International Forum on Coal, Coke and Chemicals Industry,* 15, May, 2005 (Beijing, China) pp65-72.

[4] Y. Zhang, "Research and development trends of clean coking technology", *The First China's International Forum on Coal, Coke and Chemicals Industry*, 15, May, 2005 (Beijing, China) pp104-111.

[5] L. Yang, Y. Dong, Y. Zhang, K. Xie, *Shanxi Energy Conserve.* 40, 1 (2006).

[6] ACRE Coking and Refractory Engineering Consulting Corporation, Coking Design Manual, Metallurgical Industry Press, Beijing (1978).

[7] Z. Wang, R. Han, Preprint, 2003 at http://img.hexun.com/2009-08-10/120408583.pdf.

[8] W. Chen "Development Trend of China's alcohol fuel", *The First China's International Forum on Coal, Coke and Chemicals Industry,* 15, May, 2005 (Beijing, China) pp150-156.

[9] B. Yang, "Power generation and application of coke oven gas", *The First China's International Forum on Coal, Coke and Chemicals Industry,* 15, May, 2005 (Beijing, China) pp281-288.

[10] G. Wu, Q. Li. Coke oven gas utilization technology. *The First China's International Forum on Coal, Coke and Chemicals Industry,* 15, May, 2005 (Beijing, China) pp162-170.

[11] Q. Li, *Petrochem. News,* 6, 52 (1998).

[12] Y. Zhang, K. Xie, *Chinese Patent* No. 03152797.3, April, 21, 2005.

[13] W. Ying, F. Cao, D. Fang, *The Main Products and Technology of C1 Chemical, Chemical Industry Press,* Beijing (2004).

[14] The website for the hydrocarbons industry, Term: Catalytic Steam Reforming, Definition, http://www.hydrocarbons-technology.com/glossary/catalytic-steam-reforming.html.

[15] S Rakass, H Oudghiri-Hassani, P Rowntree, N Abatzoglou, *J. Power Sources*, 158: 485 (2006).

[16] A. F. Lucrédio, E. M. Assaf, *J. power sources*, 159, 66 (2006).

[17] R. C. Vasant, C. M. Kartick, *Appl. Energy*, 83, 1024 (2006).

[18] Wikipedia, Partial oxidation, http://en.wikipedia.org/wiki/Partial_oxidation.

[19] M. Bajus, H. B. Margare, *Appl. Catal. A: Gen.* 128, 61 (1995).

[20] N. Muradov, F. Smith, C. Huang, *Catal. Today*, 116, 281 (2006).

[21] T. S. Christensen, J. H. Bak-Hansen, P. S. Christensen, "Syngas Preparation by Autothermal Reforming for Conversion of Natural Gas to Liquid Products (GTL)", Second Annual Conference "Monetizing Stranded Gas Reserves", 14, December, 1998 (San Francisco, USA) pp52-59.
[22] T. S. Christensen, P. S. Christensen, I. Dybkjaer, "Developments in autothermal reforming", the Fifth International Natural Gas Conversion Symposium, 20, Sept. 2005 (Giardini Naxos-Taormina, Italy) pp883-888.
[23] M. Ravjee, J. S. Hansen, I. I. Primdahl. "Advanced regulatory control of open flame autothermal reformers", The 45th Annual Safey in Ammonia Plants and Related Facilities Symposium, 11, Sept., 2000, (Tucson, Arizona Tucson) pp99-106.
[24] P. K. Aasberg, T. S. Christensen, I. Dybkjær, J. Sehested, M. Østberg, R. M. Coertzen, M. J. Keyser, A. P, *Stud. Surf. Sci. Catal.* 2004, 152, 258(2004).
[25] Topsoe Inc., Synthesis Gas Tech., http://www.topsoe.com/business_areas/synthesis_gas /~/media/PDF%20files/Heat_exchange_reforming/Topsoe_synthesis_gas_technology.a shx
[26] P. K. Bakkerud, J. N. GØl, K. Aasberg-Petersen, I. Dyb kjær, *Stud. Surf. Sci. Catal.* 147, 13(2001).
[27] K. S. Looc, W. S. Ernst, S. G. Thomsen, M. F. Jensen, "Improving carbon efficiency in an auto-thermal methane reforming plant with gas heated heat exchange reforming technology", The world congress of chemical engineering, 10, July, 2005 (Glasgow, UK) pp1067-1073.
[28] Wiley-VCH (ed.). Ullmann's Agrochemicals. Wiley-VCH Verlag GmbH (2007).
[29] Haber, Bosch, Mittasch. "The Haber-Bosch heritage: The ammonia production technology", 50th Anniversary of the IFA Technical conference, 25, Sept. 1997 (Seville, Spain) pp1-25.
[30] J. R. Jennings (Ed.). Catalytic Ammonia Synthesis. Plenum Press, New York(1991).
[31] A Nielsen (Ed.). Ammonia – Catalysis and Manufacture. Springer-Verlag, Berlin-Heidelberg, New York (1995).
[32] Topsoe Inc., Synthesis gas, *http://www.topsoe.com/business_areas/synthesis_gas.aspx.*
[33] F. Wang, X. Zhou, W. Guo, Z. Dai, X. Gong, H. Liu, Front. *Energy Power Eng*. China, 3, 117 (2009).
[34] A. T. Vaso, E. V. Xenophon, *Catal. Today*, 2001, 64, 83 (2001).
[35] Z.Yu, J. Yu, F. Wang, *J. Chem. Ind. Eng.*, 45, 135 (1994).
[36] S. V. Garimella, R. A. Rice, *J. Heat Trans*, 117, 871(1995).
[37] J. A. Fitzgerald, S. V. Garimella., *Int. J. Heat Mass Tran.* 41, 1025(1998).
[38] B. G. Thomas, X. Huang, R. C. Sussman, *Metall. Trans. B*, 25, 527 (1994).
[39] J. San , W. Shao, *Int. J. Heat Mass Tran.*, 49, 3477(2006).
[40] I. Dybkjaer, *Fuel Process. Technol.* 42, 85 (1995).
[41] T. S. Christensen, I. I. Primdahl, *Hydrocarbon Process.* 73, 39(1994).
[42] B. S. Bintulu, L. Schmidt, *Fuel Process. Technol.* 42, 109 (1995).
[43] Z. Akbar, B. Alireza, E. Hadi, *Eng. Technol.* 49, 247(2009).
[44] J. Pina, D. O. Borio, "Modeling and simulation of an autothermal reformer", 14, Aug. 2005 (Village Rio das Pedras, Club Med, Rio de Janeiro) pp1-10.
[45] J. Zhu, D. Zhang, K. D. King, *Fuel*, 80, 899 (2001).
[46] B. A. Albrecht, Reactor modeling and process analysis for partial oxidation of natural gas. Ph.D. Thesis, University of Twent (2004).

[47] P. Glarborg, R. J. Kee, J. F. Grcar, J. A. Miller, A fortran program for modeling well-stirred reactors, Technical Report SAND86-8209 Sandia National Laboratories (1986).
[48] J R. Rostrup-Nielsen, L. J. Christiansen, J. H. B. Hansen, *Appl. Catal.* 43, 287(1998).
[49] J. Davies, D. H. Li, *Chem. Eng. Process*, 52 71(1971).
[50] S. H. Elnashaie, A. M. Adris, M. A. Soliman, A. S. Al-Ubaid, *Can. J. Chem. Eng.* 70, 786 (1992).
[51] J. Xu, G. F. Froment, *AIChE J.* 35, 97 (1989).
[52] R. M. Quinta Ferreira, M. N. Marques, M. F. Babo, A. E. Rodrigues, *Chem. Eng. Sci.* 47, 2909 (1992).
[53] H. Ebrahimi. Modeling of top fired methane steam reforming reactors. MSc. Thesis, Sahand university of Technology, Tabriz (2004).
[54] Z. Akbar, E. Hadi, S. S. Jafar, Chem. Eng. Prog. 47, 946 (2008).
[55] J. M. Bermúdez, B. Fidalgo, A. Arenillas, J A. Menéndez, *Fuel*, 89, 2897(2010).
[56] H. Cheng, X. Lu, X. Liu, Y. Zhang, W. Ding, *J. Nat. Gas Chem.*, 18, 467(2009).
[57] Y. Zhang, Y. Liu, J. Liu, W. Ding, *Energy Technol.*, 30, 137(2009).
[58] Y. Zhang, Q. Li, P. Shen, *Int. J. Hydrogen Energy*, 33, 3311 (2008).
[59] Z. Shao, G. Xiong, H. Dong, *Sep. Purif. Technol.*, 25, 97(2001).
[60] H. Wang, T. Cristina, S. Thomas, *Catal. Commun.*, 7, 907 (2006).
[61] Y. J. Henn, M. Bouwmeester, *Catal. Today*, 82, 141(2003)
[62] X. Dong, Z. Liu, W. Jin, *J. Power Sources*, 185, 1340(2008).
[63] C. Chen, S Feng, S Ran, Angew. *Chem.Int. Edit.* 42, 5196(2003).
[64] E. P. Foster, P. J. A. Tijm, D. L. Bennett, *Stud. Surf. Sci. Catal.* 119, 8671(1998).
[65] H. Cheng, Y. Zhang, X. gang Lu, *Energy Fuels*, 23, 414 (2009).
[66] J. Shen, Z. Wang, *Energy Fuels*, 21, 3588(2007).
[67] K. Xie, Y. Zhang, W. Zhao, Shanxi energy conserv, 49, 10(2008).
[68] Yongfa Zhang, "Study of CH_4-H_2O in coke-oven gas reforming to produce syngas catalyzed by C-catalyst" 23rd Annual International Pittsburgh coal conference, 26, Sept. 2006 (Pittsburgh, PA) pp1-14.
[69] H. Zhang, Experimental study of oven gas reforming to produce synthesis gas over the coal char. MSc. Thesis, Taiyuan University of Technology, Taiyuan (2005).
[70] M. Wang, Z. Li, W. Ni, *Fron. Energy Power Eng. China*, 3, 341(2009).
[71] S. H. Seok, S. H. Han, J. S. Lee, *Appl. Catal. A: General*, 215, 31 (2001).
[72] Y. Li, R. Xiao, B. Jin, H. Zhang, F. Wang, *J. Combust. Sci. Technol.* 15, 238 (2009).
[73] W Zhang, Y Zhang, *Fron. Chem. Eng. China*, 4, 147 (2010).
[74] G. Zhang, Y. Dong, M. Feng, Y. Zhang, H. Cao, W. Zhao, *Chem. Eng. J.* 156, 519(2010).
[75] H. Zhao, Q. Chen, H. Chen, Y. Zhang, *Coal conversion*, 2009, 32, 13 (2009).
[76] Q. Zhang, "Applications and development of methanol production technology from Coke Oven Gas", The First China's International Forum on Coal, Coke and Chemicals Industry, 15, May, 2005 (Beijing, China) pp306-314.
[77] G. Liu, X. Chen, M-Sized Nitr. *Fertilizer Prog.* 3, 33(2006).

[78] X. Feng, "Discuss of non-catalytic partial oxidation Coke oven gas with pure oxygen", Applied Seminar of Using Coke Oven Gas to Produce Methanol in China Coking Industry. 15, May, 2005 (Beijing, China) pp16-25.

[79] M. Onozaki, K. Watanabe, T. Hashimoto, H. Saegusa, Y. Katayama, *Fuel*, 85, 1432(2006).

In: Syngas
Editors: Antonius Indarto and Jelliarko Palguandi

ISBN: 978-1-62100-870-5

Chapter 4

THERMODYNAMIC ANALYSIS OF SYNTHESIS GAS AND HIGHER HYDROCARBONS PRODUCTION FROM METHANE

I. Istadi*

Chemical Engineering Department, Diponegoro University, Indonesia

ABSTRACT

This chapter focused on thermodynamic chemical equilibrium analysis using method of direct minimization of Gibbs free energy for all possible methane reactions with oxygen (partial oxidation of methane), carbon dioxide (CO_2 reforming of methane), steam (steam reforming of methane), and autothermal reforming. Effects of feed ratios (methane to oxygen, carbon dioxide, and/or steam feed ratio), reaction temperature, and system pressure on equilibrium composition, conversion, and yield were studied. In addition, operating regions of carbon and no carbon formation were also considered at various reaction temperatures and feed ratios in the equilibrium system. It was found that the reaction temperature above 1100 K and CH_4/CO_2 ratio unity were favorable for synthesis gas production for methane – carbon dioxide reaction. The Carbon Dioxide Oxidative Coupling of Methane reaction to produce ethane and ethylene is less favorable thermodynamically. In addition, steam reforming of methane is the most suitable for hydrogen production from methane with low coke formation from thermodynamic point of view.

NOMENCLATURE

$\hat{\phi}$	fugacity coefficient of the species in solution
ΔC_P^o	standard heat capacity at constant pressure, J·mol^{-1}·K^{-1}
ΔG_T^o	standard Gibbs free energy change at temperature T, kJ·mol^{-1}
ΔH	heat of reaction change, kJ·mol^{-1}

* E-mail address: istadi@undip.ac.id.

ΔH^o_0	standard heat of reaction change, kJ·mol^{-1}
ΔH^o_{298}	standard heat of reaction change at 298 K, kJ·mol^{-1}
υ	stoichiometric number
ATR	Auto Thermal Reforming
b	total population of atom in the system
C	carbon coke
f	fugacity of pure species, atm
$\dot{f}$	fugacity of species of solution, atm
$\bar{g}$	partial molar Gibbs free energy, kJ·mol^{-1}
g	Gibbs free energy of the pure species, kJ·mol^{-1}
G	total Gibbs free energy, kJ·mol^{-1}
I	number of chemical species
j	sequence number of atoms
J	total number of different atoms that are present in the system
K	equilibrium constant
n	number of atom in the molecule
n	number of mol of species, mol
N	number of moles in reaction mixture, mol
OCM	Oxidative Coupling Methane
P	total pressure (atm)
R	universal gas constant, J·mol^{-1}·K^{-1}
T	temperature, K
WGS	Water Gas Shift
X	conversion, %
y	mol fraction
Y	yield, %

Superscripts and Subscripts

o	standard condition
298	standard temperature 298K
i	species *i*th
T	at temperature T

INTRODUCTION

The utilization of CH_4 conversion to produce important chemicals, especially synthesis gas (H_2 and CO), provides several advantages from the environmental and energy perspectives. Natural gas is a fuel consisting of methane, ethane, carbon dioxide, H_2S, and trace amounts of other compounds. It is highly desirable to utilize and to convert methane, typical component in natural gas, into higher value-added chemicals such as producing synthesis gas. Composition of natural gas varies widely from location to location, however, the largest component is methane.

Co-generation of synthesis gas and higher hydrocarbons from methane is important in utilization of the natural gas through reforming reactions. The reforming process yields synthesis gas with high H_2/CO molar ratio and small amount of light hydrocarbons (C_2 hydrocarbons). The synthesis gas can be converted to liquid fuels by the Fischer-Tropsch process and also to various value-added chemicals, especially methanol and gasoline via the methanol-to-gasoline (MTG) process. The synthesis gas is also the main source of hydrogen for refinery processes and ammonia synthesis. Use of CO_2 as an oxidant for the selective oxidation of methane may also be beneficial, because it is expected that the replacement of O_2 with CO_2 inhibits the gas-phase non-selective oxidation and thus increases the selectivity to higher hydrocarbons. Previously, the thermodynamic calculation on equilibrium conversion of CH_4 to C_2 hydrocarbons (C_2H_6 and C_2H_4) showed that the equilibrium conversion increased with rising temperature or CO_2/CH_4 feed ratio (Istadi and Amin, 2005). However, thermodynamic equilibrium studies on the co-generation of synthesis gas and C_2 hydrocarbons from methane using various reforming processes have been conducted separately. Accordingly, the results of thermodynamic studies can identify the constraints placed in a reacting system and can provide the recommendation of suitable operating conditions for the catalytic reacting system theoretically.

Synthesis gas production through reforming of methane can be carried out in four different ways, i.e.:

1. Methane – carbon dioxide reactions, contains:

a. Carbon dioxide Reforming of Methane (CORM)

$$CH_4 + CO_2 = 2CO + 2H_2 \qquad \Delta H^o_{298} = +247 \text{ kJ·mol}^{-1} \tag{4.1}$$

$$CO_2 + H_2 = CO + H_2O \qquad \Delta H^o_{298} = +41 \text{ kJ·mol}^{-1} \tag{4.2}$$

b. Carbon dioxide Oxidative Coupling of Methane (CO_2 OCM)

$$2CH_4 + CO_2 = C_2H_6 + CO + H_2O \qquad \Delta H^o_{298} = +106 \text{ kJ·mol}^{-1} \tag{4.3}$$

$$2CH_4 + 2CO_2 = C_2H_4 + 2CO + 2H_2O \qquad \Delta H^o_{298} = +284 \text{ kJ·mol}^{-1} \tag{4.4}$$

$$C_2H_6 = C_2H_4 + 2H_2 \quad \Delta H^o_{298} = +136 \text{ kJ·mol}^{-1} \tag{4.5}$$

2. Methane – steam reactions (Steam Reforming of Methane – SRM)

$$CH_4 + H_2O = CO + 3H_2 \qquad \Delta H^o_{298} = +206 \text{ kJ·mol}^{-1} \tag{4.6}$$

$$CO + H_2O = CO_2 + H_2 \qquad \Delta H^o_{298} = -41 \text{ kJ·mol}^{-1} \tag{4.7}$$

3. Methane – oxygen reactions, contains:

a. Partial Oxidation of Methane (POM)

$$CH_4 + ½O_2 = CO + 2\,H_2 \qquad \Delta H^o_{298} = -36\ kJ{\cdot}mol^{-1} \qquad (4.8)$$

b. Oxidative Coupling of Methane (OCM)

$$2CH_4 + ½O_2 = C_2H_6 + H_2O \qquad \Delta H^o_{298} = -80\ kJ{\cdot}mol^{-1} \qquad (4.9)$$

$$2CH_4 + O_2 = C_2H_4 + 2H_2O \qquad \Delta H^o_{298} = -144\ kJ{\cdot}mol^{-1} \qquad (4.10)$$

4. Auto Thermal Reforming (ATR)

Autothermal reforming (ATR) uses oxygen and carbon dioxide or steam in a reaction with methane to form syngas. The reaction takes place in a single chamber where the methane is partially oxidized. The reaction is exothermic due to the oxidation. The reactions can be described in the following equations, using CO_2:

$$2CH_4 + O_2 + CO_2 = 3H_2 + 3CO + H_2O \qquad \Delta H^o_{298} = -31\ kJ{\cdot}mol^{-1} \qquad (4.11)$$

and using steam:

$$4CH_4 + O_2 + 2H_2O = 10H_2 + 4CO \qquad \Delta H^o_{298} = +339\ kJ{\cdot}mol^{-1} \qquad (4.12)$$

Research studies on synthesis gas production process were mainly focused on catalyst performance studies and the output of these studies cannot be easily compared due to the big differences among the basic process variables (feed composition, pressure, and temperature). However, the effects of temperature, pressure, or feed ratio on equilibrium conversions, yields, and compositions can be analyzed via a thermodynamic analysis. The study on thermodynamic equilibrium composition has been used in investigating the feasibility of many types of reaction. Vasudeva et al. (1996) computed equilibrium compositions for steam reforming of ethanol to examine the viability of the process. Similarly, Chan and Wang (2000) examined the thermodynamic equilibrium compositions for simultaneous partial oxidation and steam reforming of natural gas. Global reaction balances and chemical equilibrium of steam reforming and partial oxidation to produce hydrogen were studied by Lutz et al. (2003; 2004) for the same objective. Thermodynamic chemical equilibrium on methane pyrolysis process were also reported by Guéret et al. (1997) using the direct Gibbs free energy minimization. Similar method was also performed by Lwin et al. (2000) on the hydrogen production from steam-methanol reforming. The method of the direct minimization of Gibbs free energy of a system was used by Chan and Wang (2000), Tang and Kitagawa (2005) and Guéret et al. (1997) for solving equilibrium thermodynamic analysis of supercritical water gasification of biomass, steam reforming of ethanol for hydrogen production, and methane pyrolysis, respectively. Meanwhile, the minimization of Gibbs free energy using Lagrange's multiplier was implemented by Douvartzides et al. (2003), and Chan and Wang (2000; 2004) for solving thermodynamic equilibrium analysis of solid oxide fuel cells, natural-gas fuel processing for fuel cell applications, autothermal methanol reformer, and catalytic combustion of methane, respectively.

This chapter focuses on thermodynamic chemical equilibrium analysis of all possible CH_4 reactions in the co-generation of synthesis gas and/or C_2 hydrocarbons through carbon dioxide reforming of methane, steam reforming of methane, partial oxidation of methane, and autothermal reforming. In this analysis, effect of various conditions, i.e. temperature, reactants feed ratio and system pressure, on chemical equilibrium (conversion, yield, and composition) are discussed. In addition, effects of reactant feed ratio and temperature on carbon formation in the reaction system at equilibrium are also investigated. Through the thermodynamic equilibrium analysis, the feasibility of methane reactions in producing synthesis gas could be addressed theoretically.

TECHNIQUE FOR THERMODYNAMIC CHEMICAL EQUILIBRIUM CALCULATION

There are two common ways to express the chemical equilibrium. One is based on equilibrium constants (K), while the other is based on the minimization of Gibbs free energy (Smith et al., 2001). In recent years, thermodynamic equilibrium results utilizing Gibbs free minimization technique have been reported for methane reforming to syngas via various routes. There are two alternative methods for solving the minimization of Gibbs free energy:

1. Direct minimization, and
2. The use of Lagrangian multiplier

The direct minimization of Gibbs free energy was reported to be effective for complicated chemical equilibrium problems. The method, which is a default method in *Chemkin*, was used to solve the chemical equilibrium system at various temperatures, feed ratios, or system pressures without requiring detailed information about the homogeneous or heterogeneous reaction rate.

In this chapter, the chemical equilibrium analysis of reactions between CH_4 and O_2, CO_2, or H_2O, and autothermal reforming are considered, in which CH_4, CO_2, O_2, CO, H_2, H_2O, C_2H_4, C_2H_6 and/or carbon (C) exist in the equilibrium system. For CORM reaction, CH_4 and CO_2 are stated as reactants, while for POM reaction, CH_4 and O_2 are defined as reactants. In addition, for SRM reaction, CH_4 and H_2O are stated as reactants, while for ATR reaction, CH_s-O_2-CO_2 and/or CH_4-O_2-H_2O are defined as reactants. The possible main reactions of CH_4 reforming of methane to produce the equilibrium products are listed in equations (4.1) - (4.12). The chemical equilibrium calculations were carried out using *Chemkin collection R3.7.1* to compute the chemical composition at specified temperature and pressure.

The practical use of the calculation method in *Chemkin* requires the following parameters:

a. The substances likely to be present at equilibrium,
b. The elements information within the system, and
c. Initial compositions of the chemical species.

Thermodynamic properties data for equilibrium calculation of the system are available readily in the *Chemkin* software.

Nevertheless, the predicted chemical compositions are still theoretical upper limits for the process, so that the optimal conditions may change in practice.

Pertaining to the direct Gibbs free energy minimization method, basic equations for chemical equilibrium calculation is considered. Consider an initial system of N_i mole of each of the i chemical species. The total Gibbs free energy G of the system is expressed as (Smith et al., 2001):

$$G = \sum_{i=1}^{I} N_i \, \bar{g}_i = \sum_{i=1}^{I} N_i \left[\left(\bar{g}_i - g_i^{\circ} \right) + g_i^{\circ} \right] \quad (4.13)$$

where $\bar{g}_i$ is the partial molar Gibbs free energy of the ith species in solution at operating conditions, g_i^o is the Gibbs free energy of the pure ith species at standard conditions, and N_i is the number of moles of each species i in the system. Using the relation (Smith et al., 2001):

$$\bar{g}_i - g_i^{\circ} = RT \ln \frac{\hat{f}_i}{f_i^{\circ}} \quad (4.14)$$

where is the fugacity of species i in solution at operating conditions, and f_i° is the fugacity of pure species i in its standard state, equation (4.13) can be written as:

$$G = \sum_{i=1}^{I} N_i \left[g_i^{\circ} + RT \ln \frac{\hat{f}_i}{f_i^{\circ}} \right] \quad (4.15)$$

If all the species are in the gaseous state and letting the reaction system pressure to be P, we can write (Smith et al., 2001):

$$\frac{\hat{f}_i}{y_i P} = \hat{\phi}_i \text{ and } \hat{f}_i = \hat{\phi}_i y_i P = \hat{\phi}_i \frac{N_i}{N} P \quad (4.16)$$

where N is total number of moles in the reaction mixture including the unreacted species; is the fugacity coefficient of the ith species in solution, and y_i is mole fraction of the ith species. Since standard state is taken as pure ideal gas state at 1 atm, f_i^o equals to 1 (Lwin et al., 2000; Smith et al., 2001). At low pressure and/or high temperature, the system can be considered to be ideal gas mixtures or ideal solutions in which $\hat{\phi}_i = 1$, equation (4.15) becomes:

$$G = \sum_{i=1}^{I} N_i \left[g_i^{\circ} + RT \ln \frac{N_i}{N} + RT \ln P \right] \quad (4.17)$$

or

$$G = \sum_{i=1}^{I} N_i RT \left[\frac{g_i^{o}}{RT} + \ln y_i + \ln P \right] \quad (4.18)$$

where R is the universal gas constant, and y_i is the mole fraction of the ith species.

The equilibrium solution at a given temperature and pressure is the distribution of N_i or y_i that minimizes the system Gibbs function, G, subject to the elemental mass balance and non-negative N_i constraints. The elemental mass balance constrain is:

$$\sum_{i=1}^{I} n_{ji} N_i = b_j \qquad j = 1,\ldots,J \quad (4.19)$$

where n_{ji} is the number of the jth atoms that appear in the ith molecule, b_j is the total population in moles of the jth atom in the system, and J is the total number of different elements/atoms that are present in the system. The standard Gibbs free energy changes (ΔG_T^{o}) at system temperature T used in analysis of the reactions is calculated by the following equation (Smith et al., 2001):

$$\frac{\Delta G_T^{o}}{RT} = \frac{\Delta G_0^{o} - \Delta H_0^{o}}{RT_0} + \frac{\Delta H_0^{o}}{RT} + \frac{1}{T}\int_{T_0}^{T} \frac{\Delta C_p^{o}}{R} dT - \int_{T_0}^{T} \frac{\Delta C_p^{o}}{RT} dT \quad (4.20)$$

Meanwhile, the equilibrium constant (K) of the reactions can be predicted by the following equation:

$$\ln K = \frac{-\Delta G_T^{o}}{RT} \quad (4.21)$$

Equilibrium conversion and yield were calculated based on the initial and equilibrium compositions resulted by the *Chemkin* calculation. The molar flow rate of reactor outlet can be determined from the mass balance developed by *Chemkin*. Conversions of CH_4, O_2, CO_2 and/or H_2O are calculated by equation (4.22).

$$X(i) = \frac{n_{i,in} - n_{i,out}}{n_{i,in}} \times 100\% \quad (4.22)$$

where i corresponds to species i and n_i is mol of species i. The yields of H_2 and CO are calculated by equations (4.23) and (4.24).

$$Y(CO) = \frac{n_{CO,out}}{\left(n_{CH_4,in} + n_{CO_2,in}\right)} \times 100\% \quad (4.23)$$

$$Y(H_2) = \frac{n_{H_2,out}}{2\,n_{CH_4,in}} \times 100\% \tag{4.24}$$

The ranges of conditions under which carbon will form in the system are presented as carbon and non-carbon region as function of feed ratio and temperature. The curve is plotted by determining the points corresponding to the first disappearance of carbon as the temperature is increased for a fixed feed ratio.

STANDARD GIBBS FREE ENERGY CHANGE ANALYSIS OF METHANE REACTIONS TO SYNTHESIS GAS AND HIGHER HYDROCARBONS

Feasibility of the reactions (equations (4.1) - (4.12)) can be studied from standard Gibbs free energy change (ΔG_T^o) which is defined as difference between Gibbs free energy change of the products and that of the reactants (weighted by their stoichiometric coefficients) at standard state (pure substance, 1 atm), and system temperature (Smith et al., 2001). The standard Gibbs free energy changes (ΔG_T^o) tabulated in Table 4.1 are calculated based on equation (4.20), while the equilibrium constants (K) of the reactions listed in Table 4.2 are calculated from equation (4.21).

In general, chemical reactions can be reversed, at which the final equilibrium composition is governed by the minimum Gibbs free energy.

The total Gibbs free energy of a closed system at constant T and P decreases during the irreversible reactions process. The condition for equilibrium is reached when Gibbs free energy attains its minimum value or in other word Gibbs free energy change equal to zero (Smith et al., 2001).

Thermodynamically, if the Gibbs free energy change (ΔG_T^o) at certain temperature decreases to a high negative value or consequently the equilibrium constant (K) attains a high positive value, the chemical reactions are said favourable (or the reaction shifts to the product side).

On the contrary, if the Gibbs free energy change (ΔG_T^o) at certain temperature increases to a high positive value or the equilibrium constant (K) tends to a lower value, the chemical reactions are said not favourable (the reaction shifts to the reactant side) (Smith et al., 2001).

From Tables 4.1 and 4.2, it can be deduced that methane reactions (equations (4.1), (4.2), (4.8), (4.9), (4.10), (4.11), and (4.12)) to form synthesis gas (H_2 and CO) are favourable reaction, while methane reactions to higher hydrocarbons (equations (4.3) and (4.4)) are not favourable reaction from thermodynamic point of view.

The high positive value of ΔG_T^o at all temperature ranges reveals that the CO_2 OCM reaction (equations (4.3) and (4.4)) are less feasible at equilibrium. During the reaction, dehydrogenation reaction (equation (4.5)) also occurs simultaneously.

From Tables 4.1 and 4.2, it is shown that the reaction in equation (4.5) is favourable at high temperature rather than the CO_2 OCM reaction, and thus reduces mole fraction of ethane at equilibrium.

Table 4.1. Standard Gibbs free energy change (ΔGTo) of the reactions at various temperatures

Reactions	Gibbs Free Energy Change, ΔG_T^o (kJ/mole)									
	298 K	373 K	473 K	573 K	673 K	773 K	873 K	973 K	1073 K	1173 K
Equation (4.1)	170.481	150.872	123.913	96.366	68.449	40.300	12.010	-16.354	-44.748	-73.136
Equation (4.2)	28.618	25.500	21.501	17.690	14.053	10.569	7.219	3.987	0.859	-2.177
Equation (4.3)	97.683	95.456	92.371	89.220	86.032	82.819	79.590	76.350	73.106	69.863
Equation (4.4)	226.616	212.013	192.180	172.156	152.058	131.949	111.865	91.832	71.866	51.980
Equation (4.5)	100.315	90.805	77.053	62.370	46.965	30.979	14.515	-2.354	-19.568	-37.084
Equation (4.6)	141.863	125.371	102.412	78.676	54.396	29.731	4.791	-20.342	-45.607	-70.959
Equation (4.7)	-28.618	-25.500	-21.501	-17.690	-14.053	-10.569	-7.219	-3.987	-0.859	2.177
Equation (4.8)	-86.709	-100.040	-119.260	-139.719	-161.121	-183.267	-206.013	-229.252	-252.904	-276.907
Equation (4.9)	-159.507	-155.195	-149.489	-143.832	-138.223	-132.663	-127.148	-121.676	-116.240	-110.836
Equation (4.10)	-287.764	-289.289	-291.541	-293.948	-296.452	-299.015	-301.610	-304.219	-306.826	-309.418
Equation (4.11)	-144.800	-174.059	-214.394	-255.682	-297.559	-339.795	-382.235	-424.772	-467.327	-509.848
Equation (4.12)	504.667	445.835	364.233	280.224	194.634	108.004	20.711	-66.976	-154.859	-242.792

Note: The calculation was conducted at pressure of 1 atm.

Table 4.2. Equilibrium constant (K) of the reactions at various temperatures

Reactions	Equilibrium constant ($K = exp(-\Delta G_T^o/RT)$)									
	298 K	373 K	473 K	573 K	673 K	773 K	873 K	973 K	1073 K	1173 K
Equation (4.1)	$1.307 x 10^{-30}$	$7.435 x 10^{-22}$	$2.068 x 10^{-14}$	$1.640 x 10^{-9}$	$4.866 x 10^{-6}$	$1.891 x 10^{-3}$	0. 1911	7.551	150.8	1807
Equation (4.2)	$9.628 x 10^{-6}$	$2.684 x 10^{-4}$	$4.222 x 10^{-3}$	$2.440 x 10^{-2}$	$8.114 x 10^{-2}$	0.1931	0.3699	0.6108	0.9082	1.250
Equation (4.3)	$7.535 x 10^{-18}$	$4.285 x 10^{-14}$	$6.293 x 10^{-11}$	$7.352 x 10^{-9}$	$2.101 x 10^{-7}$	$2.532 x 10^{-6}$	$1.729 x 10^{-5}$	$7.963 x 10^{-5}$	$2.761 x 10^{-4}$	$7.741 x 10^{-4}$
Equation (4.4)	$1.890 x 10^{-40}$	$2.036 x 10^{-30}$	$5.975 x 10^{-22}$	$2.022 x 10^{-16}$	$1.576 x 10^{-12}$	$1.212 x 10^{-9}$	$2.025 x 10^{-7}$	$1.175 x 10^{-5}$	$3.172 x 10^{-4}$	$4.844 x 10^{-3}$
Equation (4.5)	$2.605 x 10^{-18}$	$1.920 x 10^{-13}$	$3.094 x 10^{-9}$	$2.061 x 10^{-6}$	$2.263 x 10^{-4}$	$8.063 x 10^{-3}$	0.1354	1.338	8.967	44.81
Equation (4.6)	$1.358 x 10^{-25}$	$2.770 x 10^{-18}$	$4.897 x 10^{-12}$	$6.725 x 10^{-8}$	$5.997 x 10^{-5}$	$9.792 x 10^{-3}$	0.5168	12.36	166.1	1445
Equation (4.7)	$1.039 x 10^{5}$	3725	236.8	40.99	12.32	5.178	2.704	1.637	1.101	0.7999
Equation (4.8)	$1.582 x 10^{15}$	$1.023 x 10^{14}$	$1.481 x 10^{13}$	$5.460 x 10^{12}$	$3.205 x 10^{12}$	$2.424 x 10^{12}$	$2.123 x 10^{12}$	$2.031 x 10^{12}$	$2.051 x 10^{12}$	$2.145 x 10^{12}$
Equation (4.9)	$9.121 x 10^{27}$	$5.423 x 10^{21}$	$3.229 x 10^{16}$	$1.295 x 10^{13}$	$5.352 x 10^{10}$	$9.223 x 10^{8}$	$4.055 x 10^{7}$	$3.406 x 10^{6}$	$4.559 x 10^{5}$	$8.625 x 10^{4}$
Equation (4.10)	$2.769 x 10^{50}$	$3.261 x 10^{40}$	$1.573 x 10^{32}$	$6.270 x 10^{26}$	$1.023 x 10^{23}$	$1.608 x 10^{20}$	$1.114 x 10^{18}$	$2.149 x 10^{16}$	$8.652 x 10^{14}$	$6.014 x 10^{13}$
Equation (4.11)	$2.410 x 10^{25}$	$2.377 x 10^{24}$	$4.752 x 10^{23}$	$2.036 x 10^{23}$	$1.247 x 10^{23}$	$9.164 x 10^{22}$	$7.435 x 10^{22}$	$6.372 x 10^{22}$	$5.633 x 10^{22}$	$5.067 x 10^{22}$
Equation (4.12)	$3.441 x 10^{-89}$	$3.659 x 10^{-63}$	$5.962 x 10^{-41}$	$2.844 x 10^{-26}$	$7.817 x 10^{-16}$	$5.029 x 10^{-8}$	$5.764 x 10^{-2}$	3941	$3.459 x 10^{7}$	$6.488 x 10^{10}$

Note: The calculation was conducted at pressure of 1 atm.

Effect of Feed Ratio and Temperature on Equilibrium Mole Fraction, Conversion, and Yield

Methane – Carbon Dioxide Reactions

Thermodynamically, effect of temperature on the equilibrium composition can be studied from the following relations (Smith et al., 2001):

$$\frac{d\left(\Delta G^o / RT\right)}{dT} = \frac{-\Delta H^o}{RT^2} \tag{4.25}$$

or

$$\frac{d \ln K}{dT} = \frac{\Delta H^o}{RT^2} \tag{4.26}$$

For all endothermic reactions, standard Gibbs free energy change (ΔG_T^o) decreases and the equilibrium constant (K) increases with increasing temperature and therefore shifts the reaction to the product side. Meanwhile for all exothermic reactions, the ΔG_T^o increases and the equilibrium constant decreases with increasing temperature and therefore shifts the reaction to the reactant side. Effects of temperature on chemical equilibrium compositions for methane - carbon dioxide reactions are shown in Figures 4.1, 4.2, and 4.3 with respect to CH_4/CO_2 ratios of 0.5, 1, and 2, respectively. In this system, CH_4, CO_2, CO, H_2, H_2O, C_2H_6, C_2H_4 and carbon (C) exist in the system. Possible main reactions of the system are listed in equations (4.1) to (4.5). In the figures, (a) and (b) figures demonstrate the equilibrium conversion and yield as well as equilibrium compositions, respectively. Lowering CO_2 to CH_4 feed ratio from 2:1 to 1:2 (Figures 4.1a to 4.3a) enhances CO_2 conversion significantly. This is normal because for the CO_2/CH_4 feed ratio of 2/1, CH_4 acts as limiting reactant so that the CO_2 conversion is dependent of CH_4 concentration. For the CO_2/CH_4 feed ratio of 1/1 (carbon dioxide reforming of methane), CO_2 conversion is higher than that of CH_4 because simultaneously CO_2 reacts with H_2 through reverse water gas shift (RWGS) reaction as presented in equation (4.2). Ratio of H_2/CO (synthesis gas) in the product increases with CH_4/CO_2 ratio. From the equilibrium point of view, hydrogen production could be enhanced by increasing CH_4/CO_2 ratio. In term of carbon/coke formation, by increasing CH_4/CO_2 ratio, the carbon (C) concentration in equilibrium increases, particularly at high temperature. The coke formation may be mainly based on methane decomposition according to equation (4.27) beside of Boudouard reaction (equation (4.28)). This phenomenon may be true because coke formation is enhanced by increasing CH_4/CO_2 ratio as displayed in Figures 4.1b, 4.2b, and 4.3b for CH_4/CO_2 ratios of ½, 1/1, and 2/1, respectively.

$$CH_4 = C + 2\,H_2 \qquad \Delta H^o_{298} = +74\ \mathrm{kJ{\cdot}mol^{-1}} \tag{4.27}$$

$$2\,CO = C + CO_2 \qquad \Delta H^o_{298} = -172\ \mathrm{kJ{\cdot}mol^{-1}} \tag{4.28}$$

The H_2 and CO compositions at equilibrium increase with temperature. For lower CH_4/CO_2 feed ratio, both C_2H_4 and C_2H_6 mole fractions decrease as temperature increases. The C_2H_4 concentration increases and C_2H_6 decreases with temperature due to that a part of C_2H_6 is dehydrogenated to C_2H_4 and H_2 as presented in equation (4.5). The trends reveal that there is a correlation between reaction temperature and contribution of H atom from methane at equilibrium system leading to a significant effect on the ethylene and ethane products formation. Higher reaction temperature and lower CO_2/CH_4 feed ratio increase the C_2H_6 and C_2H_4 compositions at equilibrium conditions. The reaction in equation (4.2) is favoured at high temperature as evident by the ΔG_T^o value presented in Tables 4.1. Indeed, the CO yield exceeds the H_2 yield at temperature above 1000 K as revealed in Figure 4.1a at lower CH_4/CO_2 ratio.

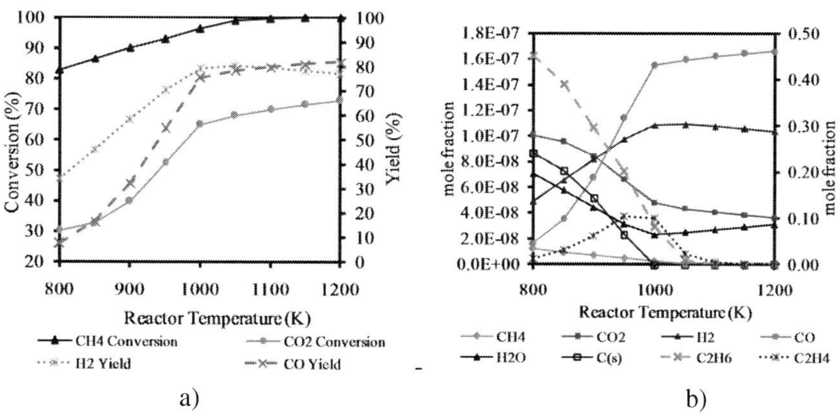

Note: for graph b., left axis defines for mole fraction of C_2H_6, C_2H_4, and right axis defines for mole fraction of CO, H_2, CO_2, CH_4, H_2O, and C.

Figure 4.1. Effect of temperature on (a) conversion, yield, and (b) mole fraction at CH_4/CO_2 ratio of ½.

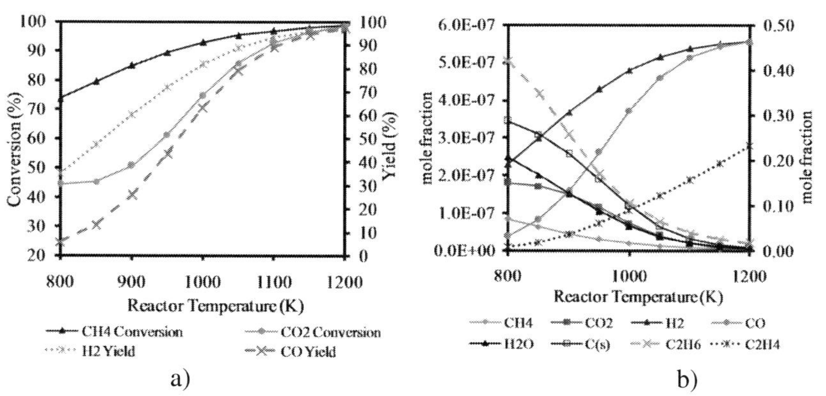

Note: for graph b., left axis defines for mole fraction of C_2H_6, C_2H_4, and right axis defines for mole fraction of CO, H_2, CO_2, CH_4, H_2O, and C.

Figure 4.2. Effect of temperature on (a) conversion, yield, and (b) mole fraction at CH_4/CO_2 ratio of 1/1.

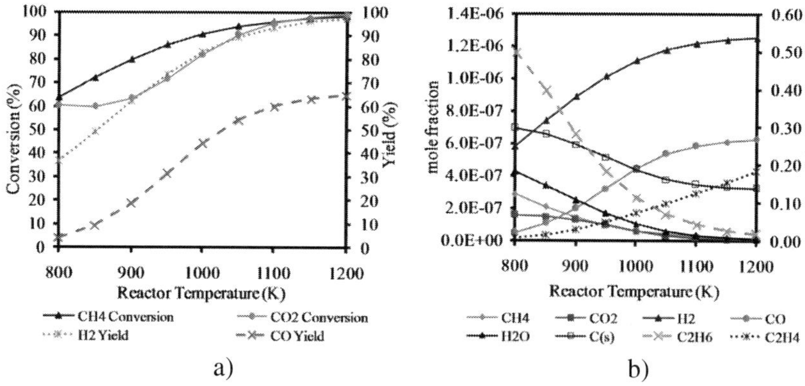

Note: for graph b., left axis defines for mole fraction of C_2H_6, C_2H_4, and right axis defines for mole fraction of CO, H_2, CO_2, CH_4, H_2O, and C.

Figure 4.3. Effect of temperature on (a) conversion, yield and (b) mole fraction at CH_4/CO_2 ratio of 2/1.

Methane – Oxygen Reactions

The possible methane – oxygen reactions are partial oxidation of methane (equation (4.8), full combustion reaction, and oxidative coupling of methane (equations (4.9) and (4.10)). Effects of temperature on chemical equilibrium compositions for methane-oxygen reactions are shown in Figures 4.4 and 4.5 for CH_4/O_2 feed ratios of 1 and 2, respectively. Components of CH_4, CO_2, O_2, CO, H_2, H_2O, C_2H_6, C_2H_4, and carbon (*C*) coexist in the equilibrium system. Possible main reactions of the system are listed in equations (4.8) to (4.10). In the Figures 4.4 and 4.5, sub-figure (a) demonstrates equilibrium conversion and yield, while sub-figure (b) presents equilibrium compositions. From the figures, O_2 conversion close to 100% for all CH_4/O_2 ratios due to oxygen is a limiting reactant. CH_4 conversion increases with temperature, while increasing CH_4/O_2 ratio decreases the conversion. That is true because at increasing oxygen content in the reactant leads to enhanced combustion reaction to produce CO_2 and H_2O as presented in Figures 4.4b and 4.5b for H_2O mole fraction. The H_2/CO ratio in the product increases with CH_4/O_2 ratio, while H_2 yield at CH_4/O_2 ratio of 2/1 is higher than that of 1/1 because the reactions shift to partial oxidation of methane. In Figures 4.4b and 4.5b, coke (*C*) formation could be reduced by decreasing CH_4/O_2 ratio or increasing oxygen content in the reactant. This is caused by oxidation of carbon formed to CO_2 indicated by increasing CO_2 mole fraction with increasing O_2/CH_4 ratio. The coke formation may be mainly based on methane decomposition according to equation (4.27) beside of Boudouard reaction (equation (4.28)). In term of higher hydrocarbons formation, C_2H_6 decreases with temperature, while C_2H_4 concentration increases which is due to that a part of C_2H_6 is dehydrogenated to C_2H_4 and H_2 as presented in equation (4.5). The dehydrogenation reaction is favoured at higher temperature which is proven by lower ΔG^o as presented in Table 4.1 or higher equilibrium constant (*K*) in Table 4.2. The reaction in equation (4.5) is favoured at high temperature as evident by the ΔG_T^o value presented in Table 4.1. The trends reveal that there is a correlation between reaction temperature and contribution of H atom from methane at equilibrium system leading to a significant effect on the ethylene and ethane products formation. Higher reaction temperature and lower O_2/CH_4 feed ratio increase the C_2H_4 mole fraction at equilibrium conditions. Indeed, H_2 yield always exceeds CO yield at all temperature tested as revealed in Figures 4.4b and 4.5b. This phenomenon is may be due to the H_2/CO ratio in the partial oxidation of methane reaction and is also favored by dehydrogenation reaction of C_2H_6 to produce C_2H_4 and hydrogen.

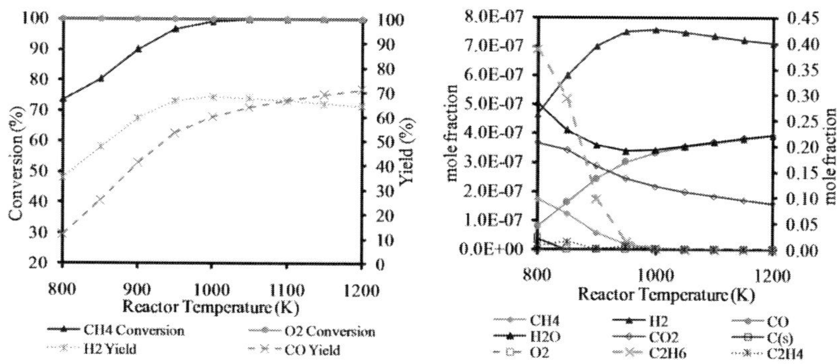

a) b)

Note: for graph b., left axis defines for mole fraction of C_2H_6, C_2H_4, and right axis defines for mole fraction of CO, H_2, CO_2, CH_4, H_2O, and C.

Figure 4.4. Effect of temperature on (a) conversion, yield, and (b) mole fraction at CH_4/O_2 ratio 1/1.

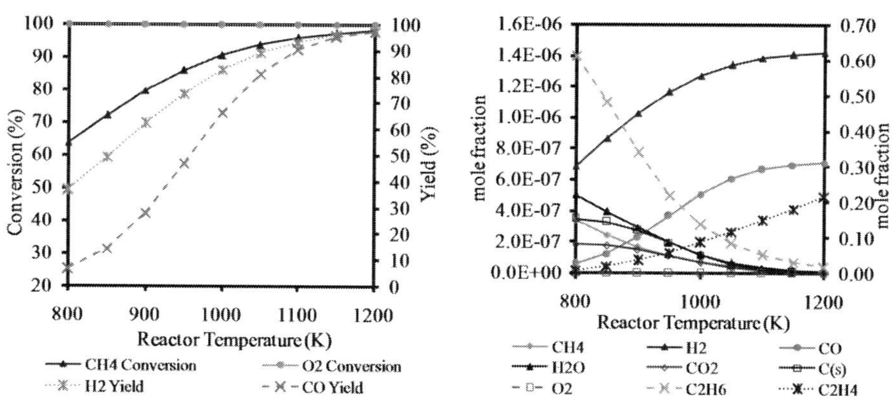

a) b)

Note: for graph b., left axis defines for mole fraction of C_2H_6, C_2H_4, and right axis defines for mole fraction of CO, H_2, CO_2, CH_4, H_2O, and C.

Figure 4.5. Effect of temperature on (a) conversion, yield, and (b) mole fraction at CH_4/O_2 ratio 2/1.

Methane – Steam Reaction

Methane and steam/water reaction has been the most applicable process to produce synthesis gas (H_2 and CO). This is truthful because the process is able to produce hydrogen-rich synthesis gas that most used in ammonia production and fuel cell application.

Stoichiometrically, H_2/CO ratio in product is about 3 – 4. Possible methane–water reaction are steam reforming of methane (SRM) process as presented in equation (4.6) and simultaneously water gas shift (WGS) reaction in equation (4.7). Combination of endothermic and exothermic reaction of both processes leads to lower energy requirement or higher energy efficiency. Effects of temperature on chemical equilibrium compositions for methane-steam reactions are shown in Figures 4.6 to 4.8 for CH_4/H_2O feed ratios of ½, 1/1, and 2/1, respectively. In this equilibrium study, components of CH_4, H_2O, CO, H_2, CO_2, and carbon (*C*) are coexisting in the system. Potential main reactions of the system are listed in equations (4.8) to (4.10). From Figures 4.6 to 4.8, H_2O conversion increases with CH_4/H_2O ratio, because at CH_4/H_2O ratio of ½, the CH_4 acts as a limiting reactant, while at CH_4/H_2O ratio of 1/1 and 2/1 the H_2O conversion closes to the CH_4 conversion particularly at higher temperature. In the steam reforming of methane, H_2 yield is always higher than CO yield, so that the process is suitable for H_2 production. H_2 content in the reaction product is higher than that of CO as presented in Figures 4.6b, 4.7b, and 4.8b. This phenomenon is caused by contribution of 3 moles H_2O and 1 mole CO in reaction stoichiometri (equation (4.6)), role of water gas shift reaction to form hydrogen and CO_2 from CO and water (equation (4.7)), as well as promoted by the hydrogen atom contribution from both methane and water. Therefore, steam reforming of methane is the most suitable for hydrogen production from methane. In Figures 4.6b to 4.8b, coke (*C*) formation could be reduced by decreasing CH_4/H_2O ratio or increasing water content in the feed. This may be due to the carbon formed is then oxidized by H_2O within the system particularly at lower CH_4/H_2O ratio. The coke formation may be mainly based on methane decomposition according to equation (4.27) beside of Boudouard reaction (equation (4.28)). It is an advantage of steam reforming of methane process that the coke formation can be reduced with the presence of water as an oxidant according to the following reaction:

$$C + H_2O = CO + H_2 \quad \Delta H^o_{298} = +131 \text{ kJ·mol}^{-1} \tag{4.29}$$

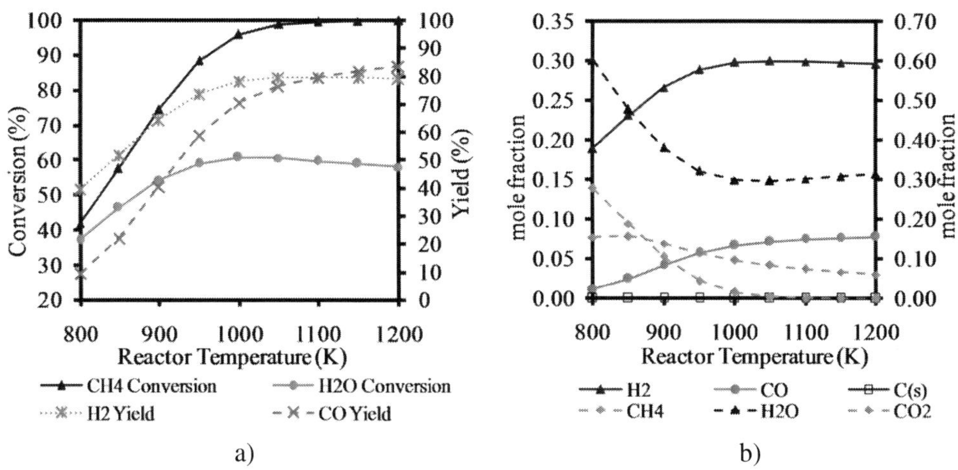

Note: for graph b., right axis defines for mole fraction of CO and H_2 while left axis defines for CH_4, H_2O, CO_2, and $C_{(s)}$.

Figure 4.6. Effect of temperature on (a) conversion, yield, and (b) mole fraction at CH_4/H_2O ratio of ½.

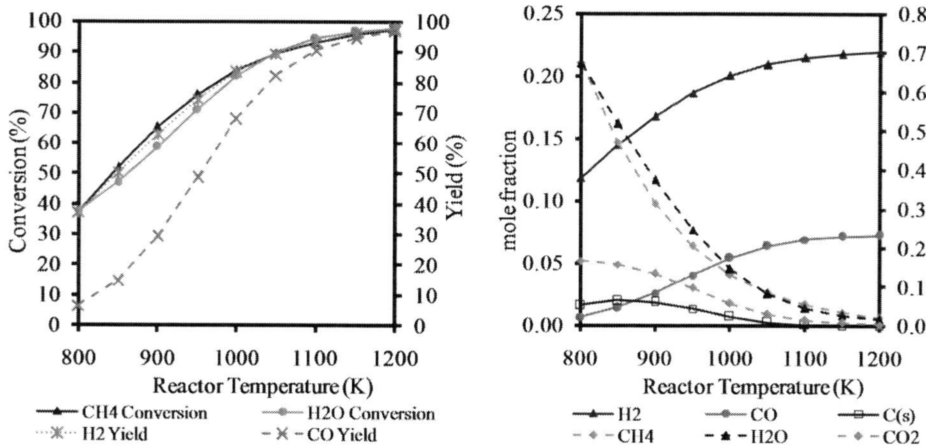

a) b)

Note: for graph b., right axis defines for mole fraction of CO and H_2 while left axis defines for CH_4, H_2O, CO_2, and $C_{(s)}$.

Figure 4.7. Effect of temperature on (a) conversion, yield, and (b) mole fraction at CH_4/H_2O ratio of 1/1.

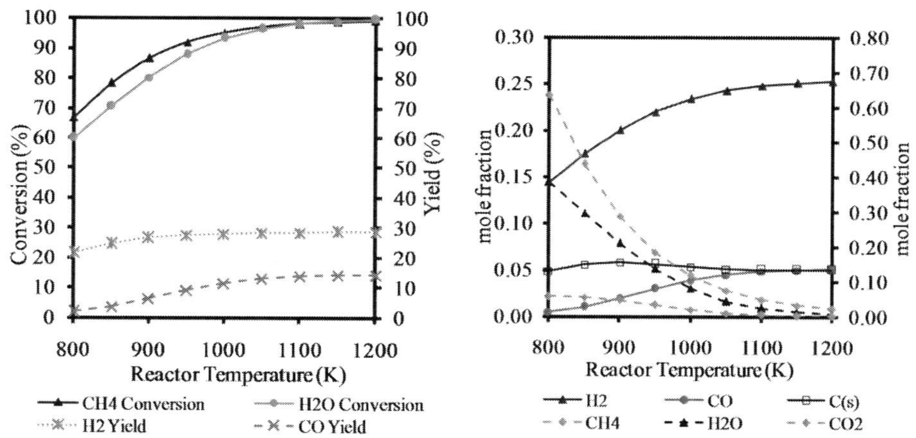

a) b)

Note: for graph b., right axis defines for mole fraction of CO and H_2 while left axis defines for CH_4, H_2O, CO_2, and $C_{(s)}$.

Figure 4.8. Effect of temperature on (a) conversion, yield, and (b) mole fraction at CH_4/H_2O ratio of 2/1.

Autothermal Reforming

In this study, autothermal reforming (ATR) focuses on addition of carbon dioxide (equation (4.11)) or steam (equation (4.12)) to methane–oxygen reactions to form synthesis gas (H_2 and CO). Due to the methane–oxygen reactions are exothermic processes, the heat released from the system could be utilized to be reacted with steam or carbon dioxide which is an endothermic reaction.

The reaction takes place in a single chamber where the methane is partially oxidized. Combination of endothermic and exothermic reaction of both process leads to lower energy requirement or higher energy efficiency.

Effects of temperature and feed ratio of CH_4-O_2-CO_2 system on chemical equilibrium compositions, conversion, and yield are depicted in Figure 4.9, while those of CH_4-O_2-H_2O system are presented in Figures 4.10. The chemical equilibrium CH_4-O_2-CO_2 reaction system was simulated for the feed ratio of 2/1/1 (Figure 4.9), while that of CH_4-O_2-H_2O reaction system was simulated for the feed ratio of 4/1/2 (Figure 4.10). In the first equilibrium study, components of CH_4, O_2, CO_2, CO, H_2, H_2O, and carbon (*C*) are coexisting in the system where the first three components as reactant. Meanwhile in the second equilibrium study, components of CH_4, O_2, H_2O, CO, H_2, CO_2, and carbon (*C*) are coexisting in the system where the first three components as reactant.

From Figures (4.9) to (4.10), CH_4 conversion is lower than that of O_2, but higher than that of CO_2 or H_2O, respectively because O_2 acts as a limiting reactant. In the first auto thermal reforming (CH_4-O_2-CO_2), CO yield increases with temperature and takes over the H_2 yield at temperature higher than 1050 K which is due to contribution of carbon atom from CO_2.

However, in the second auto thermal reforming (CH_4-O_2-H_2O), CO yield increases with temperature but only takes over the H_2 yield at temperature higher than 1200 K which is due to great contribution of hydrogen atom from H_2O molecule.

Indeed, the second auto thermal reforming is suitable for hydrogen production. It is also demonstrated in Figures 4.9b and 4.10b in which H_2 mole fraction in first system is higher than that of second system.

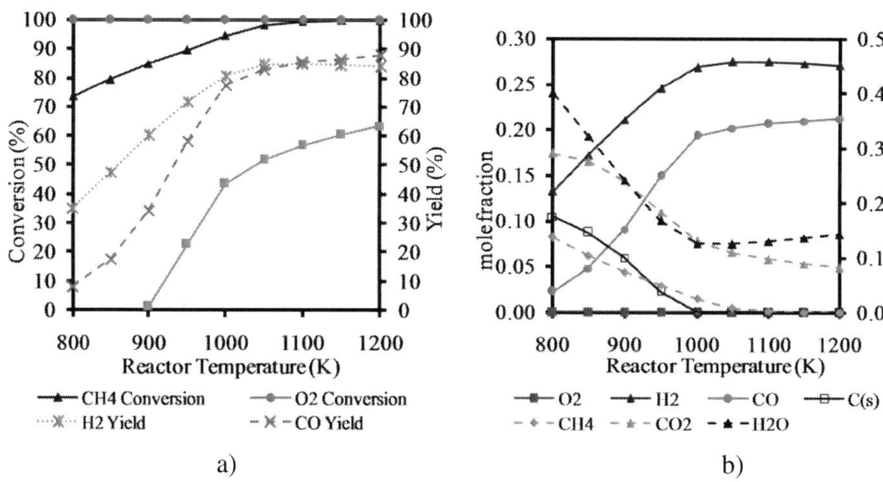

Note: for graph b., right axis defines for mole fraction of CO and H_2 while left axis defines for CH_4, H_2O, CO_2, and $C_{(s)}$.

Figure 4.9. Effect of temperature on (a) conversion, yield, and (b) mole fraction at $CH_4/O_2/CO_2$ ratio of 2/1/1.

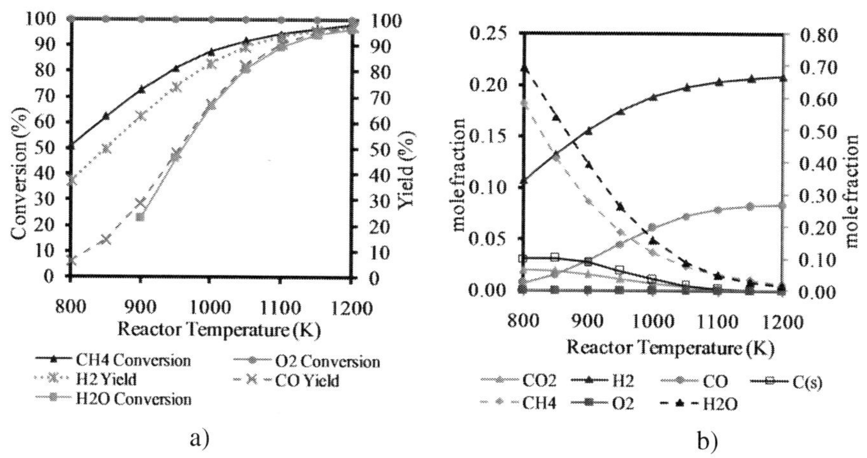

Note: for graph b., right axis defines for mole fraction of CO and H_2 while left axis defines for CH_4, H_2O, CO_2, and $C_{(s)}$.

Figure 4.10. Effect of temperature on (a) conversion, yield, and (c) mole fraction at $CH_4/O_2/H_2O$ ratio 4/1/2.

In Figures 4.9b and 4.10b, coke (C) formation in the CH_4-O_2-H_2O system is lower than that of CH_4-O_2-CO_2 system which is due to the carbon formed is then oxidized by H_2O (equation (4.29)) rather than by CO_2. The coke formation may be mainly come from methane decomposition (equation (4.27)) rather than Boudouard reaction (equation (4.28)). It is an advantage of CH_4-O_2-H_2O auto thermal system for synthesis gas (H_2 and CO) production because both hydrogen and carbon monoxide could be produced with high yield and lower energy required.

Effect of System Pressure on Equilibrium Mole Fraction, Conversion, and Yield

Influence of system pressure on performance of reaction can be expressed by equation (4.30) (Smith et al., 2001) which is agree with Le-Chatelier's principles.

$$\prod_{i=1}^{I}(y_i)^{\nu_i} = \left(\frac{P}{P^o}\right)^{-\nu} K \tag{4.30}$$

If the total stoichiometric number ν ($\equiv \sum_i \nu_i$) is negative, equation (4.30) shows that an increase in P at constant T causes an increase in implying a shift of the reaction to the right (product) side. In contrary, if ν is positive, an increase in P at constant T leads to a decrease in $\prod_{i=1}^{I}(y_i)^{\nu_i}$ exhibiting a shift of the reaction to the left (reactant) side. However, if ν is zero, the change in system pressure does not influence the reaction direction. Reactions in equations

(4.1), (4.2), (4.4), (4.5), (4.6), (4.8), (4.11), and (4.27) show positive stoichiometric number (ν), in which increasing system pressure decreases mole fractions of product components at equilibrium as depicted in Figures 4.11 - 4.14. Meanwhile, reactions in equation (4.9), oxidative coupling of methane by oxygen, shows a negative stoichiometric number (ν), so that increasing system pressure enhances mole fractions of product at equilibrium.

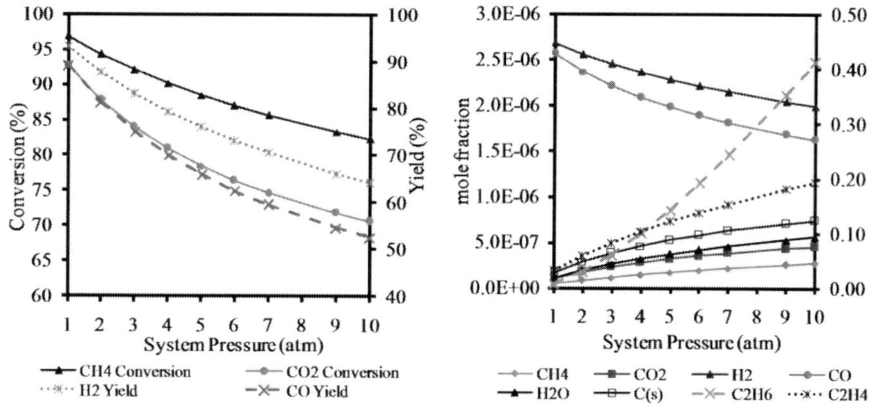

a) b)

Note: for graph b., right axis defines for mole fraction of C_2H_6 and C_2H_4 while left axis defines for CH_4, H_2O, CO, CO_2, H_2, and $C_{(s)}$.

Figure 4.11. Effect of system pressure on (a) conversion, yield, and (b) mole fraction at CH_4/CO_2 ratio 1/1 for CH_4-CO_2 reaction system.

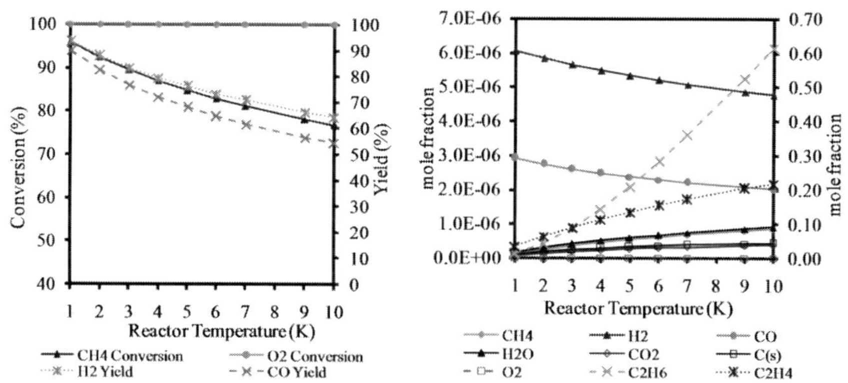

a) b)

Note: for graph b., right axis defines for mole fraction of C_2H_6 and C_2H_4 while left axis defines for CH_4, H_2O, CO, CO_2, H_2, and $C_{(s)}$.

Figure 4.12. Effect of system pressure on (a) conversion, yield, and (b) mole fraction at CH_4/O_2 ratio 2/1 for CH_4-O_2 reaction system.

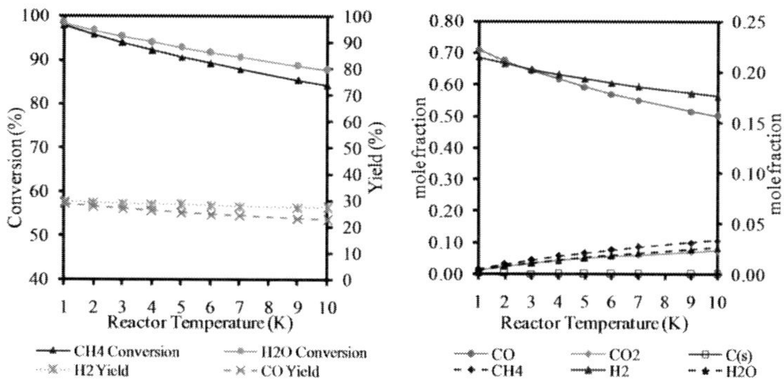

a) b)

Note: for graph b., right axis defines for mole fraction of CO and H_2 while left axis defines for CH_4, H_2O, CO_2, and $C_{(s)}$.

Figure 4.13. Effect of system pressure on (a) conversion, yield, and (b) mole fraction at CH_4/H_2O ratio of 1/1 for CH_4-H_2O reaction system.

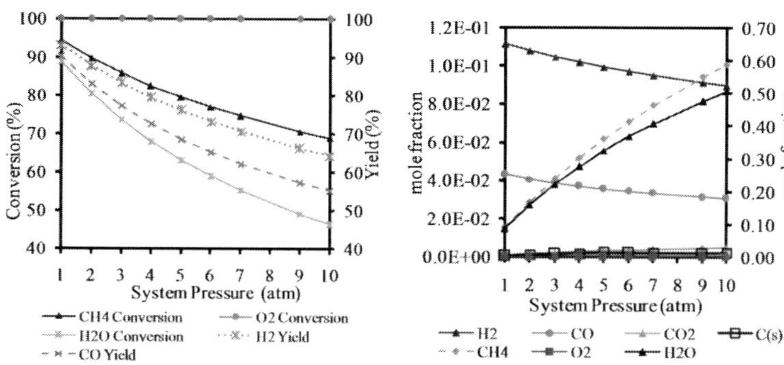

a) b)

Note: for graph b., right axis defines for mole fraction of CO, CO_2, O_2, and C(s) while left axis defines for CH_4, H_2, and H_2O.

Figure 4.14. Effect of system pressure on (a) conversion, yield, and (b) mole fraction at $CH_4/O_2/H_2O$ ratio of 4/1/2.

However, reaction in equation (4.2) for RWGS, equation (4.3) for CO_2 OCM, equation (4.7) for WGS, equation (4.28) for Boudouard reaction), and equation (4.29) for coke oxidation by steam are not influenced by the system pressure. Effect of system pressure on equilibrium conversions of CH_4 and CO_2, H_2 and CO yields, and products composition at 1100 K and CO_2/CH_4 feed ratio being 1 are exhibited in Figure 4.11. The equilibrium conversions of CH_4 and CO_2 decrease with increasing system pressure. The reaction shifted to the reactant side (negative and decreased the reaction coordinate due to positive and/or zero stoichiometric numbers (ν) of all simultaneous reactions considered (equations (4.1) - (4.5)), which in turn decreases the term in equation (4.30). Similar phenomena are found in methane – steam reactions (equation (4.6) – (4.7)), methane – oxygen reactions (equations (4.8) - (4.10)), and auto thermal reactions (equations (4.11) and (4.12)).

REGIONS OF CARBON AND NO CARBON FORMATION AT SYNTHESIS GAS AND HIGHER HYDROCARBONS PRODUCTION

In this section, carbon is present in the chemical equilibrium system in order to study conditions at which the carbon is formed. Pertaining to this system, for example in methane–carbon dioxide reaction system, components of CH_4, CO_2, CO, H_2, C_2H_4, C_2H_6, H_2O and C (carbon) exist in the equilibrium system. CH_4 and CO_2 are stated as the reactants, while CO, H_2, C_2H_4, C_2H_6, C, and H_2O are defined as products. Range of conditions under which the carbon will form in the equilibrium system is depicted in Figure 4.15. The curve was plotted by determining the points corresponding to the first disappearance of carbon as the temperature was increased for each fixed CO_2/CH_4 feed ratio. From the figure, area of carbon formation region is exhibited below the curve, while that of no carbon formation is demonstrated above the curve. The carbon may be formed via methane decomposition and/or Boudouard reaction (Froment, 2000; Ito et al., 1999) as exhibited in equations (4.27) and (4.28), respectively.

$$CH_4 = C + 2\ H_2 \tag{4.27}$$

$$2\ CO = C + CO_2 \tag{4.28}$$

From Figure 4.15, it is found that carbon is formed in the entire range of temperature at CO_2/CH_4 feed ratio below 1. At CO_2/CH_4 ratio above 1, the temperature limit for no carbon formation region decreases as the CO_2/CH_4 ratio increases. It is shown that region of no carbon formation is only found at temperature higher than 1000 K and CO_2/CH_4 ratio higher than 1. Through this figure, the operating conditions ranges of CH_4 and CO_2 reactions at which the carbon does not form at equilibrium can be recommended.

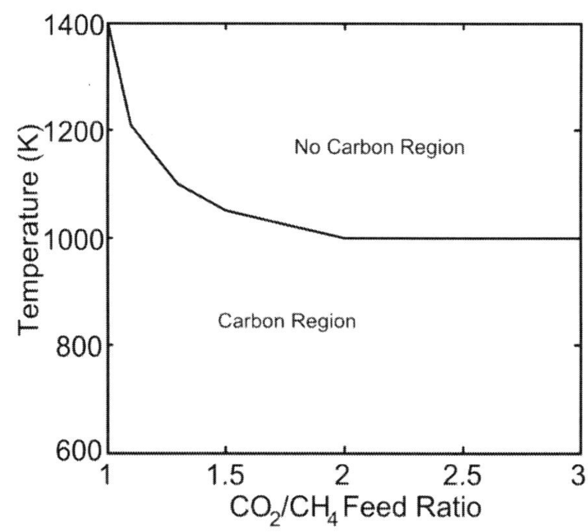

Figure 4.15. Carbon and no carbon formations region at 1 atm as function of temperature and CO_2/CH_4 feed ratio for methane–carbon dioxide reaction system.

Conclusion

Thermodynamic chemical equilibrium analysis using method of direct minimization of Gibbs free energy for all possible methane reactions with oxygen (partial oxidation of methane), carbon dioxide (CO_2 reforming of methane), steam (steam reforming of methane), and autothermal reforming were successfully studied. Feed ratio (methane to oxygen, to carbon dioxide, and/or to steam), reaction temperature, and system pressure gives significant effect on equilibrium composition, conversion, and yield as well as carbon formation possibility. It was found that the reaction temperature above 1100 K and the CH_4/CO_2 ratio unity were favorable for synthesis gas production in methane–carbon dioxide reaction, while oxidative coupling of methane with CO_2 reaction to produce ethane and ethylene is less favorable thermodynamically. In addition, steam reforming of methane is the most suitable for hydrogen production from methane with low coke formation.

Author's Biography

Dr. I. Istadi is an Associate Professor of Chemical Engineering at Department of Chemical Engineering, Faculty of Engineering, Diponegoro University. He earned his BEng. from Diponegoro University, Indonesia (1995); MEng. from Institut Teknologi Bandung, Indonesia (2000); and PhD. from Universiti Teknologi Malaysia (2006). He has published over 20 refereed articles in reputable international and national journals, book chapter at Studies in Surface Science and Catalysis, Vol.159, entitled: New Development and Application in Chemical Reaction Engineering (Page 213-216), published by Elsevier B.V. ISBN: 978-0-444-51733-3, ISSN: 0167-2991, and a book entitled: Methane–Carbon Dioxide: Conversions to Syngas and Hydrocarbons. ISBN: 978-983-52-0472-2 published by Universiti Teknologi Malaysia Press., Hybrid Catalytic – Plasma Reactor Development for Energy Conversion, ISBN: 978-979-704-839-6 (BP Undip Press), and "*Teknologi Katalis Untuk Konversi Energi: Fundamental dan Aplikasi*", ISBN: 978-979-756-734-7 (Publisher: Graha Ilmu Indonesia). He is also assigned as Peer-Reviewer in number of Elsevier's journals. He is also appointed as Editor-in-Chief of *Bulletin of Chemical Reaction Engineering and Catalysis* (ISSN 1978-2993, website: http://bcrec.undip.ac.id) that has been indexed in Scopus, DOAJ, and Ebsco. Currently, his research has focussed on hybrid plasma – catalytic reactor development for energy conversion.

References

[1] Istadi, and N.A.S. Amin. *Journal of Natural Gas Chemistry*, 14, 140, (2005).

[2] K. Vasudeva, N. Mitra, P. Umasankar, and S. C. Dhingra. *International Journal of Hydrogen Energy*, 21, 13, (1996).

[3] S.H. Chan, and H.M. Wang, *International Journal of Hydrogen Energy*. 25, 441, (2000).

[4] A.E. Lutz, R.W. Bradshaw, J.O. Keller, D.E. Witmer. *International Journal of Hydrogen Energy*, 28, 159, (2003).

[5] A.E. Lutz, R.W. Bradshaw, L. Bromberg, A. Rabinovich. *International Journal of Hydrogen Energy*, 29, 809, (2004).
[6] C. Guéret, M. Daroux, F. Billaud. *Chemical Engineering Science* 52, 815, (1997).
[7] Y. Lwin, W.R.W. Daud, A.B. Mohamad, and Z. Yaakob. *International Journal of Hydrogen Energy,* 25, 47, (2000).
[8] H. Tang, and K. Kitagawa. *Chemical Engineering Journal*, 106, 261, (2005).
[9] S. Douvartzides, F.A. Coutelieris, A.K. Demin and P. Tsiakaras, *AIChE Journal*, 49, 248, (2003).
[10] S.H. Chan, and H.M. Wang. *Journal of Power Sources*, 126, 8, (2004).
[11] J.M. Smith, H.C. Van Ness, and M.M. Abbott. *Introduction to Chemical Engineering Thermodynamics*, McGraw Hill Book, Co., New York, (2001).
[12] G.F. Froment. *Journal of Molecular Catalysis A: Chemical*, 163, 147, (2000)
[13] M. Ito, T. Tagawa, S. Goto. *Applied Catalysis A: General*, 177, 15, (1999).

In: Syngas
Editors: Antonius Indarto and Jelliarko Palguandi
ISBN: 978-1-62100-870-5

Chapter 5

SYNGAS PRODUCTION BY CO_2 REFORMING OF CH_4 UNDER MICROWAVE HEATING – CHALLENGES AND OPPORTUNITIES

B. Fidalgo and J. A. Menéndez
Instituto Nacional del Carbón, INCAR-CSIC, Oviedo, Spain

ABSTRACT

The main industrial process used to produce synthesis gas (H_2 + CO) and its resulting byproducts is the steam reforming of natural gas. Methane reforming with carbon dioxide, or dry reforming, is a promising alternative that may lead to the reduction of CO2 emissions and the production of a syngas with a lower ratio of H_2/CO. The main obstacle to the industrial implementation of CH_4 reforming with CO2 is that there are no commercial catalysts that can operate without undergoing deactivation due to carbon deposition. Consequently, new catalysts are being developed and changes are being introduced in the process in order to achieve high and steady conversions.

The microwave-assisted CO_2 reforming of CH_4 over carbon-based catalysts combines the catalytic and dielectric properties of carbonaceous materials with the advantages of microwave heating, which favours catalytic heterogeneous reactions due to, among other reasons, the generation of hot spots or microplasmas. Under certain operating conditions, the microwave-assisted dry reforming reaction can be considered as a combination of CH_4 decomposition and CO_2 gasification of carbon deposits, leading to the continuous regeneration of active centres. The most appropriate operating conditions to achieve high conversions for a long period of time are temperatures ranging between 700 and 800°C and the presence of high proportions of CO_2 in the feed (at least 50%).

The use of catalysts with a good catalytic activity gives rise to high conversions at high values of volumetric hourly space velocity, making it possible to produce large amounts of syngas. Of the catalysts evaluated the most suitable for the microwave-assisted dry reforming of methane proved to be mixtures of carbonaceous material and metal catalyst. The carbon material should be microporous and have a good reactivity towards CO_2. The metal catalyst with the best catalytic activity is Ni/alumina.

Energy consumption in the process of microwave-assisted reforming of methane with CO_2 is estimated to be 4.6 kW·h per m3 of H_2, which is a promising value compared to the energy consumption in the steam reforming of methane. Therefore, the process promises to be competitive, both from the point of view of conversion and energy consumption.

NOMENCLATURE

ΔH_{298}	Enthalpy of reaction at 298K, $kJ \cdot mol^{-1}$
ε'	Dielectric constant or real permittivity
ε''	Dielectric loss factor or imaginary permittivity
E	Electromagnetic field, $V \cdot m^{-1}$
I_D	D-band intensity in Raman spectrum
I_G	G-band intensity in Raman spectrum
RWGS	Reverse Water Gas Shift
S_{BET}	Specific BET surface area, $m^2 \cdot g^{-1}$
T	Temperature, °C
$tan\ \delta$	Dielectric loss tangent
VHSV	Volumetric Hourly Space Velocity, $L \cdot g^{-1} h^{-1}$
V_{mic}	Micropores volume, $cm^3 \cdot g^{-1}$
vol. %	Percentage in volume
WGS	Water Gas Shift
wt. %	Percentage in weight

INTRODUCTION

Synthesis gas can be obtained through various chemical and thermochemical processes from almost any carbon source, such as oil, carbon, biomass or biodegradable waste. However, natural gas is nowadays the dominant feedstock and steam reforming is the principal industrial process for the production of syngas (and hydrogen) [1-5]. Figure 5.1 shows the steps employed at the typical hydrogen production plants based on steam reforming. Steam reforming involves the endothermic conversion of methane and steam into hydrogen and carbon monoxide (reaction 5.1). It typically occurs at temperatures between 700 and 850°C, pressures between 3 and 25 bar and over Ni-based catalysts [5,6].

$$CH_4 + H_2O = 3H_2 + CO \qquad \Delta H_{298} = +206\ kJ \cdot mol^{-1} \tag{5.1}$$

Steam reforming results stoichiometrically in a H_2/CO ratio of 3:1, which is higher than the ratio needed for the synthesis of byproducts, such as methanol or derivatives from the Fischer-Tropsch reaction [2,5]. Industrially, the H_2/CO ratio is adjusted by means of the water gas shift reaction (WGS, reaction 5.2), which is usually performed in two steps: a high temperature shift (350 – 400°C) over iron oxide-based catalysts and a low temperature shift (200 – 250°C) over copper-based catalysts [2,5,7,8].

$$CO + H_2O = H_2 + CO_2 \qquad \Delta H_{298} = -41\ kJ \cdot mol^{-1} \tag{5.2}$$

Despite being a well-established process, steam reforming presents several drawbacks. The coupling of WGS for H_2/CO ratio adjustment is costly and it makes the global process more expensive [5,9,10]. In addition, an elevated heat supply is necessary in order to achieve a high methane conversion. The heat supply normally comes from combustion of part of the incoming natural gas feedstock ($\leq 25\%$) or from burning waste gases, such as purge gas [6,11]. Therefore, a large quantity of CO_2, ranging from 0.35 to0.42 m^3 of CO_2 per $m^3$of H_2 produced, is emitted due to both the reaction and the heat requirement [12].

Moreover, an excess of steam must be introduced, approximately at a H_2O/CH_4 ratio of 3 to 4, in order to avoid the deactivation of the metal catalysts due to carbonaceous deposits, and consequently, operation costs and energy consumption increase [4,8]. In addition, as the metal-based catalysts used in the steam methane reforming or the following syngas conversion processes may be poisoned by sulphur containing compounds (usually H_2S and COS), expensive desulphurization units are needed to remove these compounds [5,7].

It is for these reasons that alternative processes to steam reforming are being investigated. The catalytic decomposition of methane is the preferred alternative for hydrogen production [12-17], while partial oxidation and CO_2 reforming are the two main options for replacing steam reforming to convert methane into syngas [2,3,8,10,11].

The CO_2 reforming of methane, or dry reforming (reaction 5.3) is an endothermic reaction, like steam reforming, but it yields a syngas with a lower ratio of H_2/CO, i.e. 1:1 for a complete conversion. This ratio is preferable for the synthesis of higher hydrocarbons via Fischer-Tropsch and adequate for the production of oxygenated derivatives, which eliminates the need to adjust H_2/CO ratio by means of the WGS reaction [8-10,18-21].

$$CH_4 + CO_2 = 2H_2 + 2CO \qquad \Delta H_{298} = +247 \text{ kJ·mol}^{-1} \qquad (5.3)$$

Dry reforming is favoured by high temperatures and low pressures. Thermodynamically, it can occur at temperatures higher than 640°C but, technically and assuming a stoichiometric ratio of CH_4/CO_2 in the feedstock, temperatures higher than 800°C are needed to achieve acceptable conversions [9,21]. The process selectivity is determined by the occurrence of side reactions, which depend on the operating temperature (see Table 5.1). Thus, at temperatures lower than 800°C, the reverse water gas shift reaction (RWGS, reaction 5.4) occurs [4,9]. In addition, at temperatures lower than 700°C and a stoichiometric CH_4/CO_2 ratio, i.e. 1:1, carbon deposits may be formed from the Boudouard reaction (reaction 5.5).

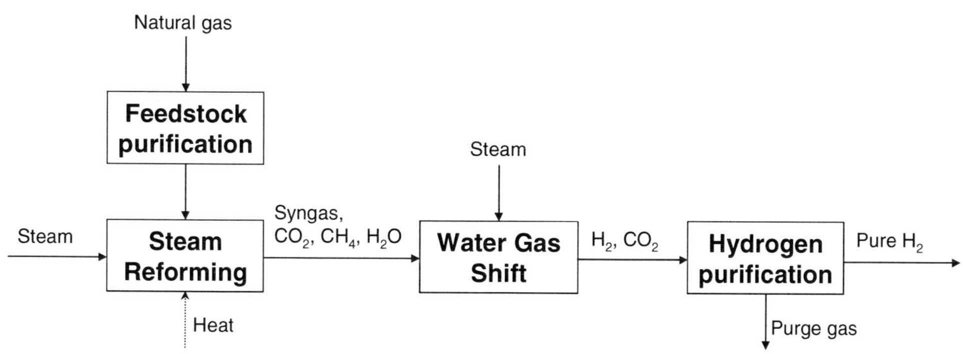

Figure 5.1. Steam reforming process for producing natural gas.

However, the reverse reaction, i.e. the CO_2 gasification of carbon, occurs at high temperatures (reaction 5.6). Carbon deposition at high temperatures is produced by decomposition of CH_4(reaction 5.7) [4,9,10].

$$H_2 + CO_2 = CO + H_2O \qquad \Delta H_{298} = +41\ kJ\cdot mol^{-1} \tag{5.4}$$

$$2CO = C + CO_2 \qquad \Delta H_{298} = -172\ kJ\cdot mol^{-1} \tag{5.5}$$

$$C + CO_2 = 2CO \qquad \Delta H_{298} = +172\ kJ\cdot mol^{-1} \tag{5.6}$$

$$CH_4 = C + 2H_2 \qquad \Delta H_{298} = +75\ kJ\cdot mol^{-1} \tag{5.7}$$

Dry reforming of methane constitutes a promising option for the conversion of natural gas into syngas mainly due to the environmental benefits that it offers. The dry reforming reaction turns two greenhouse gases (CH_4 and mainly CO_2) into a valuable feedstock and may lead to the reduction of CO_2 emissions [2,9,10,18-20]. Unless the required energy is supplied from renewable or non-fossil sources, it is unlikely that CO_2 emissions will be completely eliminated since dry reforming is an endothermic process. Nevertheless, if the heat supply is provided by the combustion of part of the methane feedstock, CO_2 emissions will be around 0.2 m^3 of CO_2 per m^3 of H_2 produced, which is lower than the quantity emitted by steam methane reforming [4,22].

Because of the high endothermicity of CO_2 reforming of CH_4, which exceeds that of steam methane reforming, this reaction has also been tested for viability in energy transmission systems (from solar energy to chemical energy), in energy storage in the form of syngas and in transportation of nuclear energy [9,18,21]. Figure 5.2 shows a schematic diagram of the chemical energy transmission system.

Table 5.1. Lower and upper limiting temperatures for reactions of the CH4/CO2 system

Reaction	Upper limiting T (°C)	Lower limiting T (°C)
$CH_4 + CO_2 = 2H_2 + 2CO$ (5.3)		640
$H_2 + CO_2 = CO + H_2O$ (5.4)	820	
$2CO = C + CO_2$ (5.5)	700	
$C + CO_2 = 2CO$ (5.6)		700
$CH_4 = C + 2H_2$ (5.7)		557

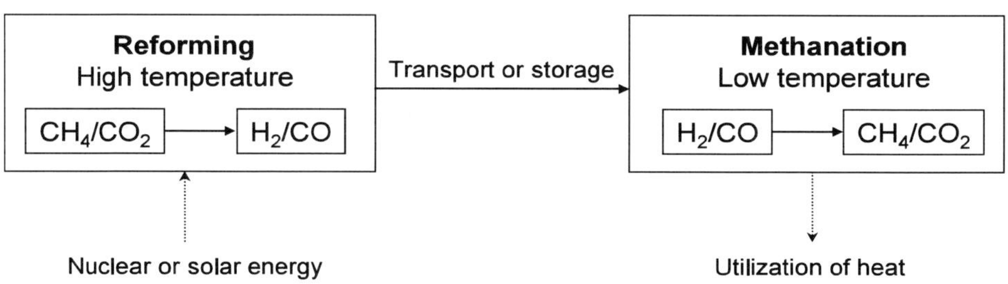

Figure 5.2. Schematic diagram of the chemical energy transmission system.

The CO_2 reforming of CH_4 is carried out by supplying nuclear or solar energy, so that the H_2 and CO produced store the incident energy. Syngas is saved for later or transported through pipelines to distant locations, where the energy is released by the reverse reaction.

The dry reforming reaction can be carried out with natural gas obtained from remote fields that contains large amounts of CO_2, without the need to separate carbon dioxide from the feed, and from petroleum fields where natural gas is flared [10,18,22,23]. In general, the dry reforming of methane may be the most effective process wherever carbon dioxide is a byproduct and available for utilization, for instance, in power plants, which emit a large amount of CO_2 at relatively high temperature, and in petrochemical industries, where effluents of light gases can be processed with waste streams of CO_2 [9,23]. In the metallurgical industry, the excess coke oven gases (COG), consisting mainly of H_2, CH_4, CO and N_2, may be turned into synthesis gas by means of dry reforming. The syngas can then be used to produce a variety of compounds but mainly methanol [24,25]. Dry reforming is also an interesting alternative for the valorisation of landfill gas or biogas from the digestion of industrial and municipal waste. Biogas is a renewable, local and low-cost raw material and it consists of approximately 55-75% of CH_4 and 25-45% of CO_2, as well as small amounts of N_2, O_2, H_2S or CO. The dry reforming of biogas to syngas for the subsequent production of hydrogen would seem to be a more efficient option than the combustion of the biogas in a turbine to produce electricity for the electrolysis of water, or the separation of methane from the biogas to turn it into hydrogen via steam reforming [20,26,27]. In addition, methane reforming with CO_2, rather than steam, is an attractive alternative in areas where water is restricted [22]. Despite the advantages afforded by the dry reforming of methane, few industrial processes utilize this reaction [9,18]. One of such processes, the SPARG process, was implemented by Sterling Chemicals Inc. in 1987 [28]. This process produces a syngas with low H_2/CO ratio by using a partially sulphur-passivated nickel catalyst. The sulphur blocks the sites for carbon nucleation while sufficient sites for the CO_2 reforming reaction are maintained. Another process, the Calcor process, was implemented by Caloric for the production of CO [29]. This is a catalytic reforming process designed to operate under low pressure and at high temperature to treat a desulphurized and CO_2-rich feedstock. The main obstacle to the industrial implementation of CO_2 reforming of CH_4 is that there are no commercial catalysts that can operate without undergoing deactivation due to carbon deposition [9,10,19-22]. Therefore, the principal objective of CO_2 reforming studies has been to develop suitable catalysts with good lifetime stability. The catalytic activity of several metals (Cu, Fe, Co, Ni, Ru, Rh, Pd, Ir or Pt) over different supports (Al_2O_3, SiO_2, TiO_2, MgO, C, ZnO, SiC, La_2O_3, ZrO_2, CeO_2 or zeolites) has been studied. The CO_2 reforming of CH_4 over noble metals was found to yield no carbon but their high cost and limited availability prevent them from being used commercially. Although the use of nickel catalysts is more feasible for industrial purposes, they deactivate more easily due to carbon deposition [8,18-20,30]. Moreover, carbon catalysts have been proposed as an alternative to metal catalysts because of their low cost, high temperature resistance and the absence of sulphur poisoning. Moreover, they have showed good catalytic activity [13,19-21,31].

In addition to the search for new catalysts, changes to the process are also being investigated in order to achieve high and steady conversions. Thus, high temperatures and CO_2/CH_4 ratios higher than unity can be used to avoid carbon deposition.

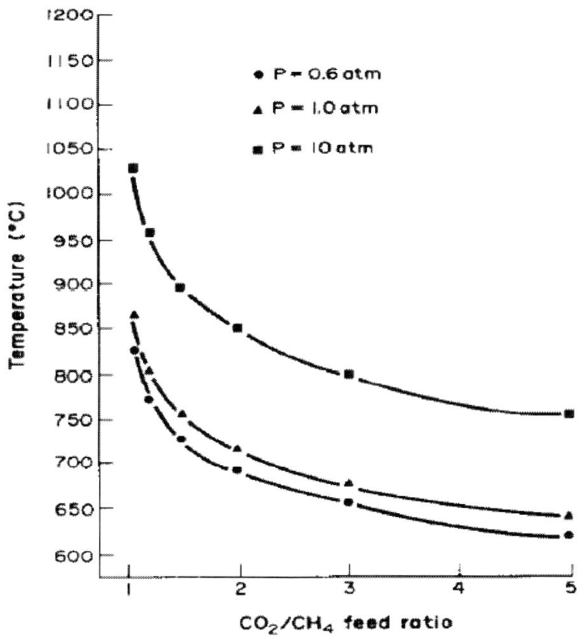

Note: The picture was obtained from Ref. [23] with the permission of Elsevier.

Figure 5.3. Effect of the CO_2/CH_4 feed ratio on the temperature below which carbon deposits form at various pressures.

Figure 5.3 shows, for different CO_2/CH_4 ratios, the temperatures above which the carbon deposition is avoided. Nevertheless, low temperatures and CO_2/CH_4 ratios of 1:1 are preferred in industrial processes in order to maintain a high selectivity towards syngas [4,9,10,22]. In certain circumstances, the simultaneous reforming of methane with CO_2, H_2O and O_2 may be advantageous. The combination of dry reforming with steam reforming and/or partial oxidation (partial combustion of CH_4 with oxygen) offers several advantages compared to the individual processes: (i) the H_2/CO ratio can be adjusted by varying the $CO_2/H_2O/O_2$ ratio in the feed; (ii) the presence of oxidant agents, such as H_2O and O_2, inhibits the process of carbon deposition; and, (iii) as partial oxidation is an exothermic reaction, the energy requirement of the process is reduced when O_2 is introduced [9,22]. In addition, alternatives to the conventional heating used in the process are being evaluated, such as, the direct conversion of CH_4 and CO_2 by using plasma [32,33] or the microwave-assisted dry reforming of methane [19,20,34], which is discussed in detail in this Chapter.

Microwave Heating in Heterogeneous Gas-Phase Catalytic Reactions

Fundamentals of Microwave Heating

Microwaves are a non-ionizing electromagnetic radiation that lies in the range of the electromagnetic spectrum limited by the frequencies between 300 MHz and 300 GHz (wavelength between 1 m and 1 mm). Domestic and industrial microwave applications generally operate at a frequency of 2.45 GHz in order to avoid interference with radar and telecommunication frequencies [35-39].

Dielectric heating is caused by high frequency electromagnetic radiation, i.e., radio and microwaves. The electric field component of the electromagnetic radiation interacts with the charged particles of a material. A current is induced when these particles are free to move. However, when the particles are linked to the material, they try to align themselves with the alternating field, as a consequence of which, the material heats up (dielectric polarization) [40]. As microwave heating is the result of the transformation of microwave energy into thermal energy, due to a molecular interaction with an electromagnetic field, it is in actual fact a question of energy conversion rather than heat transfer. Two principal dielectric polarizations, which are represented in Figure 5.4, are involved in microwave radiation [35,41]: (i) dipolar polarization, which occurs in dielectrics that have induced or permanent dipoles, such as water; and, (ii) space charge polarization, which occurs mainly in dielectric solid materials with charged particles which are free to move in a delimited region (Maxwell-Wagner polarization).

The materials which interact with microwave radiation to produce heat are called dielectrics or microwave absorbers. The ability of a material to be heated in the presence of a microwave field is described by its dielectric loss tangent $tan\delta = \varepsilon'' / \varepsilon'$. The dielectric loss tangent is composed of two parameters: the dielectric constant (or real permittivity), ε', which measures the ability to propagate microwaves into the material, and the dielectric loss factor (or imaginary permittivity), ε'', which measures the ability of the material to dissipate the energy in the form of heat [35-38]. Materials which reflect microwaves from the surface and do not heat are called conductors, and materials which are transparent to microwaves are classed as insulators.

As microwave energy is transferred directly to the material that is heated (volumetric heating), the temperature inside the material is usually higher than the temperature of the surrounding atmosphere, unlike conventional heating. In Figure 5.5 temperature gradients caused by microwave and conventional heating are compared. Microwave heating offers a number of advantages over conventional heating, such as: (i) non-contacting heating; (ii) rapid heating; (iii) selective heating of materials; (iv) quick start-up and stoppage; (v) a higher level of safety and automation; (vi) a reduction in the size of equipment and higher flexibility; and, (vii) reduced processing time [35,37,38,42].

In recent years, the industrial applications of microwave radiation have increased significantly. Microwave heating is mainly used in food processing, wood drying and rubber vulcanization. Microwave heating is also applied to organic synthesis and analytical chemistry processes, such us, drying, extractions, acid solution, decomposition or hydrolysis [43]. In addition, there is increasing interest in the use of microwaves in a wide range of areas, such as, waste management [37,44], ceramic and polymer processing [39], environmental applications [42,43], metallurgy and mineral processing [35,37,38], carbon-related processes [35], or heterogeneous gas-phase catalytic systems [34-36,41,42].

Various heterogeneous gas-phase catalytic reactions have been studied under microwave heating, such as CH_4 decomposition, CO_2 reforming of CH_4, H_2S decomposition, NO_x and SO_2 reduction or CH_4 oxidative coupling [14,19,20,34,36,41,42]. In general, in a catalytic heterogeneous system, the microwave radiation is selectively absorbed by the solid catalyst. Therefore, the catalyst should be a dielectric material and act not only as a catalyst but also as a microwave receptor [19,35]. However, some catalysts are insulator materials and, consequently, they must be used in conjunction with microwave receptors, such as carbons and certain oxides, in order to be heated easily [35,40]. Carbon materials are usually very good microwave absorbers so they can indirectly heat materials that are transparent to microwaves.

Microwave radiation is known to have the potential to increase the rate of reaction, selectivity and yield of catalytic heterogeneous reactions [14,34-36,41,42]. The improvement observed under microwave heating is normally attributed to various thermal effects [42], although the presence of hot spots within the catalyst bed, which are at higher temperature than the average temperature, may be the main reason for the improvement in gas-solid reactions. Hot spots are electric arcs which are caused by an uneven distribution of the electromagnetic fields and preferential heating, due to differences in dielectric properties, impurities or geometric defects within the catalyst.

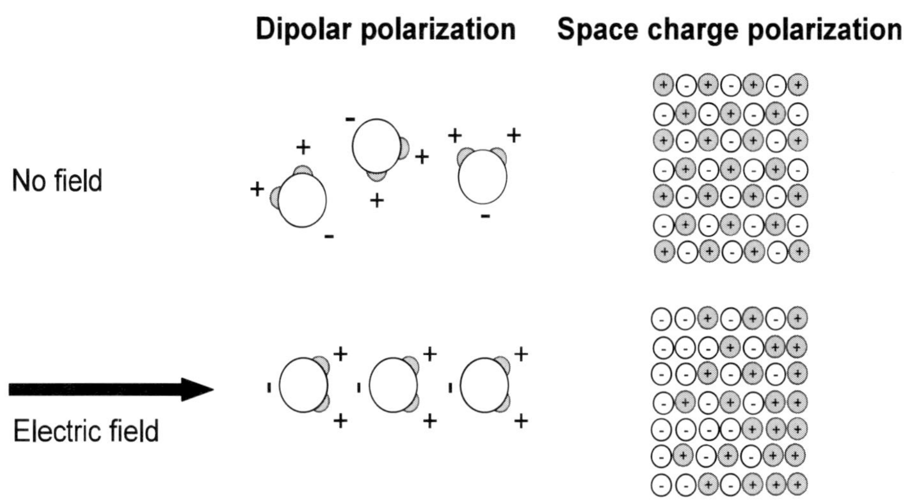

Figure 5.4. Mechanisms of dielectric polarization present under microwave heating.

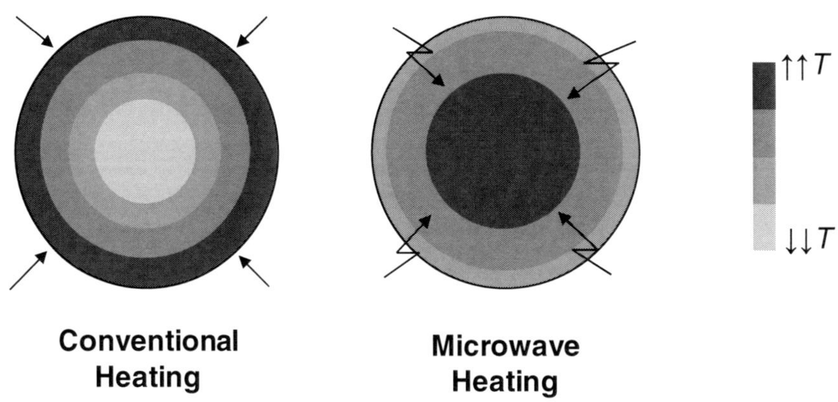

Figure 5.5. Qualitative comparison of the temperature gradients produced by microwave and conventional heating.

The electric arcs may cause the ignition of the surrounding atmosphere. Moreover, in the case of carbons, the increase in the kinetic energy of the delocalized π-electrons, which are free to move in relatively broad regions and which try to align themselves with the alternating electric component of the microwave field, may give rise to the ionization of the surrounding atmosphere. These hot spots can be therefore considered as microplasmas both from the point of view of space and time, since they are confined to a tiny region of space and last for just a fraction of a second [35,45].

Microwave Heating Technologies in Gas-Solid Systems

Various microwave technologies can be used in order to carry out gas-solid reactions. In general, microwave heating systems are made up of four basic components: (i) the power supply; (ii) the magnetron, which turns the electric energy into electromagnetic energy; (ii) the applicator or cavity for the heating of the catalyst; and, (iv) the waveguide for transporting microwaves from the magnetron to the applicator. Nevertheless, the main differences among the microwave devices arise from the method used to control the average power output of the magnetron (on/off or variable control) and the design of the applicators (unimode or multimode) [38].

The magnetrons with an on/off control operate at full power but the current is turned on and off in order to reduce the average power. The output power of the magnetrons with variable control can be altered by changing the current amplitude or the intensity of the magnetic field. Variable control is useful when continuous microwave power is required [36,39].

On the other hand, the type of applicator used in a microwave system usually depends on the materials to be processed. Common microwave applicators include unimode and multimode cavities. The unimode applicators, which have one resonant mode, display interesting features, such as precise energy input, predictable electromagnetic field distribution and the possibility of focusing the microwave field at a given location. In addition, it is possible to work with small amounts of catalyst, to determine the dielectric properties of substances and to obtain good reproducibility in the results. For these reasons, unimode cavities are preferred for most experimental set-ups, despite the fact that the effective space for the catalyst tested is limited because the applicator normally forms part of the waveguide. Multimode applicators are capable of sustaining several modes at the same time, as a result of multiple reflections from the oven walls. As multiple maxima of the electromagnetic field within the cavity are found, uniform field distribution can be achieved through changes in geometry and the use of mode stirrers. Multimode microwave ovens are usually more versatile than unimode ovens and, consequently, they are by far the most common processing devices used in industrial applications [36,39].

Temperature measurement inside microwave devices is not a straightforward issue. In order to avoid interactions with microwaves, common thermocouples can be used only when they are shielded and earthed or, in the case of fixed electric field distribution, when they are arranged perpendicular to the field. As an alternative to thermocouples, fiber optic sensors and IR-pyrometers can be employed. Fiber optics usually measure temperatures up to 400°C and are too fragile for most industrial applications. IR-pyrometers can measure higher temperatures and only record the surface temperature, which may differ from the temperature of the interior of the catalyst. Therefore, it is possible to obtain average temperatures but difficult to measure the temperature of hot spots. The use of ultrasonic temperature sensors for covering temperatures up to 1500ºC has also been proposed [36,38,41,46].

The household microwave oven represents the simplest and least expensive type of multimode on/off control device. Home microwave ovens can be modified by drilling holes on appropriate positions and simple experiments can be performed [36,43]. Figure 5.6 shows a diagram of a commercial microwave oven adapted in order to modify the surface chemistry of active carbons [47].

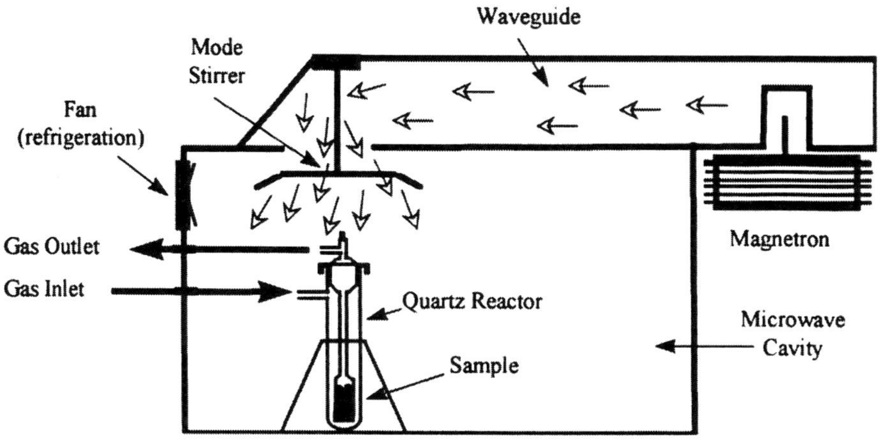

Note: The picture is obtained from Ref. [47] with the permission of Elsevier.

Figure 5.6. Schematic diagram of a commercial microwave oven adapted for modifying the chemistry of active carbons.

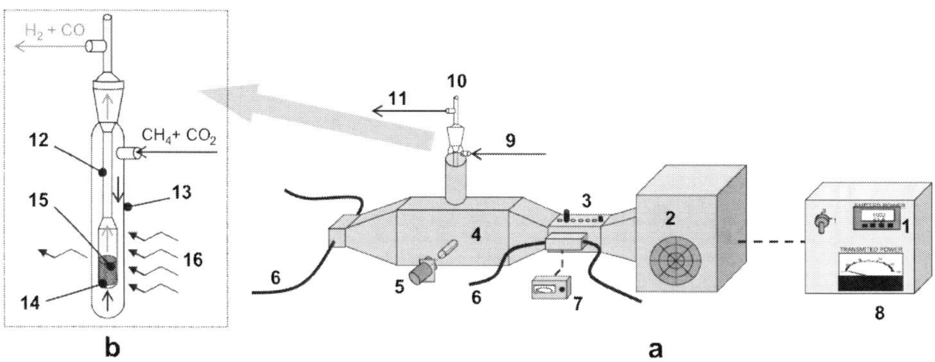

Note: The picture was obtained from Ref. [19] with the permission of Elsevier. 1. Input power control; 2. Air-cooled microwave generator or magnetron; 3. Manual 2-stub unit; 4. Waveguide; 5. Optical pyrometer; 6. Water sink;7. Reflected power control; 8. Transmitted power control; 9. Inflow gas line; 10. Set quartz reactor-jacket and catalyst/microwave receptor (C/MR); 11. Outflow gas line. 12. Quartz reactor; 13. Quartz jacket; 14. Porous plate; 15. C/MR bed; 16. Microwave radiation.

Figure 5.7. (a) Schematic diagram of the single mode microwave device.(b) Close-up view of the set quartz reactor-jacket.

Microwave devices must be specially design for most applications because proper safety precautions must be taken in order to avoid radiation leaks. These modified ovens cannot be used under explosive conditions or aggressive media.

Unimode microwave ovens have been employed for the in-lab study of various gas-solid reactions [14,19,20,34,48-50]. Figure 5.7 shows a diagram of a possible set-up for a unimode microwave device to carry out experiments of the decomposition and CO_2 reforming of methane over carbon catalysts [14,19,20]. In this case, the microwave oven can operate at a variable power from 0 to 1860 W and at a fixed frequency of 2.45 GHz. The catalyst sample is placed inside a quartz reactor, which is housed in the centre of the rectangular microwave guide that directs the microwaves from the magnetron into the sample. The non-absorbed radiation is dissipated by a water sink. The power reflected back towards the magnetron is minimized by adjusting the manual 2-stub unit in the waveguide. The catalyst temperature is measured with an IR-pyrometer. Other unimode microwave ovens with variable output power have been described elsewhere [46,48-50]. They can be illustrated by means of a similar schematic diagram to that shown in Figure 5.7, although some differences in the set-ups have been pointed out. Examples of such differences include different maximum operating powers, the use of co-axial cable instead of a rectangular microwave guide, the measurement of temperature with optic fiber and the use of mobile piston, short circuits, apertures, irises or posts, instead of stub units.

Although multimode microwaves are widely used in industrial applications, there are few examples of the use of lab-scale multimode microwave ovens in gas-solid reactions [41,51]. Figure 5.8 shows an experimental set-up used to investigate the microwave-assisted CO_2 reforming of CH_4 over carbon-based catalysts. This lab-scale pilot plant is a multimode microwave device, which operates at a frequency of 2450 MHz over a power range of 0 - 1500 W. Microwaves are generated in a magnetron and directed through a rectangular waveguide to the multimode applicator, which houses the reactor charged with the catalyst. The reactor is isolated by an insulator, which is transparent to microwaves. The catalyst temperature is measured by means of a thermocouple placed inside the catalyst bed. The insulator temperature is measured by means of another thermocouple. The power reflected back to the magnetron is minimized by adjusting a manual 3-stub unit, situated inside the waveguide.

A circulator prevents the reflected microwaves from reaching and damaging the magnetron. The reflected power is measured by a detector and dissipated by a water sink. The power needed to generate microwaves during the CO_2 reforming of CH_4 is recorded by an energy consumption meter. The instantaneous electric current and voltage are displayed on a control panel, so that the power required at any one moment can be known. In addition, the microwave pilot plant can operate in manual mode (the operating power is fixed and the C/MR is heated up to the maximum possible temperature under these conditions) and in automatic mode (the operating temperature is fixed at the desired value and controlled by a PID, which adjusts the power emitted by the magnetron in order to keep the sample temperature constant).

Another multimode microwave oven has been described to investigate the microwave-assisted heterogeneous catalytic gas-phase oxidation of hydrocarbons [41]. In this case, a commercial microwave oven with two magnetrons of 900W and rotating antennas was modified using two independent switched-mode power supplies, two metallic flanges for supporting the quartz reactor where the catalyst sample is placed and another metallic flange to support an IR-pyrometer for measuring the temperature. A thermocouple is also used to measure the catalyst temperature after the microwave power has been turned off.

The design of a microwave oven for industrial applications is very much dependent on each specific process and it must take into account critical issues such as power efficiency, uniform power distribution and consistent and reliable performance over the expected range of process conditions [38,41].

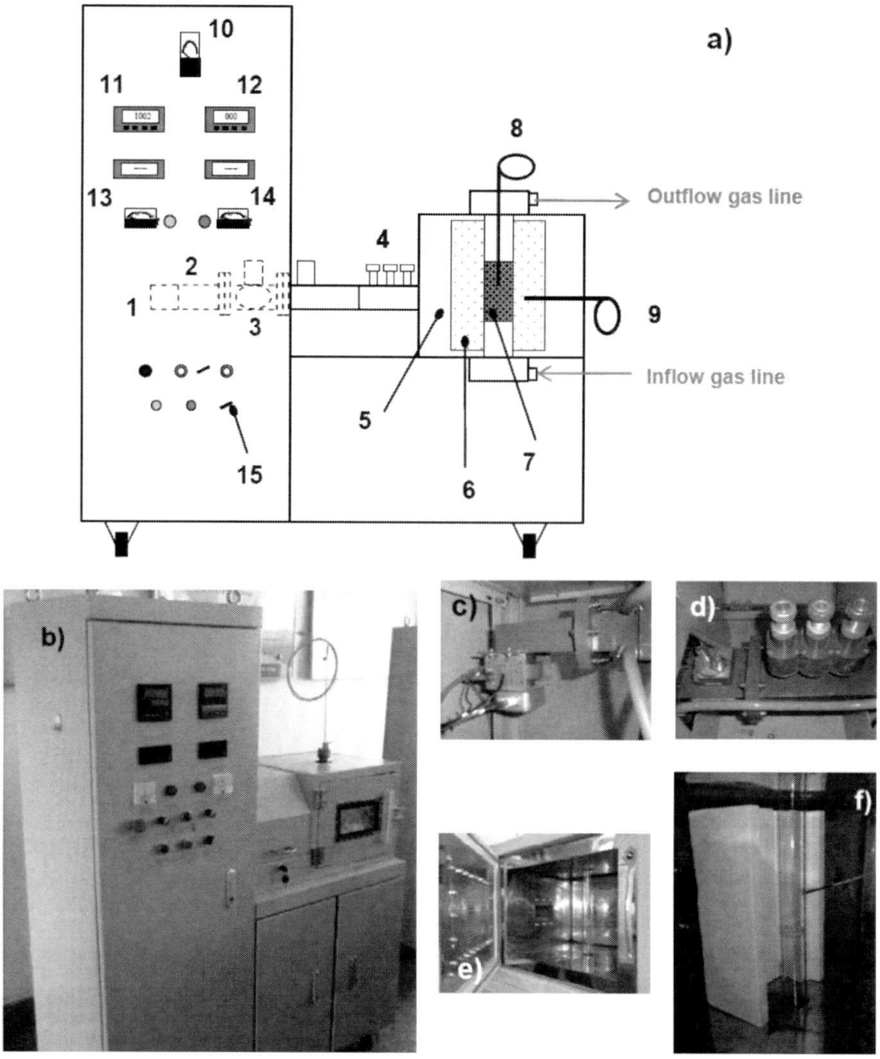

Note: 1. Magnetron; 2. Waveguide; 3. Circulator, detector of reflected power and water sink; 4. Manual 3-stub unit; 5. Multimode microwave applicator; 6. Insulator; 7. Reactor loaded with a catalyst/microwave receptor; 8. Sample thermocouple; 9. Insulator thermocouple; 10. Meter of energy consumption; 11. Controller/display of sample temperature; 12. Display of insulator temperature; 13. Display of emitted power; 14. Display of reflected power; 15. Manual/automatic mode.

Figure 5.8. (a) Schematic diagram of the multimode microwave pilot plant. (b) Multimode microwave pilot plant. (c) Microwave generator, circulator and water sink. (d) 3-stub unit.(e) Multimode applicator and reactor. (f) Reactor and insulator.

MICROWAVE-ASSISTED CO2 REFORMING OF CH4

Microwave heating has been applied to the heterogeneous catalytic reaction of CO_2 reforming of CH_4 over carbon catalysts [19,20,52] and metal catalysts [34,42,53]. Carbon catalysts are usually good microwave absorbers and, consequently, heat more easily than metal catalysts, which normally need to be mixed with microwave receptors. However, the catalytic activity of metal-based catalysts is as a rule higher, which makes it possible to work under more severe operating conditions. The operating conditions (temperature, feed gas composition or space velocity) required to ensure a good performance of microwave-assisted dry reforming may depend on the type of catalyst used.

Operating Conditions

Microwave-assisted dry reforming is an endothermic reaction and, for that reason, CO_2 and CH_4 conversions improve as the temperature increases. Normally, the H_2/CO ratio of the resulting syngas also increases when the temperature rises. An example of the evolution of the CO_2 and CH_4 conversions as a function of temperature for a series of experiments of dry reforming over a Pt-based catalyst with both microwave and conventional heating is shown in Figure 5.9. In addition, the equilibrium conversions of CO_2 and CH_4, established according to a mechanism which considers RWGS (reaction 5.4) as secondary reaction, are plotted. As well as the increase in conversion with temperature, the microwave enhancement in the dry reforming reaction is corroborated by the higher CO_2 and CH_4 conversions obtained under microwave heating compared to conventional heating, mainly below 700 °C. Moreover, under microwave heating, conversions higher than the thermodynamic equilibrium values are achieved due to the formation of random hot spots where temperatures are higher than the average temperature in a bulk metal-based catalyst [34,42]. Microwave enhancement in the CO_2 reforming of CH_4 is also observed over carbon catalysts, the CO_2 and CH_4 conversions being clearly higher under microwave heating than under conventional heating, due to the differences between the heating mechanisms and the presence of hot spots throughout the carbonaceous bed [19].

Microwave-assisted dry reforming over carbon catalyst is believed to be a combination of catalytic methane decomposition (reaction 5.7) and CO_2 gasification (reaction 5.6), as it is shown in Figure 5.10 [19,20]. Thus, the decomposition of CH_4 takes place in the active centres of the C/MR and, consequently, these are blocked by the carbonaceous deposits from methane. The nature and reactivity towards CO_2 of these carbon deposits is very heterogeneous [20,54].

Most of them are discrete deposits of irregular shape and different size that tend to physically cover the active surface. However, under microwave heating, groups of nanofilaments scattered all over the catalyst surface are formed along with the other deposits (see Figure 5.11). As a result, the highly reactive carbonaceous deposits are gasified by the CO_2, giving rise to the in situ regeneration of the active centres. Ideally, the catalytic activity of the carbon catalysts is maintained with time.

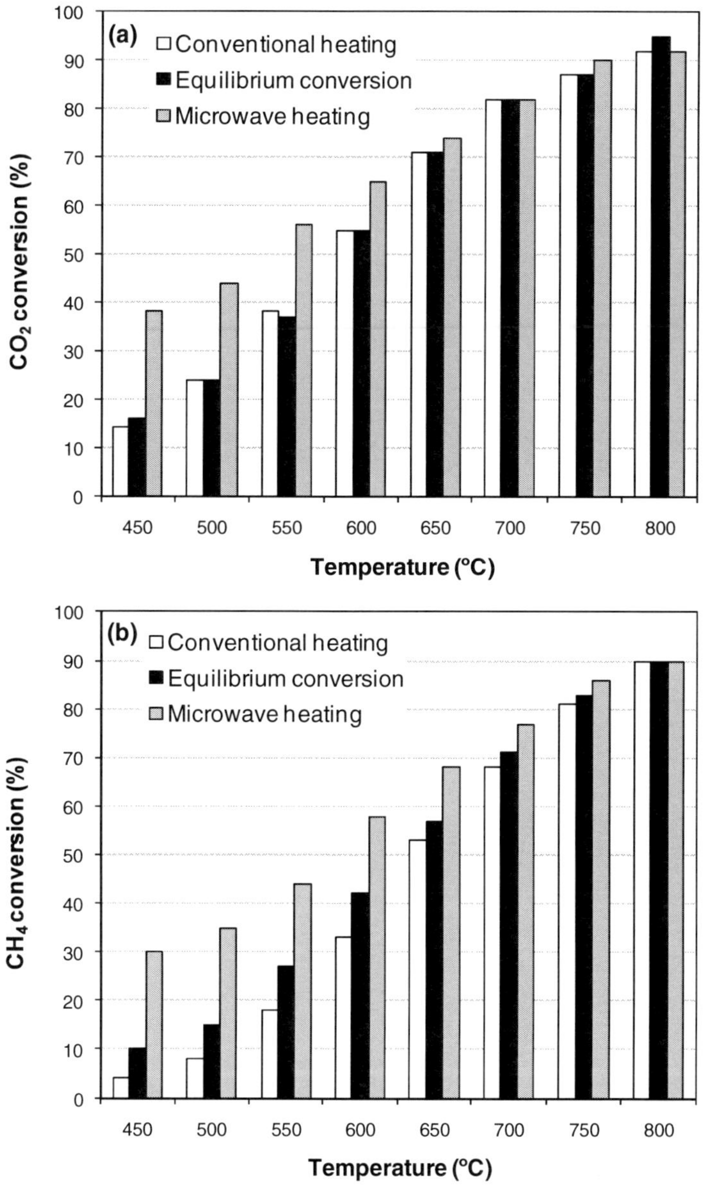

Note: Adapted from Ref. [34].

Figure 5.9. Influence of temperature on the microwave-assisted CO_2 reforming of CH_4over catalyst Pt (8 wt.%)/CeO_2 (20 wt.%)/γ-Al_2O_3 at a CH_4/CO_2 ratio of 1: (a) CO_2 conversion and (b) CH_4 conversion.

However, the conversion of CH_4 to H_2may be partiallyimpeded with time due to the existence of non-recovered active centres that are blocked by refractory carbon deposits. CO production is less restricted since CO_2 is able to react with some carbon from the original catalyst. However, the conversion of CO_2 may eventually decrease due to the fact that blockage of the pores by the refractory carbonaceous deposits may impede the access of CO_2 to the highly reactive carbon.

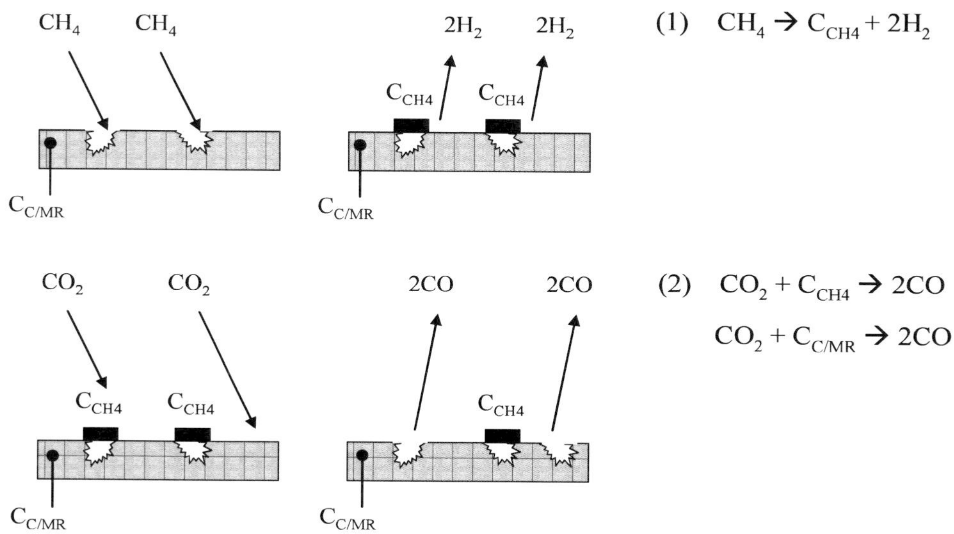

Note: C_{CH4}, carbonaceous deposits from the decomposition of CH_4 and $C_{C/MR}$, carbon of catalyst/microwave receptor.

Figure 5.10. Mechanism for the microwave-assisted CO_2 reforming of CH_4 over carbon catalysts.

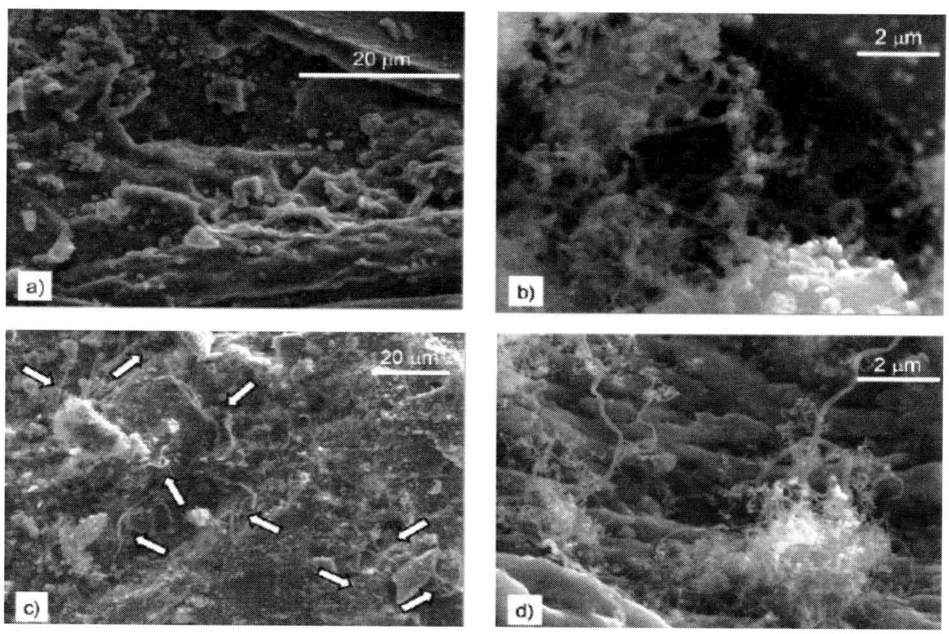

Note: Ref. [20] with the permission of ACS Publications.

Figure 5.11. SEM images of the deposits obtained after reactions at 800 °C for 2 h: (a) carbon deposits on the carbon catalyst surface, obtained after the decomposition of CH_4 under conventional heating; (b) close-up of the nanofibers obtained after the decomposition of CH_4 under microwave heating; (c) carbon catalyst surface after the CO_2 reforming of CH_4 under microwave heating (arrows indicate the location of groups of nanofibers); and (d) close-up of the nanofibers obtained after the CO_2 reforming of CH_4 under microwave heating.

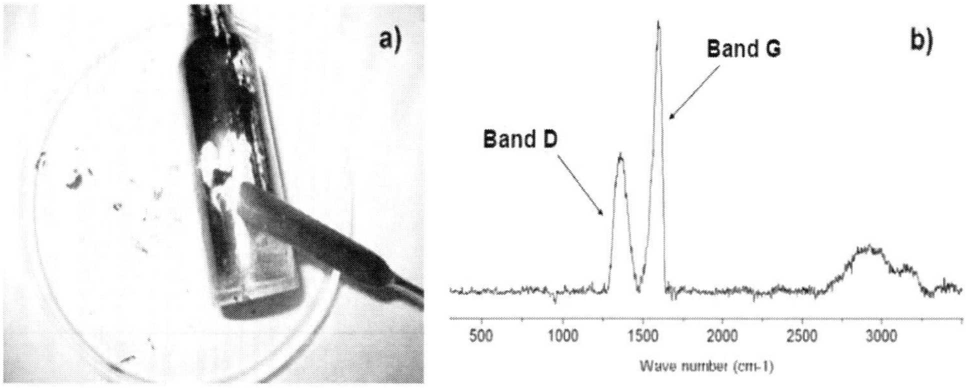

Figure 5.12. (a) Pyrolityc carbon deposits formed on the outside wall of the reactor during microwave-assisted CO_2 reforming of CH_4 at 900 °C; and,(b) Raman spectrum of the pyrolityc carbon deposits.

The mechanism proposed is verified between 700 and 800°C, which is therefore the optimum temperature range for carrying out microwave-assisted dry reforming over carbon catalysts [19].

Temperatures of around 900°C and above are discarded because they may favour the formation of a thin film of a pyrolityc carbon deposit on the outside wall of the reactor, which is shown in Figure 5.12a. This pyrolityc deposit is a mixture of non-graphitic/graphitic carbon, the latter being predominant ($I_D/I_G = 0.69$, according to the Raman spectrum shown in Figure 5.12b). Because of its graphitic nature, the deposit reflects the microwaves and prevents the C/MR from absorbing radiation, which may cause the experimental device to fail. At lower temperatures, of around 600°C, a mechanism that assumes the occurrence of the RWGS reaction as a side reaction, as in the case of microwave-assisted dry reforming over metal catalysts, is proposed, instead of the one proposed previously.

Apart from temperature, other operating conditions influence the performance of the dry reforming reaction. Thus, in the case of microwave-assisted reforming over carbon catalysts, an increase in volumetric hourly space velocity (VHSV, defined as the ratio between the feed flow rate and the catalyst charge [55]) leads to a decrease in the conversions of CO_2 and CH_4, since the contact time between the reactant gases and the catalyst is reduced.

However, the changes in VHSV may not affect the H_2/CO ratio of the synthesis gas produced [19]. The values of VHSV needed in order to achieve acceptable CH_4 and CO_2 conversions depend on the type of catalyst. For example, in the case of carbon catalysts, conversions of 80% or above may be obtained with a volumetric hourly space velocity of around 0.3 $L{\cdot}g^{-1}h^{-1}$ [19,20].

The feed gas composition, i.e. the CO_2/CH_4 ratio, also influences the microwave-assisted CO_2 reforming of CH_4 performance. In the case of carbon catalysts, the quantity of CO_2 introduced also has a key influence on the conversions achieved. Thus, as is illustrated in Figure 5.13, when the CO_2/CH_4 feed ratio is raised, the CO_2 and CH_4 conversions increase, especially in the case of CH_4. Thus, according to the mechanism explained above, when the proportion of CO_2 increases, the capacity of CO_2 for regenerating active centres increases, since there is more CO_2 available for gasifying the carbonaceous deposits which are blocking the active centres. Consequently, the conversion of CH_4 increases. At the same time, CO_2

conversion increases because there is access to the highly reactive carbon deposits and carbon of the catalyst [19].

Interestingly, in the case of the metal-based catalysts, as the ratio of CO_2/CH_4 increases, CH_4 conversion does increase whereas CO_2 decreases due to the occurrence of the secondary reaction between CO_2 and H_2 (RWGS reaction) [34]. Nevertheless, independently of the type of catalyst used, an increase in the CO_2/CH_4 ratio introduced causes the H_2/CO ratio of the resulting syngas to decrease. Thus, a H_2/CO ratio of approximately 1.4 can be obtained at a CO_2/CH_4 ratio of 0.5, while the H_2/CO ratio may diminish to 0.5 at a CO_2/CH_4 ratio of 3.0 [19,34].

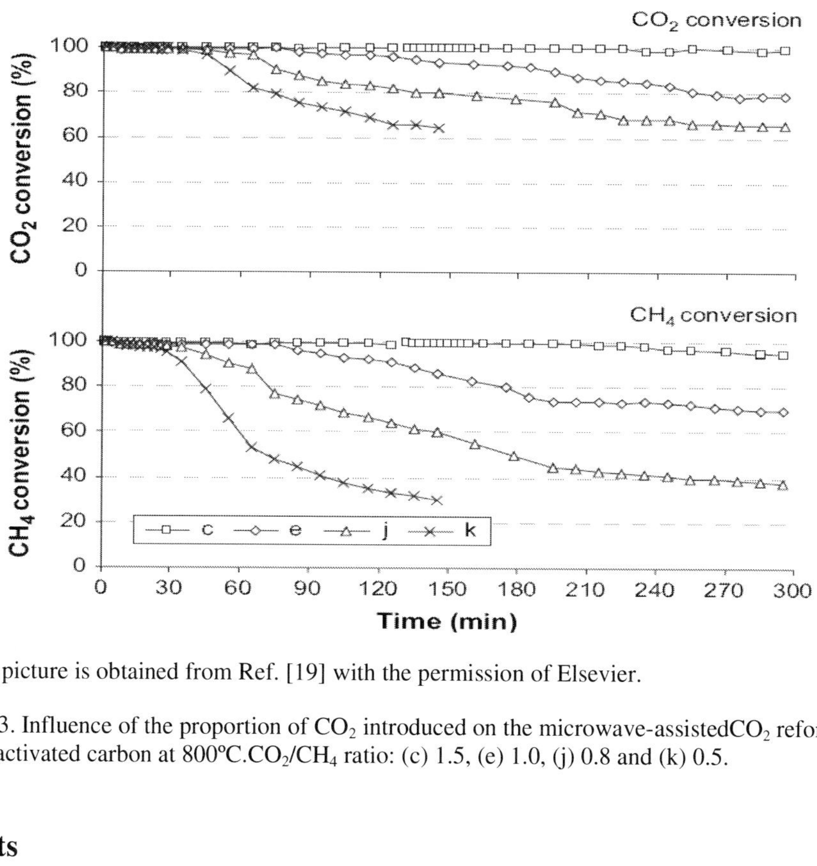

Note: The picture is obtained from Ref. [19] with the permission of Elsevier.

Figure 5.13. Influence of the proportion of CO_2 introduced on the microwave-assistedCO_2 reforming of CH_4 over activated carbon at 800ºC.CO_2/CH_4 ratio: (c) 1.5, (e) 1.0, (j) 0.8 and (k) 0.5.

Catalysts

Since one of the main obstacles for the industrial application of the dry reforming process is that there are no effective catalysts that do not undergo deactivation due to carbon deposition, numerous investigations have been carried out in order to find new catalysts. Nearly all the metals in group VIII over different supports have been studied under conventional heating, resulting in a good catalytic activity [9,10,18]. Noble metals like Rh, Pt, Ru or Ir show the best catalytic activity and the lowest susceptibility to coke formation. However, their high cost and low availability make them unattractive for implementation on an industrial scale. Ni-based catalysts are preferred since they are cheaper and, although they are more susceptible to coke formation, they exhibit a high activity in dry reforming [9,18]. In addition, the type of support used can have a big influence on the catalytic activity and deactivation of metal catalysts. As it is shown in Table 5.2, the activity, selectivity and stability of a metal catalyst can vary with the support. For this reason, several materials have been evaluated as supports, including inorganic oxides [9,10,18,23] and carbon materials [18,56,57]. Ni/Al_2O_3 is one of the catalysts that give rise to the best results [18]. To a lesser extent, the conventionally-heated CO_2 reforming of CH_4 over carbon catalysts has also been studied [19-21,31] and although they show an acceptable catalytic activity, in general, it is worse than that of metal-based catalyst.

On the other hand, microwave-assisted CO_2 reforming of CH_4 has been mainly studied over carbon catalysts [19,20,52], although the catalytic activity of some metal-based catalysts has also been investigated [34,42,58].

Table 5.2. Catalytic activity of metals on various supports

Metal activity	*T* (ºC)	Metal loading (wt. %)
Ru		
$Al_2O_3 > TiO_2 > SiO_2$	620	0.5
Pd		
$TiO_2 > Al_2O_3 >$NaY$> SiO_2 >MgO$	500	5.0
$TiO_2 > Al_2O_3 > SiO_2 >MgO$	500	1.0
Rh		
YSZ $> Al_2O_3 > TiO_2 > SiO_2 >> MgO$	650	0.5
$Al_2O_3 > SiO_2 > TiO_2 > MgO$	500	1.0
Ni		
$Al_2O_3 > SiO_2$	525-725	40
$Al_2O_3 > SiO_2$	600	10
NaY$> Al_2O_3 > SiO_2$	600	2
$SiO_2 > ZrO_2 > La_2O_3 > MgO > TiO_2$	550	4

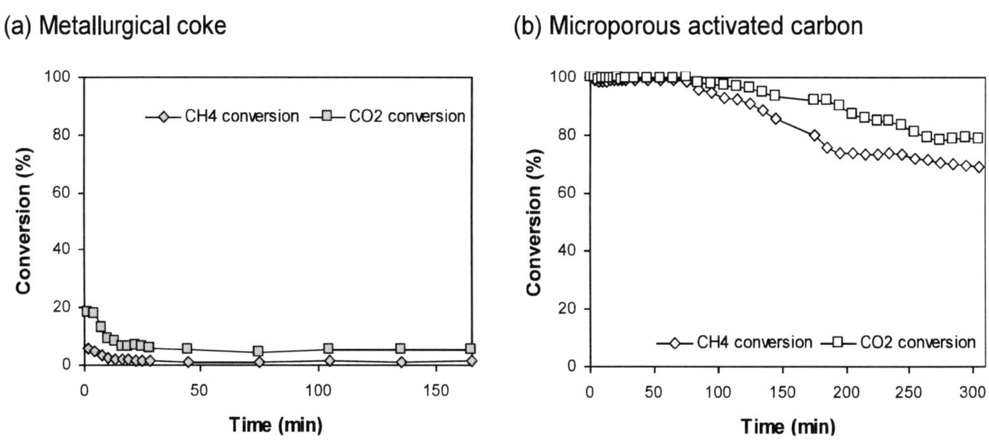

Note: Adapted from Ref. [52].

Figure 5.14. Influence of the textural properties of carbons on their catalytic activity for the microwave-assisted CO_2 reforming of CH_4: (a) Metallurgical coke – no textural development and (b) Microporous activated carbon – $S_{BET} = 821$ $m^2 \cdot g^{-1}$, $V_{mic} = 0.323$ $cm^3 \cdot g^{-1}$.

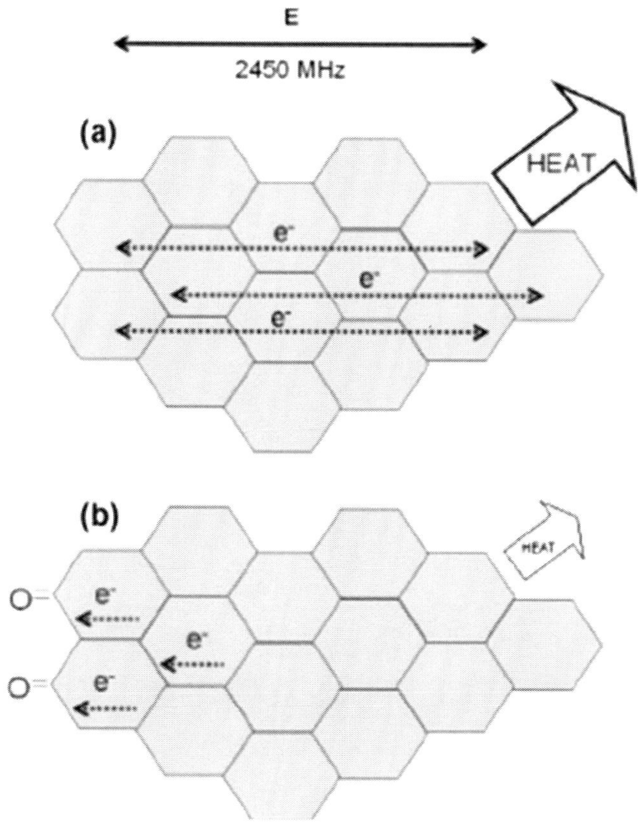

Note: The picture is obtained from Ref. [52] with the permission of Elsevier.

Figure 5.15. Schematic representation of some of the effects involved in the microwave heating of carbons: (a) When microwave heating is caused by the Maxwell–Wagner effect (Interfacial polarization), the delocalized π-electrons try to couple the changes of phase of the electric component of the electromagnetic field dissipating heat and (b) Oxygen-containing surface groups are electron-withdrawing, limiting the mobility of some of the π-electrons of the basal planes and therefore restricting the heat released.

Carbon materials are mainly used because they are usually good microwave absorbers, which make them suitable catalysts/microwave receptors for the dry reforming reaction under microwave heating [35,52]. However, many metal-based catalysts are not heated by microwave radiation because, in general, metals reflect microwaves and supports are insulators [38]. Therefore, most of the metal catalysts that exhibit good catalytic activity under conventional heating (e.g. Ni/Al_2O_3) cannot be used directly for the microwave-assisted dry reforming, they have to be mixed with or supported on microwave receptors (carbons and certain oxides) for them to be heated by microwaves [40,53,58].

The catalytic activity of carbons in the microwave-assisted CO_2 reforming of CH_4 is considered to be determined by their structural and surface properties [59], as in the case of the decomposition of CH_4, which has been widely evaluated over various carbonaceous materials (including carbon blacks, activated carbons or chars from biomass residues), mainly under conventional heating but also under microwave heating [14,17,55,60-65].

The catalytic activity of the carbons used as catalysts for microwave-assisted dry reforming of CH_4 is related to their micropores volume [52]. Carbons with a large microporosity are necessary in order to obtain acceptable CH_4 and CO_2 conversions. This is illustrated in Figure 5.14, where the negligible conversions achieved over a metallurgical coke with no textural development are compared with the high conversions obtained over a microporous activated carbon.

Moreover, as microporosity is prevented from being blocked and the catalytic activity of carbons is maintained by the CO_2 gasification of the carbon deposits (according to the mechanism described in Figure 5.10), carbon catalysts need to show good reactivity towards CO_2 [52,59]. Carbons with a low CO_2 reactivity are usually bad catalysts for the microwave-assisted CO_2 reforming of CH_4. Carbons with a high content in oxygen surface groups show a poor catalytic activity, possibly due to the fact that oxidized carbons show a lower reactivity towards CO_2 and because they are difficult to heat up [52].

Indeed, oxidized carbons heat up with difficulty under microwave radiation and the generation of microplasmas is also more difficult because oxygen, which is electronegative, shows an affinity towards electrons [66]. In other words, oxygenated functionalities are electron-withdrawing, their mobility is limited and, therefore, less density of delocalized π-electrons is available for promoting microwave heating[52,67,68]. This phenomenon is schematically represented in Figure 5.15.

Microwave-assisted dry reforming over suitable carbon catalysts can give rise to high CO_2 and CH_4 conversions for a long period of time. However, as mentioned above, this requires the use of a high CO_2/CH_4 ratio and a low VHSV, which results in a low production of synthesis gas with a H_2/CO ratio lower than unity. Consequently, carbon-supported nickel catalysts have been investigated as alternatives to carbonaceous catalysts. They have the advantages of the carbon catalysts (i.e. they heat up under microwave radiation) and the advantages of the metal catalysts (i.e. a higher catalytic activity). CO_2 and CH_4 conversions achieved over carbon-supported nickel catalysts are higher and steadier than those obtained over the non-impregnated carbons because of the good catalytic activity of metal.

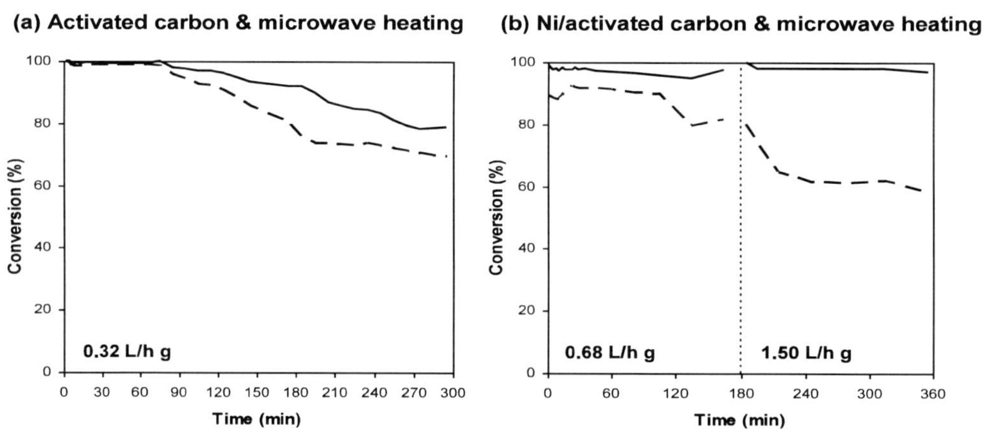

Figure 5.16. CO_2 conversion (solid line) and CH_4 conversion (dotted line) obtained by the microwave-assisted CO_2 reforming of CH_4 over different catalysts: (a) Carbon catalyst (activated carbon) and VHSV of 0.32 $L{\cdot}h^{-1}g^{-1}$, and (b) Carbon-based Ni catalyst (Ni/activated carbon) and VSHV of 0.68 and 1.50 $L{\cdot}h^{-1}g^{-1}$.

Moreover, higher conversions over carbon-supported nickel catalysts can be obtained by using higher operating times and space velocities, which gives rise to a larger syngas production. For example, as can be seen in Figure 5.16, by using Ni/activated carbon, a CO_2 conversion of 100% and a CH_4 conversion of 60% were obtained after 300 min at a VHSV of 1.50 $L{\cdot}h^{-1}g^{-1}$, whereas a CO_2 conversion of 80% and a CH_4 conversion of 70% were obtained over the non-impregnated activated carbon at a much lower VHSV (around 0.30 $L{\cdot}h^{-1}g^{-1}$). However, syngas production over Ni/activated carbon is limited by the catalyst deactivation, which is probably due to the agglomeration of Ni particles by sintering and due to the blockage of active sites by carbonaceous deposits as in the case of carbon catalysts [2,18].

As well as carbon-supported nickel catalysts, heterogeneous mixtures of carbonaceous materials and metal catalysts can be used in order to catalyze the microwave-assisted CO_2 reforming of CH_4. As in the case of carbon-supported Ni catalysts, the catalytic mixture combines the characteristics of a good microwave receptor and catalyst. In the light of previous results, the chosen carbon material needs to be microporous and have good reactivity towards CO_2, and alumina-supported nickel catalyst (Ni/Al_2O_3) is one of the preferred metal components [18,52,53]. When a heterogeneous mixture of carbon and Ni/Al_2O_3 is used as catalyst for the dry reforming reaction, the mechanism of reaction described above (Figure 5.10) is slightly modified. Thus, CH_4 decomposition tends to occur preferentially over the Ni-based fraction, whereas the CO_2 tends to gasify the carbon deposits from methane (active sites for methane decomposition are continuously being regenerated) and part of the initial carbonaceous fraction [53]. Therefore, choosing a suitable carbonaceous material for the blend is no trivial matter. Indeed, this may influence the catalytic activity of the resulting mixture of carbon fraction and metal-based fraction and the regeneration of the active centres by CO_2 gasification, since reactivity towards CO_2 varies for different carbonaceous materials [52]. In addition, the microwave heating capacity of the mixture is determined by the dielectric properties of the carbonaceous fraction, which may differ depending on the material used [35].

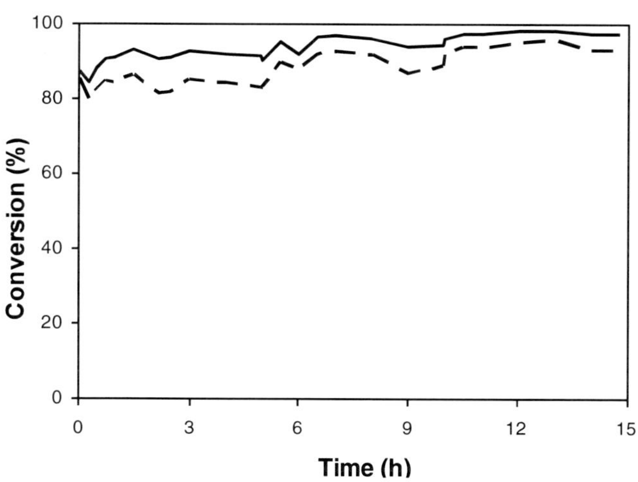

Figure 5.17. CO_2 conversion (solid line) and CH_4 conversion (dotted line) obtained by the microwave-assisted CO_2 reforming of CH_4 over a catalytic mixture of activated carbon and Ni/Al_2O_3, 800 °C, 50 *vol.%* CH_4 – 50 *vol.%* CO_2 and 3 $L{\cdot}h^{-1}g^{-1}$.

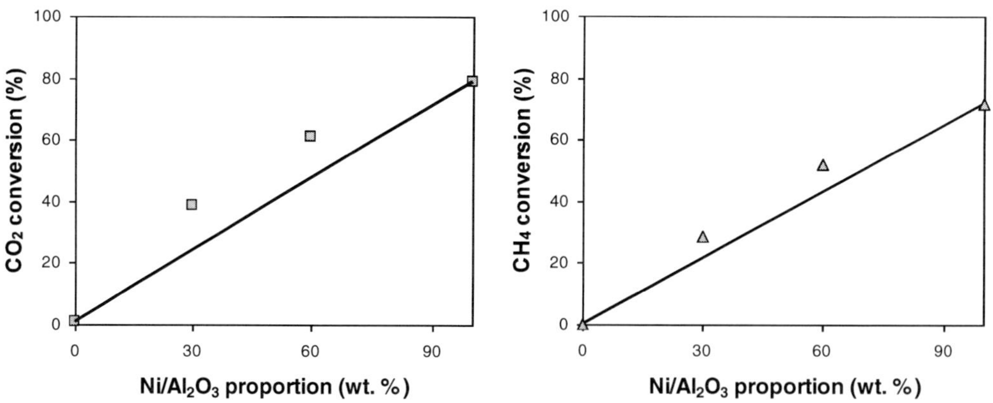

Note: Adapted from Ref. [53].Operating conditions: conventional heating at 800ºC, 50 *vol.%* CH_4 – 50 *vol.%* CO_2 and3 $L{\cdot}h^{-1}g^{-1}$.

Figure 5.18. Comparison between the CO_2 and CH_4 conversions expected by the law of mixtures (solid lines) with the experimental conversions of CO_2 (squares) and CH_4 (triangle) attained during dry reforming over different catalytic mixtures of activated carbon and Ni/Al_2O_3.

Figure 5.17 shows the profiles of the CO_2 and CH_4 conversions during microwave-assisted dry reforming over a catalytic mixture of activated carbon and Ni/Al_2O_3. Conversions close to 100% are obtained during at least 15 hours at a VHSV of 3 L/h g. This catalytic mixture therefore is preferable to Ni/activated carbon, since higher conversions are obtained at a higher VHSV (see Figure 5.16b).

According to the results presented in Figure 5.17, CO_2 and CH_4 conversions can increase with time, which may be due to the fact that the carbon deposits from the decomposition of CH_4 are more reactive than the initial activated carbon [62,69], or possibly, because the carbonaceous deposits favour changes in the metal-based fraction of the catalyst.

In addition, reduced Ni may interact with the alumina during the heating up to operating temperature (around 800ºC) and under inert atmosphere, previous to the dry reforming reaction, leading to the formation of the spinel $NiAl_2O_4$[53,70,71].In actual fact, $NiAl_2O_4$ is not active in the dry reforming reaction, but in its reduced form, which maybe continuously produced under the reaction atmosphere, it tends to inhibit the formation of carbonaceous deposits and it shows a good stability, since reduced Ni from spinel tends to be disperse, making sintering more difficult. Therefore, the interaction between the metal and the support may favour the steady conversions obtained over the catalytic mixture, although the blockage of Ni by carbon deposition and the sinterization of Ni particles may eventually lead to the deactivation of Ni/Al_2O_3 [70,71]. Moreover, a synergetic effect exists between the carbonaceous fraction and themetal-based fraction, which means that the experimental CO_2 and CH_4 conversions obtained over the mixtures of carbonaceous materials and metal-based catalysts are higher than the conversions calculated by means of the law of mixtures(i.e. by adding the weighed individual conversions) [53], as is illustrated in Figure 5.18.

The existence of a synergetic effect facilitates a compromise between the catalytic activity and the cost of the catalyst. Thus, the replacement of part of the metal-based fraction, which is probably a better catalyst for the CO_2 reforming of methane, by a carbonaceous material, which is cheaper [61], leads to a decrease in the conversions lower than that expected if there were no synergetic effect. The synergetic effect is especially advantageous in the microwave-assisted dry reforming process because, as mentioned above, part of the metal-based catalyst, which does not absorb microwaves, must be replaced by the carbonaceous material in order to be heated by microwave radiation. Due to the synergetic effect, this substitution does not involve any significant loss of catalytic activity. Moreover, any loss of catalytic activity from the use of catalytic mixtures instead of metal-based catalysts may be offset by the enhancement of the CO_2 and CH_4 conversions due to microwave heating.

The production of synthesis gas depends on the type of catalyst used. Thus, larger quantities of methane can be treated with CO_2 over the catalytic mixtures than over carbon catalysts, as a consequence of which more syngas can be produced. However, syngas composition seems to depend on the operating conditions, mainly the CH_4/CO_2 ratio and the heating device, rather than on the type of catalyst used. Thus, under similar operating conditions, the microwave-assisted dry reforming of methane produces syngas with a similar H_2/CO ratio irrespective of the carbon-based catalyst used [53].

For example, at 800 °C and 50 *vol.%*CH_4 – 50 *vol.%* CO_2, syngas with a H_2/CO ratio of 0.8 is obtained over Ni/activated carbon and of 0.9 over a mixture of activated carbon and Ni/Al_2O_3. These values are very close to the stoichiometric H_2/CO ratio of 1 and considerably higher than the values obtained from conventional dry reforming over metal catalysts, where the reverse water gas shift takes place as a side reaction (H_2/CO ratio around 0.5 or even lower) [72].

ENERGY CONSUMPTION

In this Section, preliminary estimations of energy consumption during the microwave-assisted CO_2 reforming of CH_4 over carbon-based catalysts are shown and compared with bibliographic values of energy consumption for the steam methane reforming reaction over metal catalysts in two industrial-based H2 production processes [73,74]. Table 5.3 summarizes the values of energy consumption estimated for the different processes.

In their work, Carrara et al. [73] developed a simulation tool for performance prediction in a steam reforming hydrogen production plant. A 1500 m^3/h hydrogen production plant is simulated and, although the investigation is a theoretical one, field data from an industrial plant were used to validate the model.

According to these data, the energy consumption for the steam reforming reaction over metal-based catalysts is 3.4 kW·h per m^3 of CH_4 introduced or 1.2 kW·h per m^3 of H_2 produced. The reactor outflow gases include CH_4, CO_2, H_2O, H_2 and CO, at a H_2/CO ratio of 5:1. Only the energy consumed in the reaction stage is taken into consideration in order to compare the dry and steam reforming reactions. In other words, the other stages of the global process, such as heat exchangers, WGS reactor or separation units are not taken into account.

On the other hand, Seo et al. [74] developed simulation models for the steam methane reforming, partial oxidation and autothermal reforming processes, using as a basis of calculation 1 mol s^{-1} of H_2 produced. The thermal energy required by each stage of the reforming process is evaluated.

These stages comprise a reforming reactor, a water gas shift reactor, a steam generator and a heat exchanger, but only the energy consumed in the reaction stage is taken into consideration for comparison purposes.

Table 5.3. Energy consumption for the steam reforming and microwave-assisted CO_2 reforming of methane

STEAM REFORMING OF METHANE		
Reference	Carrara, 2010 [73]	Seo, 2002 [74]
Calculation basis *	H2 production of 1500 m3·h-1	H2 production of 1 mol·s-1
Syngas H_2/CO ratio	5:1	4:1
Supplied power to reforming reactor	2094.5 kW	78.8 kW
Energy consumption respect to introduced CH_4	3.4 kW·h per m3 of CH_4	3.9 kW·h per m3 of CH_4
Energy consumption respect to produced H_2	1.2kW·h per m3 of H_2	1.2 kW·h per m3 of H_2
MICROWAVE-ASSISTED CO2 REFORMING OF METHANE		
Reference	activated carbon	Ni/Al2O3 + activated carbon
Calculation basis	1 m3·h-1 of introduced CH_4	1 m3·h-1 of introduced C CH_4
Syngas H_2/CO ratio	2:3	1:1
Supplied power to reforming reactor	84.4 kW	8.3 kW
Energy consumption respect to introduced CH_4	84.4 kW·h per m3 of CH_4	8.3 kW·h per m3 of CH_4
Energy consumption respect to produced H_2	44.4 kW·h per m3 of H_2	4.6 kW·h per m3 of H_2

Note: *Final H_2 production, after the reforming, water gas shift and purification stages.

According to this work, energy consumption for the steam reforming reaction is approximately 3.9kW·h per m^3 of CH_4 introduced or 1.2 kW·h per m^3 of H_2 produced. The reactor outflow gases are composed of CO_2, H_2O, H_2 and CO, at a H_2/CO ratio of 4:1.

The energy consumption values of the microwave-assisted CO_2 reforming of CH_4 are estimated by scaling up the experimental data obtained from the multimode device described in Figure 5.8 and using a flow rate of 1 m^3/h of CH_4as a basis of calculation. In addition, experiments over two different catalysts, an activated carbon and a mixture of activated carbon and Ni/Al_2O_3, are carried out. As can be seen in Table 5.3, the estimated energy consumption for the microwave-assisted CO_2 reforming of CH_4 over activated carbon is 84.4 kW·h per m^3 of CH_4 introduced, 44.4kW·h per m^3 of H_2 produced or 17.6kW·h per m^3 of syngas produced, the ratio of the syngas being $H_2/CO \approx 2:3$. In the case of microwave-assisted CO_2 reforming of CH_4 over a mixture of activated carbon and Ni/Al_2O_3, energy consumption is estimated as 8.3 kW·h per m^3 of CH_4 introduced, 4.6 kW·h per m^3 of H_2 produced or 2.2kW·h per m^3 of syngas produced at a H_2/CO ratio of 1:1. In effect, energy consumption during microwave-assisted dry reforming is nearly ten times lower when the catalytic mixture is used as catalyst instead of only the activated carbon. In the case of the catalytic mixture, only the carbonaceous fraction is heated by microwave radiation (the metal-based fraction cannot be heated by microwaves and is presumably heated by conduction from the particles of the carbonaceous material). However, the much lower energy consumption estimated over the catalytic mixture may be due to the fact that the reforming reaction can operate at a large volumetric hourly space velocity and still achieve high conversions, which is reflected directly in a larger syngas (or hydrogen) production per mass of catalyst.

The energy consumption value of 44.4 kW·h per $m^3$of H_2 produced estimated from the microwave-assisted CO_2 reforming of CH_4 over an activated carbon was found to be too high compared to the value of 1.2 kW·h per $m^3$of H_2 produced, obtained from the bibliographic data for the steam reforming reaction over metal-based catalysts. Under adequate operating conditions, high conversions can be obtained when the use of carbon catalysts and microwave heating are combined. However, from an energy point of view, dry reforming must be optimized for it to be able to rival the industrial process of steam methane reforming.

The energy consumption value of 4.6kW·h per $m^3$of H_2 produced estimated from the microwave-assisted CO_2 reforming of CH_4 over a catalytic mixture was also higher than the value of 1.2 kW·h per $m^3$of H_2 produced from steam reforming. However, various premises need to be considered when comparing the two reforming processes. First, the given values of energy consumption for the microwave-assisted dry reforming are estimated from experimental data obtained in a lab pilot plant, whereas the consumption values for steam reforming are established on the basis of data obtained from industrial plants of H_2 production. Presumably, energy consumption during the dry reforming would be lower with a larger scaling up and energy integration of the global process. Secondly, the dry reforming reaction is more endothermic than the steam reforming reaction (+247 kJ·mol^{-1} and +206 kJ·mol^{-1}, respectively). What is more, energy consumption during the dry reforming reaction stage could be higher than consumption during the steam reforming stage whereas consumption in the global process of microwave-assisted CO_2 reforming of CH_4 could be lower than that of the steam reforming process. For example, the dry reforming of CO_2-rich natural gas, biogas and CO_2-rich industrial residual streams can be carried out without the need for other condition stages [9,10,18,20,25]. However, previous methane separation stages would be needed in the case of the industrial steam reforming process. The same could occur with the subsequent stages. Besides, in the case of microwave-assisted dry reforming, the high CH_4 and CO_2 conversions achieved may eliminate the need for purification stages and, depending on the desired byproduct, additional stages to adjust the H_2/CO ratio could be avoided. CO_2 emissions from each process should also be taken into consideration. According to the bibliographic data, global CO_2 emissions for steam reforming are usually in the range of 0.35 to 0.42 m^3 of CO_2 per m^3 of H_2 produced [12], whereas CO_2 emissions for the dry reforming reaction can be as low as 0.2 m^3 of CO_2 per m^3 of H_2 produced [4]. Moreover, CO_2 emissions could hypothetically be zero or even negative if the electricity consumed for the generation of microwave energy in the dry reforming process were produced from renewable sources. All in all, energy consumption during the microwave-assisted dry reforming of methane over a mixture of carbon and Ni/Al_2O_3, estimated at 4.6 kW·h per m^3 of H_2 produced, can be considered a promising result compared to the value of 1.2 kW·h per $m^3$of H_2 produced in the industrial steam reforming process. Moreover, the microwave-assisted

CO_2 reforming of CH_4 over catalytic mixtures may be able to rival the steam reforming alternative, both from a point of view of conversion and energy consumption.

Challenges and Opportunities

The microwave-assisted CO_2 reforming of CH_4 offers an alternative to the well-established process of steam reforming of natural gas for the production of synthesis gas and its resulting byproducts. Dry reforming has clear environmental benefits since it turns two greenhouse gases (CH_4 and mainly CO_2) into a valuable feedstock. Moreover, microwave-assisted dry reforming could lead to the reduction of CO_2 emissions or even to their complete elimination if the electricity consumed in the generation of microwave energy in the dry reforming process were produced from renewable sources. In addition, the dry reforming of CO_2-rich natural gas, biogas and CO_2-rich industrial residual streams can be carried out without the need for previous condition stages. Also, CO_2 reforming of CH_4yields syngas with a H_2/CO ratio of 1:1 for a complete conversion, which could eliminate the need to adjust H_2/CO ratio by means of subsequent stages, such as the WGS reaction.

One of the main challenges to the industrial implementation of CO_2 reforming of CH_4 is that there are no commercial catalysts that can operate without undergoing deactivation due to carbon deposition. The microwave-assisted CO_2 reforming of CH_4 over carbon-based catalysts, which combines the catalytic and dielectric properties of the carbonaceous catalysts with microwave heating, which is known to have the potential to increase the rate of reaction, selectivity and yield of the catalytic heterogeneous reactions, has given rise to promising conversion results. However, the energy consumption during dry reforming needs to be reduced in order to be able to rival the steam reforming process. A reduction in energy consumption can be achieved by the improvement of the catalysts in order to operate at large volumetric hourly space velocity but still maintain high conversions, which would yield larger syngas production per mass of catalyst. A scaling up of the process would also enhance the energetic yield. Scaling up is in itself a considerable challenge as it entails the designing of new microwave equipment able to satisfy the requirements of a large scale process, but also it is the opportunity of implementing this type of heating in this and other analogous processes at industrial scale.

Authors' Biographies

Dr. José Angel Menéndez Díaz was born in Lugones, Asturias, Spain in 1964. He graduated from University of Oviedo, Spain, where received his M.Sc. in Chemistry and Ph.D. in Chemical Engineering in 1988 and 1994, respectively. He worked as research assistant at the Penn State University, USA, from 1995 to 1996. In 1997, he joined INCAR-CSIC, Spain, where he is currently working as a scientific researcher. His research activity is mainly focused in carbon materials and the use of microwave heating applied to carbon-related technological processes, leading various research projects on these fields. He is author and co-author of more than 100 scientific publications including various book chapters and patents. At present, he is member of the executive committee of the Spanish Carbon Group (GEC) and co-editor of the GEC bulletin.

Dr. Beatriz Fidalgo Fernández was born in Ferrol, A Coruña, Spain in 1982. She received her M.Sc. in Chemical Engineering from the University of Santiago de Compostela, Spain, in 2006 and her Ph.D. in Chemical Engineering from the University of Oviedo, Spain, in 2010. She did her doctorate research at INCAR-CSIC, Spain, from 2006 to 2010, focused on the study of microwave-assisted CO_2 reforming of CH4 over carbon-based catalysts. She is currently working as a research associate in Imperial College of London, UK, with research interest in fuel characterization and thermochemical conversion of coal and biomass. Dr. Fidalgo has been the author and co-author of 16 scientific publications and a patent since 2006.

REFERENCES

[1] J.R. Rostrup-Nielsen, *Catal. Today* 18, 305 (1993).
[2] J.R. Rostrup-Nielsen, J. Sehested and J.K. Norskov, *Adv. Catal.* 47, 65 (2002).
[3] International Energy Agency IEA (2005). Hydrogen production and storage. RandD priorities and gaps. http://www.iea.org/.
[4] J.R.H. Ross, *Catal. Today* 48, 151 (2005).
[5] I. Wender, *Fuel Process. Technol.* 48, 189 (1996).
[6] L. Barelli, G. Bidini, F. Gallorini and S. Servili, *Energy* 33, 554 (2008).
[7] E. Turpeinen, R. Raudaskoski, E. Pongracz and R.L. Keiski, *Int. J. Hydrogen Energy* 33, 6635 (2008).
[8] M.A. Peña, J.P Gómez and J.L.G. Fierro, *Appl. Catal. A* 144, 7 (1996).
[9] S. Wang, G.Q. Lu and G.J. Millar, *Energy Fuels* 10, 896 (1996).
[10] Y.H. Hu and E. Ruckenstein, *Adv. Catal.* 48, 297 (2004).
[11] J.M. Ogden, *Annu. Rev. EnergyEnviron.* 24, 227 (1999).
[12] N. Muradov, *EnergyFuels* 12, 41 (1998).
[13] N. Muradov, *Int. J. Hydrogen Energy* 26, 1165 (2001).
[14] A. Domínguez, B. Fidalgo, Y. Fernández, J.J. Pis and J.A. Menéndez, *Int. J. Hydrogen Energy* 32, 4792 (2007).
[15] R. Utrilla, J.L. Pinilla, I. Suelves, M.J. Lázaro and R. Moliner, *Fuel* 90, 430 (2011).
[16] B. Zapata, M.A. Valenzuela, J. Palacios and E. Torres-García, *Int. J. HydrogenEnergy*35, 12091 (2010).
[17] D.P. Serrano, J.A. Botas, J.L.G. Fierro, R. Guil-López, P. Pizarro and G. Gómez, *Fuel* 89, 1241 (2010).
[18] M.C.J. Bradford and M.A. Vannice, *Appl. Catal. A* 142, 73 (1996).
[19] B. Fidalgo, A. Domínguez, J.J. Pis and J.A. Menéndez, *Int. J. HydrogenEnergy*33, 4337 (2008).
[20] A. Domínguez, Y. Fernández, B. Fidalgo, J.J. Pis and J.A. Menéndez, *EnergyFuels* 21, 2066 (2007).
[21] M. Haghighi, Z. Sun, J. Wu, J. Bromly, H.L. Wee, E. Ng, Y. Wang and D. Zhang, *Proc. Combust. Inst.* 31, 1983 (2007).
[22] M.C.J. Bradford and M.A. Vannice, *Cat. Rev. Sci. Eng.* 41, 1 (1999).

[23] A.M. Gadalla and B. Bower, *Chem. Eng. Sci.* 43, 3049 (1988).
[24] J. Guo, Z. Hou, J. Gao and X. Zheng, *Energy Fuels* 22, 1444 (2008).
[25] J.M. Bermúdez, B. Fidalgo, A. Arenillas and J.A. Menéndez, *Fuel* 89, 2897 (2010).
[26] H. Purwanto and T. Akiyama, *Int. J. Hydrogen Energy* 31, 491 (2006).
[27] N. Muradov, F. Smith and A. T-Raissi, *Int. J. Hydrogen Energy* 33, 2023 (2008).
[28] N.R. Udengaard, J.-H.B. Hansen, D.C. Hanson and J.A. Stal, *Oil Gas J.* 90, 62 (1992).
[29] S. Teuner, P. Neumann and F. Von Linde, *Oil Gas European Magazine* 27, 44 (2001).
[30] Z.F. Ynag, R.G. Ding, L.H. Song and L. Qian, *Energy Fuels* 12, 1114 (1998).
[31] Q. Song, R. Xiao, Y. Li and L. Shen, *Ind. Eng. Chem. Res.* 47, 4349 (2008).
[32] N. Seyed-Matin, A.H. Jalili, M.H. Jenab, S.M. Zekordi, A. Afzali, C. Rasouli and A. Zamaniyan, *Plasma Chem. Plasma Process.* 30, 333 (2010).
[33] Q. Wang, Y. Cheng and Y. Jin, *Catal. Today* 148, 275 (2009).
[34] X. Zhang, C.S.M. Lee, D.M.P. Mingos and D.O. Hayward, *Catal. Lett.* 88, 129 (2003).
[35] J.A. Menéndez, A. Arenillas, B. Fidalgo, Y. Fernández, L. Zubizarreta, E.G. Calvo and J.M. Bermúdez, *Fuel Process. Technol.* 91, 1 (2010).
[36] H. Will, P. Scholz and B. Ondruschka, *Chem. Eng. Technol.* 27, 113 (2004).
[37] D.A. Jones, T.P. Lelyveld, S.D. Mavrofidis, S.W. Kingman and N.J. Miles, *Resour. Conserv. Recycl.* 34, 75 (2002).
[38] K.E. Haque, *Int. J. Miner. Process.* 57, 1 (1999).
[39] E.T. Thostenson and T.-W. Chou, *Composites Part A* 30, 1055 (1999).
[40] J. Jacob, L.H.L. Chia and F.Y.C. Boey, *J. Mater. Sci.* 30, 5321 (1995).
[41] H. Will, P. Scholz and B. Ondruschka, *Top. Catal.* 29, 175 (2004).
[42] X. Zhang and D.O. Hayward, *Inorg. Chim. Acta*359, 3421 (2006).
[43] A. Zlotorzynski, *Crit. Rev. Anal. Chem.* 25, 43 (1995).
[44] J.A. Menéndez, M. Inguanzo and J.J. Pis, *Water Res.* 36, 3261 (2002).
[45] J.A. Menéndez, E.J. Juárez-Pérez, E. Ruisánchez, J.M. Bermúdez and A. Arenillas, *Carbon* 49, 346 (2011).
[46] X. Zhang, D.O. Hayward and D.M.P. Mingos, *Chem. Comm.* 11, 975 (1999).
[47] J.A. Menéndez, E.M. Menéndez, M.J. Iglesias, A. García and J.J. Pis, *Carbon* 37, 1115 (1999).
[48] W.L. Perry, J.D. Katz, D. Rees, M.T. Paffet and A.K. Datye, *J. Catal.* 171, 431 (1997).
[49] J.K.S. Wan, Y.G. Chen, Y.J. Lee and M.C. Depew, *Res. Chem. Intermed.* 26, 599 (2000).
[50] C.Y. Cha and D.S. Kim, *Carbon* 39, 1159 (2001).
[51] B. Fidalgo, PhD Thesis, Universidad de Oviedo (2010). Downloadable from: http://hdl.handle.net/10261/27755.
[52] B. Fidalgo, A. Arenillas and J.A. Menéndez, *Fuel* 89, 4002 (2010).
[53] B. Fidalgo, A. Arenillas and J.A. Menéndez, *Appl. Catal. A* 390, 78 (2010).
[54] Y. Fernández, B. Fidalgo, A. Domínguez, A. Arenillas and J.A. Menéndez, *Carbon* 45, 1696 (2007).
[55] Z. Bai, H. Chen, B. Li and W. Li, *J. Anal. Appl. Pyrolysis* 73, 335 (2005).
[56] J. Matos, K. Díaz, V. García, T.C. Cordero and J.L. Brito, *Catal. Lett.* 109, 163 (2006).
[57] K. Díaz, V. García and J. Matos, *Fuel* 86, 1337 (2007).
[58] B. Fidalgo, L. Zubizarreta, J.M Bermúdez, A. Arenillas and J.A. Menéndez, *Fuel Process. Technol.* 91, 765 (2010).
[59] B. Fidalgo and J.A. Menéndez, *Chin. J. Catal.* 32, 207 (2011).

[60] H.F. Abbas and W.M.A. Wan Daud, *Int. J. Hydrogen Energy* 35, 1160 (2010).
[61] N. Muradov, F. Smith and A. T-Raissi, *Catal. Today* 102-103, 225 (2005).
[62] E.K. Lee, S.Y. Lee, G.Y. Han, B.K. Lee, T.-J. Lee, J.H. Jun, et al., *Carbon* 42, 2641 (2004).
[63] R. Moliner, I. Suelves, M.J. Lázaro and O. Moreno, *Int. J. HydrogenEnergy* 30, 293 (2005).
[64] I. Suelves, J.L. Pinilla, M.J. Lázaro and R. Moliner, *Chem. Eng. J.* 140, 432 (2008).
[65] B. Fidalgo, Y. Fernández, A. Domínguez, J.J. Pis and J.A. Menéndez, *J. Anal. Appl. Pyrolysis* 82, 158 (2008).
[66] F. Rodríguez-Reinoso, *Carbon* 36, 159 (1998).
[67] Z. Hashisho, M.J. Rood, S. Barot and J. Bernhard, *Carbon* 47, 1814 (2009).
[68] J.E. Atwater and R.R. Wheeler, *Carbon* 41, 1801 (2003).
[69] N. Muradov, *Catal. Commun.* 2, 89 (2001).
[70] S. Wang and G.Q. Lu, *Ind. Eng. Chem. Res.* 38, 2615 (1999)
[71] Z. Xu, Y. Li, J. Zhang, L. Chang, R. Zhou and Z. Duan, *Appl. Catal. A* 29, 175 (2004).
[72] M.C.J. Bradford and M.A. Vannice, *Appl. Catal. A* 142, 97 (1996).
[73] A. Carrara, A. Perdichizzi and G. Barigozzi, *Int. J. HydrogenEnergy* 35, 3499 (2010).
[74] Y.-S. Seo, A. Shirley and S.T. Kolaczkowski, *J. Power Sources* 108, 213 (2002).

In: Syngas
Editors: Antonius Indarto and Jelliarko Palguandi
ISBN: 978-1-62100-870-5

Chapter 6

EXPLOITATION OF BIOSYNGAS: HYDROGEN AND SYNTHETIC DIESEL PRODUCTION

A. R. de la Osa**, A. De Lucas, A. Romero, J. L. Valverde and P. Sánchez

Chemical Engineering Department, University of Castilla-La Mancha
Ciudad Real, Spain

ABSTRACT

In recent decades, concerns about global warming and alternative fuels have increased substantially. However, the vast world energy demand comes from fossil fuels use, due to its high energy density and conversion into heat, electricity and transport fuels which seems to be more simple and inexpensive compared to other commodities.

The use of coal as a transition feedstock for a hydrogen and synthetic fuels economy is likely to be accomplished via gasification. Coal gasification is a well established mature technology to produce a syngas (CO and H2) capable of being compatible with biomass and wastes as feedstock, reducing both fossil fuels dependency and CO2 emissions.

The hydrogen yield of such gasifier effluent may be increased by CO reacting with additional amounts of water, using the so-called water-gas shift reaction (WGSR). Although it is customarily used in industry, the reaction has not been explored at the conditions envisioned in the gasification process. However, high temperature and high pressure provide a unique scenario to increase the hydrogen yield. Water–gas shift (WGS) is then applied to increase and adjust the H2/CO molar ratio in the synthesis gas required in the synthesis of Fischer-Tropsch, FT (synthetic fuels production), and also to remove CO from the gas effluents.

Integrated Gasification Combined Cycle (IGCC) technology has clear environmental advantages over traditional coal gasification plants, being considered the cleanest and most efficient method for the production of electricity. ELCOGAS power plant, located in Puertollano (Spain), is an example of this technology, participating in two projects for generation, capture and storage of CO2 and H2 via WGSR and for obtaining "green diesel" vía FT. Thus, this chapter addresses the exploitation of an industrial coal-derived

* E-mail: AnaRaquel.Osa@uclm.es.

syngas feed, produced by gasification in an IGCC plant (ELCOGAS), via bench scale optimization of hydrogen (WGS) and synthetic diesel production (FTS) processes.

NOMENCLATURE

BTL	Biomass to Liquids
CNT	Carbon Nano Tubes
GTL	Gas to Liquids
CTL	Coal to Liquids
IGCC	Integrated Gasification Combined Cycle
WGS(R)	Water–gas shift (reaction)
FT(S)	Fischer-Tropsch (synthesis)
GHSV	Gas Hourly Space Velocity, h^{-1} for WGS and $Ncm^3 \cdot g^{-1} \cdot h^{-1}$ for FT(S)
ΔH^o	reaction enthalphy, $kJ \cdot mol^{-1}$
CCS	CO_2 capture and storage
XRD	X-Ray Diffraction
TPR	Temperature Program Reduction
TPD	Temperature Program Desorption
BET	Brunauer, Emmett and Teller
PEM	Proton exchange membrane
α	Growing chain probability

Superscribe and Subscribe

298	index for temperature of 298K

INTRODUCTION

About 80% of world energy demand comes from fossil fuels use, mainly due to its high energy density and its conversion into heat, electricity and transport fuels which seems to be more simple and inexpensive compared to other commodities. In recent decades, concerns about global warming and alternative fuels to those from petroleum have increased substantially [1]. In order to deal with this problem, some efforts are required such as reducing energy consumption, increasing energy efficiency and fossil fuel switching by renewable energy.

The use of coal as a transition feedstock for a hydrogen and synthetic fuels economy is likely to be accomplished via gasification, i.e. the conversion of solid coal into a gaseous mixture of CO, CO_2, H_2O, and H_2. Coal gasification is a well established mature technology to produce a syngas (CO and H_2) capable of being compatible with biomass and wastes as feedstock [2]. Co-gasification of biomass with coal, could contribute to reduce the fossil fuels dependency and CO_2 emissions due to additional advantages related with the low ash and sulphur content of biomass [3]. Hence, biomass is found to be an attractive alternative to produce renewable energy such as liquid fuels and hydrogen for transport [3-5].

Hydrogen is expected to play a significant role in the future energy system as a new energy carrier. The utilization of hydrogen offers significant advantages including reduction of greenhouse gas emissions, improved security of energy supply and economic competitiveness. A key aspect of hydrogen economy is the development and deployment of hydrogen production technologies whilst satisfying the sustainability criteria (competitiveness, environmental protection, etc.). It can be produced from various feedstocks, such as natural gas, oil derived products, coal and water. At present, hydrogen is used in chemical and petrochemical sectors but in the future there is hope that it can become a mainstream fuel in transport sector, for example using vehicles powered by proton exchange membrane (PEM) fuel cells.

Currently, oil is the main source for production of both chemicals and liquid fuels. A possible alternative to oil for this purpose are synthetic fuels obtained by GTL (Gas-To-Liquids), CTL (Coal-To-Liquids) and BTL (Biomass-To-Liquids) processes. Synthetic fuel can be defined as the gasoline, kerosene and diesel fuels obtained from coal, natural gas or biomass by thermochemical processes. By extension, it can also be used for other fuels (i.e. methanol, dimethyl ether or butane) and other types of unconventional raw materials (i.e. tar sands or plastic wastes). This term was invented by the German Nobel laureate scientist Friedrich Bergius in 1927. Depending on the raw material (coal, natural gas or biomass) the processes and products are respectively called CTL (Coal-to-Liquids), GTL (Gas-to-Liquids) or BTL (Biomass-to-Liquids).

GTL technology is a reality. Currently two companies have already applied this technology (Sasol in South Africa and Shell in Malaysia), and others are being implemented in the areas where natural gas reserves are important. Oil and petrochemical sector companies such as ExxonMobil, British Petroleum, Syntroleum, Chevron, etc., are devoting significant resources to the development of this process and some of them are under construction phase [6].

A combined cycle plant for the co-production of liquid fuels and electricity from coal [7] is under development in China. In addition, there is continuous production of liquid fuels and chemicals from coal-Sasol plants in South Africa which virtually ensure the sufficiency of fuel market in this country. It has also been the leading exporter of high quality waxes (long chain n-paraffins, C_{24}-C_{30}). Furthermore, some authors are investigating the energy integration issues for co-generation of hydrogen and electricity from coal gasification with carbon capture [8].

It has been highlighted in recent studies [9,10] that the use of BTL technology to obtain synthetic fuels is a promising alternative to conventional fuels from petroleum. Moreover, the European Commission, according to a 2003 directive that regulates the use of renewable transport fuels, established that biofuels should represent 5.75% of total fuels used in transport in 2010 and recommended use 20% in 2020 [11]. Since the estimated costs of producing synthetic fuels from biomass is comparable or even lower than the biodiesel or bioethanol, the development of BTL technology acquires a dimension in all energy schemes in the near future.

Current trends indicate that the future of energy use goes through the use of renewable hydrogen and energy transfer vector [12]. Up-to-date, the most widely used precursor in the production of hydrogen is natural gas.

However, assuming that live technologies of energy use are based on oil and its derivatives, it is necessary a transitional stage to accommodate the manufacture of liquid fuels from coal/biomass mixtures in order to reduce emissions with respect to middle distillates.

Hence, by means of GTL, CTL and BTL technologies, natural gas, coal and biomass, respectively, were transformed into chemicals of industrial interest, including gas (ethylene, propylene, etc.), liquid hydrocarbons, waxes and different oxygenates as shown schematically in Figure 6.1.

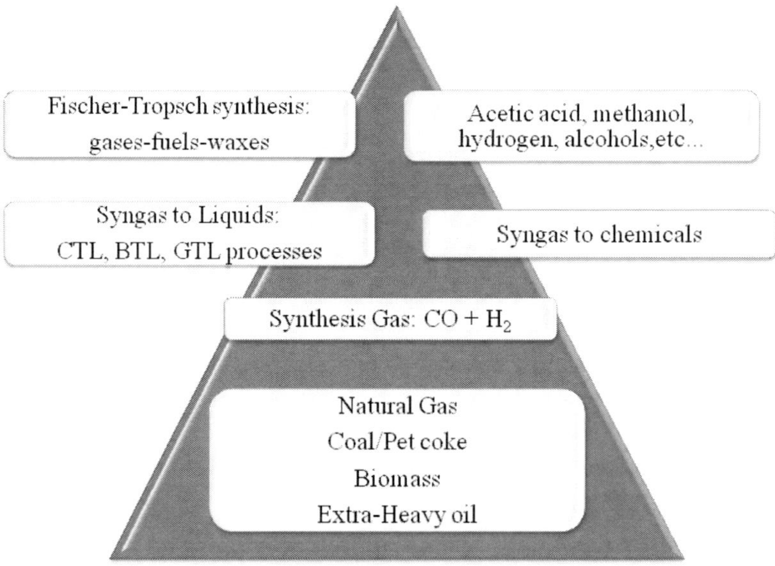

Figure 6.1. From biomass/coal blends to synthetic fuels (diesel) and hydrogen.

As commented before this chapter aims to characterize the overall process, from the gasification of biomass-coal mixtures to hydrogen and synthetic diesel production, and also assesses the feasibility of integrating such technology in the IGCC process. In this sense, the different stages involved:

1. Syngas generation
2. Syngas purification
3. Fischer-Tropsch synthesis (FTS)
4. Improvement of product quality

will be described focusing in the bench-scale optimization of WGS and FT processes (Figure 6.2).

In order to optimize both second and third stages of the overall process, the same bench scale facility was used with the corresponding and necessary modifications in each case. The set-up was fully automated and computerized and consisted of four physically separated parts: feed gas (also liquids in WGS) - where a synthetic feed stream simulated the effluent composition of ELCOGAS - and mixing supply system, reaction system, separation system and products analysis system. An Inconel fixed bed reactor (17.7 mm internal diameter (ID) and 1000 mm length) was used for both WGS and FTS experiments. The optimization was carried out at 19 bar, 300-500°C, H_2O/CO: 2.4 to 4.7 and gas hourly space velocity (GHSV): 2,757-7,877 h^{-1} for WGS and 20 bar, 210-300°C, H_2/CO: 2 and GHSV: 4,000-12,000 $Ncm^3.g^{-1}.h^{-1}$ for FTS.

Biomass Selection

The first stage of the process consists of the selection of a proper biomass for co-gasification with petroleum coke and coal mixtures. This selection is conditioned by both technical feasibility and economic integration in the Integrated Gasification Combined Cycle (IGCC) power plant of ELCOGAS without needing for additional investments to the plant, especially in regard to the grinding step. For this reason, a set of criteria was established:

a. Moisture content <12%,
b. Cost <150 €·t^{-1},
c. Size <25 mm,
d. Availability (t·$year^{-1}$) and
e. Distance to the ELCOGAS central within a 100-200 km radius.

After preliminary analysis of features, availability and costs, different types of biomass were proposed: almond shells, grape pulp, pruning vines, exhausted olive husk, olive pits and wood chips. However, exhausted olive husk was selected to be the best alternative for the co-gasification process due to its properties (Figure 6.3) [13].

Once the biomass was selected, different co-gasification tests were developed on ELCOGAS by using 6 wt. %, 8 wt. % and 10 wt. % of exhausted olive husk mixed with the common fuel (50 wt. % petcoke-50 wt. % coal), and a long-term test with 2 wt. % of biomass.

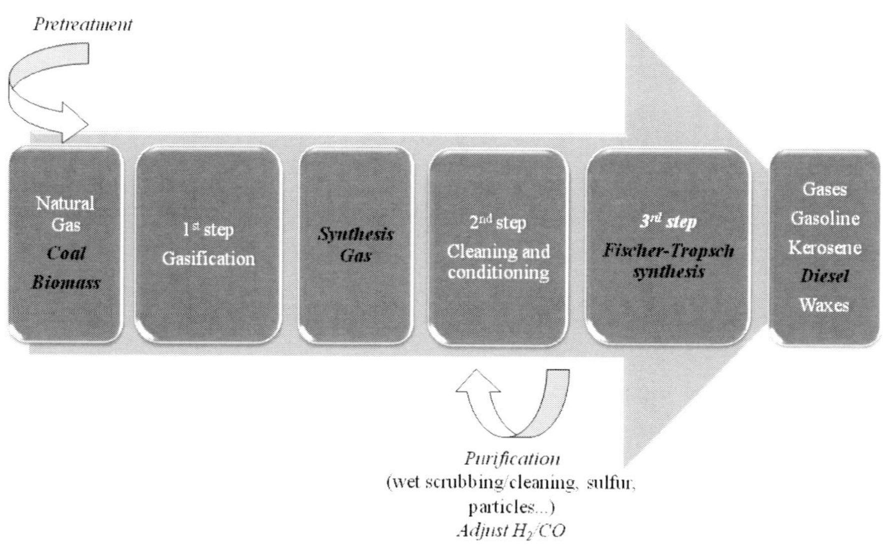

Figure 6.2. Different stages of the overall CTL/BTL process.

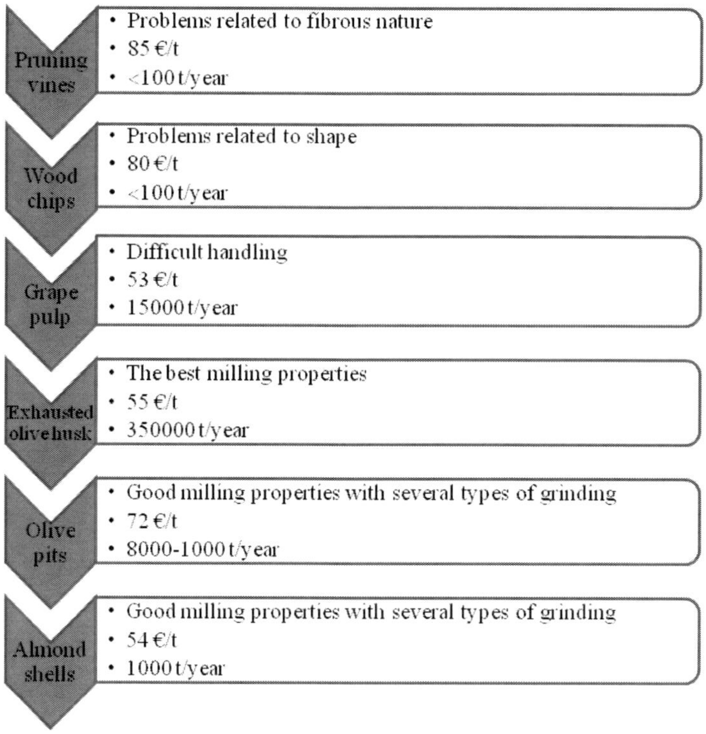

Figure 6.3. Characteristics of different kind of biomass (2008).

The recorded results evidenced that there was no technical impediments when including biomass in the feed gas since it produced a synthesis gas whose composition was within usual operating ranges obtained with the common fuel.

Parameters, such as: humidity, ashes, C, H, N and S content, were kept practically constant to those obtained operating with the common fuel whereas volatile compounds and chlorine content increased within the limits set by an Integrated Environmental Authorization set by ELCOGAS.

Syngas Production

In a first step, coal, biomass or natural gas are converted into synthesis gas, also called syngas, which is defined as a gaseous mixture of carbon monoxide and hydrogen, in which the H_2/CO molar ratio depends on the carbon source.

Over the years, the raw material for syngas production has varied depending on several factors such as the availability, economics and regulation of CO_2 emissions [14]. The most common way to obtain it starts from methane, but it can also be easily obtained from ethane, propane or butane.

It is important to note that although it can be used as fuel, it is mainly used in the production of transportation fuels and other chemicals.

Steam Methane Reforming

Nowadays, most hydrogen is produced through the process of steam reforming of methane obtained from natural gas or byproducts from petroleum refining. In this process, methane reacts with steam at pressures between 3 and 25 atmospheres, in the presence of a Ni based catalyst that operates in a temperature range of 800 - 900°C, to produce hydrogen, carbon monoxide and a relatively small amount of carbon dioxide.

$$CH_4 + H_2O \rightleftarrows CO + 3H_2 \qquad \Delta H^o_{298} = 206.3\ kJ \cdot mol^{-1} \qquad (6.1)$$

Despite being the preferred option, this route of synthesis gas production has some disadvantages. The steam methane reforming reaction is highly endothermic, i.e. it requires large amounts of energy, and the reaction rate is determined by the rate of heat from the outside of the tubes where the catalytic reaction occurs. In addition, the H_2/CO molar ratio obtained (close to 3) is too high for using directly into the conventional reactor for Fischer-Tropsch synthesis.

Partial Oxidation of Methane

The partial oxidation of methane is an exothermic catalytic process where a mixture of methane and pure oxygen, with a CH_4/O_2 molar ratio of 2, reacts on a metal catalyst (Ni or Rh) in the temperature range corresponding to 800 - 900°C [15,16]. This reaction can be carried out via purely thermal, with absence of catalyst, although required temperatures are much higher (1200 - 1500°C):

$$CH_4 + 1/2\ O_2 \rightleftarrows CO + 2\ H_2 \qquad \Delta H^o_{298} = -35.6\ kJ \cdot mol^{-1} \qquad (6.2)$$

Autothermal Reforming

Since neither methane reforming nor partial oxidation produce a synthesis gas with an ideal composition (near 2.15) for industrial application in the synthesis of hydrocarbons according to the Fischer-Tropsch reaction, a possible solution might be to combine the two processes in the autothermal reforming.

Thus, a synthesis gas of adjusted composition would be generated by mixing the effluent flows generated by each of the reactors. At present, it seems the best alternative for the industrial scale production of synthesis gas [17].

Dry Reforming

Methane reforming with carbon dioxide process, known as *dry reforming*, is a possibility for the production of synthesis gas rich in CO ($H_2/CO = 1$), which can be recommended for the production of high molecular weight hydrocarbons [18]. The reaction is, as steam reforming, a strongly endothermic process:

$$CH_4 + CO_2 \rightleftarrows 2\ CO + 2\ H_2 \quad \Delta H^{o}_{298} = 247.3\ \text{kJ·mol}^{-1} \tag{6.3}$$

This process can also be used to reduce CO_2 and CH_4 emissions, two gases that contribute to the greenhouse effect [19,20], since CO is used as a reactant in another process (carbonylation of organic compounds). However, the dry reforming provides a H_2/CO molar ratio too low to feed directly into Fischer-Tropsch synthesis.

Tri-Reforming

The tri-reforming is defined as a combination in a single reactor of dry reforming process, endothermic steam reforming and exothermic partial oxidation processes. In these terms, CO_2, H_2O and O_2 of combustion gases from fossil fuels-based power plants or gasification of solid wastes can be used as reagents for the tri-reforming of natural gas in order to produce synthesis gas. This process was first proposed by the group of professor Chunshen Song at the University of Pennsylvania [21] and, currently, it is a promising alternative to conventional reforming since it provides the following advantages: the possibility of generate a synthesis gas with appropriate H_2/CO molar ratios for the production of methanol and for Fischer-Tropsch process, mitigate the deactivation of the catalysts (based on Ni) [21-24] by deposition of coke and increase energy efficiency of the process.

Gasification

After selecting the proper biomass, the first stage to obtain "synthetic diesel" is the transformation of solid fuel (coal-biomass mixtures in this case) into synthesis gas by the gasification process. Although, firstly, this option is less desirable environmentally, economically it can be profitable in areas where there are abundant reserves of coal, such as ELCOGAS IGCC in Puertollano (Spain) or the three Sasol plants operating in South Africa [25]. Coal gasification is a well-proven technology dating back to the 18th century, although its uses have evolved significantly since then. Interest in coal gasification has wavered in the U.S. during times when the price and availability of competing fuel sources-oil and natural gas-were low. However, recent advancements in gasification technology, increasing costs of oil and gas, growing concerns about energy security, and a heightened awareness about climate change, have all led to a renewed interest in coal gasification for electric power generation in the U.S. and many other countries. This process that has been historically considered as a "dirty" technology is now emerging as a versatile process (ideal to enhance waste and by-products of other processes), allowing at the same time the development of energy efficient processes including co-production of hydrogen and electricity in addition to the capture of the emitted CO_2, so it is environmentally very attractive [26,27]. IGCC is an innovative technology that combines modern coal gasification with a gas turbine and a steam turbine to produce electric power. It is one of the most promising technologies available today to reduce the environmental impacts associated with the use of coal for electricity production. Nowadays there are just four integrated gasification combined cycle (IGCC) plants running on coal worldwide: one in Puertollano, Spain, one in Buggenum, Netherlands, one near Terre Haute, Indiana, and one in Polk County, Florida. So while the technology is not new, the experience with commercial-scale IGCC plants is limited and there is still much to learn, particularly about the economics of operating a commercial-scale plant. Coal-fueled IGCC technology offers a number of potential benefits over conventional pulverized coal plants. Depending on the final configuration of the IGCC plant, these can include:

a. *Higher efficiency.* The use of two turbines (a gas turbine and a steam turbine) leads to higher system efficiencies.
b. *Lower emissions.* The gasification process enables improved removal of naturally-occurring pollutants in coal, such as sulfur and mercury, resulting in lower emissions than conventional coal based power plants.
c. *Carbon sequestration potential.* The IGCC process makes it easier to capture carbon dioxide for carbon sequestration.
d. *Marketable byproducts.* The by-products associated with the gasification and gas clean-up process may have commercial value in nearby industries.
e. *Hydrogen as an alternative fuel source.* Hydrogen is gaining popularity as a potential clean-burning fuel source of the future for vehicles and other industries. The ability to produce hydrogen from coal for such future applications could prove to be an important benefit of IGCC technology.

Hydrogen co-production scheme requires gasification reactors with high hydrogen production efficiency, i.e., the ability to produce a syngas that consists mainly of hydrogen and carbon monoxide, with minimal amounts of hydrocarbons, tars, naphtha, ammonia, etc. The nitrogen and argon content in the syngas need also to be minimized to avoid costs and efficiency penalties for hydrogen purification step (using a Pressure swing adsorption, PSA, unit). Following the same idea, air-blown gasifiers are considered to be unsuitable for hydrogen and electricity co-production with carbon capture, mainly because of the nitrogen dilution of the syngas (this has a significant negative influence on the overall plant thermal balance and on the hydrogen purification step). Generally, the gasification reactors can be grouped into three following main categories: moving-bed gasifiers (sometimes also called fixed-bed gasifiers), fluidized-bed gasifiers and entrained-flow gasifiers [28,29]. It is not the principal aim of present section to review the characteristics of these gasification reactors and the criteria used to analyze the suitability of various gasification processes for hydrogen and electricity co-production with CO2 capture and storage (CCS), since a complete description of these criteria can be found in the literature [29,30]. Just to mention the most important criteria used when analyzing the gasification reactors for IGCC with CCS plant concept: oxidant (oxygen) purity, gasifier feeding system (dry feed vs. slurry feed), gasifier reliability and experience, cold gas efficiency (CGE), carbon conversion efficiency, syngas cooling options (gas quench vs. water quench), hydrogen production potential, downstream gas clean up issues (removing efficiently ash, hydrochloric acid, ammonia, tars, phenols, etc.), implication of gasifier reactor operating pressure on acid gas removal system, capital cost, etc. Considering the above criteria, it seems that entrained-flow gasifiers are the most promising reactors when selecting a gasifier for hydrogen and electricity coproduction with

CCS. That is the reason why among the existing technologies, ELCOGAS uses a high pressure entrained-flow gasifier. As commented before, the synthesis gas composition resulting from the co-gasification (coal, coke and biomass) stood within normal operating ranges with the usual fuel. In reference to emissions, dioxins and furans were kept similar to those obtained operating without biomass, despite the higher chlorine content of exhausted olive husk and the slight increase in SO_2 emissions within the limits set by the Integrated Environmental Authorization of ELCOGAS. Thus, tests at this stage demonstrated the technical feasibility of co-gasification of up to 10 wt. % of biomass without making changes (no change in basic operating parameters, and emissions in the operation) in the existing central.

Cleaning and Conditioning of Synthesis Gas

Once co-gasification of the ternary mixture (biomass/petcoke/coal) has been done, the next step consists of conditioning the synthesis gas from the gasification process to the feed requirements of the FTS process, which can be summarized in two: (1) the absence of substances that act as poisons for the catalyst and (2) an appropriate molar ratio H_2/CO.

Cleaning

In IGCC ELCOGAS central, the syngas was subjected to a thorough cleaning process to remove ashes, HCl, NH_3, HCN, COS and H_2S achieving a high degree of cleanness (the usual to meet the demands imposed by the turbine gas where it feeds).

The main poisons of the FTS catalysts are sulfur compounds that deactivate them irreversibly. For this reason, and even if there were specific increases of S compounds (H_2S, COS) in the effluent gas of the IGCC plant, a desulfurization reactor could be integrated to the FT process input. This reactor would consist of a packed bed absorber based on zinc oxide (ZnO) where would retain traces of H_2S and COS.

Conditioning

As mentioned previously, the second purpose of this stage is to get the suitable H_2/CO molar ratio, which depends on many factors such as type of gasifier, raw materials, reaction conditions, etc., since it is usually deficient in hydrogen. Hence, reforming of methane or Water Gas Shift reaction (WGS) (equation 2.4) are often proposed followed by CO_2 capture (CC) in order to increase the amount of H_2 (the commercial specifications of H_2/CO values ranges from 0.85 to 1.0 of the Dow and Union Carbide, respectively to 2.6 required by Synthol) without the corresponding augmentation on the amount of carbon dioxide which is discouraged in the system. The water gas shift reaction has become well known since first reported in 1888. It became one of the most important industrial catalytic reactions in 1915, when the first coal-based ammonia synthesis plant was put into operation [31,32]. The reaction is mildly exothermic and thermodynamically limited at high temperatures:

$$CO + H_2O \rightleftarrows CO_2 + H_2 \qquad \Delta H^o_{298} = -41.2 \text{ kJ.mol}^{-1} \qquad (2.4)$$

The proportion of H_2/CO in synthetic gas obtained in the IGCC ELCOGAS using biomass/common fuel mixtures is about 0.40. Therefore it is necessary to install a WGS stage to increase this ratio up to 2 [33].

WGS reaction can be catalyzed by many materials but only two classes of catalysts are used almost exclusively in industry: iron oxide-based (Fe_3O_4-Cr_2O_3 at ~500°C) and copper oxide-based (Cu-ZnO at ~200°C) catalysts [34-37]. These catalysts work extremely well in industrial application for natural gas-derived syngases where the CO concentrations are in the range 5–10%. However, in the case of coal-derived syngases where CO concentration is relatively high (40–60%), a considerably higher degree of shifting is required [38]. Moreover, the total amount of sulfur compounds (generally H_2S and carbonyl sulfide (COS)) can be significant depending upon the sulfur content present in the feed coal. In this sense, as both types are highly sensitive towards sulfur contamination of the feed, a new class of CoMo sulfur-resistant WGS based catalysts, similar to those already applied in the hydro-desulfurization (HDS) process, were proposed [32,37,39-41]. Besides the noble metals, only a limited number of elements have been used for sulfide catalyst such as Mo and W as the primary species in composing active sulfide catalyst and Fe, Ni or Co as the indispensable promoters. WGS reaction over Co-Mo catalysts has been extensively studied at laboratory conditions; however there are few papers [42] where industrial feed and bench scale set-up are used. For this reason, the effects of operating conditions on the high pressure WGS reaction as well as the catalyst pretreatment were investigated. A bench scale set-up, using a sulfur-resistant commercial catalyst and an industrial coal-derived syngas feed (from IGCC-ELCOGAS plant, Puertollano, Spain), were considered.

The WGS reaction optimization for hydrogen production was carried out at 19 bars in a bench scale fixed bed reactor constructed from Inconel (1 m in length and 17.7 mm internal diameter), operating in a temperature range between 350 and 500°C [43]. Reaction was studied further by changing the values of H_2O/CO molar ratio (from 2.4 to 4.7) and space velocity (GHSV) (from 2757 - 7877 h^{-1}). Catalytic system was selected as a sulfur-resistant based on CoMo hydrogenation catalyst supplied by ELCOGAS. Tests at this stage would make possible to verify the technical feasibility of hydrogen production and select the most suitable reaction conditions for the WGS reaction. Figure 6.4 illustrates an example of the evolution of catalytic activity with reaction temperature for different molar ratios H_2O/CO using a commercial non-sulfide CoMo catalyst and a sulfide one. The reaction system was demonstrated to be suitable for operation at elevated temperatures (up to 500°C) and pressures (19 - 20 bar) for a wide range of GHSV, involving corrosive gases. Catalytic WGS showed a strong dependence on both the reaction temperature and space velocity but seemed not to be influenced by steam-to-CO molar ratio. Activation with sulfiding agents improved catalytic results since no detectable secondary reactions were observed in the catalytic reaction tests being selectivity to H_2 close to 100% in all the experiments. However, it is important to note that despite non-sulfide CoMo catalyst have been usually reported to be non active for WGS reaction [44], this catalyst provided surprisingly similar catalytic conversion to that exposed by the sulfide one, indicating that the commercial CoMo catalyst was active for a wide range of feed compositions, with and without sulfur poison, in the industrial range of conditions studied.

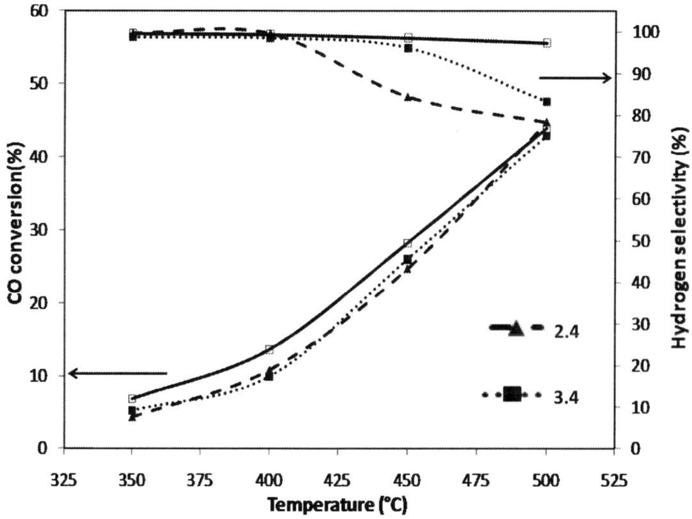

a)

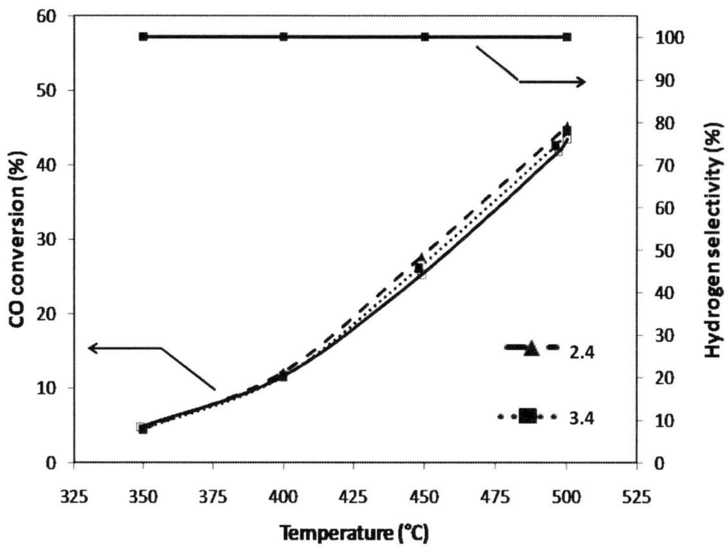

b)

Note: The data were obtained at GHSV of 7877 h-1 and under license agreement with Elsevier [43].

Figure 6.4. Influence of reaction temperature on CO conversion of (a) non-sulfide catalyst and (b) sulfide catalyst at different H_2O/CO molar ratios.

Furthermore, once the production of hydrogen was optimized and including the WGS unit after the gasification process, the H_2/CO molar ratio was demonstrated to be close to 1.8 [45], which allows the required feed (H_2/CO close to 2) for FTS to work properly.

Synthetic Diesel Production

The third stage consists of the conversion of the pretreated synthesis gas into liquid fuels (synthetic diesel) through the Fischer-Tropsch synthesis since it has been recognized as an important technology in the production of liquid fuels and chemicals from synthesis gas derived from coal (CTL), natural gas (GTL) or other materials containing carbon (BTL). Currently, there are two commercial processes: the Shell Middle Distillate Synthesis (SMDS) operating with a fixed tubular reactor and the Sasol Slurry Phase Distillate (SPD).

Synthetic Diesel or Green Diesel

Synthetic diesel or Green diesel, also known as renewable diesel, is defined as a form of diesel derived from renewable raw materials by transforming biomass into liquid fuels (Biomass to Liquid process, BTL). It is important to note that in terms of the raw material it could be classified as biodiesel, however, taking into account the processing technology and the chemical formula, green diesel (gasification) and biodiesel (transesterification) are different products.

Fischer-Tropsch Synthesis

The synthesis of hydrocarbons from synthesis gas (equation 6.5), known as Fischer-Tropsch synthesis (FT), is after more than 70 years of operation [46,47] a very attractive route for the production of clean fuels and as a source of chemicals and, therefore, is the most important step in CTL and BTL processes.

$$CO + 2H_2 \rightarrow [\text{-}CH_2\text{-}] + H_2O \qquad \Delta H_{500} = -165.0\ kJ{\cdot}mol^{-1} \qquad (6.5)$$

The Fischer-Tropsch synthesis is a complex network of primary and secondary reactions and can be compared to a polymerization reaction in which the surface monomer $-CH_2-$ is formed from carbon monoxide. Based on the monomer $-CH_2-$, alkanes, alkenes and oxygenated hydrocarbons were formed. The formation of the $-CH_2-$ group, methane and the Boudouard reaction (the decomposition of CO to elementary carbon and carbon monoxide) can be considered as primary reaction steps; all others are secondary ones. The relative velocity of each reaction step depends strongly on catalyst type and chosen reaction conditions. The Water Gas-Shift reaction, a negligible pathway on cobalt catalysts, increases in importance for secondary reactions on iron based catalysts. Thus, the FT process results in a complex mixture of straight and branched chain, and oxygenated products (alcohols, aldehydes and esters). The majority of them are linear paraffins and α-olefins in the range of high quality gasoline and diesel characterized by no heteroatoms (S, N), no aromatic structures and a high cetane number.

Since the FT synthesis is not selective to a single reaction product or a specific range of number of carbons, being methane the only exception, the objective of this section was the optimization of both the operating conditions and the catalytic system in order to obtain a hydrocarbon product distribution centered on C_{13} - C_{18} fraction, typical of diesel.

Iron-, nickel-, ruthenium- and cobalt- based materials are the classic Fischer-Tropsch catalysts, but only iron and cobalt appeared to be economically feasible at an industrial scale and, depending on each active metal, a different product distribution is obtained. On Fe-catalysts, low-boiling hydrocarbons and alcohols are the predominant products besides a great tendency towards the water-gas shift reaction. High-boiling hydrocarbons can be obtained on nickel catalysts in addition to oxygenated carbon products. On ruthenium, the formation of polyethylene was found by Pilcher [48]. In the present chapter cobalt catalysts were the subject of interest. The advantages of cobalt catalysts are the absence of the WGS reaction, the favored formation of high-boiling, unbranched hydrocarbons (waxes), barley oxygenated by-products and, finally, the good availability and the relatively low price in order to scale-up to industrial scale [49,50].

It is generally accepted that several parameters affect the performance of cobalt catalysts for FTS, i.e., the catalyst support, the nature and amount of added promoters as well as the cobalt dispersion. The literature data concerning these subtle influences are resumed below. Moreover, between the existing modes, the research was carried out under 20 bars, using an H_2/CO molar ratio of 2 (a WGS unit was supposed to be included as commented before), operating in a temperature range between 210 and 300°C, as has been shown to favor the primary diesel court that provides a good diesel performance [51,52].

Influence of Promoters over Supported Co-Based Catalysts

The goal of promotion is to enhance cobalt FT catalysts by the addition of small amounts of noble metals and metal oxides. Numerous studies have shown that introduction of a noble metal (Ru, Rh, Pt and Pd) has a strong impact on the structure and dispersion of cobalt species, FT reaction rates and selectivities. The promoting metal is typically introduced via co-impregnation or subsequent impregnation. Analysis of the literature data [53] suggests that introduction of noble metals could result in the following phenomena:

a. Easing cobalt reduction;
b. Enhancement of cobalt dispersion;
c. Better resistance to deactivation;
d. Formation of metallic alloys;
e. High concentration of hydrogen activation site;
f. Modification of intrinsic activity of surface sites.

However, due to high prices, their industrial application is restricted. Thus, promotion with less expensive metal oxides has also been one of the methods to improve activity and hydrocarbon selectivity of FT catalysts. Among the oxide promoters, ZrO_2, La_2O_3, MnO and CeO_2 have been most often employed. Addition of oxide promoters could [53]:

1. Modify the catalyst texture and porosity;
2. Reduce formation of hardly reducible cobalt mixed oxides;
3. Increase cobalt dispersion, reducibility and fraction of different cobalt metal crystalline phases;
4. Enhance mechanical and chemical attrition resistance of cobalt FT catalysts;
5. Improve chemical stability of the support.

More information about the effect of promotion on cobalt based catalysts is available in recent reviews [53,54]. There are numerous papers studying the influence of addition of alkali promoters on FTS over iron-based catalysts [55,56]. However, literature related to alkali promoter effect on cobalt catalysts is scarce [57] and there are only some separately studies on the magnesium [58] or calcium [59] promoter effect. It is believed that alkali addition to cobalt leads to an enhanced adsorption of carbon monoxide [60] and that the growth probability (α) of Fischer-Tropsch synthesis increases up to $\alpha = 0.87$ in the presence of alkali promoter [61].

As there was not any study of comparison between alkali-earth and alkali metal promoter on FTS activity and selectivity over cobalt-based catalyst, the aim of this section was to study the effects of the addition of these promoters over a common cobalt-alumina based catalyst on the bench scale FTS activity and product distribution.

Characterization results allow comparing between alkali-earth and alkali metal promoter effect on FTS activity and selectivity over cobalt-based catalyst. X-ray diffraction (XRD) patterns, H_2-chemisorption and Temperature Programmed Reduction (TPR) results showed that both alkali and alkali-earth oxide suppresses the interaction between cobalt oxide and support resulting in an increased cobalt reducibility and more cobalt active sites. Moreover, a positive correlation between basicity and particle size was observed with the exception of K promoted sample. An example of TPR profiles is depicted in Figure 6.5.

Furthermore, CO conversion (Figure 6.6) and C_5^+ selectivity (Table 6.1) were found to be also influenced by the addition of promoters. It was found that the activity increased by addition of promoters, attributed to the enhanced dispersion and reducibility that provided abundant active sites.

The addition of alkali metal caused an increase in 1-alkene selectivity (not shown here). Thus, the addition of alkali metal promoter was suggested to suppress the hydrogenation activity better than alkali-earth metal promoters. Moreover, the addition of Ca to the cobalt based catalyst greatly improved the diesel selectivity. However Ca promoted catalyst yielded a higher amount of methane at 242°C than the Na sample due to hydrogen spillover (not shown here).

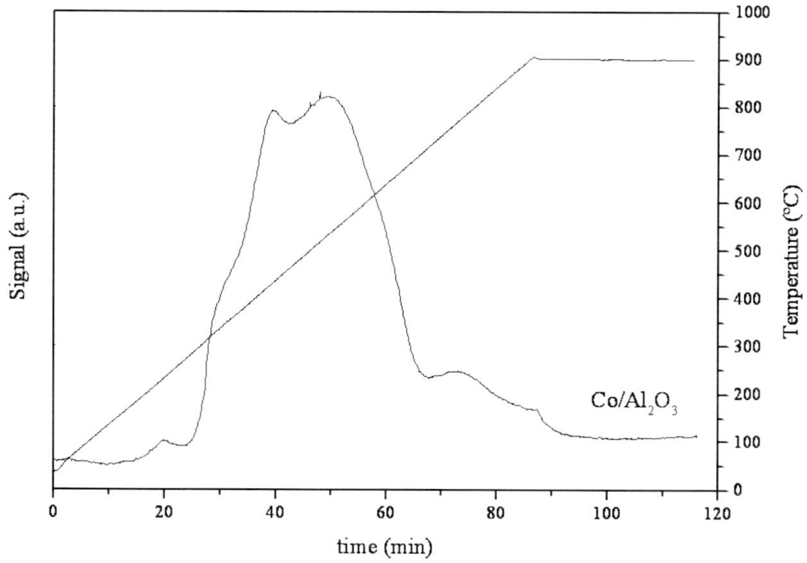

Figure 6.5. (Continued).

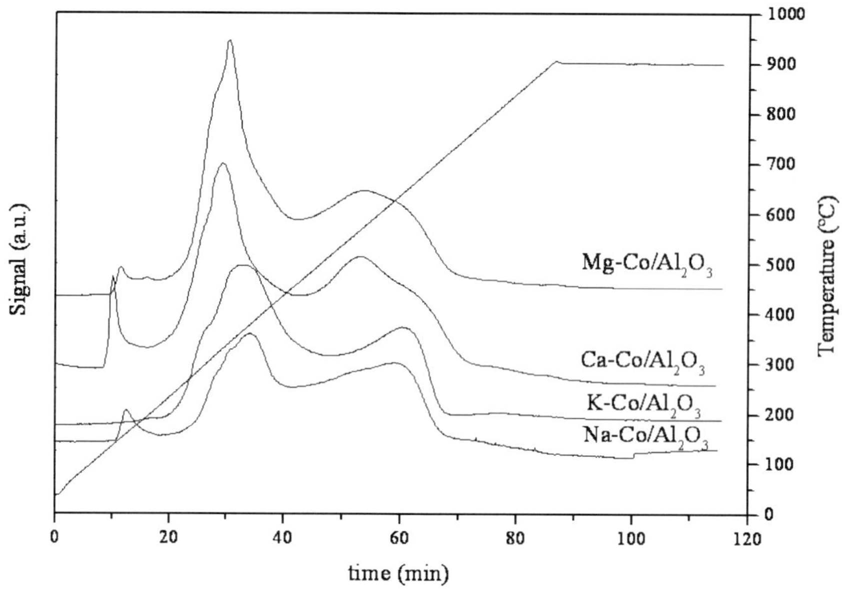

Note: The figures are obtained under the license agreement with Elsevier [62].

Figure 6.5. TPR profiles of alkali-promoted and unpromoted Co/Al_2O_3 catalysts.

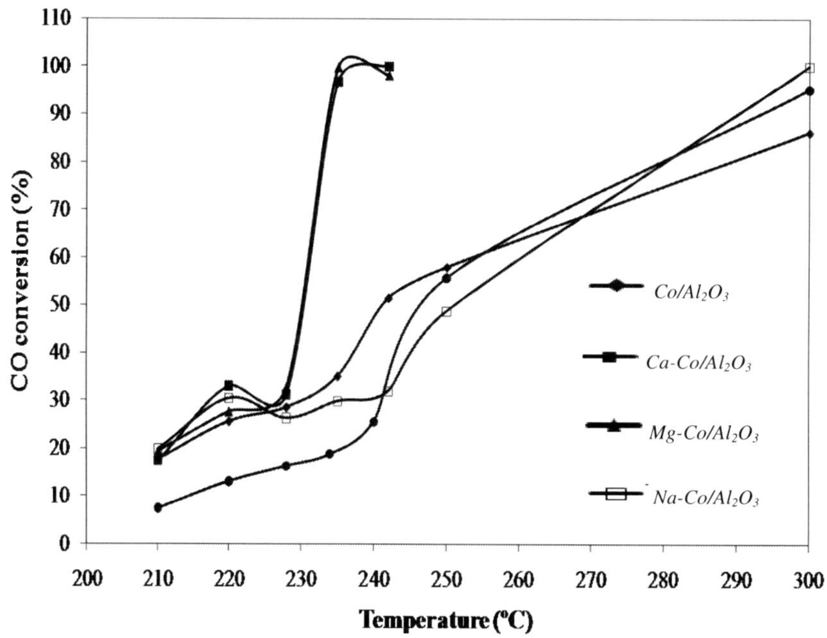

Note: The figure is obtained under the license agreement with Elsevier [62].

Figure 6.6. Catalytic activity of alkali-promoted and unpromoted catalysts.

Table 6.1. C5+ hydrocarbon product distribution

Sample	Gasoline (C_7-C_{10}) (%)	Kerosine (C_{11}-C_{14}) (%)	Diesel (C_{15}-C_{18}) (%)	Lubricants (C_{19}-C_{20}) (%)	Waxes (C_{20}^+) (%)	Total Diesel Selectivity (%)
Co/Al_2O_3	37.5	47.0	6.6	0.7	10.8	6.2
Ca-Co/Al_2O_3	0.3	10.9	49.7	16.8	22.4	19.5
Mg-Co/Al_2O_3	14.8	66.8	15.2	1.2	2.1	4.6
K-Co/Al_2O_3	11.1	35.1	26.2	8.8	18.8	22.5
Na-Co/Al_2O_3	13.8	52.8	21.5	4.6	7.2	20.4

From these results, it can be concluded that the CO conversion improved in presence of alkali/alkali-earth promoter due to an enhancement of dispersion with consequently increase of active sites at the same time reducibility is favored, with exception of K-Co/Al_2O_3. Moreover, it can be concluded that the addition of Ca to the cobalt based catalyst greatly improved diesel selectivity [62].

Influence of Reaction Conditions

Once the catalyst has been properly selected, the synthesis of high molecular weight hydrocarbons is the main objective of FT synthesis [62]. It is well known that reaction conditions strongly influence the performance of the catalyst and the selectivity of the products [63,64]. Several studies have been conducted to elucidate the effects of reaction conditions on the performance of both iron and cobalt catalysts [64,65].

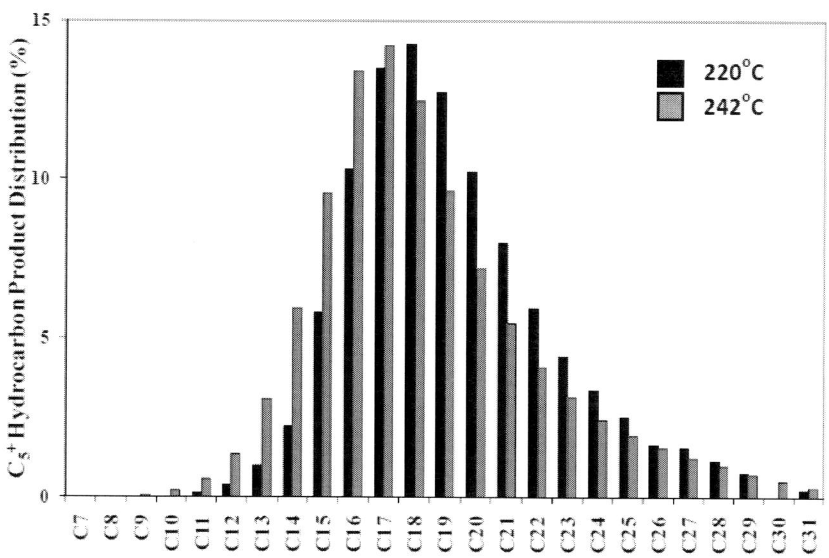

Note: The figure is obtained under the license agreement with Elsevier [71].

Figure 6.7. Influence of reaction temperature on C_5^+ hydrocarbon distribution over Co-Ca/Al_2O_3 catalyst.

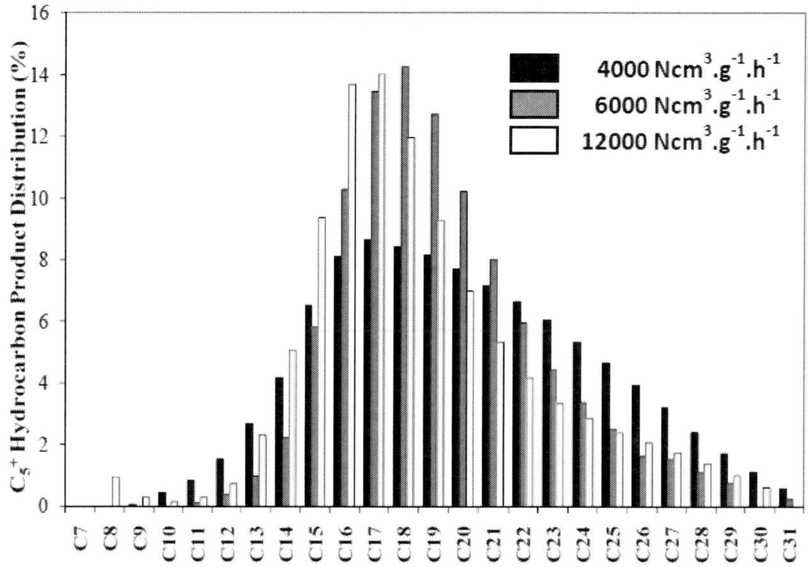

Note: The figure is obtained under the license agreement with Elsevier [71].

Figure 6.8. Influence of GHSV on C_5^+ hydrocarbon distribution over Co-Ca/Al_2O_3 catalyst.

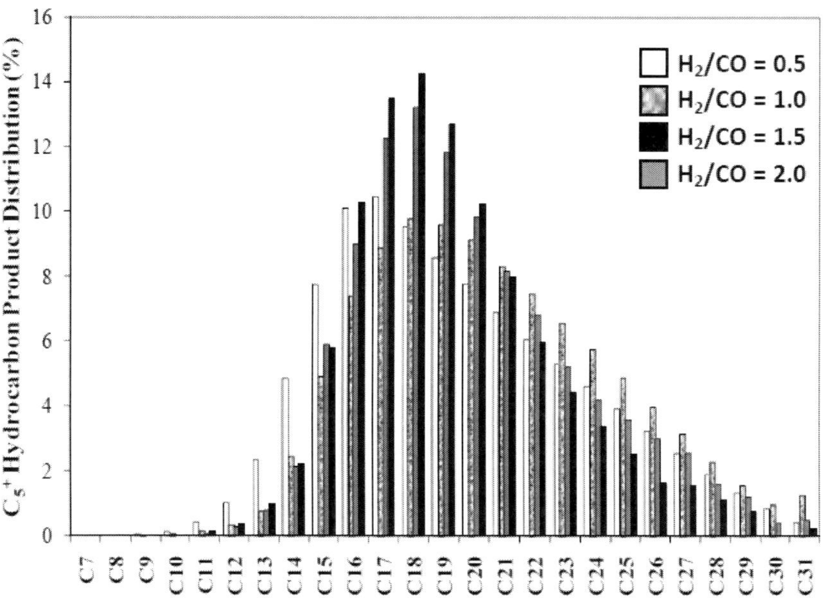

Note: The figure is obtained under the license agreement with Elsevier [71].

Figure 6.9. Influence of H_2/CO molar ratio on C_5^+ hydrocarbon distribution over Co-Ca/Al_2O_3 catalyst.

It was found that the selectivity to low molecular hydrocarbons is basically improved with increasing both reaction temperature and inlet H_2/CO ratio, while the selectivity to high molecular hydrocarbons is favored by decreasing total pressure [66]. The addition of a strongly basic alkaline promoter has been demonstrated to be required to shift the selectivity towards the more useful long-chain hydrocarbons [67].

However, the literature related to the study of the influence of the reaction conditions using alkali-earth promoted-cobalt materials as FTS catalysts appears to be scarce. Thus, several studies should be conducted to elucidate the effects of reaction conditions on the FTS performance of calcium-promoted cobalt based catalyst and the differences with the unpromoted one. Experimental tests showed that despite being an exothermic reaction, the conversion of CO and the selectivity to light hydrocarbons, not desirable in this case, were favored by increasing reaction temperature. Figure 6.7 presents an example of CO conversion and product selectivity versus temperature. Distributions of liquid products of more than five carbon atoms (C_5^+) obtained, were in agreement with those published by other authors [68-70], suggesting that increasing the reaction temperature would favor the displacement of the distribution of oil products to lower molecular weight. It is important to note that the variation of the reaction conditions seemed to influence product distribution in a lesser extent than unpromoted catalyst.

However, a displacement of hydrocarbon distribution to higher molecular weight with decreasing space velocity (Figure 6.8) and temperature was also observed. Moreover, the addition of calcium to the cobalt based catalyst was found to greatly maintain selectivity to C_5^+ for a wide range of H_2/CO molar ratios (Figure 6.9).

From these results it was considered necessary to adopt a compromise between conversion and selectivity in order to obtain a majority of diesel distribution. Hence, choosing 2.0 as a suitable inlet H_2/CO ratio, 6000–12000 $Ncm^3.g^{-1}.h^{-1}$ as the space velocity and 242°C as reaction temperature can result in optimal catalytic performance of the catalyst, achieving an optimal diesel distribution [71].

Influence of Support

Catalytic support has significant and multiple roles in the design and catalytic performance of cobalt FT catalysts since the main function of catalyst support is to disperse cobalt and to produce stable cobalt metal particles after catalyst reduction and activation. Efficient control of metal dispersion is often a challenge in the design of efficient metal supported catalysts. Porous structure of the support could control the sizes of supported cobalt particles. A “negative” effect of support could be related to the formation of cobalt support mixed compounds (aluminate and silicate) that should be avoided since they do not produce active sites for FT synthesis. In addition, the catalytic support could modify diffusion of reagents and products inside the catalyst grains, capillary condensation of the reaction products in the catalyst pores, heat dissipation and mechanical strength. In FT process, some inorganic supports with high surface area such as silica and alumina have been used to increase the surface area of the cobalt. Also, other supports such as titania or carbon in the form of activated carbon and carbon nanotubes (CNTs) have been investigated in the FTS [72-75]. However, silicon carbide has hardly ever been used for this purpose although it is one of the most advanced ceramic materials for its remarkable chemical and thermomechanical properties [76]. It was found that the support material influenced dramatically the performance of cobalt catalysts, i.e., the carbon monoxide conversion and the chain growth probability (α-value). As reported by Iglesia [77] and Reuel and Bartholomev [78] the catalyst activity over some of them decreased in the following order: Co/TiO_2 > Co/Al_2O_3 > Co/SiO_2 > Co/C > Co/MnO > Co/MgO. In these terms, the aim of this section was to study and compare the influence of different supports, including the common alumina, titania, and the newer sodium-bentonite and silicon carbide, on the bench-scale activity and diesel product selectivity of cobalt-based Fischer-Tropsch catalysts. It was reported that both FTS rate and CO conversion are strongly dependent and proportional to the number of surface reduced active cobalt sites. Catalysts Co/Bentonite and Co/TiO_2 showed a wide single TPR reduction peak which was related to the complete reduction of Co_3O_4 to Co^0. However, multiple reduction peaks were observed for catalysts Co/Al_2O_3 and Co/SiC that were more difficult to reduce (Figure 6.10). Lower reducibility of small cobalt particles in narrow pore supports appears [79] to be a cause of their different catalytic performance. It was also reported that Co particle size could be related to the basic behaviour of catalyst, suggesting a direct proportionality between surface basicity and hydrogenation rate. . However, in agreement with XRD and CO_2-Temperature Programmed Desorption (TPD-CO_2) (basicity) measurements (Table 6.2) no clear correlation between particle size and basic behaviour of catalysts was observed. Thus, textural properties of the support could be responsible for the differences on metal deposition. Particle size (XRD) depended on nature of support and specific area. In particular, the pore characteristics of the support have a significant effect on the Co_3O_4 crystallite size measured after impregnation and calcination [49]. Catalytic results (Figure 6.11) indicated that the activity increased in the order of Co/TiO_2 < Co/Al_2O_3 < Co/Bentonite < Co/SiC at temperatures higher than 235ºC. Moreover, catalyst Co/Bentonite showed good catalytic activity at low temperatures, although this support enhanced the formation of oxygenated compounds. Hydrocarbon distribution also varied with the support showing a positive correlation with pore diameter (Figure 6.12). It is important to note that Co/SiC not only improved CO conversion and FTS rate, but also shifted to higher molecular weight hydrocarbons products distribution than Co/Al_2O_3.

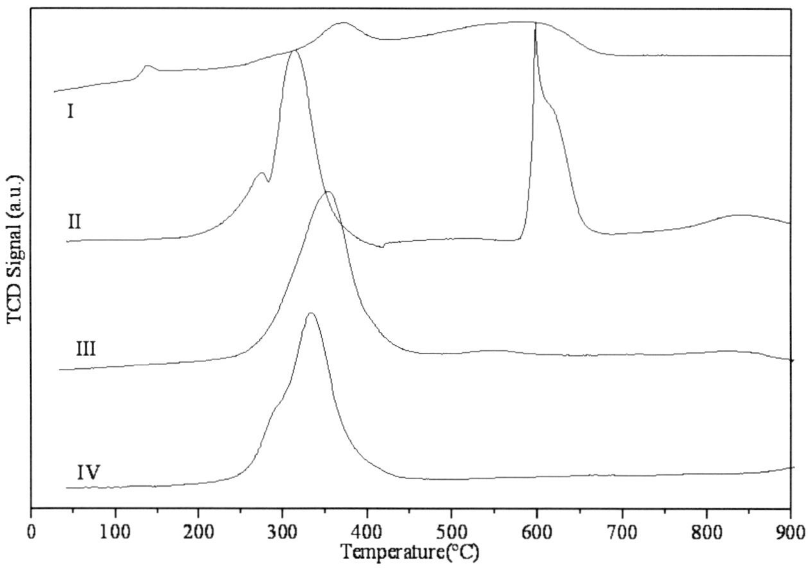

Note: The figure is obtained under the license agreement with Elsevier [80].

Figure 6.10. TPR profiles of (I) Co/Alumina; (II) Co/SiC; (III) Co/Bentonite; (IV) Co/TiO_2.

Table 6.2. Chemical composition and physical properties of FTS catalysts

Catalyst	wt. % Co	BET area ($m^2.g^{-1}$)	D_{pore} (nm)	Total Pore Volume ($cm^3.g^{-1}$)	dCo_3O_4 (nm) (XRD)	Degree of reduction (O_2 pulse)	D (%) (XRD)	Weak basicity ($mmol.g^{-1}$)	Strong basicity ($mmol.g^{-1}$)
Co/Al_2O_3	11	90.1	6.0	0.136	30.7	48.6	4.2	0.165	0.066
Co/Bentonite	13	9.6	19.1	0.046	106.1	40.4	1.2	0.077	0.086
Co/TiO_2	13	9.4	8.6	0.020	85.6	45.6	1.5	0.026	0.016
Co/SiC	10	21.5	7.2	0.039	97.2	79.0	1.3	0.002	0.029

Note: D_{pore} = pore diameter; dCO_3O_4 = cobalt oxide particle size; D = dispersion=96/dCo^0; dCo^0 = metallic cobalt particle size = 0.75· dCO_3O_4.

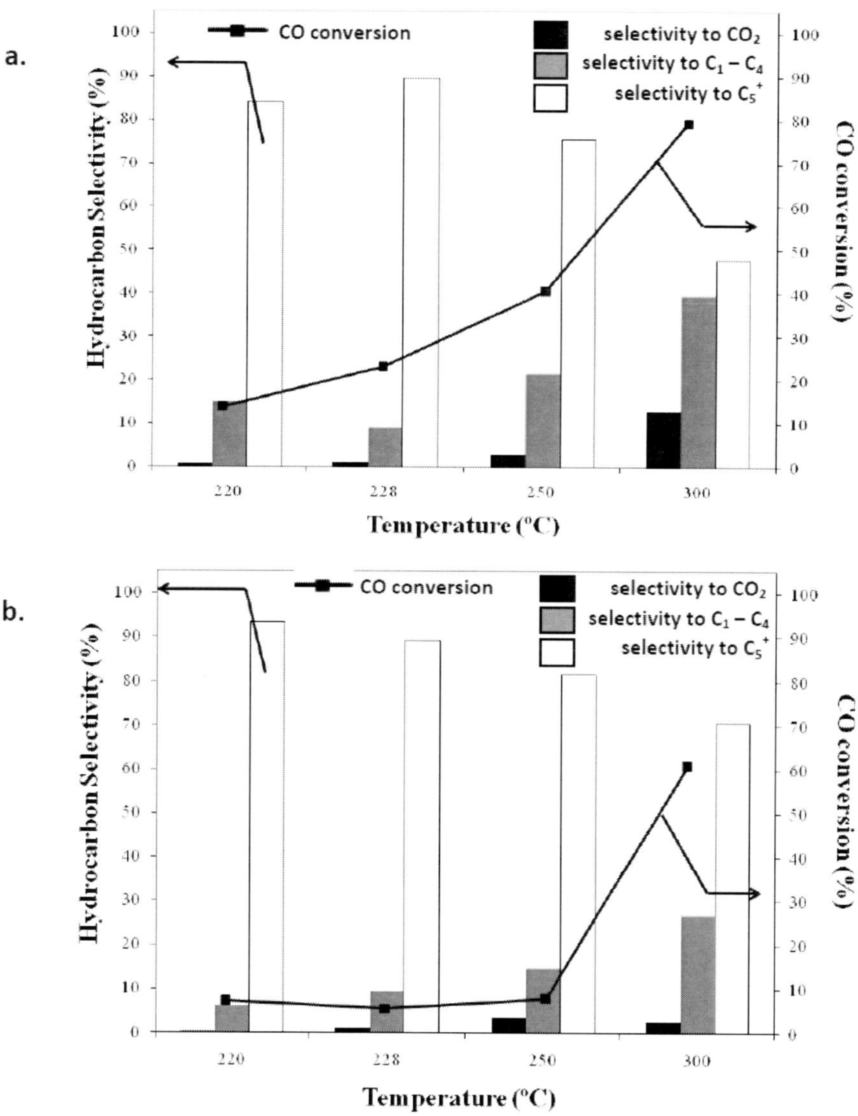

Figure 6.11. (a-b). Influence of supports, (a) Alumina and (b) TiO_2, on FTS catalytic performance.

c.

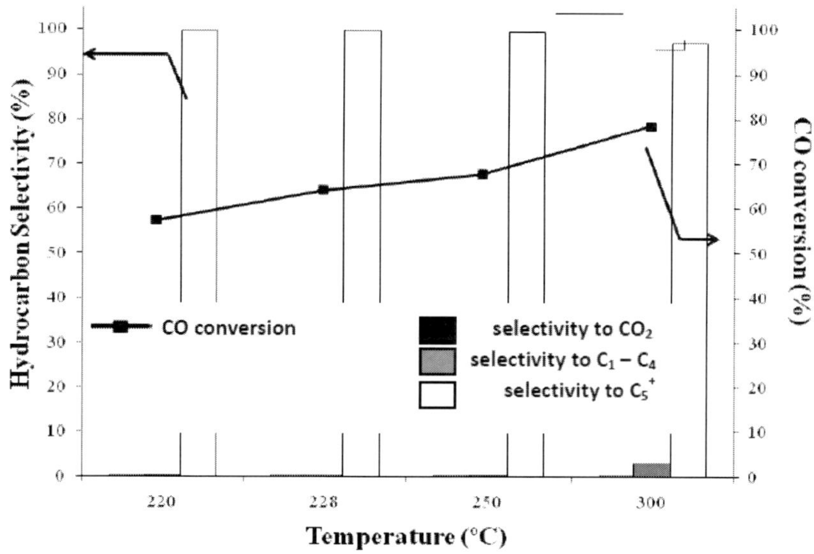

d.

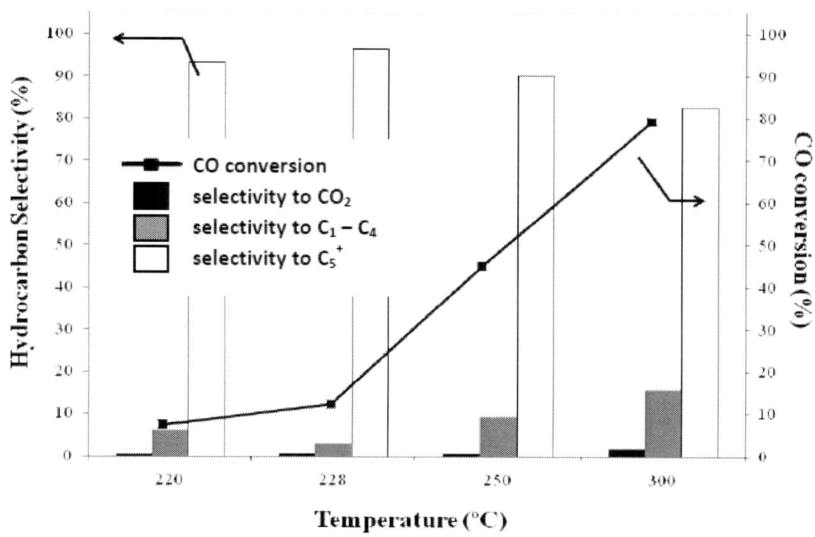

Figure 6.11. (c-d). Influence of supports, (c) Bentonite and (d) SiC, on FTS catalytic performance.

Moreover, alumina, bentonite and titania supports provided higher amounts of C_{21+} whereas silicon carbide showed negligible amounts of these hydrocarbons, resulting a very selective catalyst.

Thus, silicon carbide has been found to be a promising FTS catalyst since it provides better catalytic results toward the diesel fraction than common alumina support [80]. In fact, a more complex study about the influence of preparation method (influence of cobalt precursor, influence of metal loading, influence of basic promoter) of Co/SiC catalyst is now being developed.

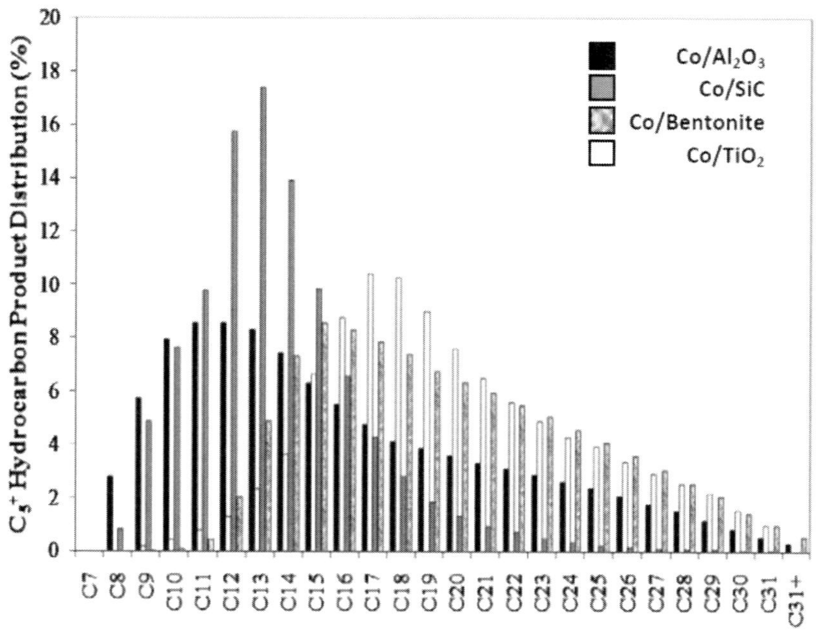

Note: The data were obtained at reaction temperature of 250°C, H_2 to CO ratio of 2, and GHSV of 6000 $Ncm^3.g^{-1}.h^{-1}$. The presented material is under license agreement with Elsevier [80].

Figure 6.12. Influence of support on C_5^+ product distribution.

CONCLUSION

The goal of the present chapter was to deal with the possibility of exploitation of an industrial coal-gasification derived syngas by means of the bench scale H_2 and synthetic diesel production processes.

In this chapter, the use of up to 10% of exhausted olive husk has been demonstrated to be technically feasible as biomass for co-gasification with the common industrial fuel (coke/petcoke) without producing significant changes in the effluent synthesis gas composition. Moreover, the optimization of both bench scale WGS and FTS catalytic processes was accomplished. WGS reaction was carried out operating at high pressure in a wide range of GHSV, temperature and feed composition (with and without sulfur compounds). A commercial sulfur-resistant catalyst and an industrial coal-derived feed were considered. It have been demonstrated the need of a WGS unit for hydrogen production with a selectivity of 100% to increase the H_2/CO molar ratio from 0.4 to 1.8-2, required for the proper FTS development. Finally, the feasibility of synthetic diesel production from synthesis gas from the co-gasification of biomass-coal blends using a cobalt-based catalyst prepared by impregnation was also demonstrated. A series of different Co-based catalysts (different supports, presence or absence of promoters, etc.) was prepared in order to study the effect on the Fischer–Tropsch synthesis activity and selectivity resulting in a typical diesel fraction. Catalytic FTS performance not only showed a strong dependence on reaction conditions but also on basicity, particle size, pore diameter and metal-support interactions. From these results it was considered necessary not only to adopt a compromise between conversion and selectivity, but to optimize the composition of the catalyst (support, promoters) in order to obtain a majority in distribution of diesel.

ACKNOWLEDGMENTS

Financial Supports from the Ministry of Science and Innovation of Spain (The National Plan for Scientific Research, Development and Technological Innovation 2008-2011) (CENIT-PiIBE project) and Regional Government of Castilla-La Mancha (Project PSE 2-2005) and ELCOGAS S.A. are gratefully acknowledged.

AUTHORS' BIOGRAPHIES

Ana Raquel de la Osa is assistant professor of Chemical Engineering at Castilla-La Mancha University in Ciudad Real (Spain). She received the Chemical Engineering degree, focused on synthesis and characterization of Carbon Nanofibers, from the University of Castilla-La Mancha, in 2006. Currently, she is finishing the PhD degree focused on bench scale conventional catalysis (water-gas shift, Fischer-Tropsch) which participates in two National projects with ELCOGAS S.A. company. She joins the Prof. J.H. Bitter group at the University of Utrecht (The Netherlands) during her predoctoral research. She is author of 6 international and 2 national scientific papers and around fourteen communications to congresses. Her current research interest is focused on: conventional catalysis: Water-Gas Shift and Fischer-Tropsch.

Antonio de Lucas Martínez is Professor of Chemical Engineering (since 1990) at Castilla-La Mancha University in Ciudad Real (Spain) although he was also a teacher at Complutense (1974-90), San Pablo CEU (1976-77) and Rey Juan Carlos (1997) universities of Madrid. She received the PhD degree (Chemical Engineering), from the Complutense University of Madrid and PPD from IESE of Navarra University, respectively. He also currently works as an scientific advisor of Caja Rural of Ciudad Real (since 2008). He is author of 224 scientific papers, 6 patents, 18 books and around 242 communications to congresses and 28 conferences. He also directed 29 PhD thesis and 44 investigation proyects (European, National and Regional ones). His current interest is focused on relationship between research and companies, participating in around ten administrative councils and elaborating more than 100 technical reports.

Jose Luis Valverde completed his graduated and PhD studies at the Complutense University of Madrid. His research activities developed as a Full Professor of the University of Castilla-La Mancha have been focused on different subjects related to conventional catalysis, application of the electrochemical promotion of catalysis (EPOC), synthesis and uses of carbon nanostructures (carbon nanotubes, carbon nanofibers and carbon nanospheres) as catalysts, and synthesis of microcapsules containing PCMs. He co-authored several book chapters and more than 145 research papers. He has been the principal investigator of more than 35 research projects with European, National and Regional institutions and private companies.

Amaya Romero is Associate Professor of Chemical Engineering at Castilla-La Mancha University in Ciudad Real (Spain). She received the Chemical Engineering degree and the PhD degree, focused on synthesis and characterization of Pillared Clays, from the University of Castilla-La Mancha, in 2000 and 2003, respectively. She joins the Prof. Krijn de Jong group in the University of Utrecht (The Netherlands) during her postdoctoral research. She is author of 54 scientific papers, 6 book chapters and around seventy communications to congresses. Her current research interest is focused on: conventional catalysis (water-gas shift, Fischer-Tropsch, selective hydrogenation and oxidation reactions); application of the electrochemical promotion of catalysis (EPOC); synthesis and uses of carbon nanostructures (carbon nanotubes, carbon nanofibers and carbon nanospheres) as catalysts in fine chemical processes and in hydrogen storage; synthesis of microcapsules containing PCMs.

Paula Sánchez is Professor of Chemical Engineering (since January 2011) at Castilla-La Mancha University in Ciudad Real (Spain). She received the Chemistry degree and the PhD degree (Chemical Engineering), focused on synthesis and characterization of zeolites, from the University of Castilla-La Mancha, in 1991 and 1997, respectively. She joins the Prof. van Hoff group in the University of Eindhoven (The Netherlands) during her postdoctoral research. She is author of 70 scientific papers, 1 patent, 10 book chapters and around a hundred of communications to congresses. Her current research interest is focused on: conventional catalysis (water-gas shift, Fischer-Tropsch, hydrocarbons and alcohols steam reforming, selective hydrogenation and oxidation reactions); application of the electrochemical promotion of catalysis (EPOC); synthesis and uses of carbon nanostructures (carbon nanotubes, carbon nanofibers and carbon nanospheres) as catalysts in fine chemical processes and in hydrogen storage; synthesis of microcapsules containing PCMs.

References

[1] http://www.ipcc.ch/pdf/assessment-report/ar4/syr/ar4_syr.pdf.

[2] S. Lögdberg. Development of Fischer-Tropsch Catalysts for Gasified Biomass. Licentiate Thesis in Chemical Engineering Stockholm, Sweden 2007.

[3] A. Demirbas. *Prog. Energy Combust. Sci.* 33, 1 (2007).

[4] M.J.A. Tijmensen, A.P.C. Faaij, C.N. Hamelinck and M.R.M. van Hardeveld. *Biomass Bioenerg.* 23, 129 (2002).

[5] G.W. Huber, S. Iborra and A. Corma. *Chem. Rev.* 106, 4044 (2006).

[6] T.H. Fleisch, R.A. Sills and D.M. Briscoe. *J. Nat. Gas Chem.* 11, 1 (2002).

[7] A.P. Steynberg and H.G. Nel. *Fuel.* 83, 765 (2004).

[8] C.C. Cormos. *Int. J. Hydrogen Energy.* 35, 7485 (2010).

[9] C.N. Hamelinck, A.P.C. Faaij, H. den Uil and H. Boerrigter. *Energy.* 29, 1743 (2004).

[10] K.W. Jun, H.S. Roh, K.S. Kim, J.S. Ryu and K.W. Lee. *Appl. Catal. A: Gen.* 259, 221 (2004).

[11] Directiva 2003/30/CE del Parlamento Europeo y del Consejo relativa al fomento del uso de biocarburantes u otros combustibles renovables en el transporte. 2003.

[12] E.F. Sousa-Aguilar, L.G. Appel and C. Mota. *Catal. Today.* 101, 3 (2005).

[13] A. Romero, P. Sánchez, N. Mota, F. García and P. Casero. *Ing. Quim.* 469, 100 (2009).

[14] D.J. Wilhelm, D.R. Simbeck, A.D. Karp and R.L. Dickenson. *Fuel Process. Technol.* 71, 139 (2001).

[15] M.A. Peña, J.P. Gómez and J.L.G. Fierro. *Appl. Catal. A: Gen.* 144, 7 (1996).

[16] M. Fernández-García and J.A. Anderson. Supported metals in Catalysis, Catalytic Science Series 5, Imperial College Press, London (2005).

[17] J.R. Rostrup-Nielsen. *Catal. Today.* 71, 243 (2002).

[18] A.K. Prabhu, R. Radhakrishnan and S.T. Oyama. *Appl. Catal. A: Gen.* 183, 241 (1999).

[19] J. Rostrup-Nielsen. *Appl. Catal. A: Gen.* 255, 3 (2003).

[20] J.H. Edwards. *Catal. Today.* 23, 59 (1995).

[21] C. Song and W. Pan. *Catal. Today.* 98, 463 (2004).

[22] J. Lee, O. Joo, Y. Baek, Y. Yu and K. Jung. *B. Kor. Chem. Soc.* 24, 1623 (2003).

[23] J.S. Kang, D.H. Kim, S.D. Lee, S.I. Hong and D.*J. Moon. Appl. Catal. A.* 332, 153 (2007).

[24] H.T. Jiang, H.Q. Li and H.B. Xu. *Fuel Process. Technol.* 88, 988 (2007).

[25] M.E. Dry. *Catal. Today.* 71, 227 (2002).

[26] J.C. Van Dyk, M.J. Keyser and M. Coertzen. *Int. J. Coal Geol.* 65, 243 (2006).

[27] K. Yamashita and L. Barreto. *Energy.* 30, 2453 (2005).

[28] C. Higman and M. van der Burgt. Gasification; Elsevier Science – Gulf Professional Publishing: Burlington, 2008.

[29] C.C. Cormos. Decarbonizarea combustibililor fosili solizi prin gazeificare; Cluj University Press: Cluj-Napoca, 2008.

[30] C.C. Cormos. *Int. J. Hydrogen Energy.* 34, 6065 (2009).

[31] K. Kochloefl. Handbook of Heterogeneous Catalysis, G. Ertl, Wiley-VCH, Weinheim, (1997).

[32] Y. Lian, H. Wang, Q. Zheng, W. Fang, and Y. Yang. *J. Nat. Gas Chem.* 18, 161 (2009).

[33] S.T. Chaudhari, S. K. Bej, N.N. Bakhshi and A.K. Dalai. *Energy Fuel.* 15, 736 (2001).

[34] L. Lloyd, D.E. Ridler and M.V Twigg. In The Water Gas Shift Reaction in Catalysis Handbook; Twigg, M. V.; 2nd Ed.; Manson Publising Ltd.: Frome, 1996; p. 283.

[35] J.L. Yu, F.J. Tian, L.J. Mckenzie and C.Z. Li. *Process. Saf. Environ.* 84, 125 (2006).

[36] M. Laniecki, M. Malecka-Grycz and F. Domka. *Appl. Catal. A Gen.* 196, 293 (2000).

[37] I.D. González, R.M. Navarro, M.C. Álvarez-Galván, F. Rosa and J.L.G. Fierro. *Catal. Commun.* 9, 1759 (2008).

[38] S.S. Hla, D. Park, G.J. Duffy, J.H. Edwards, D.G. Roberts, A. Ilyushechkin, L.D. Morpeth and T. Nguyen. *Chem. Eng.* J. 146, 148 (2009).

[39] D.S. Newsome. *Catal. Rev. Sci. Eng.* 21, 275 (1980).

[40] X. Xie, H. Yin, B. Dou and J. Huo. *Appl. Catal. A Gen.* 77, 187 (1991).

[41] J.F. Pasquier, S. Komarneni and R. Roy. *J. Mater. Sci.* 26, 3797 (1991).

[42] J. Abbasian, R.P. Bachta, J.R. Wangerow, W. Mojtahedi and K. Salo. *Ind. Eng. Chem. Res.* 33, 91 (1994).

[43] A.R. de la Osa, A. de Lucas, J.L. Valverde, A. Romero, I. Monteagudo, P. Sánchez. *Int. J. Hydrogen Energy.* 36, 44 (2011).

[44] C. Ratnasamy and J.P. Wagner. *Catal. Rev.* 51, 325 (2009).

[45] J. Fermoso, B. Arias, M.V. Gil, M.G. Plaza, C. Pevida, J.J. Pis and F. Rubiera. *Bioresource Technol.* 101, 3230 (2010).

[46] O.P.R. Van Vliet, A.P.C. Faaij and W.C. Turkenburg, W.C. *Energy Conversion and Management.* 50, 855 (2009).

[47] T. Hanaoka, Y. Liu, K. Matsunaga, T. Miyazawa, S. Hirata and K. Sakanishi. *Fuel Process. Tech.* 91, 859 (2010).

[48] H. Pichler. Adv. Catal. 4, 271 (1952).

[49] Ø. Borg, S. Eri, E.A. Blekkan, S. Stosater, H. Wigum, E. Rytter and A. Holmen. *J. Catal.* 248, 89 (2007).

[50] S.Z. Li, S. Krishnamoorthy, A.W. Li, G.D. Meitzner and E. Iglesia. *J. Catal.* 206, 202 (2002).

[51] M.E. Dry. *Appl. Catal. A Gen.* 189, 185 (1999).

[52] A. de Klerk. *Catal. Today.* 130, 439 (2008).

[53] A.Y. Khodakov, W. Chu and P. Fongarland. *Chem. Rev.* 107, 1692 (2007).

[54] A.Y. Khodakov. *Catal. Today.* 144, 251 (2009).

[55] Y. Yang, H.-W. Xiang, Y.-Y. Xu, L. Bai and Y.-W. Li. *Appl. Catal. A Gen.* 266, 181 (2004).

[56] J. Yang, Y. Sun, Y. Tang, Y. Liu, H. Wang, L. Tian, H. Wang, Z. Zhang, H. Xiang and Y. Li. *J. Mol. Catal. A Chem.* 245, 26 (2006).

[57] J. Gaube and H.-F. Klein. *Appl. Catal. A Gen.* 350, 126 (2008).

[58] Y. Zhang, H. Xiong, K. Liew and J. Li. *J. Mol. Catal. A Chem.* 237, 172 (2005).

[59] A. Bao, K. Liew and J. Li. *J. Mol. Catal. A Chem.* 304, 47 (2009)

[60] D.A. Wesner, G. Linden and H.B. Bonzel. *Appl. Surf. Sci.* 26, 335 (1986).

[61] S. Eri, Jr. J.G. Goodwin, G. Marcelin and T. Riis. US Patent 4,880,763, 1989.

[62] A.R. de la Osa, A. de Lucas, J.L. Valverde, A. Romero, I. Monteagudo, P. Coca and P. Sánchez. *Catal. Today.* 167, 96 (2011).

[63] D.B. Bukur, X. Lang, A. Akgerman and Z. Feng. *Indus. Eng. Chem. Res.* 36, 2580 (1997).

[64] Y. Liu, B.-T. Teng, X.-H. Guo, Y. Li, J. Chang, L. Tian, X. Hao, Y. Wang, H.-W. Xiang, Y.-Y. Xu and Y.-W. Li. *J. Mol. Catal. A Chem.* 272, 182 (2007).

[65] D. Tristantini, S. Lögdberg, B. Gevert, Ø. Borg and A. Holmen. *Fuel Process. Technol.* 88, 643 (2007).

[66] L. Tian, C.F. Huo, D.B. Cao, Y. Yang, J. Xu, B.S. Wu, H.-W. Xiang, Y.-Y. Xu and Y.-W. Li. *J. Mol. Struct. Theochem.* 941, 30 (2010).

[67] M.E. Dry and G.J. Oosthuizen. *J. Catal.* 11, 18 (1968).

[68] S.-H. Kang, K.-J. Woo, J.W. Bae, K.-W. Jun and Y. Kang. *Korean J. Chem. Eng.* 26, 1533 (2009).

[69] A. Tavasoli, R.M. Malek Abbaslou, M. Trepanier and A.K. Dalai. *Appl. Catal. A Gen.* 345, 134 (2008).

[70] H. Jung, J.-I. Yang, J.H. Yang, H.-T. Lee, D.H. Chun and H.J. Kim. *Fuel Process. Technol.* 91, 1839 (2010).

[71] A.R. de la Osa, A. de Lucas, A. Romero, J.L. Valverde and P. Sánchez. *Fuel.* 90, 1935 (2011).

[72] G.L. Bezemer, P.B. Radstakea, U. Falke, H. Oosterbeek, H.P.C.E. Kuipers, A.J. van Dillen and K.P. de Jong. *J. Catal.* 237, 152 (2006).

[73] G.L. Bezemer, J.H. Bitter, H.P.C.E. Kuipers, H. Oosterbeek, J.E. Holewijn, X. Xu, F. Kapteijn, A.J. van Dillen and K.P. de Jong. *J. Am. Chem. Soc.* 128, 3956 (2006).

[74] A. Tavasoli, K. Sadaghiani, A. Nakhaeipour and M. Ghalbi Ahangari. *Iran. J. Chem. Chem. Eng.* 26, 1 (2007).

[75] M.C. Bahome, L.L. Jewell, K. Padayachy, D. Hildebrandt, D. Glasser, A.K. Datye and N.J. Coville. *Appl. Catal. A Gen.* 328, 243 (2007).

[76] A. Balbo, D. Sciti, A.L. Costa and A. Bellosi. *Mater. Chem. Phys.* 103, 70 (2007).

[77] E. Iglesia. *Stud. Surf. Sci. Catal.* 107, 153 (1997).

[78] R.C Reuel and C.H. Bartholomew. *J. Catal.* 85, 63 (1984).

[79] A.Y. Khodakov, A. Griboval-Constant, R. Bechara and V.L. Zholobenko. *J. Catal.* 206, 230 (2002).

[80] A.R. de la Osa, A. de Lucas, A. Romero, J.L. Valverde and P. Sánchez. *Catal. Today* (2011) doi:10.1016/j.cattod.2010.12.010.

In: Syngas
Editors: Antonius Indarto and Jelliarko Palguandi
ISBN: 978-1-62100-870-5

Chapter 7

PLASMA GASIFICATION AS A PROMISING WAY OF WOOD ENERGY USE

Ph. G. Rutberg, A. N. Bratsev, V. A. Kuznetsov, V. E. Popov and A. A. Ufimtsev*

Institute for Electrophysics and Electric Power RAS, Saint-Petersburg, Russia

ABSTRACT

The process of high temperature plasma gasification of wood is fully considered in the paper. The efficiency of use the wood as an energy feedstock is discussed. The comparative analysis of wood with other types of feedstock is carried out. The general advantages of plasma gasification as compared with autothermal gasification and pyrolysis are described. The dependence of the process parameters versus wood moisture, plasma forming gas composition and energy consumption value is studied. The syngas efficiency for electric power and synthesis process and hydrogen production is estimated. The large-scale experimental installation for plasma gasification is described. The experimental results which agree well with the calculated data are presented. The recommendations on the process optimization are indicated.

NOMENCLATURE

ρ	Density, $kg{\cdot}m^{-3}$
χ	Volume fraction
η	Efficiency
AC	Alternating Current
DC	Direct Current
E	Power inputs per unit of feedstock mass, $J{\cdot}kg^{-1}$
G	Specific flow-rate, $kg{\cdot}kg^{-1}$
LHV	Low Heating Value, $J{\cdot}kg^{-1}$

* E-mail: rc@iperas.nw.ru, lab10@iperas.nw.ru.

P	Specific energy yield, $J{\cdot}kg^{-1}$
Q	Yield of chemical or thermal energy, $J{\cdot}kg^{-1}$
RDF	Refuse Derived Fuel
T	Temperature, °C, K

Superscript and Subscript

c	Corrected, calculated value
CC	Combined Cycle
ch	Chemical
CH_2	Hydrocarbon
CO	Carbon oxide
el	Electricity
in	Introduced
M	Calculated by Mendeleev equation
net	Net energy
SC	Steam Cycle
SF	Synthetic fuel
SG	Syngas
th	Thermal

INTRODUCTION

Global energy consumption in 2009 amounted to ~$4.7{\times}10^{11}$ GJ/year [1]. From them: coal, oil and natural gas constituted ~88%, and the remaining portion of ~12 % came from Hydro and Nuclear stations. In terms of 1 person the annual consumption of primary energy makes ~70 GJ, whereas from the general power of power stations on 1 person falls ~0.3 kW [1,2]. It is easy to calculate that the primary energy is not very efficiently used by mankind.

Global economic and world population growth leads to increased demand for power resources. It is predicted that by 2100 the energy consumption will increase approximately in 1.5 times [3]. Thus oil and gas consumption will drop, and consumption of solid fuel (mainly of coal) and other energy sources will increase [4,5]. On the other hand power use of fossil hydrocarbons leads to a climb in carbon dioxide concentrations in the atmosphere which is the most dangerous greenhouse gas. Promising approaches decreasing the demand of fossil energy resources and technogenic load on biosphere are increase in efficiency of use of fuel combustion energy and involving of renewed energy sources in the energy balance including waste.

The most effective industrial method of primary energy conversion into electricity is a combined Brayton and Steam cycle (hereinafter combined cycle). This process uses only a gaseous fuel, therefore it is necessary to gasify solid fuels for electric power generation transforming them in syngas, a mixture mainly consisting of hydrogen and carbon monoxide. It allows increasing the efficiency of use of solid fuel primary energy in comparison with the direct combustion process. Besides, syngas produced at gasification can be a raw material for production of liquid fuel, hydrogen, and other substances of chemical-technological application.

Gasification processes differ on autothermal and allothermal after the manner of energy supply. In autothermal process thermal energy comes from combustion of a part of raw material for achievement of required temperature level, and in allothermal, it is brought from the outside. Syngas is contaminated by ballast impurities and hazardous substances in the autothermal process due to low temperatures and combustion that reduces efficiency of its further use.

Low temperature dense plasma application for allothermal gasification allows production of pure syngas with reduced content of admixtures, which combustion energy is practically completely equivalent to the combustion energy of solid fuel.

ALTERNATIVE FUEL

Types and Properties

All carbonaceous waste and resources, which cost mainly depends on the transportation distance, can be referred to substances capable to be involved in the general power balance. The approximate list of these substances is presented in Table 7.1 and their key properties are specified. Sludge is a solid component of sewage water separated by mechanical methods and precipitation at the sewage cleaning stations. Depending on sources of formation they can contain a considerable quantity of mineral inclusions and have high moisture. Thus a mineral component, as a rule, is hardly separated from deposits. The moisture, containing in pores of deposit particles, can be separated by traditional methods (squeezing, evaporation).

In the second half of the XX century the rate of formation of household waste per capita steadily grew together with rise of living standards. Today this parameter is at level of ~0.85 kg/day of solid municipal waste. Landfill is the most common method of waste disposal. Thus environmental contamination factors are inevitable penetration of hazardous substances into subsoil waters, and also releasing of gaseous products of waste biodecomposition (in general methane - one of the main greenhouse gases). At the same time part of waste is already used as an energy source. A lot of incineration plants have been built in the world.

Direct power use of waste is not always possible because of high variability of morphological composition and mechanical properties of this type of feedstock. Besides, the process efficiency will be decreased due to the energy consumption for warming up of the inorganic components of fuel and excess moisture decomposition. Therefore before burning or gasification wastes should be subjected to preliminary preparation. Refuse Derived Fuel (RDF) is the product of this preparation. RDF possesses lower moisture and ash content and higher heat of combustion, in comparison with the primary waste. RDF is widely used in power engineering in developed countries. For example RDF production in Europe increased from 1.4 million of ton in 2001 to 12.4 million of ton in 2005 [15]. Cost of RDF production is ~10 €/t [16,17].

Table 7.1. Characteristics of waste and renewable resources

Feedstock	SmpNo.	Moisture	Ash	Composition of organic mass, %						LHV ($MJ \cdot kg^{-1}$)
				C	H	O	N	S	Cl	
Sludge [6]	1	8.90	67.60	6.51	1.59	13.16	1.08	1.16	0	2.94
	2	78.10	8.08	8.17	1.27	3.00	1.20	0.18	0	1.45
	3	6.10	36.06	28.11	3.80	21.96	2.98	0.99	0	12.92
Municipal Solid Waste (MSW) [7]	1	34.76	31.09	17.40	2.24	13.85	0.55	0.11	0	5.34
	2	40.25	26.65	19.36	1.95	11.11	0.47	0.21	0	5.95
	3	24.20	43.27	21.32	1.45	9.27	0.22	0.27	0	7.25
	4	27.03	25.36	25.53	3.18	18.37	0.42	0.12	0	8.58
Refuse Derived Fuel (RDF) [8-10]	1	30.73	11.59	28.30	4.20	24.25	0.61	0.32	0	9.50
	2	7.83	8.71	41.70	5.58	35.29	0.80	0.09	0	15.01
	3	3.20	11.99	43.06	7.69	32.07	0.90	0.30	0.80	19.47
	4	10.00	10.80	53.80	9.20	14.90	0.30	0.30	0.70	25.00
Wood [11]	1	50.00	0.40	25.10	3.21	21.19	0.05	0.05	0	2.96
	2	60.00	0.50	19.39	2.53	16.02	1.43	0.13	0	4.03
	3	20.00	0.80	39.80	4.82	34.33	0.16	0.08	0	13.90
	4	12.00	0.50	43.79	5.39	38.14	0.09	0.09	0	15.63
Used car tires [12]	1	1.20	9.88	76.67	10.28	0.00	0.00	1.98	0	24.08
	2	1.14	13.35	73.65	5.93	2.97	0.49	1.48	0.99	33.57
Peat [13, 14]	1	16.10	21.40	35.44	3.44	21.81	1.13	0.69	0	11.58
	2	9.40	16.87	42.44	4.50	24.98	1.60	0.18	0.03	16.46

Note: LHV = Low Heating Value; Smp No = sample number.

Approximate power inputs on primary waste milling are no more than 0.013 kWh/kg, on secondary 0.033 kWh/kg [18]. RDF can also be produced from landfill wastes. At this to the cost of production should be added the cost of extraction ~0.7 kg of oil per ton of waste [19].

Wood waste which cannot be used in industry is secondary energy resources. They are of special interest for power engineering of such countries, as Russia, USA, Brazil and Canada possessing the greater part of world wood reserves [20]. The estimations show that in the USA about half of the consumed oil can be replaced by biofuel from wood waste, agriculture and bioenergy crops [21]. It is possible to increase the existing biomass reserves at the expense of planting of fast-growing bioenergy crops on vacant lands unusable for agriculture. Wood waste are of different types: green wood (sample 1), garden plants knots (sample 2), mixed wood waste (sample 3), hardwood (sample 4), etc.

The basic source of rubber waste is the world car park which quantity already makes more than billion transport facilities [22]. Average global consumption of rubber is ~3.8 kg/man·year approximately half of it is used car tires [23,24]. That is the city with the million population consumes about 434 kg/h of rubbers which basic part is used for manufacture of car tires. Unlike the previous representatives of this group of a feedstock source the used car tires on composition and heat of combustion are close to coals, that are they possess a high heat of combustion, carbon content and low oxygen content.

Peat is a product of plant decomposition and is formed at swamps. According to different estimations the global peat amount is about 270 - 6000 billion tons [25]. The average cost of peat production in Russia is ~$10 per ton. This kind of fuel does not widely used in power engineering due to comparatively low combustion heat. Interest in developing the use of peat for energy purposes has diminished because it is a renewable energy source.

WAYS OF ENERGY USE OF SOLID FUEL

The process of direct combustion of solid fuel with utilization of released heat for steam production which rotates the steam turbine and generator rotor (hereinafter steam cycle) is the most widely spread method in power engineering. Efficiency (share of primary energy converted into electricity) of such power stations greatly depend on fuel characteristics. For coal the efficiency can reach ~44%, whereas the average efficiency of coal power stations in the USA is ~33% [26]. Use of dry biomass gives efficiency of about 30-34%, and at operation on municipal solid waste ~22% [27].

The combined cycle is the most efficient technology of electricity generation nowadays. It uses only gaseous fuel combusted in a special gas turbine at high temperatures. The turbine rotor rotates the generator, and thermal energy of burned gases is utilized by the steam cycle. The process efficiency is up to ~60 % [28-30], and mainly depends on maximum temperature rating in a gas turbine and technology of gas burning. The rest 30 - 35% of primary energy can be used as low-potential heat.

The combined cycle is widely used in solid fuel power engineering due to high efficiency. The solid fuel is preliminary subjected to gasification, at which the organic component of fuel under the influence of temperature and blast agents transforms into a gas phase, and mineral part forms ash or slag. The process is carried out at oxidizing agent deficit therefore the formed gases contain combustible components CO, H_2, CH_4, etc. Carbon monoxide and hydrogen are the two primary components of syngas, methane and other gaseous hydrocarbons are presented in a kind of impurity. In addition the produced gas can be ballasted by nitrogen, carbon dioxide, steam, and other gases. The composition and heat of combustion of such gas essentially depend on properties of the primary fuel and from the gasification process parameters. The optimization problem consists in increase of yield of primary components at decrease of ballast gases and impurities.

PLASMA GASIFICATION ADVANTAGES

Traditional methods of control of a heterogeneous process of gasification are variation of blast gases flow rate and pressures in the reactor. Pressure growth leads to increase of the reaction rate due to increase of the oxidizer concentration. For example, pressure increase from 1 to 40 atm at 1500 K results in reduction reaction rate of CO_2 on carbon in ~5.3 times. As a result the module output increases. But in this case the specific yield of end products (H_2 and CO) and process energy efficiency decrease. For example let us consider the variation of composition of products of stochiometric gasification (when all oxygen of the oxidizer and feedstock is used for single oxidizing of all carbon of oxidizer and feedstock) of graphite by steam under specified conditions and 100% conversion. Equilibrium mole composition of gasification products will vary as follows: CO concentration decreases from 50.0 to 49.3%, and H_2 — from 49.9 to 46.2%, whereas CH_4 concentration increases from 0.1 to 2.2%, H_2O — from 0.0 to 1.6%, CO_2 — from 0.0 to 0.7%. The specified pressure will require increase in power inputs on ~2.74 MJ/kg and results in decrease of energy efficiency on ~2.3%.

The module output can be increased by increasing the blast gases flow rate. With its growth the thermal stress of reactionary volume and process temperature increase that leads to decrease of thermal losses (in relative magnitudes) and decrease of specific yield of end products (H_2 and CO) and energy efficiency of the process. This method can be used for both autothermic and allothermic approaches but they have different output upper limit.

Plasma application for increase the process temperature is an alternative method of increasing productivity. The same CO_2 reduction rate on hydrogen increases in ~7.8 times if the temperature increases from 1500K to 1900K under atmospheric pressure. At this the power inputs for temperature increase are ~0.5 MJ/kg, whereas pressure increase requires ~1.6 MJ/kg. In other words use of plasma for productivity raise in ~4.7 times is more effective than increase in pressure and it does not lead to displacement of balance towards ballast gases (H_2O and CO_2). On the contrary, raising energy density, for example in air plasma, it is possible to lower essentially the content of ballast nitrogen in gasification products.

Plasma use is alternative to use of pure oxygen as a blast agent which application raises level of danger of manufacture and also results in additional expenses for its production.

Calculations of equilibrium composition of gasification products, for example, wood waste with 20% moisture at 1,500K and 101,325 Pa with the use of air blast show that if air plasma with calorific value 1 - 9 MJ/kg is used instead of air, the specific yield of gas decreases on 13.3 - 34.4%, the share of incombustible components in syngas composition decreases to 61.0 - 23.2%, power inputs for heating of gasification products decreases to 17.4 - 46.5%, and the specific yield of chemical energy of gasification products increases by 39.7 - 106%. It will allow increasing productivity by 18.4 - 72.5%.

Thus, the basic advantages of plasma application plasma in gasification are the increase in gasification rate and syngas heat of combustion. Plasma gasification technology allows treatment of both traditional kinds of fuel and various waste into syngas. Some kinds of feedstock can be treated both for power generation and for environment protection.

Comparative Estimation of the Processes for Different Fuels

The comparative analysis of the basic process parameters should be carried out for reasons of availability of use of this or that feedstock in power engineering. Calculation of specific energy yield is realized by equations (7.1) and (7.2),

$$P_{SC} = LHV \cdot \eta_{SC} \tag{7.1}$$

$$P_{CC} = Q_{ch} \cdot \eta_{CC} - E_{in} \tag{7.2}$$

where, P_{SC} is specific energy yield at use of direct combustion process and steam cycle (MJ/kg); LHV is low heating value of fuel (MJ/kg); η_{SC} is efficiency of electricity generation at use of direct combustion and steam cycle (in decimal fractions); P_{CC} is specific electricity yield at use of a plasma gasification process and combined cycle (MJ/kg); Q_{ch} is yield of chemical energy (LHV of syngas) on unit of wood mass (MJ/kg); η_{CC} is efficiency of electricity generation from syngas in combined cycle (in decimal fractions), E_{in} is power inputs per unit of feedstock mass (MJ·kg^{-1}). Data on fuel LHV represented in references are not always well agreed with calculated once, for example by equation (7.3):

$$LHV_M = 34.09 \cdot C + 110.46 \cdot H - 11.99 \cdot (O + N) + 6.84 \cdot S - 1.53 \cdot A - 2.44 \cdot W \qquad (7.3)$$

where, LHV_M is low heating value of the working mass of feedstock (MJ/kg); C, H, O, N, S, A, W are mass fraction of carbon, hydrogen, oxygen, nitrogen, sulphur, inorganic component and water on working mass (in decimal fractions). Magnitudes E_{in}, P_{SC} and P_{CC} were calculated on the basis of reference data. Formulas (7.4) and (7.5) were used for correction of values of specific yields.

$$P_{SC}^{c} = P_{SC} \cdot LHV_M / LHV \qquad (7.4)$$

$$P_{CC}^{c} = P_{CC} + LHV_M - LHV \qquad (7.5)$$

where, P_{SC}^{c} is corrected value P_{SC}; P_{CC}^{c} is corrected value P_{CC}. Efficiency of electricity generation on steam cycle η_{SC} in calculations was taken as:

- Sludge 15%,
- Solid municipal waste 22%,
- RDF 22%,
- Wood waste 25%,
- Car tires 20%,
- Peat 20%,

Combined cycle efficiency was taken as constant (η_{CC} = 60%), because at evaluation of plasma gasification parameters the mode with maximum possible combustion heat of the produced gas was chosen. The calculation results are tabulated in Table 7.2.

It should be considered that real efficiency of both processes can essentially differ from the accepted at comparison of results of calculation of specific yields the electric power for processes on the basis of burning and plasma gasification. As, the parameters of plasma process are not optimized, and the efficiency of combustion depends not only on a fuel heat of combustion, but also from development production engineering of its burning.

The first and second sludge samples are of little use for burning because of low adiabatic combustion temperature, and for plasma gasification - because of high power inputs. Taking into account reference data and correction on a heat of combustion the specific yield of the useful electric power at use of the system on the basis of plasma gasification and the combined cycle for the third sample can make 2.45 - 5.22 MJ·kg^{-1}, against 1.52 - 1.94 MJ·kg^{-1} for the process on a basis of combustion. It is obvious that plasma use in this case is much more effective than combustion.

From 1 kg of solid municipal waste in the system on the basis of combustion and steam cycle can be generated 1.18-1.96 MJ. At use of a plasma method the specific yield of the electric power for samples 3 and 4 makes 0.50-1.65 MJ/kg, and for samples 1 and 2 it appears less than zero.

Table 7.2. Indices of waste treatment and renewable resources

Feedstock		LHV	LHV_M	Air + Steam	E_{in}	Q_{ch}	P_{SC}	P_{SC}^c	P_{CC}	P_{CC}^c	Increment	
		$(MJ{\cdot}kg^{-1})$		$(kg{\cdot}kg^{-1})$	$(MJ{\cdot}kg^{-1})$						P_{CC}/P_{SC}	P_{CC}^c/P_{SC}^c
Sludge	1	2.94	1.10	0 + 0	1.66	2.58	0.44	0.16	-0.11	-1.95	-	-
	2	1.45	1.66	0 + 0	5.73	4.44	0.22	0.25	-3.07	-2.85	-	-
	3	12.92	10.15	0.417 + 0	1.98	12.00	1.94	1.52	5.22	2.45	~2.7	1.6
MSW	1	5.34	5.36	0 + 0	5.03	8.01	1.18	1.18	-0.23	-0.21	-	-
	2	5.95	5.99	0 + 0	5.40	8.92	1.31	1.32	-0.05	-0.01	-	-
	3	7.25	6.50	0 + 0	4.51	9.61	1.60	1.43	1.25	0.50	-	-
	4	8.58	8.92	0 + 0	5.94	12.09	1.89	1.96	1.31	1.65	-	-
RDF	1	9.50	10.40	0 + 0	6.74	13.58	2.09	2.29	1.40	2.31	-	-
	2	15.01	15.74	0.552 + 0	5.91	17.67	3.30	3.46	4.69	5.42	~1.4	1.7
	3	19.47	18.98	0.931 + 0	7.75	19.94	4.28	4.18	7.43	6.94	~1.7	1.3
	4	25.00	26.29	1.986 + 0	11.98	25.40	5.50	5.78	8.92	10.2	~1.6	1.8
Wood	1	2.96	8.33	0 + 0	11.10	11.34	0.74	2.08	-4.30	1.07	-	-
	2	4.03	5.85	0 + 0	7.50	8.79	1.01	1.46	-2.22	-0.41	-	-
	3	13.90	14.27	0.038 + 0	6.57	17.85	3.48	3.57	4.14	4.50	~1.2	1.3
	4	15.63	16.00	0.395 + 0	5.90	18.46	3.91	4.00	5.18	5.55	~1.3	1.4
Used car tires	1	24.08	37.44	4.191 + 0	16.41	31.23	4.82	7.49	2.33	15.69	-	-
	2	33.57	31.12	3.811 + 0.025	0.00	25.56	6.71	6.22	15.34	12.88	~2.3	2.1
Peat	1	11.58	12.46	0.460 + 0	6.12	14.67	2.32	2.49	2.68	3.56	~1.2	1.4
	2	16.46	15.78	0.962 + 0	4.07	16.83	3.29	3.16	6.03	5.35	~1.8	1.7

Note: Increment means the increase of electrical yield.

Data on RDF energy content well agreed with formula (7.3) only for the second, third and fourth samples (divergence from -2.5 to 5.2%), for the first sample the heat of combustion differs from the calculated one by ~10%. Taking into account reference and calculated data from 1 kg of RDF (depending on composition) it is possible to produce 1.40-10.2 MJ of electricity by plasma method, against 2.09 - 5.78 MJ by combustion process.

Data on wood waste energy content for samples 1 and 2 are in rather poor agreement with formula (7.3) (divergence by 181% and 45% respectively), and for samples 3 and 4, those are well agreed. Taking into account reference and correction for the third sample the specific electricity yield can be 2.64 - 3.48 MJ/kg at use of the combustion process and 3.57 -4.14 $MJ{\cdot}kg^{-1}$ at use of the plasma gasification and combined cycle system, for the fourth sample, 2.38 - 3.91 $MJ{\cdot}kg^{-1}$ for combustion and 4.00 - 5.18 $MJ{\cdot}kg^{-1}$ for plasma gasification. It is clear that in this case, the plasma method allows significant increase of the energy efficiency of wood waste use. From 1 kg of used car tires can be produced about 6.22 - 7.49 MJ by combustion method or 12.9 - 15.7 MJ by plasma method. Data on peat energy content are satisfactory agreed with formula (7.3) (divergence of 4.1% - 7.6%). According to the preliminary estimations the specific yield of electricity produced by plasma technologies is 2.68 - 6.03 $MJ{\cdot}kg^{-1}$, versus 2.32 - 3.29 $MJ{\cdot}kg^{-1}$ the combustion process. Plasma gasification application can result in increase of electricity yield only for some kinds of feedstock. The final decision on applicability of plasma technology for treatment of this or that feedstock can be taken on the basis of economical analysis taking into account all aspects of the process.

THEORETICAL ANALYSIS OF WOOD WASTE PLASMA GASIFICATION PROCESS

Average moisture of technically dry wood waste is 20% [11]. The number of oxygen atoms in a mass unit is close to number of atoms of carbon in wood waste with such humidity. That is, for gasification of such feedstock the oxidizing agent is not practically required for carbon transformation in carbon monoxide. In traditional methods it is required for compensation of endothermic effect of gasification. At plasma gasification the endothermic effect of gasification can be compensated by the external energy source. Besides, wood with 20% moisture has sufficient heat of combustion for autothermic process conducting. Plasma gasification of such raw material is promising, as energy of plasma can be used directly on feedstock conversion in hydrogen and carbon monoxide without increase in power inputs for decomposition H_2O or CO_2. CO_2 yield per unit of thermal energy (~94.43 $g{\cdot}MJ^{-1}$) at wood combustion is compared with the similar index for coal (85 - 100 $g{\cdot}MJ^{-1}$), and chemical underburning, as a rule, is significantly lower. Wood use in power engineering does not contribute to CO_2 emission as the quantity of CO_2 absorbed at tree growth is equal to CO_2 quantity formed as a result of its combustion. It is known, that in an hour the Earth absorbs more solar energy, than is consumed by mankind in a year [31]. Therefore, despite rather low general efficiency of transformation of solar energy in energy of biomass (no more than 5 - 6% [32]), vegetative biomass is one of the most promising renewed energy sources.

Fundamental Tenets and Assumptions

The most important aspect at development of technologies of syngas use is the connection between its composition and electricity cost. The content of the basic components (H_2 and CO), macro-and microimpurities in syngas influences on the efficiency of its treatment technologies. Numerical modeling was carried out of efficiency of syngas use, operating conditions of plasma gasification and wood moisture to study the interconnection of efficiency of syngas use. Two methods of syngas use were considered: as a fuel for electric power generation and as a feedstock for synthesis of liquid fuels. During modeling the following assumptions were used:

1. Ashless composition of products of wood waste gasification is taken as thermodynamically equilibrium composition of mixture of plasma and ashless mass of wood waste calculated at temperature 1500K and atmospheric pressure (101,325 Pa).
2. Inorganic component of wood waste does not participate in chemical reactions, and energy inputs on its warming up from 298K to 1500K are ~1.367 $MJ{\cdot}kg^{-1}$.
3. There is no change in composition of gasification products during their cooling to 298K.
4. Mass composition of wood waste: C – 50.25, H – 6.09, O – 43.35, N – 0.2, S – 0.1% (for dry ashless mass), ash content – 1% (for dry ashless mass), moisture – 20% (for working mass); *LHV* – 13.90 $MJ{\cdot}kg^{-1}$ (for working mass), *LHV* of absolutely dry wood ~17,99 $MJ{\cdot}kg^{-1}$ [11].

5. All gases which have already found application in traditional gasification are used as oxidizing agents (air, O_2, CO_2 and H_2O).
6. For comparison of use of passive oxidizing agents in autothermic gasification with other methods their mixtures with active oxidizing agents are used, in this case the oxidizing agent flow rate (G_{in}) is registered as the sum of flow rates of all oxidizing agents.
7. Mass composition of air: N_2 – 74.43; O_2 – 22.81; Ar – 1.27; CO_2 – 0.04; H_2O – 1.45%.
8. The specific electricity yield at use of the produced syngas as a fuel for the combined cycle was estimated on the basis of dependence of the combined cycle efficiency versus temperature over the range 1400 - 2000K [33]. It has been approximated by cubic polynomial:

$$\eta_{CC} = 0{,}133 + 4{,}3\cdot10^{-5}\cdot T + 2{,}6\cdot10^{-7}\cdot T^2 - 7{,}9\cdot10^{-11}\cdot T^3 \tag{7.6}$$

where T is temperature (K); efficiency of the combined cycle at temperatures above 2000K is taken to be constant and equal to the efficiency at 2000K.

9. Key parameters of the system for production of thermal and electric energy were estimated by equations (7.7 – 7.9):

$$P_{el} = Q_{ch}\cdot\eta_{CC} \tag{7.7}$$

$$P_{net} = P_{el} - E_{in} \tag{7.8}$$

$$P_{th} = LHV + E_{in} - P_{el} \tag{7.9}$$

where, P_{el} is specific electricity yield ($MJ\cdot kg^{-1}$); P_{net} is specific yield of useful electric power ($MJ\cdot kg^{-1}$), P_{th} is specific yield of thermal energy ($MJ\cdot kg^{-1}$).

10. Value of a specific yield of liquid fuels was estimated on the basis of stoichiometry of Fisher-Tropsch process (7.10), without taking into account the influence of impurities:

$$CO + 2\,H_2 \rightarrow CH_2 + H_2O \tag{7.10}$$

At volumetric relation $H_2/CO \geq 2$ potentially possible specific yield of liquid fuel is determined by equation (7.11):

$$G_{SF} = G_{SG}\cdot\chi_{CO}\cdot\rho_{CH2} \tag{7.11}$$

where G_{SF} is yield of liquid fuel (kg/kg), G_{SG} is yield of dry syngas (m^3/kg), χ_{CO} is volume fraction CO in syngas (in decimal fractions), ρ_{CH2} is density (at 25°C and 101,325 Pa) conditional gas with chemical formula of CH_2 (kg/m^3).

At volumetric relation $H_2/CO < 2$ potentially possible specific yield of liquid fuel is determined by equation (7.12):

$$G_{SF} = G_{SG} \cdot (\chi_{H2} + \chi_{CO}) \cdot \rho_{CH2} / 3 \quad (7.12)$$

where χ_{H2} – volume fraction of H_2 in syngas (in decimal fractions); in formula (7.12) was taken into account catalytic conversion CO in H_2 on shift reaction (7.13):

$$CO + H_2O \rightarrow CO_2 + H_2 \quad (7.13)$$

Comparative Assessment of Various Plasma Forming Media

Search of energy efficient approach of high-temperature wood waste treatment is carried out by a method of comparison of processes of pyrolysis, autothermic and plasma gasification (see Table 7.3 and 7.4). The estimation of the maximum yield of liquid fuels is made approximately. If the gasification process is designed as a component of chemical synthesis it will be necessary to consider in addition the influence of syngas composition on the synthesis parameters. Requirements to syngas composition will depend in this case on the applied process.

For example, for synthesis of liquid fuels by Fisher-Tropsch process, and also for catalytic synthesis of methanol the relation of concentrations H_2 : CO should be ~2, and for synthesis of dimethyl ether it is sufficient H_2 : CO ~ 1.

Gasification by passive medium differs by high level of power inputs, because the main reactions in this case are endothermic. Autothermic gasification by pure carbon dioxide or steam is impossible.

To show the effect of use of these media, oxygen and air have been added in composition of blast gases for compensation of endothermic losses. If we compare these results with the results of autothermic gasification by air and oxygen, it is obvious that specific yields of chemical energy of gas decrease.

Increase of passive oxidizer flow rate at steam-air and steam-oxygen gasification leads to decrease in specific yield of carbon monoxide, increase in the specific yield of carbon dioxide and thereafter decrease of heat of combustion of gas.

Use of carbon dioxide as the secondary blasting decreases the specific yield of hydrogen and increases the specific yield of carbon dioxide, that results in decrease of the of heat of combustion of gas.

Therefore, CO_2 and H_2O use for autothermic gasification is promising only for regulation of relation H_2 : CO. There is little point to use steam and carbon dioxide as a secondary blasting for air autothermic gasification, because high content of ballast gases results in decrease of efficiency of synthesis or combustion.

Plasma gasification by pure CO_2 and H_2O, as a whole, has similar characteristics as oxygen gasification that is high heat of combustion of syngas and low content of impurities (see Figure 7.1).

The growth of the specific power inputs (in terms of 1 kg of feedstock) at steam-plasma gasification leads to the decrease of CO content in a dry syngas and increase of H_2, and at carbon dioxide gasification concentration of H_2 decreases.

Table 7.3. Key parameters of high temperature processes of wood processing with 20% moisture

Oxidizer	E_{in} (MJ/kg)	Oxidizer specific flow rate G_{in} (kg·kg^{-1})	G_{SG} (m^3·kg^{-1})	Composition of syngas (%vol.)			
				H_2	CO	CO_2	N_2
no	6.57	0	1.65	51.7	48.1	0.02	0.08
Air	0	2.80	2.93	9.40	14.6	13.0	62.2
	1.98	1.98	2.54	16.7	22.3	9.61	50.8
	2.76	1.66	2.39	20.4	25.8	8.17	45.1
	5.29	0.59	1.92	37.6	39.2	3.01	19.9
O_2	0	0.44	1.23	34.2	47.0	18.6	0.11
	0.42	0.42	1.26	35.4	47.5	16.8	0.11
	2.50	0.28	1.39	41.4	49.1	9.30	0.10
CO_2	10.1	2.02	2.31	16.4	56.3	27.3	0.06
	7.35	0.37	1.75	41.8	54.5	3.6	0.08
Air + CO_2	0	3.11 + 0.48	3.28	4.80	13.4	19.4	61.6
	0	2.86 + 0.11	3.03	9.23	13.4	15.4	61.3
O_2 + CO_2	0	0.54 + 1.41	1.52	13.2	43.1	43.6	0.09
	0	0.46 + 1.52	1.34	31.0	42.1	26.7	0.11
H_2O	13.3	2.66	2.08	61.0	19.8	19.2	0.07
	7.53	0.38	1.79	54.6	39.2	6.02	0.08
Air + H_2O	0	3.30 + 0.51	3.30	9.51	6.64	18.0	65.1
	0	2.88 + 0.11	3.06	11.4	10.8	15.8	61.3
O_2 + H_2O	0	0.61 + 1.03	1.34	39.2	17.4	43.3	0.11
	0	0.47 + 0.20	1.35	39.9	31.5	28.4	0.10

Table 7.4. Estimations of the key parameters of application of the syngas produced from wood with 20% moisture in corresponding technologies

Oxidizer	E_{in} (MJ·kg^{-1})	G_{in} (kg·kg^{-1})	Q_{ch} (MJ·kg^{-1})	P_{el} (MJ·kg^{-1})	P_{th} (MJ·kg^{-1})	G_{SF} (g·kg^{-1})
no	6.57	0	17.7	11.3	8.96	315
Air	0	2.80	7.70	4.03	9.86	134
	1.98	1.98	10.8	6.62	9.25	189
	2.76	1.66	12.0	7.61	9.03	211
	5.29	0.59	15.9	10.1	9.07	282
O_2	0	0.44	10.9	6.92	6.97	191
	0.42	0.42	11.3	7.23	7.08	200
	2.50	0.28	13.6	8.64	7.75	240
CO_2	10.1	2.02	18.8	12.0	12.0	321
	7.35	0.37	18.2	11.6	9.63	322
Air + CO_2	0	3.11 + 0.48	6.65	3.07	10.8	114
	0	2.86 + 0.11	7.46	3.79	10.1	131
O_2 + CO_2	0	0.54 + 1.41	9.58	6.10	7.79	164
	0	0.46 + 1.52	10.6	6.75	7.14	187
H_2O	13.3	2.66	17.3	11.0	16.2	236
	7.53	0.38	17.8	11.3	10.1	321
Air + H_2O	0	3.30 + 0.51	5.63	2.41	11.5	102
	0	2.88 + 0.11	7.26	3.63	10.3	130
O_2 + H_2O	0	0.61 + 1.03	7.87	4.99	8.90	134
	0	0.47 + 0.20	10.3	6.55	7.34	184

Content of CO in a dry syngas at plasma gasification of wood waste by CO_2 plasma increases with the increase of power inputs up to ~8.7 $MJ \cdot kg^{-1}$, and then decreases.

Relation H_2/CO varies in the design range within the limit 0.18 - 1.07 at use of CO_2 plasma and from 1.07 to 3.65 at use of steam plasma.

Differences in trends of behavior of concentration of syngas components are caused by the character of changes brought in by the oxidizing agent into the element balance of a mixture of fuel with the oxidizing agent. Fractions of oxygen and carbon increase at increase in CO_2 flow rate. Fractions of oxygen and hydrogen increase at increase in steam flow rate. The relation H_2/CO can be varied at plasma gasification over wide limits, changing the composition of the plasma forming gas and compensating power inputs for the process at the expense of outside energy supply.

Power inputs increase with the increase of the oxidizer flow rate (see Figure 7.2). This is because plasma energy of passive oxidizers (H_2O or CO_2) goes not only for compensation of the endothermic effect of gasification, but also for destruction and heating of oxidizing gas up to the process temperature.

With the increase of plasma calorific value power inputs decrease (in the limit they tends to the power inputs on pyrolysis). Use of passive oxidizer plasma with high calorific value allows production of a syngas containing low amount of impurities, which is favorable for its application for the liquid fuel synthesis [34].

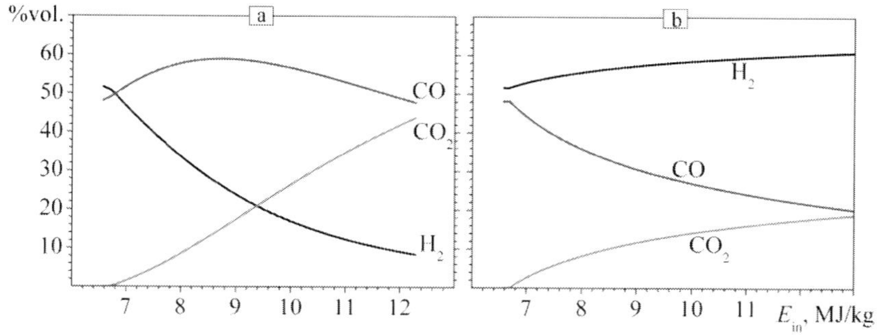

Figure 7.1. Influence of power inputs on volumetric composition of dry gas at plasma gasification of wood waste by (a) carbon dioxide and (b) steam.

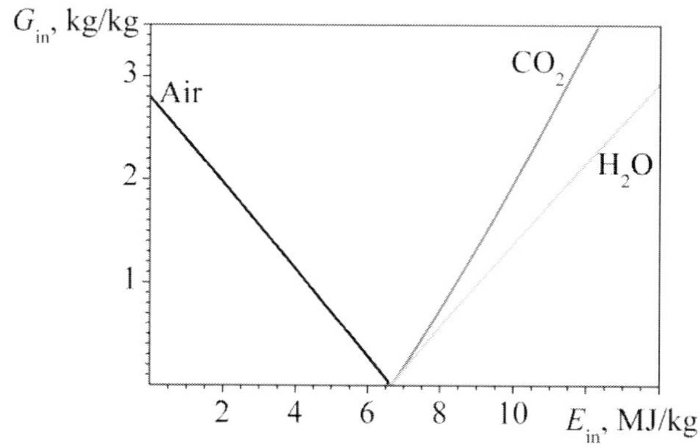

Figure 7.2. Flow rates of oxidizers at plasma wood gasification.

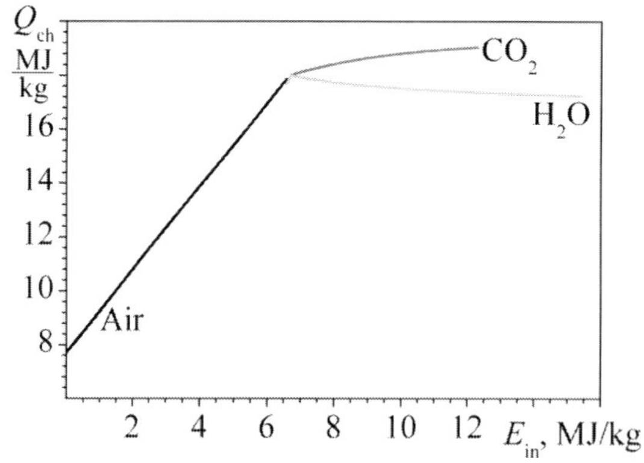

Figure 7.3. Gasification of wood by various kings of plasma.

Despite of high syngas heat of combustion, there is no point in application of wood plasma gasification technology by passive oxidizers for power engineering. At increase of power inputs the specific yield of chemical energy varies insignificantly (see Figure 7.3).That is why the mode with minimum oxidizer flow rate and therefore maximum plasma calorific value will be the most efficient. The efficiency of electric energy conversion into thermal energy decreases at increase of plasma calorific value.

Specific yield of chemical energy increases with the increase of power inputs at use of CO_2 plasma, and for steam plasma, it decreases. It is caused by balance displacement (at 1500K) towards hydrogen oxidation and reduction of carbon dioxide. It should be noted that energy efficiency of gasification at use of CO_2 in blast gas composition is higher than at use of steam blasting. It is true both for autothermic and for plasma gasification.

In pyrolysis of wood, its own reserved energy and one supplied from the outside are transformed into chemical and thermal energy of the syngas. Thus the specific yield of gasification products is minimum and their heat of combustion is maximum. The estimations show that if the produced gas is used in the modern combined cycles ~11.3 MJ of electricity can be generated from 1 kg of wood, and ~6.57 MJ will return to the process, that is from 1 kg of wood waste with 20 % moisture can be obtained ~4.69 MJ of electricity. The rest heat energy ~8.96 MJ also can be used: "high potential" part (hot syngas) for high pressure steam production, and "low potential" part (after combined cycle) for the preliminary drying or heating of feedstock. The produced gas is suitable for dimethyl ether production on the amount of impurities and H_2/CO relation.

Effective implementation of this process without application of the heat-transfer agent contacting with feedstock, is now impossible. Use of warmed walls by analogy to the processes of low temperature pyrolysis or steam conversion of methane for high-temperature processes is ineffective. Ignition of the discharge in the reactor, is of some interest from the point of view of raise of energy concentration in a mass unit of pyrolysis products, however, due to complexity of the discharge stabilization this process has low efficiency. Part use of syngas or inert gases as plasma forming ones can be alternative to these methods. Then, depending on used plasma forming gas, at plasma temperature of 5000 - 6000K power inputs on pyrolysis will increase to 7.23 - 7.07 MJ/kg or 8.30-7.46 MJ/kg for syngas and nitrogen, respectively. Syngas application does not change the composition of gasification products, and in case of nitrogen use its content will be ~40.1-25.6%.

Use of active oxidizing agents takes the part of energy used for the process from raw materials, therefore power inputs on their implementation are more lower.

In the process of autothermic gasification of wood waste by air, syngas has the high content of impurities and a low calorific value. About 44.6% of energy of feedstock is spent for heating of products of gasification to the process temperature. Content of impurities in syngas essentially decreases at use of oxygen blasting, and its heat of combustion increases. However, production, storage and supply of oxygen into the reactor is a complicated and expensive process. Hence, power cost of syngas will be determined by the method of oxygen production and maintenance costs.

Moreover the use of oxygen blasting at plasma gasification is complicated by high chemical activity of oxygen. In this case at a stage of plasma generation even cooled metal electrodes will be exposed to considerable erosion. On the other hand generation of air plasma is well mastered process and the parameters of air-plasma gasification tabulated in Table 7.3 are quite achievable.

Power inputs increase with the decrease of the oxidizer flow rate (see Figure 7.2). The reason is that at feedstock oxidizing by smaller quantity of the oxidizing agent, less energy releases, and the plasma energy consumption on production and heating of wood gasification products is higher than expenses for heating of air ballast nitrogen.

Figure 7.4 shows that at gasification of wood waste by air plasma the growth of specific power inputs (in terms of 1 kg of feedstock) leads to increase in concentration of combustible components of syngas (H_2 and CO) and decrease in concentration of incombustible gases (CO_2 and N_2). Relation H_2/CO in a design range changes from 0.64 to 1.07.

Increasing plasma calorific value we can approximate the characteristics of air-plasma gasification to the plasma pyrolysis characteristics. Thus the energy efficiency of the process will increase with the growth of plasma calorific value. Yield of chemical energy increases linearly with the increase of power inputs (see Figure 7.3): for 1 MJ of input energy the yield of chemical energy of syngas increases by ~1.55 MJ. Power inputs for wood pyrolysis are ~6.57 MJ·kg^{-1}. Maximum of the specific yield of chemical energy is reached near the pyrolysis point.

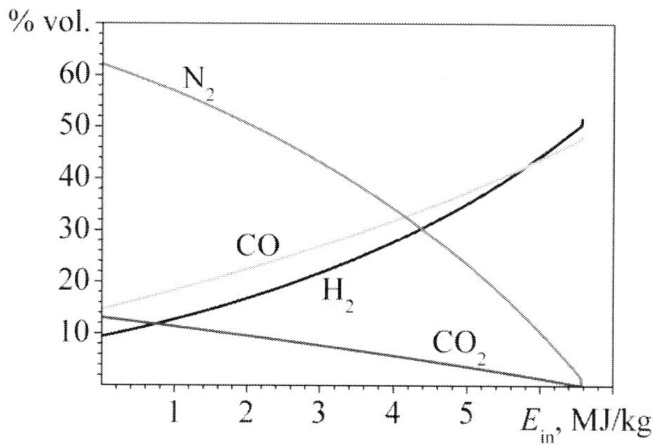

Figure 7.4. Influence of power inputs on the volumetric composition of dry syngas at plasma gasification of wood waste by air.

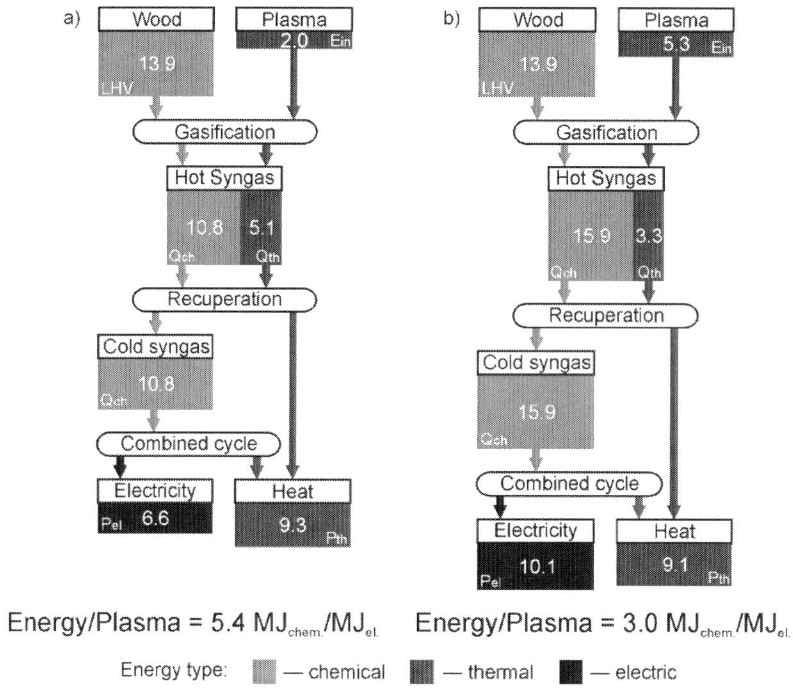

Note: All values are given in MJ in the terms of 1 kg of wood waste.

Figure 7.5. Comparison of energy balances of generation of electricity and thermal energy from wood waste using air plasma with enthalpy: (a) 1 MJ·kg^{-1}; (b) 9 MJ·kg^{-1}.

Figure 7.5 shows two variants of implementation of air-plasma gasification process of wood waste with 20% moisture in the modern system for electricity generation (without taking into account heat losses). The schematic of the process is as follows: mixture of wood waste and air plasma in the gasifier transforms into syngas, which thermal energy is recuperated, then part of chemical energy (amount depends on gas composition) transforms into electricity in the combined cycle and thermal energy recuperates.

Modes "a" and "b" (Figure 7.5) differ by plasma calorific value used in the process. In mode "b" the yield of chemical energy from the unit of feedstock mass and syngas quality increases at the expense of high plasma calorific value, at the same time power inputs for heating of the gasification products decrease to the process temperature due to decrease of their specific yield. Thus the specific yield of the electric power from a wood mass unit increases by magnitude (~3.5 MJ·kg^{-1}), comparable to increase in power inputs (~3.3 MJ/kg). The difference between a specific yield of the electric power and specific power inputs changes slightly (4 MJ·kg^{-1}), no less than specific yield of thermal energy (9 MJ·kg^{-1}). That is, in power aspect these regimes are comparable.

On the other hand, if both processes are carried out in equal volumes, nonreversible energy losses in "a" mode will be higher, because of the big Reynolds number.

Figure 7.6 shows the estimation of influence of specific power inputs on specific yield of useful electric power from the system using plasma gasification of wood by various oxidizers and combined cycle. It is obvious that air plasma use allows ~20% increasing of electricity yield from the unit of feedstock mass and application of steam plasma and carbon dioxide makes sense only if plasma have high calorific value (above 14 MJ·kg^{-1} for CO_2 plasma, above 24 MJ·kg^{-1} for H_2O plasma).

Figure 7.7 shows the estimation of influence of power inputs on the specific yield of synthetic liquid fuel from the system using plasma gasification and Fisher-Tropsch process. It is obvious that air plasma use allows increasing of liquid fuel from the unit of feedstock mass more than in 2 times, and use of steam and carbon dioxide plasma (with low calorific value) can raise the cost of end product. It should be noted that the values of specific yield of liquid fuels for gasification by air plasma with low calorific value will be even less than Figure 7.7 shows. It is caused by decrease of efficiency of Fisher-Tropsch synthesis at increase of ballast gases content (N_2 and CO_2). It can be concluded that air plasma gasification is the most simple and promising method of improvement of the technology of combustible gas production from wood for its energy application. This technology allows increasing of the plant output and improvement of the gas quality, which is important for the process control. Estimations show that the specific yield of net electricity increases with the increase of plasma calorific value. It is conditioned by the growth of efficiency of the combined cycle with the increase of the adiabatic temperature of gas combustion.

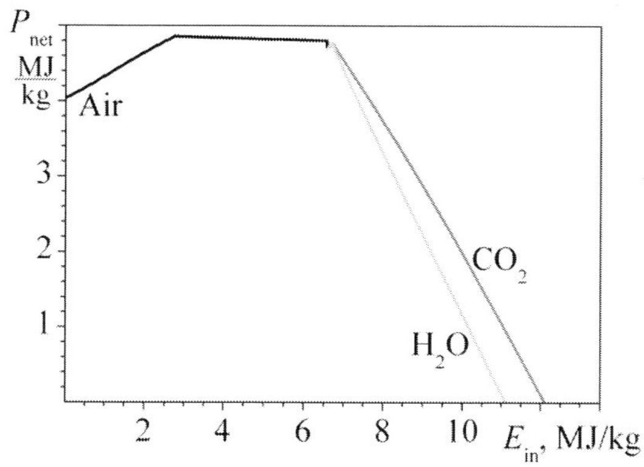

Figure 7.6. Electricity generation from wood in the process of its gasification by different types of plasma.

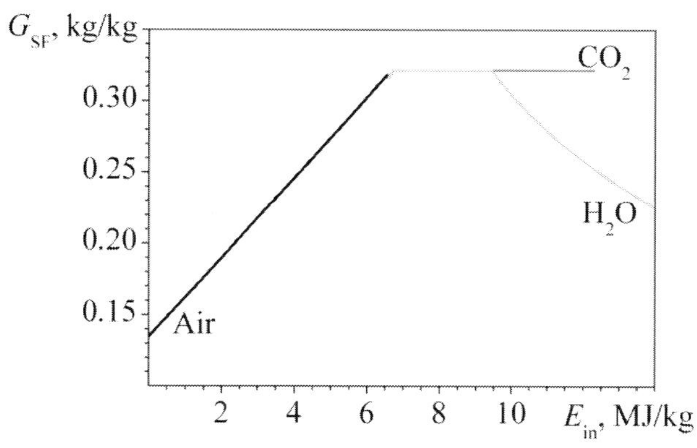

Figure 7.7. Production of liquid fuel from wood in the process of its gasification by different types of plasma.

Influence of Wood Moisture

For complete gasification of 1 kg of absolutely dry wood it is required not less than 969 g air, and for technically seasoned wood (moisture ~20%) — no less than 38.2 g. However, for gasification of technically dry wood with the specified specific air flow rate it is necessary to supply ~6.56 MJ/kg. If this energy is supplied with air, calorific value of plasma will be about 165 MJ/kg. Practically such parameters in low-temperature plasma generators are unattainable, therefore the energy deficiency is compensated at the expense of excess of the oxidizer. Use of part of syngas or inert gases as plasma forming ones can be an alternative way. Then, depending on used plasma forming gas, at temperature of plasma of 5000 - 6000K power inputs for pyrolysis increase to 7.23 - 7.07 MJ·kg^{-1} and 8.30 - 7.46 MJ·kg^{-1} for syngas and nitrogen, respectively. Use of syngas does not essentially change the composition of gasification products, and in case of nitrogen use its content will be ~40.1 - 25.6%.

The disadvantage of syngas use is loss of its thermal energy that increases power inputs. When nitrogen plasma is used, its content in syngas becomes high, rather than total content of CO_2 and N_2 at air plasma gasification with the same temperature.

Calculation of key parameters of air-plasma gasification has been made for wood with moisture 0, 10, 20 and 30%. Figure 7.8 shows the influence of power inputs and wood moisture on specific yields of the basic gaseous products of gasification.

Curves in Figure 7.8 have dependence on power inputs close to linear one. With increase of wood moisture the maximum values of specific yield of hydrogen increase up to humidity of ~20%, then — decrease, and specific yield of carbon monoxide monotonously decreases.

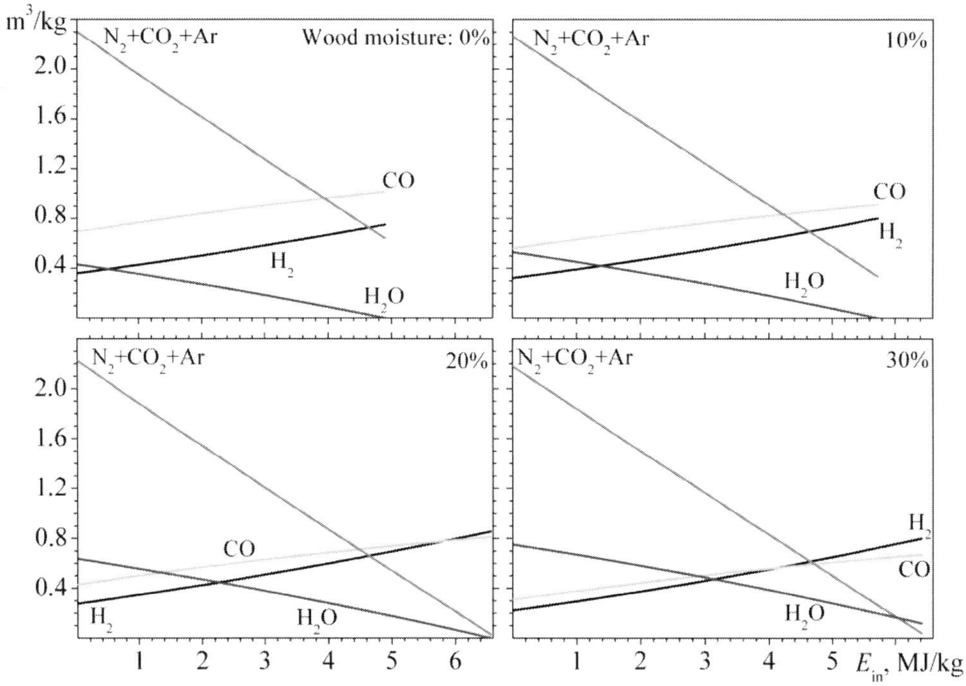

Figure 7.8. Specific yields of the main gaseous products in air-plasma gasification of wood.

This results from the fact that at increase in wood moisture the hydrogen content in it increases, and carbon content decreases. Decrease of the maximum values of the specific yield of hydrogen after ~20% moisture is connected with so-called carbon boundary of gasification (number of carbon atoms in a mixture is equal to the number of oxygen atoms). If the carbon boundary (on moisture) is crossed, it is not necessary to feed the oxidizer for complete allothermic gasification of wood because all carbon is oxidized by oxygen of water and organic part of wood. For wood specified composition the carbon boundary on moisture is ~20.82%.

As a whole, with moisture growth the specific yield of CO and sum of H_2+CO decreases, steam — increases, and for other uncondensed gases — it slightly decreases. At autothermic gasification with moisture growth H_2O specific yield increases, and H_2, CO and sums N_2+CO_2+Ar — decreases. Dependences of a specific yield of chemical and thermal energy in a gas phase from power inputs and moisture are very close to the linear (Figure 7.9). Approximations of calculated values of yield of chemical (7.14) and thermal energy (7.15) repeat the calculated values with precision $\leq 0.05\%$.

$$Q_{ch} = 11{,}584 - 19{,}442 \cdot M + 1{,}546 \cdot E_{in} \tag{7.14}$$

$$Q_{th} = 6{,}3910 - 0{,}9774 \cdot M - 0{,}5454 \cdot E_{in} \tag{7.15}$$

where Q_{th} – yield of thermal energy (without taking into account heat of condensation of steam) on unit of wood mass ($MJ \cdot kg^{-1}$), M – moisture of wood for working mass (in decimal fractions).

To compare the degree of moisture influence and power inputs on yield of chemical and thermal energies it is necessary to reduce one magnitude to another (because M is taken as dimensionless value and E_{in} in dimension), that is to care out handy scaling. It is natural to measure energy input in units of wood *LHV* (17.99 $MJ \cdot kg^{-1}$). Then the dimension factor at $E_{in}/17.99$ in equation 7.14 will be significantly higher in ~10 times than the factor prior to moisture, and in equation 7.13 this factors are comparable. That is, wood moisture influences more significantly on the specific yield of chemical energy and rather weaker on specific yield of thermal energy than power inputs.

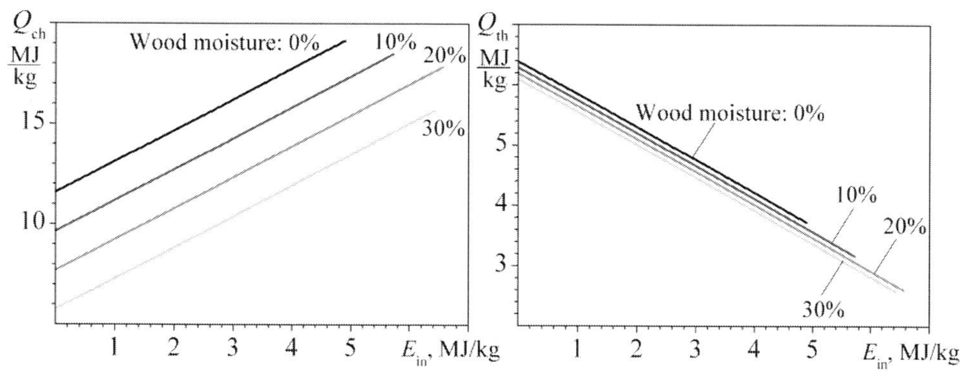

Figure 7.9. Specific yield of chemical Q_{ch} and thermal Q_{th} energy for various values of wood moisture.

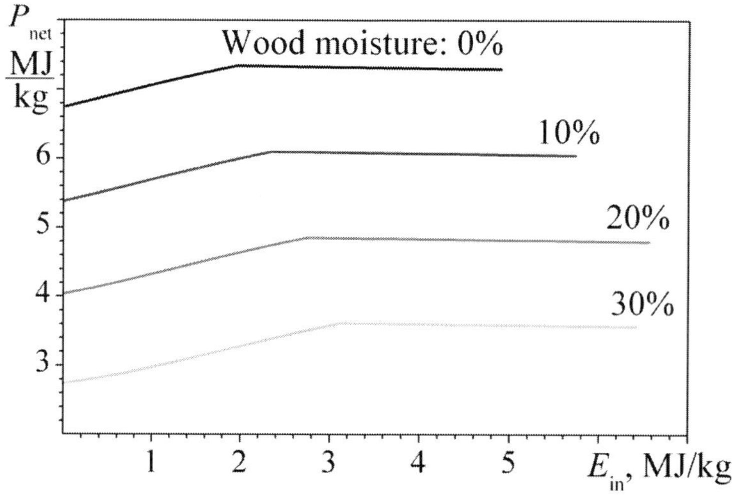

Figure 7.10. Specific yield of electricity at use of a system on the basis of plasma gasification of wood and combined cycle for various values of the primary feedstock moisture.

Recalculation of the equation 7.13 on dry mass shows that the derivative on a fraction of moisture from yield of chemical energy from unit of dry wood mass changes from -7.86 to -16.03 MJ·kg^{-1} over the range of moistures from 0 to 0.3 of mass fractions. Power inputs on removal of 1 kg of water should be less than ~7.86 MJ of thermal energy (in recalculation on dry wood mass) in order to make the process energetically effective over the whole range of moistures. Energy of evaporation of water (taking into account heating to 100°C is taken as ~2.58 MJ·kg^{-1}). That is the efficiency of drying should be no less than 33%.

If we use the equation 7.13 for extrapolation, it is possible to determine the limit value of moisture for down draft autothermic gasification of wood (~59.6 %).

At recalculation of equation 7.13 on dry mass of wood the factor prior to the specific power inputs will not change, as the relation of power inputs on working mass to a share of dry mass is equal to power inputs on dry mass. Thus, without dependence from wood moisture by means of plasma technology it will be possible to increase the specific yield the electric power if efficiency of its generation from syngas is above ~64.7%.

However, in practice the increase in a specific yield of the electric power is achieved at the expense of increase in efficiency of syngas use syngas with growth of its adiabatic combustion temperature. Influence of this factor on specific yield of the useful electric power is shown in Figure 7.10.

Derivative discontinuity of the first derivatives at curves in Figure 7.10 does not have physical sense. They are caused by achievement of adiabatic combustion temperature of syngas of 2000K (on the accepted model the efficiency of the combined cycle remains constant at temperatures above 2000K).

Maxima in breakpoints of specific yield of the electric power and values of power inputs matching them and specific yield of air approximated by linear dependence from wood moisture:

$$E_{in} = 1{,}9595 + 3{,}9535 \cdot M \qquad (7.16)$$

$$P_{net} = 7{,}3432 - 12{,}434 \cdot M \tag{7.17}$$

$$G_{in} = 2{,}1996 - 0{,}026936 \cdot M \tag{7.18}$$

Error of approximation does not exceed 0.03%. Recalculation of equation (7.16) on dry wood mass shows that yield of useful electricity decreases with moisture growth. The derivative on fraction of moisture from the yield of useful electric power from the unit of dry wood mass changes from -5.09 to -10.39 $MJ \cdot kg^{-1}$ in the range of moistures from 0 to 0.3 mass fractions. Power inputs on removal of 1 kg of water from wood should be less than ~5.09 MJ, in order to have the energy efficient process over the whole range of moistures. That is, efficiency of a heat transfer at resistance heating should be more than ~51%.

Dependence of specific yield of chemical energy from plasma calorific value is fractionally linear. This nonlinearity is caused by that calorific value of plasma necessary for gasification is defined as the relation of specific power inputs to specific air flow rate, and relation between any two of the following magnitudes: specific yield of chemical energy, specific power inputs and specific air flow rate can be approximated by the linear equation.

Linear connection of these magnitudes is observed, when the specific air flow rate is in the range from stochiometric gasification to stochiometric incineration. In this range part of combustible products of gasification oxidizes proportionally to increase in the specific air flow rate and, hence, the specific yield of chemical energy decreases proportionally to a share of the oxidized products of gasification. At the expense of the released energy the specific power inputs decrease.

At autothermic gasification of bone-dry wood a calorific value of the drained gas is on ~0.83 MJ/m^3 higher than at gasification of wood with 20% moisture. At transition to allothermic process this difference decreases with growth of power inputs.

In the range of wood moisture between 0 and 20% the composition of dried syngas at constant value of power inputs slightly depends on wood moisture (concentration of the principal components vary no more than on 6%), however with increase of wood moisture its specific yield decreases. The increase of moisture leads to growth of CO_2 concentration and decrease in CO concentration CO, influence of moisture on H_2 and N_2 content in syngas is less considerable and mainly depends on magnitude of power inputs.

Review of Modern Plasma Gasification Technology Development

The considerable promise of the presented method is indirectly proved to be true by all-around activity observed recently in the field of research, and engineering works. Tens of experimental, pilot, and trial installations have been created.

Brief Review of the Implemented Processes

Westinghouse Plasma Corp. (Department of AlterNrg) has developed the technology of one stage plasma gasification (Figure 7.11) [35].The petroleum coke, coal, biomass, household or industrial waste can be used as a feedstock. Updraft gasification process is used. The feedstock is loaded from the top, and the oxidizing agent — air-oxygen blasting — moves from the bottom. Plasma is introduced into the bottom part of the reactor, part of its energy is used for fusion of slag which then is released in liquid state.

Syngas produced as a result of feedstock gasification is removed in the top part of the reactor. The metals containing in slag, go to secondary use, and the remained slag is used in building. Pyrogenesis Canada Inc. developed the technology of two-stage plasma gasification of waste [36].

At the first stage the high-temperature pyrolysis on molten bed in the chamber type reactor based on the metallurgical furnace is carried out. The produced gases polluted by tars go to the second stage where under the action of a plasma stream their complete decomposition takes place.

After that the gas is cooled and treated and goes to the combustion chambers of the gas-piston aggregate. The firm states that produced energy is enough for compensation of own expenses. The firm constructed a mobile installation by productivity ~500 $kg{\cdot}h^{-1}$ for US Air Force. Advanced Plasma Power (APP) developed a pilot plant for RDF gasification using a plasma converter [37].

It is a two stage process. Preliminary prepared waste move into the fluidized bed gasifier with an oxygen-steam blast where it is partially gasified transforming in two products: crude syngas, polluted by small particles of feedstock and tars, and the solid residual. Both these of products arrive in the plasma converter with a melt at the bottom part, and a rod electrode at the top part. The electric arc burns between the melt and electrode. Besides, the oxygen-steam blast is supplied into the converter.

In the plasma converter tars are subjected to cracking, parts of feedstock are gasified, and inorganic components turn to melt. Further syngas is cleared and moves on gas-piston engine for electric power and heat production Its stated that ~40 % of the electric power generated in gas-piston engine are used for work of the installation and, that the offered technology has the negative contribution to emission CO_2 emission.

Figure 7.11. AlterNRG plasma-gasifier.

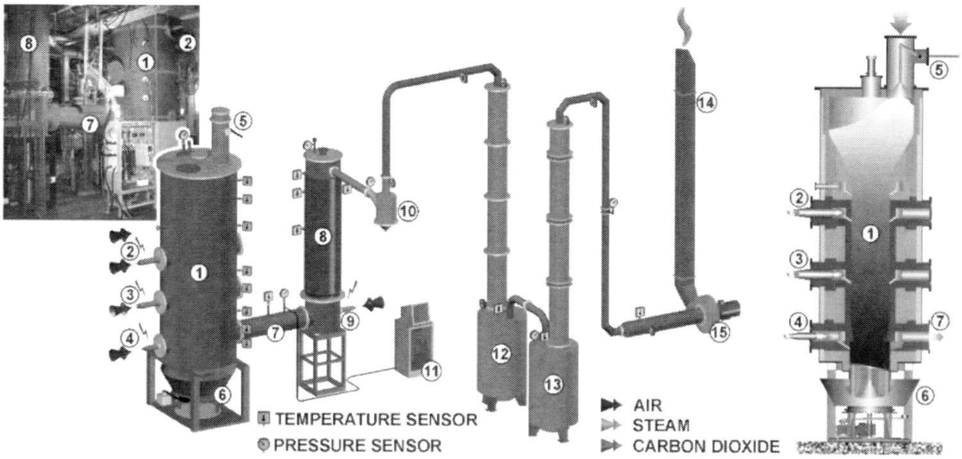

Note: 1. Reactor-gasifier; 2. Main plasma generator; 3. Additional plasma generator (plasma-forming media: air, steam or carbon dioxide); 4. Additional plasma generator; 5. Feeder; 6. Device for slag removal and cooling; 7. Gas duct; 8. Afterburner; 9. Emergency torch; 10. Cyclone; 11. Gas-analytical complex; 12. Spraying scrubber; 13. Packed bed scrubber; 14. Stack; 15. Exhaust fan.

Figure 7.12. General view and schematic diagram of the experimental unit for plasma gasification of waste.

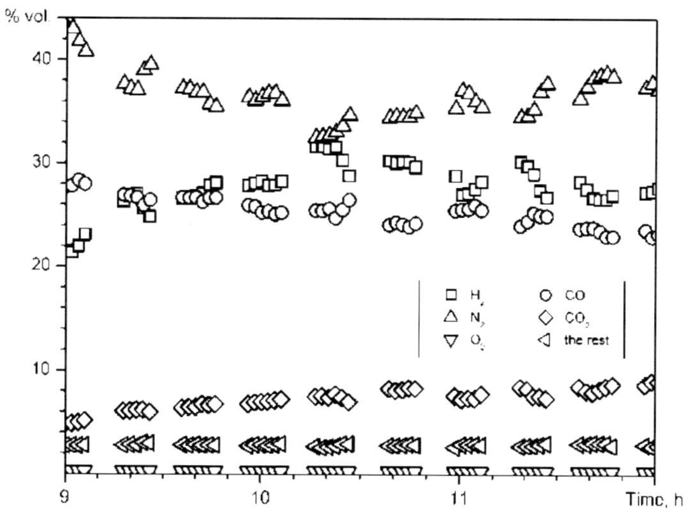

Figure 7.13. Change of syngas composition during the experiment on air-plasma gasification of wood waste.

The experimental unit for investigation of the plasma gasification processes for various kinds of solid hydrocarbon waste has been created at the Institute for Electrophysics and Electric Power of Russian Academy of Sciences [38,39].

In 2010 design of plasma-gasifier has been changed for enrich the ability of experimental researches (Figure 7.12). Calculated output of the down-draft gasifier is 50 kg/h, however, as the experiments have shown, it can be increased to 150 - 200 kg·h^{-1}.

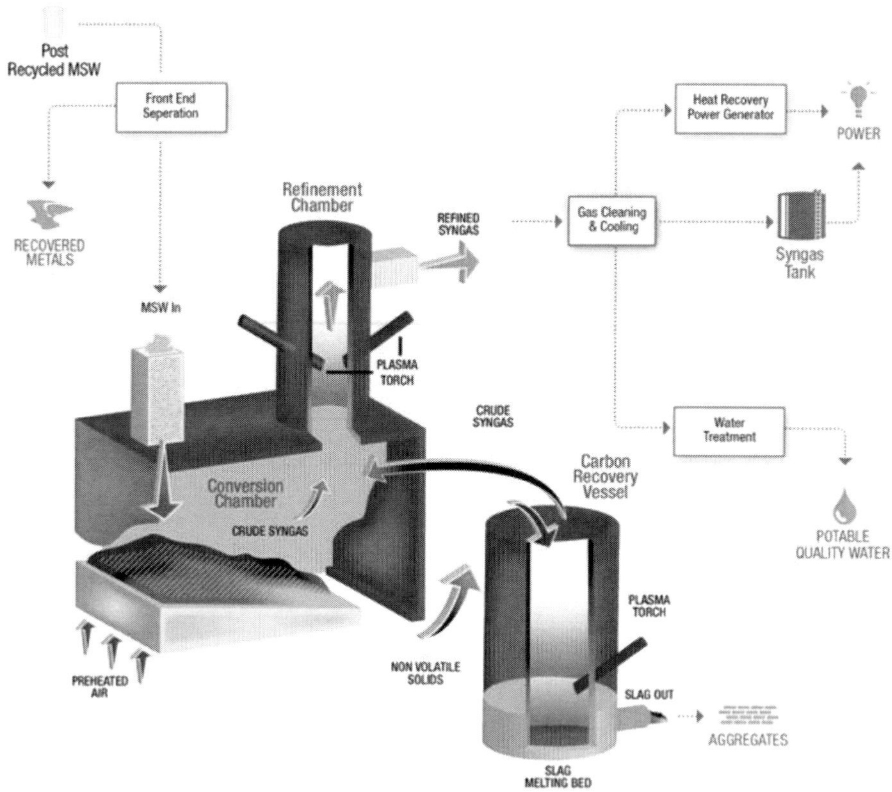

Figure 7.14. Schematic of the installation for plasma gasification Plasco.

One or several three-phase high-voltage plasma generators with rod electrodes are used [40]. Air is the basic plasma forming environment, there is a possibility to use carbon dioxide and steam. Continuous monitoring of syngas composition syngas on an exit from the gasifier is carried out during experiments by means of a complex of mass spectrometer and optical devices.

Temperatures, pressures, flow rates of blast gases, and also electric parameters of plasma generator operation are controlled and displayed. Experimental researches of the process of air-plasma gasification of solid organic fuels including RDF and wood waste are carried out at the installation [41-46].

Figure 7.13 shows the characteristic composition of the produced syngas in the course of experiment on gasification of wood waste with moisture of ~20%.

Plasco Energy Group proposes the system for plasma conversion of crushed waste in syngas (Figure 7.14) [47].

The raw materials move in the conversion chamber from the top and form a flat inclined layer through which from the bottom the air heated by recuperated energy is passed. Crude syngas, produced as a result of air gasification of feedstock by hot air moves to the cleaning chamber.

In the cleaning chamber the syngas mixes up with plasma and is partially cleaned. The solid residuals from the conversion chamber get to the high-temperature chamber of carbon recuperation. In this chamber the inorganic substances turn to liquid slag at the expense of energy of plasma fed into the chamber, and organic substances are gasified.

Crude syngas, received in the carbon recuperation chamber, goes to the conversion chamber. Slag is removed in a liquid state and will be converted to small granules. Syngas from the cleaning chamber goes to the system of cooling and cleaning secondary cleaning. Then it is used for burning in internal-combustion engines for generation of thermal and electric energy. It is stated that from 1 kg of waste (with calorific value 14.2 $MJ \cdot kg^{-1}$) treated by the Plasco conversion system is produced 1 kWh of electricity.

Plasma Generators

The success of implementation of these technologies depends on type of a design of the plasma reactor, its operating modes, character of heat exchange etc. However, the most important is selection of a system of plasma generation which supplies the energy necessary for the organization of process into the reactor and also partially formed the composition of produced gases. There are two basic methods of plasma generation.

In the first one — the electric arcs generating plasma are ignited directly in the reactor between the system of electrodes, which is mechanically its part [48], and are free burning. As a cathode of a direct current arc is used a graphite or metal electrode, and the anode, as a rule, is a conducting melt or a conducting rod injected into melt.

In case of use of an alternating current the polarity of electrodes constantly varies. The second way is use of plasma generators, when one or several arcs burn in the discharge chamber and the received stream of plasma with the help of nozzle apparatus is injected into the reactor volume.

Stabilization of arcs is carried out gas-dynamically (the special organization of a stream of working gas), or by superposition of an external magnetic field. Key parameter of classification of plasma generators is the current: direct or alternative. Electrode systems of DC and AC plasma generators can be rod, toroidal, ring, tubular, and the electrode system in some cases can be simultaneously the discharge chamber.

High power systems both DC [49-51], and AC [52-55] are of the primary interest for implementation of the described technologies. The in-depth analysis of processes in the discharge chambers and design features both of plasma generators and their power supplies shows the essential advantage of application of powerful alternating current plasma generators in comparison with direct current plasma generators, and also arc systems working on melt on a variety of factors. It is possible to familiarize in more details with these systems in the special literature.

Conclusion

Treatment of wood and other vegetative biomass is the most promising direction in development of plasma gasification technologies for the countries possessing the corresponding resources as it is ecologically pure renewed energy source with low content of inorganic components. Many types of solid low-grade fuels, including different kinds of wastes, can be included into the energy balance due to the plasma gasification technology. Efficiency of the electric power generation from them with use of plasma gasification increases with growth of hydrogen content in a feedstock at constant oxygen content. Waste processing by the newest methods will essentially decrease the ecological load on the environment.

AUTHORS' BIOGRAPHIES

Philip Rutberg was born on 22.09.1931

Leningrad State University and Leningrad Polytechnic Institute — MD.

Doctor of Science, Professor, Member (Academician) of Russian Academy of Sciences

Institute for Electrophysics and Electric Power RAS (IEE RAS) — director.

The major interests are concentrated in physics of dense low temperature plasma, discharges in gas flows, powerful gas discharges in dense media, pulsed discharges in liquids, methods of water purification without chemical reagents, as well as constructions of different types of dense plasma generators and their electrical supply systems, plasma technologies for waste treatment of different types, renewable energy generation and synthetic liquid fuels production. He is the author and co-author of more than 450 papers and inventions, 40 patents, and 4 monographs.

Alexander Bratsev was born on 16.06.1948.

Leningrad Electromachine Manufacturing College — BD.

Leningrad Polytechnical Institute — MD.

Ph. D

IEE RAS — Head of Laboratory

His specialty is electrical machines and devices. Area of scientific interests: electrophysics, development and creation of various electrophysical devices using the electric arc discharge, high pressure, high temperature. He is the author and co-author of more than 40 scientific works and of 7 patents.

Vadim Kuznetsov (29.03.1983)

Baltic State University — MD.

IEE RAS — senior staff scientist.

His specialty is aircraft and space rocket construction. Area of scientific interests: plasma processing of solid fuels and waste, calculations, modeling and experimental researches.

He is the author and co-author of 15 scientific works.

Victor Popov (28.04.1974)

Baltic State University — MD.

Ph.D

IEE RAS — Head of Laboratory

His specialty is aviation and space rocket heat engineering. Area of scientific interests: electrophysics, plasma processing of solid fuels and waste, development and creation of plasma technologies. He is the author and co-author of more than 40 scientific works.

Alexander Ufimtsev (15.10.1948)
Novosibirsk State University — MD.
IEE RAS — senior staff scientist.

His specialty is physics. Area of scientific interests: aerohydrodynamics, thermal physics, turbulence, transport phenomenon in continua, physics of low temperature plasma, chemical physics. He is the author and co-author of more than 50 scientific works.

REFERENCES

[1] BP Statistical Review of World Energy, http://www.bp.com (2010)

[2] World Population Data Sheet, http://www.prb.org (2010)

[3] V.V. Klimenko, A.G. Tereshin, O.V. Mikushina, *Russ. Chem. J.*, 52, 11 (2008).

[4] B. Williams, *Oil and Gas J.*, 101, 18 (2003).

[5] C.J. Campbell, *Oil and Gas J.*, 101, 38 (2003).

[6] P. Thipkhunthod, V. Meeyoo, P. Rangsunvigit, B. Kitiyanan, K. Siemanond, T. Rirksomboon, *Fuel*, 84, 849 (2005).

[7] I.S. Masaev *Fuel from Household and Vegetable Waste*, JSC "Neftyanik", Moscow (2002).

[8] M.R. DeLallo, R. Zahkchuk, R.B. Reuther, D.L. Bonk, "Performance and economics of co-Firing a CoaWaste slurry in advanced fluidized-bed combustion" *The 21st Int. Technical Conf. on Coal Utilization and Fuel Systems (Clearwater, Florida USA)* March 18-21 1996.

[9] K. Raghunathan, K.R. Bruce, "Control of emissions from cofiring of coal and RDF" (1997), http://www.nrel.gov/docs/legosti/fy97/26036.pdf

[10] Cioni M., C. La Marca, J. Riccardi, "RDF gasification in a circulating fluidized bed gasifier: characterization of syngas and ashes" *The Clean Choice for Carbon Management (Noordwijk, Netherlands)* April 08-10 2002.

[11] D.W. Green, R.H. Perry, *Perry's Chemical Engineers' Handbook: 8th ed.*, McGraw-Hill, USA (2007).

[12] CalRecovery, Inc., Stanwell Drive, Concord, CA 2454, http://www.calrecycle.ca.gov/publications/Tires/62095001.pdf

[13] F.D. Raniere, L.P. Combs, A.Y. Falk, Experimental investigation of peat hydrogasification, Canoga Park, California, http://www.anl.gov/PCS/acsfuel/preprint%20archive/Files/24_3_WASHINGTON_09-79_0064.pdf

[14] S. Stournas, M. Papachristos, G.B. Kyriakopoulos. Preprint 1987, at *Papers of American Chemical Society, Division of Fuel Chemistry*, 32, 227 (1987).

[15] B. Bilitewski State of the art and new developments of waste to energy technologies *Venice 2006 Biomass and Waste to Energy Symp. Proc. (Venice, Italy)* 29 Nov – 01 Dec 2006 (IWWG and CISA).

[16] A.C. Caputo and P.M. Pelagagge, *Applied Thermal Engineering*, 22, 439 (2002).

[17] J. Nithikul, *Potential of Refuse Derived Fuel Production from Bangkok Municipal Solid Waste*, Asian Institute of Technology School of Environment, Thailand, (2007).

[18] RDF processing systems for alternative fuels, http://www.l-rt.com/index.php?id=60andno_cache=1andL=1andcid=640anddid=534andsechash=aff925cf

[19] D. Chen, X. Zhai, G. Zhou, *Int. Conf. on Sustainable Solid Waste Management (Chennai, India)* Sept 5-7 2007 pp406-414.

[20] A. Kumar, P. Flynn, S. Sokhansanj, *Renewable Energy*, 33, 1354 (2008).

[21] G.W. Huber, B.E. Dale, *Sci. Am.*, 301, 52 (2009).

[22] D. Sperling, D. Gordon, *TR News Magazine*, 259, 3 (2008).

[23] World rubber and tire, industry study with forecasts for 2011 and 2016. Study #2282 February 2008, http://www.freedoniagroup.com/brochure/22xx/2282smwe.pdf.

[24] Monthly world population figures, http://www.census.gov/ipc/www/popclockworld.html

[25] S.G. Maslov, L.I. Inisheva, *Chemistry of Vegetable Feedstock*, 4, 5 (1998).

[26] D.K. Bellman, *Working Document of the NPC Global Oil and Gas Study* (2007), http://www.npc.org/Study_Topic_Papers/4-DTG-ElectricEfficiency.pdf

[27] *Biomass for power generation and CHP* (2007), http://www.iea.org/techno/essentials3.pdf

[28] Ladwig M., Lindvall K., Conzelmann R., *VGB Conf. Gas Turbines and Operation of Gas Turbines* (*Friedrichshafen, Germany*) June 5-6 2007.

[29] T. Ichinose, M. Hishida, E. Ito, *Mitsubishi Heavy Industries Technical Review*, 45, 7 (2008).

[30] E. Ito, K. Tsukagoshi, A. Muyama, J. Masada, T. Torigoe, *Mitsubishi Heavy Industries Technical Review*, 47, 19 (2010).

[31] N. Lewis, G. Crabtree, *Basic Energy Sciences Workshop on Solar Energy Utilization* (2005).

[32] X.G. Zhu, S.P. Long, D.R. Ort, *Current Opinion in Biotechnology*, 19, 153 (2008).

[33] M. Ishikawa, M. Terauchi, T. Komori, J. Yasuraoka, *Mitsubishi Heavy Industries Technical Review*, 45, 15 (2008).

[34] A.N. Bratsev, V.A. Kuznetsov, V.E. Popov, A. Ph. Rutberg, A.A. Ufimtsev, S.V. Shtengel, *High Temperature Material Processes (New York)*, 13, 147 (2009).

[35] Alter NRG, 215, 4000 - 4th Street SE Calgary, Alberta T2G 2W3, http://alternrg.com

[36] Pyrogenesis Canada Inc., 1744 William St. Suite 200 Montreal Quebec Canada H3J 1R4, http://www.pyrogenesis.com

[37] Advanced Plasma Power Ltd., B3 Marston Gate South Marston Business Park Swindon Wiltshire SN3 4DE United Kingdom, http://www.advancedplasmapower.com

[38] A. N. Bratsev, V. E. Popov, A. F. Rutberg and S. V. Shtengel', *High Temp.*, 44, 823 (2006).

[39] A.N. Bratsev, V.E. Popov, S.V. Shtengel, A.Ph. Rutberg, *High Temp. Mater. Processes (New York)*, 10, 549 (2006).

[40] F. G. Rutberg, A. V. Pavlov, S. D. Popov, A. I. Sakov, E. O. Serba, V. A. Spodobin and A. V. Surov, *High Temp.*, 47, 175, (2009).

[41] A.N. Bratsev, I.L. Glezin, V.B. Kovshechnikov, I.I. Kumkova, V.A. Kuznetsov, V.E. Popov, S.V. Shtengel, A.A. Ufimtsev *Proc. XXVIII Intern. Conf. on Phenomena in Ionized Gases (Prague)* July 15-20 2007 (Institute of Plasma Physics AS CR, Czech Republic) pp 1848-1851 (2007).

[42] V.A. Kuznetsov, A.N. Bratsev, V.B. Kovshechnikov, I.I. Kumkova, V.E. Popov, S.V. Shtengel, A.A. Ufimtsev, *Pulsed Power Conf. 2007 16th IEEE International (Albuquerque)* June 17-22 2007 vol 2 (Madison, WI USA) pp 1223-1226 (2007).

[43] V.E. Popov, A.N. Bratsev, V.A. Kuznetsov, S.V. Shtengel, A.A. Ufimtsev, *J. Phys. Conf. Ser.* 275. 012015 (2011).

[44] A.N. Bratsev, I.I. Kumkova, V.A. Kuznetsov, V.E. Popov, S.V. Shtengel', A.A. Ufimtsev, *IOP Conf. Ser.: Mater. Sci. Eng.* 19, 012004 (2011).

[45] Ph.G. Rutberg, A.N. Bratsev, V.A. Kuznetsov, V.E. Popov, A.A. Ufimtsev, S.V. Shtengel', *Biomass Bioenergy*, 35, 495 (2011).

[46] A. N. Brattsev, V. A. Kuznetsov, V. E. Popov and A. A. Ufimtsev, *High Temp.*, 49, 244 (2011).

[47] Plasco Energy Group Inc., 1000 Innovation Dr., Suite 400 Ottawa, ON K2K 3E7 http://www.plascoenergygroup.com

[48] N.A. Barcza 1987 Application of Plasma technology to Steel processing *Plasma Technology in Metallurgical Processing* ed. J. Feinman (Iron and Steel Society, Warrendale, Pa.) chapter 11 pp 131-148.

[49] D.A. Maniero, P.F. Kienast, C. Hirayama, *Westinghouse Engr.,* 26, 66 (1966).

[50] M.F. Jukov, I.M. Zasipkin, A.N. Timoshevskiy et al. *Electric arc thermal plasma generators*, Nauka, Novosibirsk (1999).

[51] Westinghouse Plasma Corp. Plasma Center P.O. Box 410 I-70 Exit 54 Madison, Pa. 15663 http://www.westinghouse-plasma.com

[52] Ph.G. Rutberg, A.A. Safronov, S.D. Popov, A.V. Surov and Gh.V. Nakonechny, *Plasma Phys. and Control. Fusion*, 47, 1681 (2005).

[53] Ph. Rutberg, *Physics and Technology of High Current Discharges in Dense Gas Media and Flows*, Nova Science Publishers, New York, (2009).

[54] P.G. Rutberg, *Three-phase plasma torch // Some aspects of gas discharge plasma research and creation of strong magnetic fields*, Nauka, Leningrad (1970).

[55] J. Anderson, *Gasodynamic Lasers: Introduction*, Mir, Moscow (1977).

In: Syngas
Editors: Antonius Indarto and Jelliarko Palguandi

ISBN: 978-1-62100-870-5

Chapter 8

Role of Electrolysis in Regenerative Syngas and Synfuel Production

Qingxi Fu*
European Institute for Energy Research (EIfER), Karlsruhe, Germany

Abstract

In the long term, hydrocarbon fuels, characterised by their compatibility with existing infrastructures and engine technologies and their extremely high energy densities, need to be produced from regenerative energy sources to realise a sustainable transport especially a sustainable aviation. As most renewable energies are harvested in the form of electricity, a system to convert electrical energy to chemical energy stored in hydrocarbon fuels is therefore to be developed. Electrolysis, a technology that has been applied for more than 100 years, is believed to be able to play an import role in such an energy conversion system. Starting from electrolysis, various routes can be followed to produce liquid fuels from H_2O, CO_2 and renewable electricity. Syngas is considered as the key intermediate energy carrier that can be produced either by water electrolysis followed by reverse water gas shift (RWGS) reaction or directly by co-electrolysis of CO_2 and H_2O. In the present chapter, various conversion routes are described. Individual technologies that can be integrated into such an energy conversion system are briefly reviewed, with a focus on H_2O and CO_2 electrolysis. Economic perspective of the combined high-temperature CO_2/H_2O co-electrolysis-FT (Fischer-Tropsch) process, which is identified to be potentially the most efficient pathway, is presented. Finally research and development needs towards the realisation of regenerative syngas and synfuel production are proposed.

8.1. Introduction

Petroleum-derived hydrocarbon fuels in liquid form (e.g. gasoline, diesel and jet fuel) provide the major part of the energy for the transport sector. While other energy carriers such

* E-mail address: fu@eifer.org.

as batteries, ultra capacitors and hydrogen could play potentially an increasing role for short-range and light-duty transport, liquid hydrocarbon fuels, featured by their high energy densities, will continuously be needed for heavy-duty trucks, ships and especially airplanes.

As petroleum is depleted with time [1], alternative means of producing hydrocarbon fuels from other energy sources need to be developed. For a certain limited period of time, other fossil fuels (natural gas and coal) can be converted into liquid fuels using the GTL (gas-to-liquids) and CTL (coal-to-liquids) processes, which have already been commercialised since many decades. However, there is a need for decarbonising the transport as the GHG (greenhouse gas) emission from the transport sector is increasing rapidly (projected to increase by nearly 50% by 2030 and more than 80% by 2050 given current trends [2]), which makes the GTL and CTL options highly unfavourable. Therefore, in the long term, hydrocarbon fuels need to be produced from regenerative energy sources to realise a sustainable transport especially a sustainable aviation.

One option is to convert biomasses to biofuels. Conventional or first-generation biofuels are made from sugar, starch, and vegetable oil and are on the market in considerable amounts today, e.g. bioethanol in USA and Brazil and biodiesel in Europe. The increasing criticism of the sustainability of first-generation biofuels leads to the development of advanced biofuels (second- and third-generation) that can be produced from non-food biomass feedstocks such as woody and cellulosic plants and waste materials via either bio-chemical or thermo-chemical pathways. The so-called biomass-to-liquids (BTL) process is based on thermo-chemical conversion of biomass into syngas followed by Fischer-Tropsch (FT) synthesis to yield the high-quality "designer fuel". Advanced biofuel technologies including the BTL technology is still under development and have not yet been commercialised. Today, biofuels provide around 3% of total road transport fuel globally and considerably higher shares are achieved in certain countries. Brazil, for instance, met about 21% of its road transport fuel demand in 2008 with biofuels. In the United States, the share was 4% of road transport fuel and in the European Union (EU) around 3% in 2008. According to projections of IEA (International Energy Agency), globally biofuels could provide 8% of the total road transport fuel demand in 2035 [3] and even 27% in 2050 [4].

Another option is to convert renewable electrical energy into chemical energy stored in liquid fuels. To reduce the carbon footprint and for the sake of sustainability in the power generation sector, the utilisation of renewable resources, especially wind and solar energies, has been strongly scaled up during the last decade. Globally, the cumulative installed wind power capacity has grown at an average rate of around 30% per year since 2000 (18 GW in 2000 and around 195 GW at the end of 2010) [5-6]. Solar PV (Photovoltaic) was the fastest-growing renewable power technology worldwide. Estimates suggest that cumulative installed capacity of solar PV reached roughly 40 GW at the end of 2010, up from 1.5 GW in 2000 [5]. The inherent intermittent feature of wind and solar power, however, leads to grid stability problem, as evidenced for instance by more frequent negative electricity price on the spot market in Germany as more and more wind power is brought online [7]. Massive Electricity Storage (MES) (or grid-scale energy storage) with multi-MW capacity and multi-hours of application duration is thus considered critical to further increase the penetration of renewable power [8-9]. However, currently available MES technologies are not ready for full-scale deployment. Pumped hydro and compressed-air energy storage (CAES) are well established but are limited by the availability of suitable geologic sites. Three technologies based on the conversion between electrical and chemical energy – the sodium-sulfur battery, vanadium redox flow battery and zinc bromide flow battery – are potential MES candidates.

Table 8.1. Gravimetric and volumetric energy densities of various on-board energy carriers (based on lower heating values)

Type	Energy density without package		Energy density with package	
	MJ kg^{-1}	MJ l^{-1}	MJ kg^{-1}	MJ l^{-1}
Diesel	42.8	35.8	35.0	28.7
Gasoline	43.4	32.4	34.4	25.4
Compressed hydrogen, 700 bar	120	4.7	5.8	2.9
Liquid hydrogen	120	8.5	6.1	4.8
Compressed natural gas (CNG), 200 bar	48.6	8	7	3
Liquefied natural gas (LNG)	48.6	20.8		
Liquefied petroleum gas (LPG)	46.6	23.7		
Ammonia	18.6	14.1		
Methanol	20.1	16.0	16.3	12.8
Ethanol	27.0	21.3	22.0	17.3
Dimethyl ether (DME)	28.9	19.2		
Battery, lead-acid			0.14	0.36
Battery, NiMH, high power design			0.25	0.49
Battery, Li-ion			0.54-0.9	0.9-2.2
Ultracapacitor			0.02	0.026

Note: main data sources: [11-13].

Electrolysis offers another attractive alternative to convert efficiently the electrical energy into chemical energy stored in fuels. For instance, water electrolysis can be used to produce hydrogen fuel using renewable electricity. It is possible to produce again electricity in a fuel cell using hydrogen, although the round-trip efficiency is relatively low mainly due to the large efficiency loss in the electricity-regeneration step. Recently a reversible solid oxide cell (SOCs) operating with methane-oxygen has been proposed for an improved round-trip efficiency [10]. In view of the need of sustainable fuels for transportation, however, a back conversion of fuel to electricity is irrelevant. Such a one-direction conversion of electrical energy into chemical energy can already contribute to the grid stabilisation through prompt consumption of surplus renewable power, which is beyond the digestion capability of the grid.

Water electrolysis is a well established technique for large-scale electricity-to-fuel conversion, producing hydrogen fuel and oxygen as by-product. The utilisation of hydrogen as a transportation fuel, however, faces several challenges such as relatively low net volumetric and gravimetric energy density (due to the need of heavy pressure vessels or cryogenic tanks) (Table 8.1, [12-13]), big energy penalty for hydrogen compression (10-20% of higher heating value (HHV) of hydrogen for compression to 800 bar) or liquefaction (ca. 40% of HHV of hydrogen) [14-15] and a lack of distribution infrastructure that requires huge investment to be established. Synthetic liquid fuels (hydrocarbons and alcohols) with high energy densities, in contrast, are completely compatible with existing engine technology and distribution infrastructure and are thus strongly preferred.

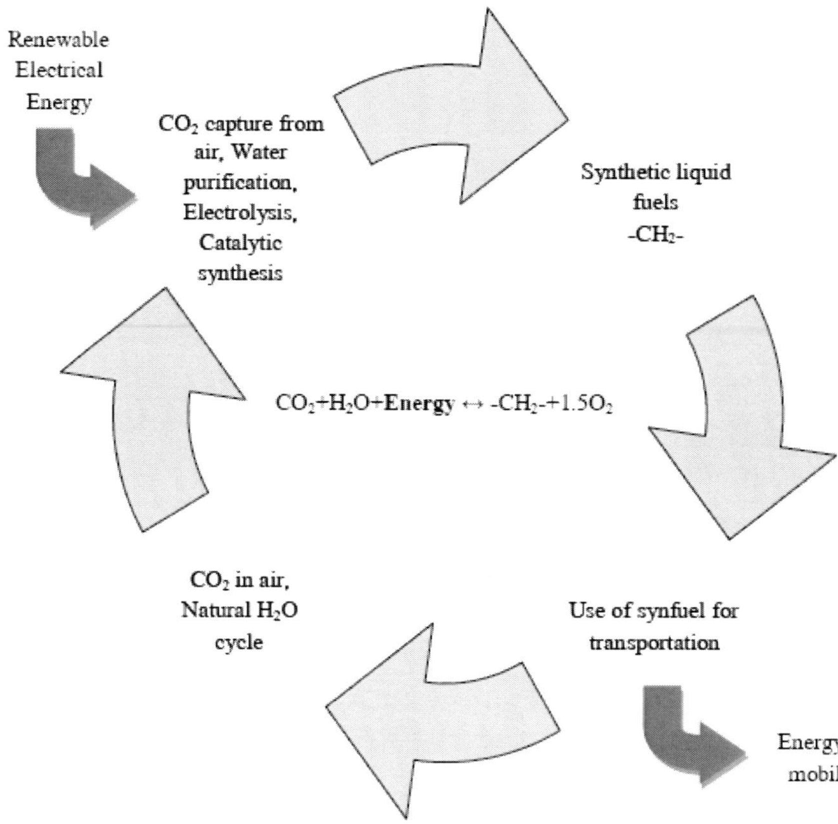

Figure 8.1. Energy concept for a sustainable transportation (especially a sustainable aviation) in the future: renewable electricity powers mobility through a synthetic liquid fuel cycle.

While alcohols can also find application for the road transport, it is probably hard to find an alternative to hydrocarbon fuels (jet fuels) for the aviation application. To form hydrocarbon or alcohol fuels, hydrogen produced by water electrolysis should be combined in some way with carbon feedstock. Considering that the burning products of hydrocarbon or alcohol fuels are CO_2 and H_2O, CO_2 can be used as the carbon feedstock to close the material loop.

Figure 8.1 shows schematically the as-described synthetic liquid fuel cycle for powering sustainable transportation (especially sustainable aviation) in the future using renewable electricity. Through electrolysis and subsequent catalytic synthesis process, CO_2 and H_2O are recycled into synthetic liquid fuels. The consumption of liquid fuels in motor engines releases again CO_2 and H_2O, forming a closed loop. In this sense, CO_2 and H_2O serve only as the medium to absorb the renewable energy and transform it into a practically useful form.

In the present chapter, various routes to produce liquid fuels from H_2O, CO_2 and renewable electricity are described. Individual technologies that can be integrated into such an energy conversion system are briefly reviewed, with a focus on H_2O and CO_2 electrolysis. Economic perspective of the combined high-temperature CO_2/H_2O co-electrolysis-FT process, which is identified to be potentially the most efficient pathway, is then presented. Finally research and development needs towards the realisation of regenerative syngas and synfuel production are proposed.

8.2. Electrolysis-Based Conversion Routes and Processes

Starting from electrolysis, various routes can be followed to convert H_2O and CO_2 into synthetic liquid fuels using renewable electricity as the energy source. Some routes have been depicted in Figure 8.2. H_2 derived from H_2O electrolysis can be reacted with CO_2 through the reverse water gas shift (RWGS) reaction, forming syngas ($CO+H_2$), which can be converted into methanol or hydrocarbon liquid fuels using technologically mature processes (methanol synthesis and FT synthesis, respectively). Alternatively, direct reaction of CO_2 and H_2 can also yield methanol that can be further converted to gasoline using the methanol-to-gasoline (MTG) process. Dimethyl ether (DME), as a good diesel substitute, can be produced by methanol dehydration. Alternatively, syngas can be produced through the H_2O/CO_2 co-electrolysis process, which has the potential to simplify the reaction system by avoiding a separate RWGS reactor.

Before the conversion processes, the feedstock (H_2O and CO_2) with required purity must be acquired. In principle, H_2O and CO_2, which are released into the atmosphere upon the consumption of liquid fuels in vehicles, have to be collected from the atmosphere, as dictated by a closed material loop (Figure 8.1). In the case of H_2O, thanks to the natural cycling in the biosphere, it can be taken conveniently from seas, lakes, rivers or underground. It needs only to be purified before entering the electrolyser.

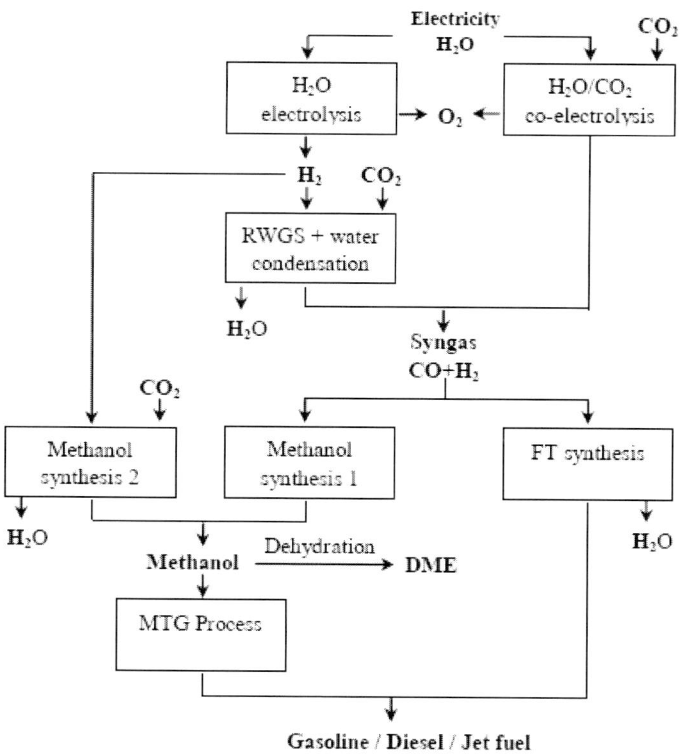

Note: RWGS: Reverse Water Gas Shift; FT: Fischer-Tropsch; MTG: Methanol-to-Gasoline; DME: Dimethyl Ether.

Figure 8.2. Various routes for liquid fuel production from renewable electricity, H_2O and CO_2.

In the case of CO_2, ideally it should be extracted from the atmosphere at any place. Alternatively, it can be derived from the use of biomass, taking advantage of the natural photosynthesis process for CO_2 recycling. For a certain period of time, CO_2 can be recycled from the use of fossil fuels in power plants or industries (e.g. cement production, steel production and ammonia production), which supply concentrated CO_2 sources and hence lower capture cost in comparison to extraction from ambient air. It should be noted that integrating fossil fuel-derived CO_2 in the synfuel cycle results in net CO_2-emission to the atmosphere and is thus not sustainable in the long term.

In the following paragraphs individual processes involved in the above-mentioned conversion routes are described.

8.2.1. H_2O Purification and CO_2 Collection

Generally deionised water is required to avoid damage to the electrolyser system (e.g. corrosion or scaling of the steam generator for the high-temperature electrolyser due to presence of minerals in water). Large-scale implementation of renewable synfuels requires large quantities of water, which might necessitate the use of seawater in light of limited fresh-water resources on the Earth. The most commonly used seawater desalination technologies are reverse osmosis and thermal processes such as multi-stage flash and multi-effect distillation. Due to its much lower energy consumption than its thermal alternatives, reverse osmosis has gained much wider acceptance. Desalinated water is typically produced with a cost of less than 1 US$ per cubic meter [16], which translates into a very small fraction of the water purification cost in the final total production cost of synfuel (less than one percent, see below).

In terms of CO_2 collection, concentrated point sources can be first exploited in the early-phase adoption of renewable electricity-derived synfuels. CO_2 can be captured from flue gases of coal- or natural gas-based power plants using mature separation technologies mainly based on absorption through liquid solvents. Depending on the type of the power plant and the CO_2 concentration in the gas stream, the capture cost ranges from 25 to 60 US$/t$CO_2$ (tonne of CO_2) [17]. For some industrial processes (ammonia, ethylene oxide and fermentation) that generate nearly pure CO_2 streams, only compression cost (6-12 US$/t$CO_2$) is needed to facilitate transport and storage (ca. 20 US$/t$CO_2$ inclusive transport cost). As mentioned above, CO_2 capture from ambient air can completely close the carbon loop to achieve fully carbon-neutral and sustainable synfuels. Although more energy is needed to separate CO_2 from air than from flue gases due to the big concentration difference (0.04% vs. 3-15%), air capture decouples the processes of CO_2 emission and capturing in time and location. An air capture plant can be constructed at any favorable location. The practice of removing CO_2 from air dates back to the 1940s in the context of producing CO_2-free air and has received in recent years more attention in the context of CO_2 emission compensation [18]. Typically hydroxide solutions (e.g. NaOH) have been used as CO_2 absorbents to form carbonates [19]. With currently available technologies, the cost of CO_2 capture from air has been estimated to be around 100-200 US$/t$CO_2$ [20-21]. Recently a new CO_2 air capture method with significantly reduced energy demands has been developed [18,22]. It is based on an anionic exchange resin that readily absorbs CO_2 from air when it is dry and releases the CO_2 again when it is exposed to moisture. The CO_2 capture cost with this new humidity swing sorbent is estimated to be around 250 US$/t$CO_2$ today and has a potential to be reduced to as low as 30 US$/t$CO_2$ as the technology matures [18, 22].

8.2.2. Electrolysis of H_2O and CO_2

Currently three main technologies are used to realise H_2O/CO_2 electrolysis: alkaline electrolysis with a liquid electrolyte, PEM (polymer electrolyte membrane) electrolysis with a solid acidic polymer electrolyte and high-temperature solid oxide electrolysis (SOE) with a solid oxide electrolyte (mainly yttria-stabilised zirconia, YSZ). Due to their low operation temperatures, alkaline and PEM electrolysis are also called low-temperature electrolysis. Beside solid oxide electrolysis, another type of high-temperature electrolysis with a molten carbonate (Li_2CO_3) electrolyte, a Ti cathode and a graphite anode has been recently demonstrated to be able to electrolyse CO_2 into CO and O_2 at 900°C [23]. Here the molten carbonate acts also as a reactant. Since this technique is still at a very early stage of development, it will not be further discussed in the present section.

8.2.2.1. Thermodynamics

The dissociation of H_2O and CO_2 can be expressed by the following equations:

$$H_2O_{(l/g)} + \Delta H \rightarrow H_{2(g)} + \tfrac{1}{2}\, O_{2(g)} \qquad \Delta H^o = 286 \text{ kJ/mol} \tag{8.1}$$

$$CO_{2(g)} + \Delta H \rightarrow CO_{(g)} + \tfrac{1}{2}\, O_{2(g)} \qquad \Delta H^o = 283 \text{ kJ/mol} \tag{8.2}$$

Both reactions are endothermic. The total energy needed to split H_2O or CO_2 corresponds to the enthalpy of reaction (ΔH), which consists of two parts according to the expression:

$$\Delta H = \Delta G + T\Delta S \tag{8.3}$$

where ΔG is the free energy of reaction and corresponds to the minimum share of ΔH which has to be applied as work (here electricity for electrolysis), and the term $T\Delta S$ is the product of the thermodynamic temperature T and the entropy of reaction ΔS and represents the maximum share of ΔH which can be applied as thermal energy to the process.

According to Faraday's law, the energy demand for the electrolysis of H_2O or CO_2 can be correlated to the cell voltage. More specifically, the Nernst voltage (V_N) of the cell can be expressed by

$$V_N = \Delta G / 2F \tag{8.4}$$

where F is the Faraday's constant (96485 C mol^{-1}). The Nernst voltage represents the voltage at which the system is in equilibrium (zero current). A cell voltage higher than V_N is required to draw current through the cell so as to start the electrolysis process. The cell voltage related to the enthalpy of reaction is called thermoneutral voltage (V_{tn}):

$$V_{tn} = \Delta H / 2F \tag{8.5}$$

At thermoneutral voltage, the electricity input to the system exactly matches the total energy need of the electrolysis reaction. The electrolyser works under so-called thermoneutral mode. When the cell operating voltage (V_{op}) is lower than the thermoneutral voltage but higher than the Nernst voltage ($V_N < V_{op} < V_{tn}$, endothermic mode), external thermal energy is needed for energy balance. When the operating voltage is even higher than the thermoneutral voltage ($V_{op} > V_{tn}$, exothermic mode), extra thermal energy has to be removed from the electrolyser (cooling) to avoid overheating.

Figure 8.3 shows the temperature dependence of the energy demand and corresponding cell voltage for the electrolysis of H_2O and CO_2. It should be noted that here the partial pressure for all species involved in the electrolysis reactions (1) and (2) is assumed to be 1 bar. Under real electrolysis conditions, the partial pressures of gaseous species are often different so that the actual Nernst voltages (more practically open circuit voltages) deviate from the values shown in this figure. In the case of H_2O electrolysis, the phase change at 100°C leads to a discontinuity of the enthalpy curve. At a first glance it seems that steam electrolysis needs less energy than liquid water electrolysis. This is only true when steam instead of liquid water is already available as a feedstock without any energy penalty. In fact, in most cases additional energy is needed for the evaporation of water. Taking into account the water evaporation enthalpy also for the steam electrolysis, the total energy need for H_2O electrolysis is marked in Figure 8.3 with a dashed line.

A notable feature for both H_2O and CO_2 electrolysis is the rapid decrease of the electrical energy demand (ΔG) as temperature increases, while the total energy demand (ΔH) remains essentially unchanged.

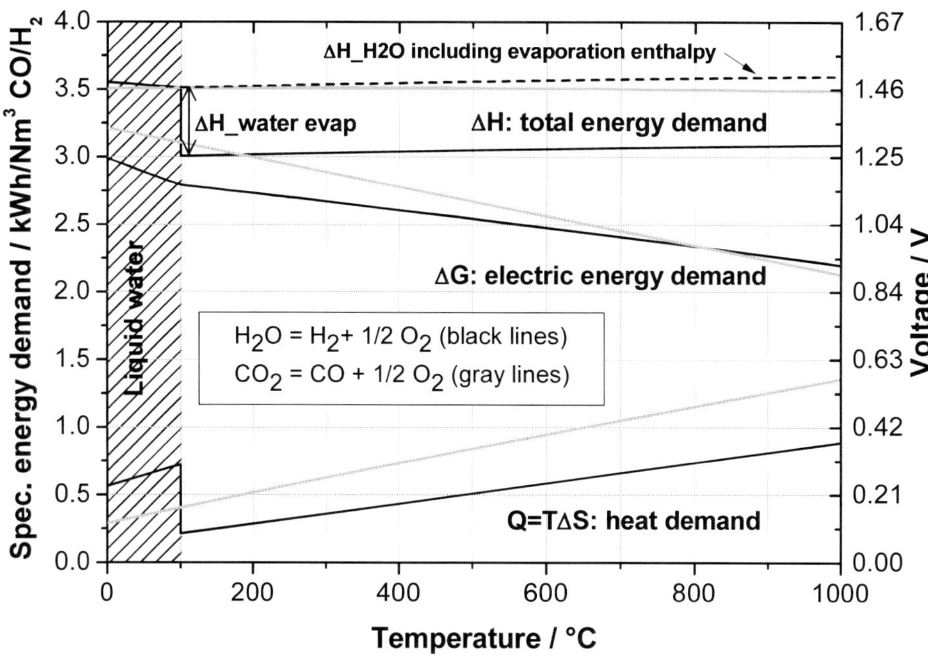

Note: data calculated using HSC Chemistry 5.11 at a pressure of 1 bar.

Figure 8.3. Temperature dependence of the thermodynamic data for H_2O and $\mathbf{CO_2}$ splitting.

Table 8.2. Thermoneutral voltages and Nernst voltages (in Volt) for the electrolysis of H_2O and CO_2 at different temperatures (extracted from Figure 8.3)

Parameter	H_2O electrolysis			CO_2 electrolysis		
	25 °C	80 °C	800 °C	25 °C	80 °C	800 °C
V_{tn}	1.48	1.47	1.29 (1.50[a])	1.47	1.47	1.46
V_N	1.23	1.18	0.98	1.33	1.31	0.98
max. heat share [b]	17%	20%	34%	10%	11%	33%

Note: [a] including the water evaporation enthalpy. [b] calculated using $(1\text{-}V_N/V_{tn\_25\,°C})\times 100\%$, where $V_{tn\_25\,°C}$ is the V_{tn} at 25 °C representing the Higher Heating Value (HHV) of H_2 or CO.

This provides a chance to replace a significant part of the electrical energy with thermal energy, which represents the main advantage of high-temperature electrolysis since in most cases heat has a lower economic value than electricity. Table 8.2 lists the V_{tn} and V_N values for the electrolysis of H_2O and CO_2 at three temperatures as a measure of total energy demand and minimum required electrical energy, respectively. The maximum share of thermal energy is also calculated. The necessary electrical energy for H_2O electrolysis is only 0.98 V at 800 °C instead of 1.18 V at 80 °C. Correspondingly, up to 34% of the energy can be supplied as heat at 800 °C in comparison to 20% at 80 °C. The effect is more pronounced for CO_2 electrolysis, with the maximum heat share increasing from 11% to 33% as the temperature increases from 80 to 800 °C.

The global energy conversion efficiency of an electrolyser system is defined by

$$\eta_{sys} = \frac{output\ chemical\ energy}{total\ input\ energy} = \frac{output\ chemical\ energy}{electrical\ input + heat\ input} \tag{8.6}$$

The output chemical energy can refer to either the lower heating value (LHV) or the higher heating value (HHV) of H_2/CO. While the LHV of CO is the same as its HHV (3.51 kWh/Nm^3), there is a significant difference between the LHV (3.00 kWh/Nm^3) and the HHV (3.54 kWh/Nm^3) of H_2 (LHV/HHV=0.845). Considering the fact that in most cases liquid water is used as the feedstock, the energy needed for water evaporation (either electricity or heat input to the system) should be taken into account so that the HHV of H_2 should be used for a proper measure of the system efficiency. Regarding the energy input term, besides electrical energy, heat energy may also be integrated in the high-temperature electrolyser for the evaporation of water, the preheating of feedstock, the heating of the stack (for isothermal operation) or for compensating for unavoidable thermal insulation losses due to radiation and convection.

For the low-temperature electrolysis, normally only electrical energy is needed. In an electrolyser system, the electricity input is used to power not only the stack, but also the peripheral devices such as circulation and feed water pumps, solenoid valves and power converters (rectifier).

The energy use outside the electrolyser stack represents typically an additional efficiency loss of 5-10% [24].

If only electrical energy is consumed in the stack and the current efficiency or so-called Faraday efficiency is 100%, the stack efficiency for H_2O electrolysis can then be expressed as a function of cell voltage:

$$\eta_{stack,HHV} = \frac{V_{tn}^{o}}{V_{cell}} = \frac{1.48\,(V)}{V_{cell}\,(V)} \tag{8.7}$$

In a real stack, the current efficiency will be slightly lower than 100% due to the existence of stray currents, H_2 diffusion through the electrolyte membrane, possible partial electronic conduction of the electrolyte and gas leakage.

8.2.2.2. Brief History

Some important historical events related to water electrolysis are listed in Table 8.3 with references. The phenomenon of water electrolysis was discovered more than 220 years ago. As a well-established technology, alkaline water electrolysers have been used commercially for more than 100 years. In the 1920s and 1930s several large 100-MW size plants were built in Canada, Norway and elsewhere, mainly for the production of ammonia fertilizers based on cost-effective hydro-power. In 1951 Lurgi (Germany) presented the first commercially available high-pressure alkaline water electrolyser at 30 bar. Advanced alkaline electrolysers with improved energy efficiency have been developed since the introduction of the "zero gap" concept in 1967 by Costa and Grimes. Advanced electrolysers are at a pre-commercial stage.

In comparison to alkaline electrolysis, PEM electrolysis, also known as Solid Polymer Electrolyte (SPE) water electrolysis, is a fairly young technology. In 1966 the first PEM electrolysis system was built by General Electric and in 1987 the first 100-kW system was delivered by BBC.

The development of high-temperature electrolysis using stabilised-zirconia electrolyte for the dissociation of CO_2 and H_2O was started in 1960s mainly under NASA (National Aeronautics and Space Administration) contracts for the purpose of O_2-regeneration from respiratory CO_2 and H_2O in manned spacecrafts. From the middle of 1970s to the end of 1980s, as initiated by the so-called oil crisis in 1973, development work unrelated to space exploration was conducted worldwide, notably by Dornier and Lurgi in Germany in the framework of the "HOT ELLY" (*High operating temperature* electrolysis) project and by Westinghouse, both using tubular solid oxide cells operating at ca. 1000 °C. By the end of the HOT ELLY project, a 2-kW electrolysis module consisting of 100 electrolysis tubes (each tube comprising 10 cells) connected in series and in parallel was built up in 1987 and tested under elevated pressure (3.5 bar). Each 10-cell electrolysis tube produced H_2 at a rate of ca. 6.8 Nl/h, translating to ca. 0.68 Nm^3/h H_2 for the 2-kW module. As oil became relatively cheap again in the middle of 1980s, the development of SOE was largely terminated. In the past decade, renewed interest on SOE appeared based on the progress achieved with planar SOFC technology operating at reduced temperatures (ca. 800 °C). For instance, the nuclear hydrogen program in USA was started in 2003 and the Hi2H2 project (Highly efficient, high temperature, hydrogen production by water electrolysis) and then the RELHY (Innovative solid oxide electrolyser stacks for efficient and reliable hydrogen production) project in Europe were started in 2004 and 2008 respectively.

Table 8.3. Some historical events related to water electrolysis

Year	Events
1789	van Troostwijk and Deiman first discovered the phenomenon of water electrolysis using an electrostatic generator [25-26]
1800	Volta invented "electric pile" (battery), an inexpensive and easily reproducible source of nearly constant voltage with which electrochemistry took off [25]
1834	Michael Faraday used the term "electrolysis" for the first time [27]
1899	Nernst developed the Nernst mass – 15 wt% yttria-stabilised zirconia (YSZ) [28-29] – basis for solid oxide fuel cell and solid oxide electrolysis cell
1900	The first industrial electrolyser based on a filter-press design was presented by O. Schmidt in Zurich [27]
1902	Already more than 400 industrial electrolysers in operation [30]
1920s-1930s	Several large 100 MW size plants were built worldwide, mainly using low-cost hydropower for production of ammonia fertilizers [31]
1948	First pressurised electrolyser was built by Zdansky/Lonza [30]
1951	First commercial high-pressure electrolyser (30 bar) was presented by Lurgi [27]
1957	First Raney nickel cathode was introduced by Winsel and Justi [27]
1962-1966	Gemini-Apollo program, first PEM fuel cell using polymer electrolyte
1966	First solid polymer electrolyte system was built by General Electric [30]
1967	"Zero gap" electrode design was introduced by Costa and Grimes; a cornerstone for advanced alkaline electrolyser designs
1960s	Development of SOEC began mainly under NASA contracts [32-35]
1987	First 100-kW Membrel electrolyser (PEM electrolyser) by BBC (later ABB) [36]
1970s-1980s	SOEC development (tubular cells, operation temperature ~1000 °C) by Dornier and Lurgi (HOT ELLY project in Germany) and by Westinghouse [37-39]
2000s	Renewed interest on SOEC development (planar cells, reduced operation temperature ~ 800 °C)

8.2.2.3. Status of Technology

While low-temperature electrolysis has already been commercialised, the high-temperature SOE is still under development. Most SOE testing to date has focused on the performance and durability of cells and stacks. Recently Idaho National Laboratory (INL) has developed a 15 kW test facility, termed the Integrated Laboratory Scale (ILS) high temperature electrolysis test facility, aiming at addressing balance-of -plant (BOP) issues such as thermal management, hydrogen recycle and others. The system contains three 5-kW solid oxide modules (720 cells total, each of 64 cm^2 active area). The testing of the INL facility was started in September 2008 and lasted for 1080 h, with a peak H_2 production rate of 5.7 Nm^3/h, corresponding to a peak electrolyser power of 18 kW [42].

Table 8.4. Comparison of different water electrolysis technologies

	Alkaline electrolyser	PEM electrolyser	Solid oxide electrolyser
Electrolyte	25-30% KOH solution	Sulfonated (fluoro-) polymer	Y-/Sc-/Ce-substituted ZrO_2
Charge carrier	OH^-	H^+	O^{2-}
Operating temperature / °C	40-90	20-100	700-1000
Status of technology	Commercial	Initial commercial	RandD

Table 8.4. (Continued)

	Alkaline electrolyser	PEM electrolyser	Solid oxide electrolyser
Current density / A cm^{-2}	0.2-0.5	0.5-2.0	0.5-1.0
Cell voltage / V	1.7-2.0	1.7-2.1	~ 1.3 (V_{tn} @ 800 °C)
Power consumption at stack level / d.c. kWh/Nm^3 H_2	4.1-4.7	4.1-5.1	~ 3.1
Energy efficiency at stack level (HHV)	75-86%	69-86%	~ 100%
Power consumption at system level / a.c. kWh/Nm^3 H_2	4.6-5.2	4.6-5.7	~ 3.4
Energy efficiency at system level (HHV) [a]	68-77%	62-77%	~ 90%
Commercially available unit capacity and pressure	760 Nm^3/h H_2 @ 30 bar (ELT, IHT)	30 Nm^3/h H_2 @ 30 bar (Proton Onsite)	-
Typical stack lifetime	10-20 years (ELT, IHT)	5000-20000 h	under evaluation
Advantages	Established Long durability No precious metal catalyst High pressurisation possible	High current density High pressurisation possible High purity H_2 Simple system configuration	Low electricity demand Potentially highest efficiency No precious metal catalyst Potential for high current density Co-electrolysis of CO_2/H_2O
Disadvantages	Relatively low current density Corrosive electrolyte Complex system	Precious metal catalyst High investment cost Insufficient durability	Still RandD phase Unproved durability Complex heat management

Note: [a] The efficiency at system level is assumed to be 10% lower than the efficiency at stack level due to unavoidable energy losses on system components outside the stack. Data sources: Stolten (2010) [27], Smolinka (2010) [40], Genovese (2009) [24], Graves (2011) [41] and internet survey.

The three types of electrolysers are compared in Table 8.4, including their advantages and disadvantages. The most important aspects are discussed below.

8.2.2.3.1. Energy Consumption and Efficiency

The power consumption is directly related to the operating voltage of the electrolysis cell. For an electrolyser system, an optimum cell operating voltage exists for the most economic production of H_2/CO. For this purpose a trade-off has to be found between capital cost and operating cost. The capital cost can be reduced through increased current density and hence higher operating voltage. The operating cost, however, will rise with increasing operating voltage as more power is needed. For conventional alkaline water electrolysers such a trade-off results in a typical cell voltage in the range of 1.7-2.0 V, which is significantly higher than the thermoneutral voltage (1.48 V). At lower cell voltages, the benefit due to reduced power consumption will be overweighed by the increased capital cost. The stack needs cooling due to the exothermal operation mode. According to Eq. (8.7) the stack HHV efficiency is only 74% at an operating voltage of 2.0 V_{DC}. Commercially available alkaline water electrolysis stacks typically have a specific power consumption of 4.1-4.7 d.c. kWh/Nm^3 H_2 and an HHV efficiency of 75-86%. At the system level, the power consumption is increased to 4.6-5.2 a.c. kWh/Nm^3 H_2 and the HHV efficiency is correspondingly reduced to 68-77%, given 10% energy loss on system components. As an example, a HySTAT^{TM}-60 unit (24-60 Nm^3/h H_2, 10 bar, indoor solution) from Hydrogenics [43] has a specific power consumption of 4.9 a.c.-kWh/Nm^3 H_2 and an HHV efficiency of 72% at the system level. Although not yet commercialized, advanced alkaline electrolysers were demonstrated to operate at 1.6-1.65 V_{DC} at 300 mA cm^{-2} and 80 °C, which means a specific power consumption of 3.85-3.95 d.c. kWh/Nm^3 H_2 and an HHV efficiency of 90-92% at the stack level [44].

Commercial PEM electrolysis cells operate at slightly higher voltages than conventional alkaline electrolysis cells, but with much higher current densities (up to 2 A cm^{-2}) in order to compensate for the inherent high cost of the electrolyser stack as precious metal catalysts have to be used. The specific power consumption is in the range of 4.1-5.1 d.c. kWh/Nm^3 H_2 at the stack level and estimated to be 4.6-5.7 a.c. kWh/Nm^3 H_2 at the system level. As a reference, a HOGEN® C30 unit (30 Nm^3/h H_2, 30 bar) from Proton Onsite [45] has a specific power consumption of 5.8 a.c.-kWh/Nm^3 H_2 and an HHV efficiency of 61% at the system level.

As discussed in the previous section, the power consumption of SOE can be significantly reduced thanks to the high operating temperature. The high operating temperature allows, on one hand, a lower Nernst potential of the cell (thermodynamic benefit), and on the other hand, a reduced cell internal resistance and overpotential (kinetic benefit), both contributing to a reduced cell operating voltage without compromising the current density. For instance, already in the HOT ELLY program as mentioned above, a current density of 0.6 A cm^{-2} was achieved at a cell voltage of 1.32 V, which is slightly higher than the thermoneutral voltage at the operating temperature (1.3 V at 995 °C) [37]. The state-of-the-art hydrogen-electrode-supported planar solid oxide cells achieved a current density of 2.6 A cm^{-2} and 950 °C [46] and 1.4 A cm^{-2} at 810 °C [47] at 1.3 V at the cell level. At the stack level, 1 A cm^{-2} appeared achievable at 800 °C and 1.3 V [48]. It should be noted that reducing the operating temperature from 1000 °C to 800 °C is beneficial in that cost-effective metallic interconnect materials can be used. Operating SOE cells at thermal neutral voltage means that all the input electrical energy is converted into chemical energy so that the energy conversion efficiency is 100% at the stack level. It should be noted here that the V_{tn} at 800°C for water electrolysis (1.3 V) is lower than the V_{tn} at 25°C (1.48 V), which represents the HHV of H_2. As discussed above, such a difference reflects mainly the water evaporation enthalpy (equivalent to 0.21 V). At the system level, unavoidable energy loss on power converter and thermal insulation losses will reduce the efficiency. It is considered that for an up-scaled plant with minimized thermal insulation losses, such an efficiency loss should not exceed 10%. Therefore the system efficiency of a high-temperature SOE should be able to reach 90%, which is ca. 10% higher than low-temperature electrolysers. Nevertheless, so high an efficiency at the system level is still to be demonstrated in the next stage of development.

From an economic point of view, the possibility of replacing a significant part of electrical energy with heat energy represents another advantage for the high temperature SOE. For instance, heat from geothermal sources or from concentrated solar power (CSP) (or from nuclear reactors) can be directly used for steam generation, thus reducing the energy cost since electricity derived from these energy sources are much more expensive due to the low thermal-to-electricity conversion efficiency (typically 35%). In addition, it is in principle also possible to couple a SOE with a high-temperature gas-cooled nuclear reactor [49-50] in order to supply high-temperature heat to the electrolyser stack when it is operated under endothermic mode. When the electrolyser capital cost is low, this would offer another economic benefit. Furthermore, considering the downstream fuel synthesis processes, which are exothermal processes, it is possible to thermally couple the SOE unit with the fuel synthesis unit, using the waste heat for steam generation, thus improving the overall efficiency of the whole energy conversion chain.

8.2.2.3.2. Durability under Static Operation Mode

As a well proved technology, conventional alkaline electrolyser has a very good durability, with the cell/stack lifetime in the range of 10-20 years. As an example, the IHT S556 unit (up to 760 Nm^3/h H_2, 30 bar) [51] requires replacement of old cell blocks every 10-15 years.

Regarding the lifetime of PEM electrolyser, considerable progress has been made over the past decade. Degradation or failure mechanisms include membrane thinning, titanium embrittlement by H_2, corrosion/dissolution/agglomeration of the catalyst and its support, and deterioration of seals. A well-designed and correctly operated PEM electrolysis stack can reach a lifetime of up to several tens of thousands of hours in industrial and military applications [40].

High-temperature SOE is currently under intensive development. So far almost all the cell/stack testing work has been carried out using simply currently existing solid oxide fuel cells (SOFCs). During the last decade, considerable progresses have been achieved on the SOFC development in terms of system-level efficiency, durability, robustness and cost reduction [52]. The degradation rate in the stack level has been reduced to 0.4-1.0% / kh (voltage degradation in percent per thousand hours) of static operation, which is very close to the target 0.3% / kh for stationary applications. The longest SOFC stack operation with planar cells has reached nearly 4 years with an average degradation of 1%/kh [53]. The test of SOE cells and stacks, however, has been performed so far only in a time scale of less than 10 000 hours. The degradation rate in the electrolysis mode observed so far tend to be higher than in the fuel cell mode. Reported voltage degradation rate can be as high as 2-40% / kh at both cell and stack levels [42, 48, 54-57].

Major degradation mechanisms that have been discussed or identified include:

i. electrode poisoning by impurities such as silicon coming from glass sealants [58] and sulfur from feed gases [57, 59-60],
ii. high pressure development and oxygen formation in the electrolyte grain boundaries just near the oxygen electrode/electrolyte interface, leading to oxygen electrode delamination especially under high current density conditions (> 1 A cm^{-2}) [61-65],

iii. Cr-migration from the interconnect, leading to Cr-substitution of the oxygen electrode bond layer and hence reduced electrical conductivity of the bond layer material [55,63] and
iv. microstructural change of the hydrogen electrode (Ni-relocation) [54].

An important feature to be noted is the effect of current density, which is critical for the economic competitiveness of the SOE technology. Recent results [64,66] indicate that state-of-the-art hydrogen-supported cells are not able to operate at a current density higher than ca. 1 A cm^{-2} due to the fast degradation rate. It is considered that the oxygen evolution reaction kinetics at the oxygen electrode is the limiting factor. Further improvement of the oxygen electrode should lead to a higher allowable current density. Nevertheless, it has been shown very recently that, through cleaning of the inlet gases to the hydrogen electrode, a TOFC (Topsoe Fuel Cells A/S, Denmark) stack has been operated for ca. 1200 h without notable degradation under CO_2 and H_2O co-electrolysis conditions at current densities of up to 0.75 A cm^{-2} at 850 °C [67]. The durability of SOE stacks in a time scale of longer than 10 000 hours is yet to be demonstrated.

8.2.2.3.3. Flexibility under Partial Loads and Dynamic Operation

In order to be able to couple with renewable but intermittent power sources, the electrolyser should have the flexibility to operate under partial loads and be capable for dynamic operation.

Alkaline electrolyser can operate with partial loads in a typical range of 25-100% of maximum capacity (e.g. ELT, Germany). Gandía et al [68] tested a commercial alkaline water electrolyser of 5 kW (model H_2 IGen 300/1/25 formerly from Vandenborre and Stuart Energy, now HySTAT from Hydrogenics) under emulated wind conditions during a period of 14 h. The electrolyser worked reliably with satisfactory gas purities and fast dynamic response to the varying conditions imposed. Another test [69] with a 6 kW alkaline electrolyser (Stuart Energy Systems) has shown that maintaining a minimum electrolyser current is critical to avoid performance decline induced by dynamic operation. At the island of Utsira in Norway a 10 Nm^3/h H_2 alkaline electrolyser has been used in an autonomous wind/hydrogen energy demonstration system. NEL Hydrogen [70] (originally Norsk Hydro, Norway) has developed a new model (60 Nm^3/h H_2, 15 bar) for integration with renewable energy sources. It has an extended dynamic operation range of 5-100% of maximum capacity and a load change response time of less than 1 second.

PEM eletrolyser stack can follow the load in the full range of 0-100% of maximum capacity. In practice, however, the range is limited to 5-100% due to the power consumption of auxiliary components [40]. Due to their inherent simplicity and potential for quick startup, PEM electrolyser may prove to be more suitable for wind/hydrogen-systems than alkaline electrolysers.

The operability of SOE under partial loads has not yet received much attention. The high operating temperature in combination with the strong thermal effect induced by the variation of the cell voltage probably means less flexibility of SOE under dynamic conditions. Thermodynamic calculation shows that under adiabatic conditions, a variation of the cell voltage by only 46 mV would tend to heat up or cool down the gases by approximately 150 °C [71]. Therefore the stack would be subjected to thermal cycling under dynamic conditions. The issue of coupling an SOE with intermittent power sources is being addressed for instance in the framework of a European project called ADEL (Advanced electrolyser for hydrogen production with renewable energy sources).

8.2.2.3.4. Pressurised Operation

When pressurised H_2/CO is needed for application (e.g. H_2 storage and transportation), pressurised operation of the electrolyser system is advantageous because it can save the energy that would otherwise be needed by an external compressor. Thermodynamically, increasing the operation pressure would lead to increased Nernst potential and thus increased electricity demand. Kinetically, however, high operation pressure would accelerate the electrode reaction rates (e.g., by decreasing the size of bubbles in the case of alkaline electrolyser), thus reducing the electrode overpotentials, which may compensate for the unfavourable thermodynamic effect. For example, for a 5-kW high-pressure alkaline electrolyser stack developed at Forschungszentrum Jülich, there is no statistically significant shift in power demand as the pressure is varied in the range of 30-120 bar [27]. Despite of the advantage of energy saving, pressurised operation poses engineering challenges such as necessary safety provisions and difficult pressure control especially under intermittent operation mode. Most commercially available alkaline electrolysers are pressurised at 2-10 bar, just a few operate at ca. 30 bar and above [27]. In the case of commercial PEM electrolysers, most are pressurised at 15-30 bar.

Pressurised operation of high-temperature co-electrolysis of CO_2/H_2O for syngas production is also favoured in view of the downstream fuel synthesis step operated normally under pressure (e.g. 10-40 bar for FT synthesis). Recently test of SOEC under pressure has been started at Risoe National Laboratory. A cell was tested at a pressure of up to 10 bar and its performance was weakly affected by the pressure [72], evidencing again the counterbalance between the thermodynamic effect and the kinetic effect.

8.2.2.3.5. Capability of CO_2/H_2O Co-Electrolysis

While SOE is still under development, it does have another important feature in the context of syngas and synfuel production – it is capable for CO_2 electrolysis [38-39,57,73-74] and CO_2/H_2O co-electrolysis [32,66,74-77]. CO_2/H_2O co-electrolysis is a one-step process to produce syngas, eliminating the need of a RWGS reactor when comparing with the two-step process (H_2O electrolysis plus RWGS). Under high-temperature SOE conditions, the RWGS reaction can easily reach equilibrium at the presence of Ni-catalyst, which is widely used as a catalyst in the hydrogen electrode of the cell. This fact makes it possible to predict the product gas composition according to the feed gas composition, the current and the operating temperature.

Figure 8.4 shows the test result of an electrolyte-supported cell for CO_2/H_2O co-electrolysis at an oven temperature of 850 °C. It can be seen that the predicted product gas compositions agree well with the measured ones. Figure 8.4 also shows that high CO_2/H_2O conversion can be achieved to obtain syngas with low concentration of residual CO_2 and H_2O (conversion 93% at 1.5 A cm^{-2}).

In the case of RWGS reaction, however, multi-stage water condensation is normally needed to reach high CO_2 conversion, leading to increased system complexity and efficiency loss [78].

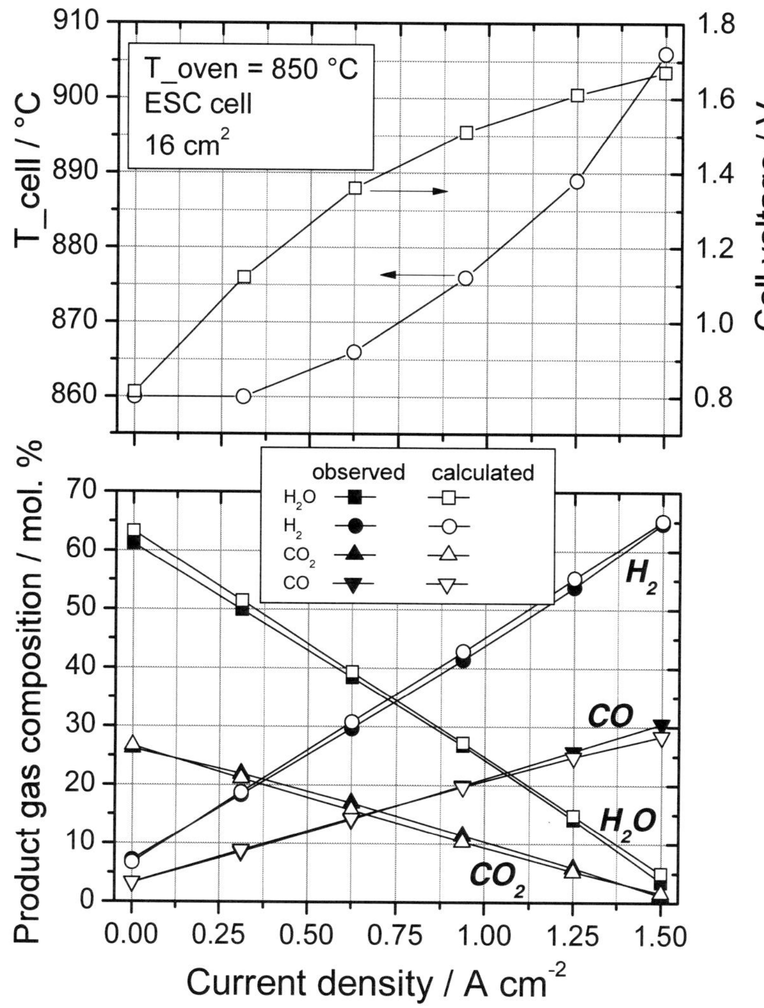

Note: The cathode feed gas is composed of 120 sccm H_2O(g) (60 mol %), 60 sccm CO_2 (30 mol %) and 20 sccm H_2 (10 mol %). Closed symbols indicate experimentally observed results and open symbols indicate calculated results by assuming that the water gas shift reaction reaches equilibrium. At a current of 24 A (1.5 A cm^{-2}), the CO_2/H_2O conversion reaches 93%.

Figure 8.4. Product gas composition, cell temperature and cell voltage as a function of current at an oven temperature of 850 °C.

Although low-temperature CO_2/H_2O co-electrolysis for syngas production is also possible, it is still far away from application due to its low selectivity, low current density or low energy conversion efficiency [41]. For instance, Delacourt et al. [79] developed an electrochemical cell to realise CO_2/H_2O co-electrolysis at 25 °C. It is based on the insertion of a PH-buffer layer (aqueous $KHCO_3$) between the Ag-based cathode catalyst layer and Nafion membrane. A CO/H_2 ratio of 1/2, suitable for methanol synthesis, was successfully obtained. However, the current density was only 80 mA cm^{-2} at a cell voltage of as high as 3.9 V, meaning an energy efficiency of only 35%. In addition, the product selectivity is unstable upon long-term electrolysis.

8.2.3. Catalytic Synthesis Processes

8.2.3.1. RWGS Reaction

The water gas shift (WGS) reaction has been widely applied for industrial production of hydrogen, ammonia and other bulk chemicals utilising synthesis gases with H_2/CO molar ratios higher than can be produced by coal gasification or by steam reforming of hydrocarbons [80]. The RWGS reaction, however, has received much less attention due to little industrial demand. In the context of human exploration of Mars, the RWGS process, in combination with water electrolysis, has been studied as a viable process for oxygen production under the In Situ Propellant Production project supported by NASA [81-83]. In the so-called CAMERE process (Carbon dioxide hydrogenation to form methanol via a reverse water gas shift reaction) [84], a RWGS reactor has been developed. With the elimination of water by the RWGS reaction, the purge gas volume was minimized and the methanol yield was increased from 69% to 89%.

The RWGS reaction is needed to reduce CO_2 to CO in order to obtain syngas when only H_2 is first produced by water electrolysis.

$$CO_2 + H_2 \leftrightarrow CO + H_2O \qquad \Delta H^{o}_{600\,K} = 39 \text{ kJ mol}^{-1} \qquad (8.8)$$

As an endothermic, reversible reaction, the conversion of CO_2 increases with temperature. The equilibrium constant K_p reaches 1 at 816 °C (HSC Chemistry 5.11). Catalysts such as Ru/CeO_2-ZrO_2 [82-83], $ZnAl_2O_4$ [85-86] and Ni-CeO_2 [87] have been tested. An effective solution to drive the RWGS reaction to the right and achieve a high CO_2 conversion would include H_2O separation by condensation in a multi-stage process [78]. For instance, by adding a second reactor in series with water removal between the two reactors, the maximum conversion was increased from 45% to 60% using Ru/CeO_2-ZrO_2 catalyst [83]. The realisation of multi-stage condensation with gas-to-gas recuperators with high efficiency is challenging. Doty [88] proposed a process that should permit over 90% efficiency.

8.2.3.2. Methanol and DME Synthesis

Methanol is one of the top ten organic chemicals manufactured in the world (about 40 million tonnes produced in 2007). It is mainly used for the production of formaldehyde, methyl tert-butyl ether (MTBE), acetic acid, methyl methacrylate (MMA) and as a solvent or gasoline extender. Almost all methanol is synthesized from syngas containing 2-8 vol% of CO_2 ($H_2/CO/CO_2$) derived predominantly by steam reforming of natural gas [89]. The synthesis of methanol from syngas is one of the technically very well-developed industrial processes. Modern industrial methanol synthesis is largely carried out on Cu-based catalysts ($Cu/ZnO/Al_2O_3$), which are extremely active and selective.

Methanol is catalytically synthesised by the reactions

$$CO + 2\,H_2 \leftrightarrow CH_3OH \qquad \Delta H^{o}_{600\,K} = -100 \text{ kJ mol}^{-1} \qquad (8.9)$$

$$CO_2 + 3\,H_2 \leftrightarrow CH_3OH + H_2O \qquad \Delta H^{o}_{600\,K} = -61 \text{ kJ mol}^{-1} \qquad (8.10)$$

The WGS reaction occurs simultaneously with methanol synthesis.

$$CO + H_2O \leftrightarrow CO_2 + H_2 \qquad \Delta H^{0}_{600\ K} = -39\ kJ\ mol^{-1} \qquad (8.11)$$

Since they are all exothermic and involves a decrease in the number of moles, the maximum conversion is achieved at low temperature and high pressure. Regarding the reaction mechanism, it is widely accepted today that methanol is formed from CO_2 in the $H_2/CO/CO_2$ feed. Copper is the active catalyst component. The role of CO is to keep the cooper in a highly reduced state. Although methanol can be made from H_2/CO feeds on $Cu/Zn/Al_2O_3$, the rate is about 100 times slower than when CO_2 is present. The presence of CO_2 enhances the durability of the catalyst [89].

For methanol synthesis, a nearly stoichiometric composition of the synthesis gas is desired. The "modulus" $(H_2\text{-}CO_2)/(CO+CO_2)$ should be equal or slightly above 2. Under-stoichiometric gas (modulus < 2) should be avoided since it leads to high formation of byproducts and loss of synthesis gas as increased purge. In addition, a high CO to CO_2 ratio should be used since a high concentration of CO_2 leads to unfavourable equilibrium, high water concentration in the raw product, low conversion per pass, low reaction rate and increased rate of catalyst deactivation. Furthermore, the inert (e.g. CH_4, N_2 or Ar) concentration in the syngas should be low [90].

It is interesting to note that despite the above mentioned feedstock criteria for conventional syngas-based methanol synthesis, CO_2 and H_2 (without CO) can be also used to produce methanol. For instance, Lurgi AG developed and thoroughly tested a high activity catalyst for methanol production from CO_2 and H_2 [91]. In Japan, a pilot plant with a capacity of 50 kg day^{-1} of methanol, which was equipped with facilities for recycling unreacted gases and gaseous products, was successfully operated [92]. In Iceland the company Carbon Recycling International is currently building the first contemporary commercial CO_2 to methanol recycling plant using locally available cheap geothermal energy after successful pilot plant scale operation. There the CO_2 is a byproduct accompanying local geothermal energy sources or industrial sources (aluminium production), and H_2 is produced by water electrolysis using geothermal-derived electricity. Dimethyl ether (DME) can be used as spray propellant, chemical intermediate or as a clean and economical fuel alternative (e.g. LPG substitute, diesel substitute, power generation, etc.). Its production through methanol dehydration using e.g. alumina catalyst is also a well-proved industrial process.

$$2\ CH_3OH \leftrightarrow CH_3OCH_3 + H_2O \qquad \Delta H = -23\ kJ\ mol^{-1}\ [89] \qquad (8.12)$$

Alternatively, DME can be produced through direct synthesis, using a dual catalyst system that permits both methanol synthesis and dehydration in the same process unit, with no methanol isolation and purification. The combination of reactions results in synergistic effect that eases the thermodynamic constraints for the methanol synthesis [89].

8.2.3.3. Methanol to Gasoline (MTG) Process

The MTG process was invented by Mobil over 30 years ago. In this process, methanol is converted to gasoline with a zeolite catalyst, ZSM-5. Essentially three steps are involved in the MTG process: conversion of methanol to DME, initial formation of a C-C bond, and finally aromatisation with hydrogen transfer [89]. The first (and only) commercial MTG plant came on stream in 1985 in New Zealand, originally producing 14 500 bpd of gasoline from natural gas. The plant was later closed for economic reasons. A second generation MTG fluid bed reactor was built and operated in Wesseling, Germany with a productivity of 100 barrel per day. It gave a higher yield of gasoline than the fixed bed MTG reactor. During the last decade, other methanol-to-hydrocarbon (MTH) processes, such as methanol-to-olefins (MTO), methanol-to-propene (MTP) have been developed [93].

8.2.3.4. Fischer-Tropsch (FT) Synthesis

The FT synthesis is the catalytic hydrogenation of CO to give a range of products, which can be further processed to yield high-quality diesel fuel, gasoline and linear chemicals such as 1-alkenes, alkanes and oxygenated hydrocarbons. The process was developed by two German chemists Franz Fischer and Hans Tropsch in 1923 and was applied during the period of 1935-1945 in Germany for liquid fuel production from coal. The productivity in 1944 reached a maximum of about 0.65 million tons per year [89]. Sasol (South Africa) started producing synthetic fuels using the FT process since 1955 to make the country less dependent on imported oil. Today Sasol is producing over 7 million tons synfuel annually. The FT process has also been used to a less extent in other countries. New plants are under construction e.g. in Qatar and Nigeria to produce synfuel from natural gas [94].

The FT reaction yields a wide spectrum of hydrocarbons (mainly paraffins and olefins) and oxygenated compounds (mainly primary alcohols). A large number of reactions occur during the FT reaction. The overall reaction to form hydrocarbons can be represented by:

$$CO + 2\,H_2 \rightarrow -CH_2- + H_2O \tag{8.13}$$

where $-CH_2-$ is a part of a hydrocarbon chain. The reaction is highly exothermic: the heat released per $-CH_2-$ is about 145 kJ mol^{-1} [94]. It is therefore important to ensure efficient transfer of heat from the catalyst particles to the surrounding medium.

The chain growth depends on parameters like reactor temperature and pressure, syngas composition and catalyst. Regarding the catalyst, currently either Fe-based or Co-based catalysts are used. For instance, all the FT plants in South Africa use iron catalysts while the Shell plant in Malaysia uses cobalt. In the Sasol plants, the high-temperature FT (HTFT) reactors are operated at about 340 °C, yielding large amounts of olefins, a lower boiling range and very good gasoline. Diesel fuel can be produced readily by oligomerisation of olefins. Substantial amounts of oxygenates are also produced. The low-temperature FT (LTFT) reactors are operated at 220-270 °C, yielding much more paraffins and linear products and can be adjusted to very high wax selectivities. The primary diesel cut and wax cracking products can give excellent diesel fuels [89].

8.3. Economic Analysis of High-Temperature CO_2/H_2O Co-Electrolysis and FT Synthesis for Syngas and Synfuel Production

For all of the energy conversion pathways shown in Figure 8.2, the energy conversion efficiency of each step needs to be maximised for the economic competiveness of the whole system. Targeting the synthetic jet fuel for the aviation application, the high-temperature CO_2/H_2O co-electrolysis process combined with the FT process is potentially the most efficient pathway due to the high efficiency of the high-temperature SOE and the avoidance of an intermediate RWGS step. In order to assess the marketing potential of such a combined conversion process, an economic analysis has been performed through process modelling and sensitivity analysis [71]. The main findings are summarised in this section.

As the FT process is an established technology with published techno-economic parameters, the analysis was focused on a syngas plant based on the high-temperature CO_2/H_2O co-electrolysis process. A syngas plant with a designed productivity of 5 tonne per day was simulated with Aspen Plus®. A process flow sheet containing all of the components that would be present in an actual high-temperature co-electrolysis plant was first created, and then the parameters for each component were established and optimised in order to maximise the energy conversion efficiency. A key issue in the design of the flow sheet was to maximise the recovery of the heat energy contained in the hot outlet gases for the preheating of feed stocks. A large part of the heat was used for steam generation. Recycling of the cathode product gas was considered in order to keep the stack inlet zone always under reducing conditions to avoid cathode re-oxidation. If redox-stable hydrogen-electrode materials can be implemented, such a gas recycling step would become unnecessary. In recent years progress has been made on the development of alternative SOFC hydrogen-electrode materials with improved redox stability (for instance [95-98]). The targeted H_2/CO ratio in the syngas is 2.12. Detailed procedures for the simulation of the co-electrolysis stack and reasonable assumptions were given in [71].

Table 8.5 lists the parameters for the baseline study and ranges for the sensitivity analysis. Justifications for the choice of all parameter values were supplied in [71]. In the baseline case, electricity from the grid is used to power the system under static operation mode. The use of renewable electricity, operating the stack under intermittent conditions (higher electricity price, lower capacity factor), is included in the sensitivity analysis.

In the baseline case, the syngas production cost was calculated to be 0.775 €/kg. Figure 8.5 shows the breakdown of the cost. The cost of syngas is energy and feedstock intensive, and is not capital intensive. The capital and maintenance contributes only 9.3% to the total cost. The major contribution comes from electricity consumption (57.3%). Due to relatively high cost of CO_2 (160 €/t assumed for the baseline study), feedstock contributes about one-third (31.6%) to the total cost.

The results of sensitivity analysis are shown in Figure 8.6. The high sensitivity of the syngas production cost to the electricity price is evident. It is also apparent that a high-performance stack (here a typical CSC stack with cathode-supported cells) offers great advantages over a low-performance stack (here a typical ESC stack with electrolyte-supported cells) due to the lower capital cost (smaller cell surface area is needed for the designed productivity). A low-performance stack would need to have either very long lifetime or very low acquisition cost to keep the syngas production cost low. On the contrary, with a high-performance stack, the syngas production cost is not any more very sensitive to the lifetime and acquisition cost. If the CO_2 price can be reduced to 20 €/t, the syngas cost can be reduced by 27.5%. Although reducing the cell operating voltage would allow partial replacement of electrical energy with cheaper heat energy, the syngas production cost would not be reduced due to the counter effect of increased stack acquisition cost (lower voltage leads to lower current and bigger cell surface area).

Table 8.5. Parameters used for the baseline study and ranges for the sensitivity analysis

	Parameter		Unit	Baseline value	Range
Invest. and Maint.	Electrolyser	Cost	€/m² cell	1500	500 - 5000
		Lifetime	h	20000	5000 - 80000
		ASR @ 800 °C	Ω cm²	0.25	0.25 - 1.25
		Operating voltage	V	1.343	1.100 - 1.343
		Maintenance	%	4	-
	Power converter	Efficiency	%	92	92 - 96
		Cost	€/kW	130	-
		Lifetime	h	100000	-
		Maintenance	%	2	-
	Heat exchanger	Average cost	€/unit	6000	-
		Average lifetime	h	15000	-
		Maintenance	%	4	-
	Compressor	Cost	€/kW	700	-
		Lifetime	h	80000	-
		Maintenance	%	4	-
	Other components	Total cost	€	20000	-
		Lifetime	h	50000	-
		Maintenance	%	2	-
	Interest rate	%	9	-	
	Capacity factor		-	1.0	0.2 - 1.0
Feed stock	CO_2 cost		€/t	160	20 - 160
	Deionised water cost		€/t	1.15	-
Energy	Electricity cost		€/MWh	56	0 - 200
	Low temperature heat cost		€/MWh	20	33.3% or 100 % of electricity
	High temperature heat cost		€/MWh	56	33.3% or 100 % of electricity
O_2 sale credit	O_2 sale price		€/t	0	0-50

If the byproduct O_2 can be sold at a price of 50 €/t, the syngas cost would be reduced by 10%. Using intermittent energy such as wind and solar electricity would require higher capital expense due to the largely increased nominal capacity of the plant operated under partial loads, resulting in higher syngas production cost. It shows here that a capacity factor 0.2 instead of 1.0 would increase the syngas cost by as much as 38%.

Based on the syngas production cost, the FT diesel production cost can be calculated using known economic parameters for the FT process. Figure 8.7 shows the calculated production cost of FT diesel as a function of electricity price. Effects of CO_2 price and credit from O_2 sale are also illustrated.

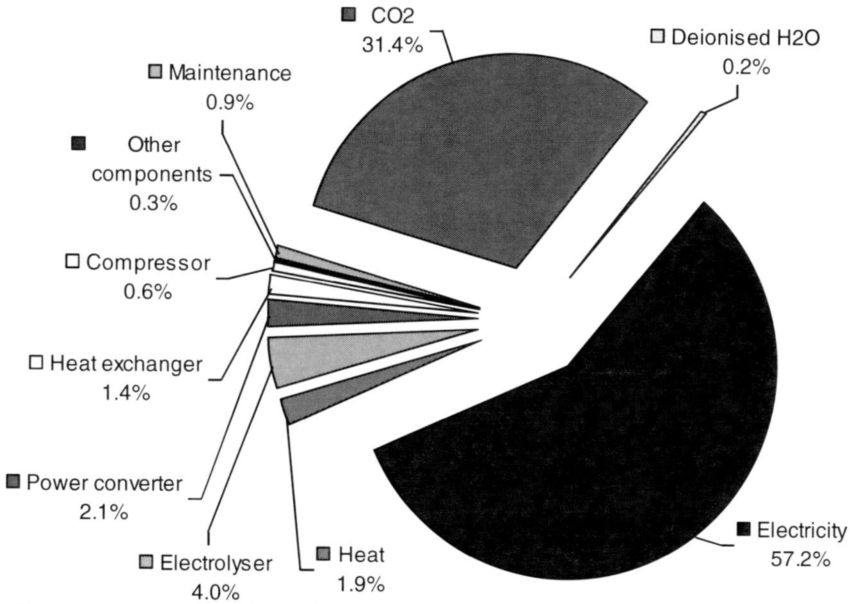

Note: baseline case, syngas cost 0.775 €/kg.

Figure 8.5. Breakdown of the syngas production cost.

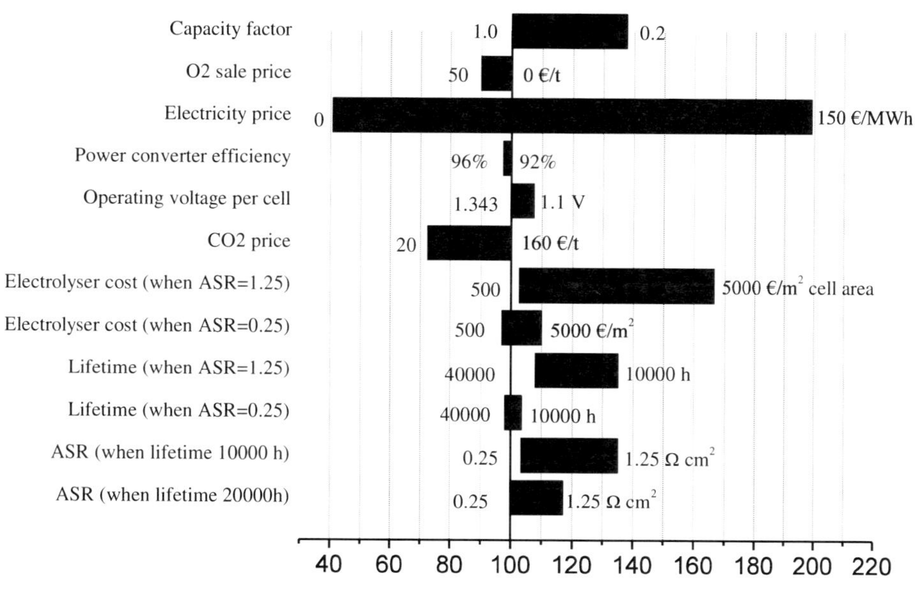

Note: ASR: area specific resistance of the repeating unit in the stack. ASR=0.25 Ω cm^2 represents the performance of a CSC stack with cathode-supported cells, and ASR=1.25 Ω cm^2 represents the performance of an ESC stack with electrolyte-supported cells. Capacity factor of 0.2 represents the coupling of the electrolyser with renewable electricity (e.g. on-shore wind power).

Figure 8.6. Sentivitity analysis results.

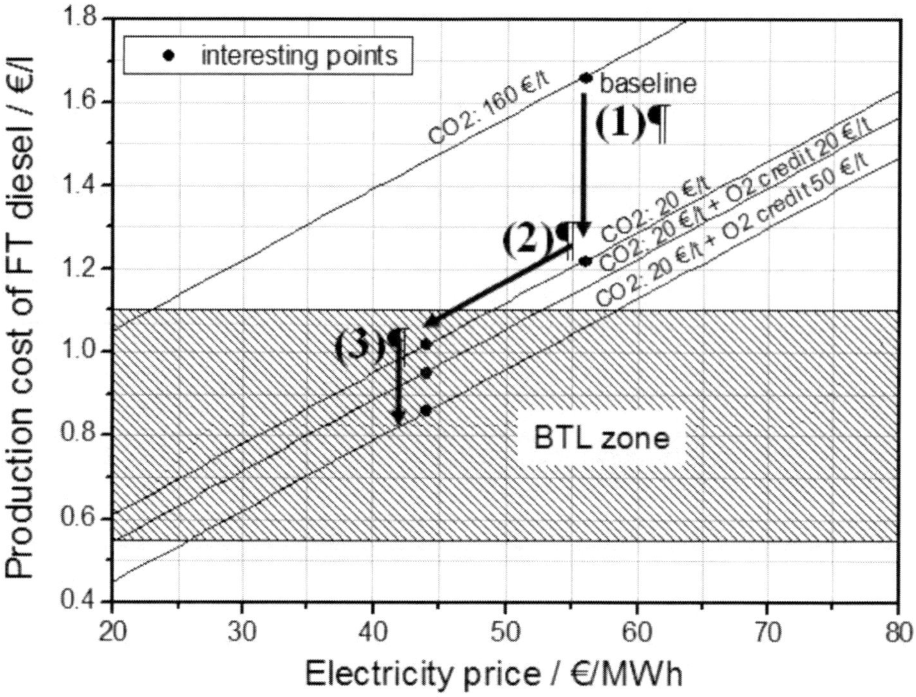

Note: The production cost of BTL fuels estimated in the literature is indicated with a patterned zone (0.55 – 1.1 €/l). The market price (without taxes, source: Eurostat) of conventional petroleum-derived diesel is 0.57 €/l in 2008 in both France and Germany. Arrows show favourable scenarios: 1) reduced cost of CO_2 from concentrated sources; 2) cheaper electricity price with up-scaled productivity; 3) O_2 credit considered.

Figure 8.7. Production cost of FT diesel as a function of electricity price, CO_2 cost and credit from O_2 sale.

In the baseline case (CO_2 price 160 €/t, electricity price 56 €/MWh, no O_2 sale credit), it would cost 1.66 €/l. It should be noted that the CO_2 price of 160 €/t is a quotation from a gas handler and may not represent the real production cost. Compression and transport cost of CO_2 from concentrated sources such as ammonia plants is only about 20 €/t. If this price is used, the production cost of FT diesel would be remarkably reduced to 1.22 €/l. Considering the up-scale effect, a large plant can benefit from a much lower electricity tariff than in the baseline case. At an electricity price of 44 €/MWh (in France) and CO_2 price of 20 €/t, the diesel production cost is reduced to 1.02 €/l, which is already comparable with BTL technology. When the pure O_2 by-product can be sold, the diesel production cost can be further reduced to 0.95 and 0.86 €/l at an O_2 sale price of 20 and 50 €/t, respectively. At such a price it could compete with BTL process. To compete with petroleum diesel today (0.57 €/l in 2008), electricity price needs to be around 20 €/MWh. Surplus or off-peak wind power may be sold at a price lower than this value (in extreme cases even negative electricity price on the spot market [7]), thus enabling an immediate economic competitiveness of the process. It should be mentioned that in their "WindFuels" concept, Doty et al. [99-100] proposed to convert off-peak wind power to synthetic liquid fuels through low-temperature water electrolysis, RWGS reaction and FT synthesis, i.e. one of the pathways shown in Figure 8.2 of the present chapter.

To have a sense on the scale of implementation of the as-described co-electrolysis-FT process for synfuel production from renewable energies, the wind energy in Germany is taken as an example. In 2009, 38.6 TWh electricity was produced from wind in Germany [101]. The wind energy generation is projected to increase to 150 TWh/a in 2020. If 30% of such amount of wind energy (assumed here to be off-peak cheap electricity) is converted into synfuel with an energy conversion efficiency of 59% (the overall efficiency of the present co-electrolysis-FT process: from electricity to LHV of FT-synfuel), 26 TWh/a fuel (equivalent to ~2.2 Mt/a kerosene) can be produced, which is 25% of the total fuel consumption for aviation in Germany in 2009 (102 TWh).

CONCLUSION

In the long term, hydrocarbon fuels, characterised by their compatibility with the existing infrastructures and engine technologies and their extremely high energy densities, need to be produced from regenerative energy sources to realise a sustainable transport especially a sustainable aviation. As most renewable energies are harvested in the form of electricity, a system to convert electrical energy to chemical energy stored in hydrocarbon fuels is therefore to be developed. Electrolysis, a technology that has been applied for more than 100 years, is believed to be the most realistic technology that can be integrated in such an energy conversion system.

Syngas is an important intermediate to produce synthetic fuels using well-established catalytic synthesis processes, e.g. the FT process. Starting from low-temperature alkaline or PEM water electrolysis, such an energy conversion system can be readily realised, if the bridging step, the RWGS process can be demonstrated to produce syngas efficiently. In comparison to the low-temperature electrolysis, the high-temperature SOE has the potential to increase the efficiency by around 10% and offers additionally the possibility to avoid the RWGS step, which further simplifies the system. Economic analysis shows the potential economic competitiveness of the combined high-temperature co-electrolysis-FT process. Intensive RandD on the high-temperature SOE technology, however, is still needed to address the following issues among others:

a. The durability of the SOE stack under static operating conditions needs to be improved to reach a lifetime of at least 10 000 – 20 000 h depending on the performance and cost of the stack. The biggest challenge would be the durability of high-performance cells under high current densities (> 1 A cm^{-2}) in order to reduce the capital cost. For that purpose performance limiting factors and degradation mechanisms need to be first identified. If necessary, specific cell/stack materials for SOE need to be developed.
b. A high-temperature SOE system with integrated balance of plant components (e.g. heat exchangers, stream recycling) needs to be developed and demonstrated. Efficient thermal management is crucial for the efficiency of the whole system.
c. Robustness of the cell under dynamic operating conditions simulating the coupling with intermittent renewable energies needs to be evaluated and improved if necessary.

d. System operational and control strategy when coupling with renewable energies needs to be developed by taking into account the operation limitations of SOE stacks.
e. The feasibility of pressurised operation of SOE cells/stacks needs to be investigated. Real benefits of pressurised operation need to be quantified.

Regarding the conversion routes starting from low-temperature electrolysis, it seems necessary to enhance the RandD regarding the coupling of renewable electricity with the electrolyser. Operation range of the alkaline electrolyser under partial loads should be extended. For PEM electrolysers, the lifetime needs to be extended. In addition, an efficient RWGS reactor needs to be developed and demonstrated. Finally, the air capture technology should be further developed to supply cost-effective CO_2 in the long term.

Author's Biography

Dr. Qingxi Fu is a researcher and project manager at European Institute for Energy Research (EIfER). After receiving his Ph.D. in materials physics and chemistry at University of Science and Technology of China (USTC) in 2002, he had worked at Forschungszentrum Jülich (Germany) for five years, focusing on new materials development and component processing for solid oxide fuel cells (SOFCs). As a holder of an EU Marie Curie Incoming International Fellowship during the period of 2004-2006, he had developed alternative ceramic-based SOFC anode materials with improved reduction-oxidation stability. Since 2008 he has been working at EIfER, extending his research interest to high-temperature steam and carbon dioxide electrolysis using solid oxide cells for the production of hydrogen and synthesis gas. He is author and co-author of more than 20 scientific articles in peer reviewed journals.

References

[1] I.S. Nashawi, A. Malallah, and M. Al-Bisharah, Forecasting World Crude Oil Production Using Multicyclic Hubbert Model. Energy and Fuels, 2010. 24(3): p. 1788-1800.
[2] IEA, Tranport, energy and CO_2 - moving toward sustainability. IEA/OECD, 2009.
[3] IEA, World Energy Outlook 2010 - executive summary. OECD/IEA, Paris, 2010.
[4] IEA, Technology roadmap - biofuels for transport. OECD/IEA, Paris, 2011.
[5] IEA, Clean energy progress report. OECD/IEA, Paris, 2011.
[6] IEA, Technology roadmap - wind energy. OECD/IEA, Paris, 2009.
[7] M. Nicolosi, Wind power integration and power system flexibility-An empirical analysis of extreme events in Germany under the new negative price regime. *Energy Policy,* 2010. 38(11): p. 7257-7268.
[8] B. Lee and D. Gushee, Massive electricity storage. an AIChE White Paper, *AIChE's Government Relations Committee*, June 2008.
[9] D. Rastler, New Demand for Energy Storage. *Electric Perspectives.* Sept 2008: p. 30-47.

[10] D.M. Bierschenk, J.R. Wilson, and S.A. Barnett, High efficiency electrical energy storage using a methane-oxygen solid oxide cell. *Energy and Environmental Science*, 2011. 4(3): p. 944-951.

[11] GREET Transportation Fuel Cycle Analysis Model, GREET 1.8b, developed by Argonne National Laboratory, Argonne, IL, released September 5, 2008.

[12] R.J. Pearson, J.W.G. Turner, and A.J. Peck, Gasoline-ethanol-methanol tri-fuel vehicle development and its role in expediting sustainable organic fuels for transport. *Low Carbon Vehicles Conference, Institution Mechanical Engineers Conference,* London UK, May 20-21, 2009.

[13] U. Eberle, GM's Research Strategy - Towards a Hydrogen-Based Transportation System. Presentation on FuncHy Workshop, Hamburg, Sept. 2006.

[14] U. Bossel, B. Eliasson, and G. Taylor, The Future of the Hydrogen Economy: Bright or Bleak? Proc. European Fuel Cell Forum, 2003; with revised foreword 26 February 2005. www.efcf.com/reports/E08.pdf.

[15] U. Bossel, Does a Hydrogen Economy Make Sense? *Proceedings of the IEEE*, 2006. 94(10): p. 1826-1836.

[16] C. Fritzmann, J. Löwenberg, T. Wintgens, and T. Melin, State-of-the-art of reverse osmosis desalination. *Desalination*, 2007. 216(1-3): p. 1-76.

[17] J.J. Dooley, R.T. Dahowski, C.L. Davidson, M.A. Wise, N. Gupta, S.H. Kim, and E.L. Malone, Carbon Dioxide Capture and Geologic Storage - a Core Element of a Global Energy Technology Strategy to Address Climate Change. A technology report from the second phase of the global energy technology stragety program, 2006.

[18] K.S. Lackner, Capture of carbon dioxide from ambient air. *European Physical Journal-Special Topics*, 2009. 176: p. 93-106.

[19] F. Zeman, Energy and Material Balance of CO_2 Capture from Ambient Air. *Environmental Science and Technology*, 2007. 41(21): p. 7558-7563.

[20] D. Keith, M. Ha-Duong, and J. Stolaroff, Climate Strategy with CO_2 Capture from the Air. *Climatic Change*, 2006. 74(1): p. 17-45.

[21] F.S. Zeman and D.W. Keith, Carbon neutral hydrocarbons. *Philosophical Transactions of the Royal Society A: Mathematical, Physical and Engineering Sciences*, 2008. 366(1882): p. 3901-3918.

[22] K.S. Lackner, Capturing Carbon Dioxide with a Humidity Swing Sorbent. Presentation on the workshop "Capture and conversion of CO_2 into sustainable hydrocarbon fuels" April 2011, Roskilde, Denmark.

[23] V. Kaplan, E. Wachtel, K. Gartsman, Y. Feldman, and I. Lubomirsky, Conversion of CO_2 to CO by Electrolysis of Molten Lithium Carbonate. *Journal of the Electrochemical Society*, 2010. 157(4): p. B552-B556.

[24] J.E. Genovese, K. Harg, M. Paster, and J.A. Turner, Current (2009) State-of-the-Art Hydrogen Production Cost Estimate Using Water Electrolysis. Independent Review, Published for the U.S. Department of Energy Hydrogen Program, NREL/BK-6A1-46676, 2009.

[25] R. de Levie, The electrolysis of water. *Journal of Electroanalytical Chemistry*, 1999. 476: p. 92-93.

[26] S. Trasatti, Water electrolysis: who first? *Journal of Electroanalytical Chemistry*, 1999. 476: p. 90-91.

[27] D. Stolten and D. Krieg, Alkaline Electrolysis - Introduction and Overview, in Hydrogen and Fuel Cells - Fundamentals, Technologies and Applications, D. Stolten, Editor. 2010, Wiley-VCH: Weinheim. p. 243.

[28] H.-H. Möbius, On the history of solid electrolyte fuel cells. *Journal of Solid State Electrochemistry*, 1997. 1(1): p. 2-16.

[29] W. Nernst, Material for electric-lamp glowers, US Patent 685730, application filed on Aug. 24, 1899. 1901.

[30] W. Kreuter and H. Hofmann, Electrolysis: The important energy transformer in a world of sustainable energy. *International Journal of Hydrogen Energy*, 1998. 23(8): p. 661-666.

[31] R.L. LeRoy, Industrial water electrolysis: Present and future. *International Journal of Hydrogen Energy*, 1983. 8(6): p. 401-417.

[32] J. Weissbart and W.H. Smart, Study of electrolytic dissociation of carbon dioxide-water using a solid oxide electrolyte, NASA report CR-680. 1967.

[33] H.S. Spacil and C.S. Tedmon, Electrochemical Dissociation of Water Vapor in Solid Oxide Electrolyte Cells .I. Thermodynamics and Cell Characteristics. *Journal of the Electrochemical Society,* 1969. 116(12): p. 1618-1626.

[34] J. Weissbart, W.H. Smart, S.H. Inami, C.M. McCullough, and S.A. Ring, Development of a carbon dioxide-water solid oxide electrolyte electrolysis system: first annual report, NASA report CR-73358. 1969.

[35] L. Elikan and J.P. Morris, Solid electrolyte system for oxygen regeneration, NASA report CR-1359. 1969.

[36] S. Stucki, G.G. Scherer, S. Schlagowski, and E. Fischer, PEM water electrolysers: evidence for membrane failure in 100kW demonstration plants. *Journal of Applied Electrochemistry*, 1998. 28(10): p. 1041-1049.

[37] W. Dönitz, E. Erdle, and R. Streicher, High temperature electrochemical technology for hydrogen production and power generation, in Electrochemical hydrogen technologies : electrochemical production and combustion of hydrogen, H. Wendt, Editor. 1990, Elsevier: Amsterdam [u.a.]. p. 213.

[38] W.D. E. Erdle, R. Schamm, A. Koch, Reversibility and polarization behaviour of high temperature solid oxide electrochemical cells. *International journal of Hydrogen Energy*, 1992. 17(10): p. 817-819.

[39] A.O. Isenberg, Energy conversion via solid oxide electrolyte electrochemical-cells at high-temperatures. *Solid State Ionics*, 1981. 3-4: p. 431-437.

[40] T. Smolinka, S. Rau, and C. Hebling, Polymer Electrolyte Membrane (PEM) Water Electrolysis, in Hydrogen and Fuel Cells - Fundamentals, Technologies and Applications, D. Stolten, Editor. 2010, Wiley-VCH: Weinheim. p. 271.

[41] C. Graves, S.D. Ebbesen, M. Mogensen, and K.S. Lackner, Sustainable hydrocarbon fuels by recycling CO_2 and H_2O with renewable or nuclear energy. *Renewable and Sustainable Energy Reviews*, 2011. 15(1): p. 1-23.

[42] A. Brisse, J. Schefold, C. Stoots, and J. O'Brien, Electrolysis using fuel cell technology, in Innovations in fuel cell technologies, R. Steinberger-Wilckens and W. Lehnert, Editors. 2010, Royal Society of Chemistry: Cambridge. p. 267.

[43] Hydrogenics HySTAT™ Alkaline Electrolyser, www.hydrogenics.com/hydro/hystat_a, accessed in July 2011.

[44] G. Schiller, R. Henne, P. Mohr, and V. Peinecke, High performance electrodes for an advanced intermittently operated 10-kW alkaline water electrolyzer. *International Journal of Hydrogen Energy*, 1998. 23(9): p. 761-765.

[45] Proton Onsite PEM Electrolyser, www.protononsite.com/resources/commercial-products-brochures/, accessed in July 2011.

[46] S.H. Jensen, P.H. Larsen, and M. Mogensen, Hydrogen and synthetic fuel production from renewable energy sources. *International Journal of Hydrogen Energy*, 2007. 32(15): p. 3253-3257.

[47] M. Zahid, J. Schefold, and A. Brisse, High-Temperature Water Electrolysis Using Planar Solid Oxide Fuel Cell Technology: a Review, in Hydrogen and Fuel Cells - *Fundamentals, Technologies and Applications*, D. Stolten, Editor. 2010, Wiley-VCH: Weinheim. p. 227.

[48] A. Brisse, J. Schefold, M. Zahid, J.U. Nielsen, and P. Noyé. Long Term Operation of Solid Oxide Fuel Cell Stacks in High Temperature Water Electrolysis. in 9th European Solid Oxide Fuel Cell Forum, p.14-49. 2010. Lucerne / Switzerland: European Fuel Cell Forum 2010.

[49] J.E. O'Brien, M.G. McKellar, C.M. Stoots, J.S. Herring, and G.L. Hawkes, Parametric study of large-scale production of syngas via high-temperature co-electrolysis. *International Journal of Hydrogen Energy*, 2009. 34(9): p. 4216-4226.

[50] R. Hino, H. Aita, K. Sekita, K. Haga, Y. Miyamoto, and T.-o. Iwata, Present status of RandD on hydrogen production by high temperature electrolysis of steam, Japan Atomic Energy Research Institute report, *JAERI-Research* 95-057. 1995.

[51] X. Song and Z. Guo, Technologies for direct production of flexible H_2/CO synthesis gas. *Energy Conversion and Management*, 2006. 47(5): p. 560-569.

[52] Solid Oxide Fuel Cells 12 (SOFC XII) (ESC Transactions, Vol. 35, No. 1). in Twelfth International Symposium on Solid Oxide Fuel Cells (SOFC XII). 2011. Montreal, Canada: The Electrochemical Society.

[53] L.G.J. de Haart and I.C. Vinke, Long-Term Operation of Planar Type SOFC Stacks. *ECS Transactions*, 2011. 35(1): p. 187-194.

[54] A. Hauch, S.D. Ebbesen, S.H. Jensen, and M. Mogensen, Solid oxide electrolysis cells: Microstructure and degradation of the Ni/yttria-stabilized zirconia electrode. *Journal of the Electrochemical Society*, 2008. 155(11): p. B1184-B1193.

[55] V.I. Sharma and B. Yildiz, Degradation Mechanism in $La_{0.8}Sr_{0.2}CoO_3$ as Contact Layer on the Solid Oxide Electrolysis Cell Anode. *Journal of the Electrochemical Society*, 2010. 157(3): p. B441-B448.

[56] J.E. O'Brien, C.M. Stoots, J.S. Herring, K.G. Condie, and G.K. Housley, The High-Temperature Electrolysis Program at the Idaho National Laboratory: Observations on Performance Degradation. INL report: INL/CON-09-15564, International Workshop on High Temperature Water Electrolysis Limiting Factors, Karlsruhe, 2009.

[57] S.D. Ebbesen and M. Mogensen, Electrolysis of carbon dioxide in Solid Oxide Electrolysis Cells. *Journal of Power Sources*, 2009. 193(1): p. 349-358.

[58] A. Hauch, S.H. Jensen, J.B. Bilde-Sorensen, and M. Mogensen, Silica segregation in the Ni/YSZ electrode. *Journal of the Electrochemical Society*, 2007. 154(7): p. A619-A626.

[59] S.D. Ebbesen, C. Graves, A. Hauch, S.H. Jensen, and M. Mogensen, Poisoning of Solid Oxide Electrolysis Cells by Impurities. *Journal of the Electrochemical Society*, 2010. 157(10): p. B1419-B1429.

[60] S.D. Ebbesen and M. Mogensen, Exceptional Durability of Solid Oxide Cells. *Electrochemical and Solid State Letters*, 2010. 13(9): p. D106-D108.

[61] T. Jacobsen and M. Mogensen, The Course of Oxygen Partial Pressure and Electric Potentials across an Oxide Electrolyte Cell, M. Mogensen, T. Armstrong, T. Gur, and H. Yokokawa, Editors. 2008, ECS. p. 259-273.

[62] A.V. Virkar, Mechanism of oxygen electrode delamination in solid oxide electrolyzer cells. *International Journal of Hydrogen Energy*, 2010. 35(18): p. 9527-9543.

[63] J.R. Mawdsley, J.D. Carter, A.J. Kropf, B. Yildiz, and V.A. Maroni, Post-test evaluation of oxygen electrodes from solid oxide electrolysis stacks. International *Journal of Hydrogen Energy*, 2009. 34(9): p. 4198-4207.

[64] R. Knibbe, M.L. Traulsen, A. Hauch, S.D. Ebbesen, and M. Mogensen, Solid Oxide Electrolysis Cells: Degradation at High Current Densities. *Journal of the Electrochemical Society*, 2010. 157(8): p. B1209-B1217.

[65] M.S. Sohal, J.E. O'Brien, C.M. Stoots, J.S. Herring, J.J. Hartvigsen, D. Larsen, S. Elangovan, J.D. Carter, V.I. Sharma, and B. Yildiz, Critical Causes of Degradation in Integrated Laboratory Scale Cells During High-Temperature Electrolysis. INL report n° INL/EXT-09-16004, 2009.

[66] C. Graves, S.D. Ebbesen, and M. Mogensen, Co-electrolysis of CO_2 and H_2O in solid oxide cells: Performance and durability. *Solid State Ionics,* 2011. 192(1): p. 398-403.

[67] S.D. Ebbesen, J. Høgh, K.A. Nielsen, J.U. Nielsen, and M. Mogensen, Durable SOC stacks for production of hydrogen and synthesis gas by high temperature electrolysis. *International Journal of Hydrogen Energy*, 2011. 36(13): p. 7363-7373.

[68] L.M. Gandia, R. Oroz, A. Ursua, P. Sanchis, and P.M. Dieguez, Renewable hydrogen production: Performance of an alkaline water electrolyzer working under emulated wind conditions. *Energy and Fuels*, 2007. 21(3): p. 1699-1706.

[69] A. Bergen, L. Pitt, A. Rowe, P. Wild, and N. Djilali, Transient electrolyser response in a renewable-regenerative energy system. *International Journal of Hydrogen Energy*, 2009. 34(1): p. 64-70.

[70] NEL Hydrogen, brochure "If the future could choose", www.nel-hydrogen.com/home/?Brochures, accessed in July 2011.

[71] Q.X. Fu, C. Mabilat, M. Zahid, A. Brisse, and L. Gautier, Syngas production via high-temperature steam/CO_2 co-electrolysis: an economic assessment. *Energy and Environmental Science*, 2010. 3(10): p. 1382-1397.

[72] S.H. Jensen, X. Sun, S.D. Ebbesen, R. Knibbe, and M. Mogensen, Hydrogen and synthetic fuel production using pressurized solid oxide electrolysis cells. *International Journal of Hydrogen Energy*, 2010. 35(18): p. 9544-9549.

[73] F. Bidrawn, G. Kim, G. Corre, J.T.S. Irvine, J.M. Vohs, and R.J. Gorte, Efficient reduction of CO_2 in a solid oxide electrolyzer. *Electrochemical and Solid State Letters*, 2008. 11(9): p. B167-B170.

[74] Q. Fu, J. Dailly, A. Brisse, and M. Zahid, High-Temperature CO_2 and H_2O Electrolysis with an Electrolyte-Supported Solid Oxide Cell. *ECS Transactions*, 2011. 35(1): p. 2949-2956.

[75] C. Stoots, J. O'Brien, and J. Hartvigsen, Results of recent high temperature coelectrolysis studies at the Idaho National Laboratory. *International Journal of Hydrogen Energy*, 2009. 34(9): p. 4208-4215.

[76] C.M. Stoots, J.E. O´Brien, J.S. Herring, and J.J. Hartvigsen, Syngas production via high-temperature coelectrolysis of steam and carbon dioxide. *Journal of Fuel Cell Science and Technology* 2009. 6: p. 011014-1.

[77] S.D. Ebbesen, C. Graves, and M. Mogensen, Production of Synthetic Fuels by Co-Electrolysis of Steam and Carbon Dioxide. *International Journal of Green Energy*, 2009. 6(6): p. 646-660.

[78] F.D. Doty, L. Holte, and S. Shevgoor, Securing our transportation future by using off-peak wind energy to recycle CO_2 into fuels. *Proceedings of the ASME 2009 3rd International Conference of Energy Sustainability*, ES2009, July 19-23, 2009, San Francisco, California, USA.

[79] C. Delacourt, P.L. Ridgway, J.B. Kerr, and J. Newman, Design of an electrochemical cell making syngas ($CO+H_2$) from CO_2 and H_2O reduction at room temperature. *Journal of the Electrochemical Society*, 2008. 155(1): p. B42-B49.

[80] K.-O. Hinrichsen, K. Kochloefl, and M. Muhler, Water Gas Shift and COS Removal, in Handbook of Heterogeneous Catalysis, G. Ertl, H. Knözinger, F. Schüth, and J. Weitkamp, Editors. 2008, Wiley-VCH Verlag GmbH and Co. KGaA.

[81] J.E. Whitlow and C. Parrish, Operation, Modeling and Analysis of the Reverse Water Gas Shift Process. 2001 NASA/ASEE summer program, JFK Space Center, 2001.

[82] J.D. Holladay, K.P. Brooks, P. Humble, J. Hu, and T.M. Simon, Compact reverse water-gas-shift reactor for extraterrestrial in situ resource utilization. *Journal of Propulsion and Power*, 2008. 24(3): p. 578-582.

[83] J.D. Holladay, K.P. Brooks, R. Wegeng, J. Hu, J. Sanders, and S. Baird, Microreactor development for Martian in situ propellant production. *Catalysis Today*, 2007. 120(1): p. 35-44.

[84] O.-S. Joo, K.-D. Jung, I. Moon, A.Y. Rozovskii, G.I. Lin, S.-H. Han, and S.-J. Uhm, Carbon Dioxide Hydrogenation To Form Methanol via a Reverse-Water-Gas-Shift Reaction (the CAMERE Process). *Industrial and Engineering Chemistry Research*, 1999. 38(5): p. 1808-1812.

[85] S.-W. Park, O.-S. Joo, K.-D. Jung, H. Kim, and S.-H. Han, Development of ZnO/Al_2O_3 catalyst for reverse-water-gas-shift reaction of CAMERE (carbon dioxide hydrogenation to form methanol via a reverse-water-gas-shift reaction) process. *Applied Catalysis A: General*, 2001. 211(1): p. 81-90.

[86] O.S. Joo and K.D. Jung, Stability of $ZnAl_2O_4$ catalyst for reverse-water-gas-shift reaction (RWGSR). *Bulletin of the Korean Chemical Society*, 2003. 24(1): p. 86-90.

[87] L. Wang, S. Zhang, and Y. Liu, Reverse water gas shift reaction over co-precipitated Ni-CeO_2 catalysts. *Journal of Rare Earths*, 2008. 26(1): p. 66-70.

[88] F.D. Doty, Hydrocarbon and alcohol fuels from variable, renewable energy at very high efficiency. PCT WO 2008/115933, 2008.

[89] I. Wender, Reactions of synthesis gas. *Fuel Processing Technology*, 1996. 48(3): p. 189-297.

[90] J.B. Hansen and P.E. Højlund Nielsen, Methanol Synthesis, in Handbook of Heterogeneous Catalysis, G. Ertl, H. Knözinger, F. Schüth, and J. Weitkamp, Editors. 2008, Wiley-VCH Verlag GmbH and Co. KGaA.

[91] G.A. Olah, A. Goeppert, and G.K.S. Prakash, Chemical Recycling of Carbon Dioxide to Methanol and Dimethyl Ether: From Greenhouse Gas to Renewable, Environmentally Carbon Neutral Fuels and Synthetic Hydrocarbons. *The Journal of Organic Chemistry*, 2009. 74(2): p. 487-498.

[92] M. Saito, M. Takeuchi, T. Fujitani, J. Toyir, S. Luo, J. Wu, H. Mabuse, K. Ushikoshi, K. Mori, and T. Watanabe, Advances in joint research between NIRE and RITE for developing a novel technology for methanol synthesis from CO_2 and H_2. *Applied Organometallic Chemistry*, 2000. 14(12): p. 763-772.

[93] S. Kvisle, T. Fuglerud, S. Kolboe, U. Olsbye, K.P. Lillerud, and B.V. Vora, Methanol-to-Hydrocarbons, in *Handbook of Heterogeneous Catalysis*, G. Ertl, H. Knözinger, F. Schüth, and J. Weitkamp, Editors. 2008, Wiley-VCH Verlag GmbH and Co. KGaA.

[94] M.E. Dry, The Fischer–Tropsch (FT) Synthesis Processes, in Handbook of Heterogeneous Catalysis, G. Ertl, H. Knözinger, F. Schüth, and J. Weitkamp, Editors. 2008, Wiley-VCH Verlag GmbH and Co. KGaA.

[95] Q. Fu and F. Tietz, Ceramic-based anode materials for improved redox cycling of solid oxide fuel cells. *Fuel Cells*, 2008. 08(5): p. 283-293.

[96] Q. Fu, F. Tietz, D. Sebold, S. Tao, and J.T.S. Irvine, An efficient ceramic-based anode for solid oxide fuel cells. *Journal of Power Sources*, 2007. 171(2): p. 663-669.

[97] Q. Fu, F. Tietz, and D. Stoever, $La_{0.4}Sr_{0.6}Ti_{1-x}Mn_xO_{3-delta}$ perovskites as anode materials for solid oxide fuel cells. *J. Electrochem. Soc.*, 2006. 153: p. D74-D83.

[98] Q. Ma, F. Tietz, A. Leonide, and E. Ivers-Tiffée, Anode-supported planar SOFC with high performance and redox stability. *Electrochemistry Communications*, 2010. 12(10): p. 1326-1328.

[99] G.N. Doty, F.D. Doty, L.L. Holte, D.L. McCree, and S.K. Shevgoor, Securing Our Energy Future by Efficiently Recycling CO_2 into Transportation Fuels – and Driving the Off-peak Wind Market. *Proceedings of WindPower* 2009 May 4-7, 2009, Chicago IL USA, 2009.

[100] F. Doty, Overview of the Energy Challenge, and the Science and Engineering of WindFuels: Economically Producing Fuels and Chemicals from CO_2 and Off-peak Wind. Presentation on web http://www.windfuels.com/, 2010.

[101] BMU, Erneuerbare Energien in Zahlen (renewable energies in figures). BMU report, 2010.

In: Syngas
Editors: Antonius Indarto and Jelliarko Palguandi
ISBN: 978-1-62100-870-5

Chapter 9

DECARBONIZED ENERGY VIA SYNGAS ROUTES

Wojciech M. Budzianowski*
Wrocław University of Technology, Poland

ABSTRACT

The chapter focuses on organic fuel-based power generation processes without atmospheric CO2 emissions via chemical routes involving syngas steps. The role of syngas in such low-carbon processes, i.e. in power cycle, is presented and main technological benefits of syngas based fuel processing are expounded. Three promising low-carbon power generation cycles are examined with emphasis put on their energy efficiency, thermodynamics, thermal integration and the feasibility of low-cost CO2 capture. The first power cycle - biogas fuelled oxy-reforming fuel cell (OR-FC) cycle - enables to achieve negative net CO2 emissions from the utilization of renewable biogas by its oxy-reforming to syngas followed by water gas shift, CO2 capture and final H2 oxidation in a fuel cell. Its in-depth modeling and simulation analysis is provided. The second zero-carbon solid biomass fuelled multi-step gasification gas turbine (MSG-GT) power cycle can achieve improved energy efficiency by applying multi-step solid biomass conversion, which is a more reversible process than one-step biomass combustion. The third zero-carbon coal fuelled oxy-gasification steam chemical looping combustion (OG-SCLC-GT) cycle benefits from (i) improved cycle efficiency due to the increased reversibility of a steam chemical looping combustion (SCLC) process, (ii) cycle mass and energy integration due to several recirculation loops involved and (iii) extremely high CO2 capture rate due to purity of CO2/H2O mixture achieved at the outlet from a syngas reactor.

NOMENCLATURE

A	pre-exponential factor, mol $m^{-3}\cdot s^{-1}\cdot Pa^{-2}$
C/H	C to H ratio in a mixture (molar basis)
C/O	C to O ratio in a mixture (molar basis)

* E-mail: wojciech.budzianowski@pwr.wroc.pl.

CCS	Carbon Capture and Sequestration
CLC	Chemical Looping Combustion; A - air, S - steam
c_p	specific heat, $J \cdot kg^{-1} \cdot K^{-1}$
D^{EFF}	effective diffusivity in the porous phase, $m^2 \cdot s^{-1}$
D^G	bulk molecular diffusivity in the gaseous phase, $m^2 \cdot s^{-1}$
DMR	Dry Methane Reforming
E	activation energy, $J \cdot mol^{-1}$
H^{CAT}	thickness of the catalytic layer, m
HT-WGS	High Temperature-Water Gas Shift
I	unit diagonal matrix
ID	internal diameter of the reactor channel, m
IGCC	Integrated Gasification Combined Cycle
K^{EQ}	thermodynamic equilibrium constant (Pa basis)
k^{MS}	constant in equation (9.17)
L^{CAT}	length of the catalytic zone, m
LT-WGS	Low Temperature-Water Gas Shift
$\dot{m}$	mass flow rate, $kg \cdot s^{-1}$
M	molecular mass of species, $kg \cdot mol^{-1}$
m, n, a, b	stoichiometric coefficients
MGD	Membrane Gas Desorption
MSG-GT	Multi Step Gasification-Gas Turbine
NC	number of components, NC = 7
NR	number of reactions, NR = 5 in oxy-reformer and 1 in WGS reactors
OG-SCLC-GT	Oxy Gasification-Steam Chemical Looping Combustion-Gas Turbine
OR-FC	oxy-reforming fuel cell
P, P^C	pressure, partial pressure, partial pressure in the catalytic phase
P^{REF}	reference pressure $P^{REF} = 0.1013 \times 10^6$ Pa
PSA	pressure swing adsorption
r	radial coordinate, m
R	reaction rate, $mol \cdot m^{-3} \cdot s^{-1}$
R^G	universal gas constant = 8.314 $J \cdot mol^{-1} \cdot K^{-1}$
RQ	reaction quotient, -
SMR	steam methane reforming
SOFC	solid oxide fuel cell
T	temperature, K
v^D	diffusion volume, $m^3 \cdot mol^{-1}$
w	velocity, $m \cdot s^{-1}$
WGS	water-gas shift
WSTP	Wrocław Sewage-Treatment Plant "Janówek"
x	molar fraction
y	mass fraction
z	axial coordinate, m

Greek Letters

ΔH	heat of reaction, $J{\cdot}mol^{-1}$
ε	porosity
κ	permeability of the porous medium, m^2
λ	thermal conductivity, $W{\cdot}m^{-1}{\cdot}K^{-1}$
μ	viscosity, $kg{\cdot}m^{-1}{\cdot}s^{-1}$
v	stoichiometric coefficient in a reaction
$v^{b'}$	reaction order for a backward reaction
$v^{f'}$	reaction order for a forward reaction
ρ	density, $kg{\cdot}m^{-3}$

Subscripts and Superscripts

C	catalytic phase
G	gas, gaseous phase
i,j	indexes for species
IN	inlet
OUT	outlet
r	index for reactions
SUB	substrates

INTRODUCTION

Decarbonized power generation has recently received increasing interest due to climate warming concerns. The final product of most power generation technologies involving organic fuels is CO_2, which is released to the atmosphere. Power generation-derived anthropogenic CO_2 accumulates in the atmosphere. This accumulation is rather irreversible because nowadays, photosynthesis-based natural processes are very slow and due to industrialization and deforestation plants' potential for atmospheric CO_2 binding is now substantially lower than in the past centuries. Consequently, anthropogenic CO_2 emissions affect equilibria of natural carbon cycles and CO_2 concentration in the atmosphere increases at a rate of around 2 ppm/year reaching now the level of around 390 ppm. At this rate the Copenhagen Accord target equal to 450 ppm CO_2/+2°C will be reached before 2040. Therefore, low-carbon power generation technologies are necessary in order to limit the emissions of greenhouse gases into the atmosphere. CO_2 generation intensity of renewable and nuclear electricities ranges from 10 to 50 kg CO_2 MWh^{-1}. In contrast, fossil fuel-fired power generators are characterized by substantially larger CO_2 emission amounting to around 400 kg CO_2 MWh^{-1} for natural gas, 850 kg CO_2 MWh^{-1} for oil and 900 kg CO_2 MWh^{-1} for coal. Thus decarbonization of fossil fuel-based power generation necessitates CO_2 capture and sequestration (CCS). CO_2 capture can be realized by using a variety of processes such as reactive absorption or membranes. CO_2 sequestration can be achieved via e.g. geological storage or via some much more interesting options such as CO_2 utilization in chemical industries [1], fuel synthesis [2], algae cultivation, enhanced gas and oil recovery or enhanced methane coal-bed recovery. Sustainable and renewable energy sources such as hydro, wind, solar and geothermal can offer low-carbon power generation, but they have limited energy capacities and produce relatively costly electricity. On the other hand, fossil fuels are available in large quantities and are used to produce relatively cheap electricity. The main problem, however, are their atmospheric CO_2 emissions due to combustion of carbonaceous fossil fuels. It is thus desirable to use primary energy of fossil fuels, such as coal or hydrocarbons, but with limited CO_2 emissions. Alternatively, power can be produced from renewable biomass-derived fuels. However, in this case additional technical measures are needed in order to decrease the costs of electricity from biomass-based power cycles. Those two important topics are thus addressed in this chapter. Main idea behind pre-combustion CO_2 capture techniques in relation to carbonaceous fuels is to oxidize carbon to CO_2 while transferring all chemical energy of the fuel into H_2, which can be further oxidized without any atmospheric release of CO_2. This chapter will show that syngas routes can be very useful in the realization of this pre-combustion CO_2 capture concept. Moreover, based on thermodynamic considerations it will be investigated how power cycles involving a syngas step can achieve improved process integration and increased energy efficiency under the constraints of low-carbon operation.

ROLE OF SYNGAS IN DECARBONIZED ENERGY

Figure 9.1. presents the role of syngas in decarbonized power generation cycles.

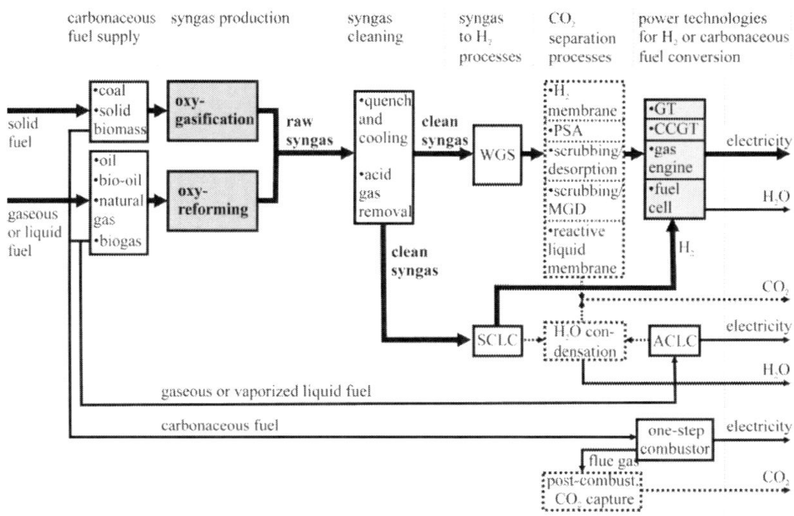

Note: WGS: Water Gas Shift Reaction, PSA: Pressure Swing Adsorption, MGD: Membrane Gas Desorption, SCLC: Steam Chemical Looping Combustion, ACLC: Air Chemical Looping Combustion, GT: Gas Turbine, CCGT: Combined Cycle Gas Turbine.

Figure 9.1. Role of syngas in decarbonized power generation cycles.

Namely, carbonaceous fuels are gasified/reformed in controlled oxygen/steam atmosphere to syngas. The syngas is then shifted via water gas shift reaction to a CO_2/H_2 mixture. The CO_2/H_2 mixture is then separated to capture CO_2 and a resulting H_2 stream is sent to a power unit for electricity production.

Alternatively, syngas is directed to a steam chemical looping combustion (SCLC) section in which H_2 is generated in a steam reactor and simultaneously CO_2/H_2O mixture is produced. The CO_2/H_2O mixture is then separated by simple H_2O condensation via cooling below the dew point. For comparison it is shown, that a conventional chemical looping combustion (CLC) process uses direct conversion of gaseous or vaporized liquid fuels in an air reactor (an ACLC process) and thus, similarly as in one-step fuel combustion, the syngas route is omitted.

SYNGAS PRODUCTION FROM CARBONACEOUS FUELS

Carbonaceous fuels are chemical energy carriers. They can be either fossil or renewable organic fuels. Fossil fuels include coal, oil and natural gas while renewable fuels include solid biomass, bio-oil and biogas. A substantial amount of chemical energy in all organic fuels is bound in the form of carbon and thus the release of this chemical energy by carbon oxidation is linked with CO_2 generation.

According to the concept presented in Figure 9.1, syngas is produced via oxy-gasification of solid fuels and oxy-reforming of gaseous or vaporized liquid fuels. Oxy-gasification is a process of reacting carbonaceous solid fuels at high temperatures with controlled amount of oxygen and steam. Oxy-gasification is thus a method for converting chemical energy of any carbonaceous fuels into the form of syngas. In an oxy-gasifier solid carbonaceous fuel such as coal or biomass is pyrolized to volatile tars and to char. The char is then simultaneously oxidized and reformed to produce syngas. Chemistry of gasification is dominated by irreversible carbon oxidation reactions, mostly reversible reforming reactions and the reversible water gas shift reaction which usually enables to maintain near equilibrium conditions between CO/H_2O and CO_2/H_2 in the reaction system [3]. Gasifiers can be a fixed-bed, a fluidized bed and an entrained flow-type unit. Recent research activities address novel designs of oxygen-blown coal gasifiers directed at efficient integrated gasification combined cycles power plants integrated with CCS (IGCC-CCS) [4]. Similarly, oxy-reforming is a process of reacting carbonaceous gaseous or vaporized fuels again at high temperatures with controlled amount of oxygen and steam. In an oxy-reformer hydrocarbons undergo the same chemical reactions but the contribution of solid phase reactions is avoided or at least limited. Due to controlled oxygen availability CO and H_2 are not fully oxidized via relevant irreversible oxidation reactions of those two species and thus syngas is formed.

SYNGAS CONDITIONING AND CLEANING

Syngas is a mixture of CO and H_2 usually in different proportions. Besides, depending on the process, it can comprise particulate matter, sulfur and other noxious substances. Therefore, syngas cleaning step is usually needed.

Syngas should be desulfurized in an acid gas removal unit. Also particulate matter must be removed. It is desirable to realize cleaning processes under high-temperature, however, such processes are still emerging concepts.

SYNGAS TO H_2 PROCESSES

Clean syngas is further used for hydrogen production. It is subjected to either water gas shift (WGS) reaction or steam chemical looping combustion (SCLC). In a WGS reactor the syngas is shifted to CO_2/H_2 mixture usually via catalytic reactions. SCLC uses two reactors, i.e. a fuel reactor and a steam reactor. The former is used for reducing the oxygen carrier while the latter is designed to produce H_2 utilizing energy from oxidation of the oxygen carrier [5].

It needs to be noted that a fuel reactor in air chemical looping combustion (ACLC) can utilize any gaseous or liquid vaporized hydrocarbon fuels. The steam reactor is replaced by an air reactor which extracts oxygen from air and do not produces hydrogen. Thus the ACLC system schematically shown in Figure 9.1 omits the syngas route. The syngas route is also omitted in one-step fuel combustion.

The role of syngas routes becomes clearer in comparison with those two well-known fuel conversion processes.

CO_2 SEPARATION PROCESSES

H_2/CO_2 mixture produced in the WGS reactor can be separated via several mature techniques such as H_2-membranes, pressure swing adsorption (PSA), solvent scrubbing/desorption as well as some emerging processes such as scrubbing/membrane gas desorption (MGD) [6] or reactive liquid membranes. CO_2/H_2O mixture produced in the SCLC fuel reactor can be separated via much more simple H_2O condensation via cooling below the dew point.

POWER TECHNOLOGIES FOR THE CONVERSION OF H_2 INTO ELECTRICITY

The advantage of using hydrogen as an energy carrier is such that its combustion is potentially more thermodynamically efficient than one-step fuel combustion route because the hydrogen can be burned at higher temperatures than solid organic fuels or even in fuel cells thus the Carnot efficiencies are improved or even overcome in the case of fuel cells. Hydrogen is oxidized in gas turbines (GT) which can operate in a more efficient combined cycle (CCGT).

For small scale applications gas engines can be used. Further, fuel cells can be used including solid oxide fuel cells (SOFCs) [7].

POWER CYCLE I: BIOGAS FUELLED OXY-REFORMING FUEL CELL (OR-FC)

Biomass includes carbon entirely assimilated from atmospheric CO_2 during its growth via photosynthesis with contribution of solar energy since plants are unable to assimilate carbon from any other sources. Hence, bioenergy processes that involve utilization of biomass-derived biogas with simultaneous CO_2 capture are characterized by 'negative net CO_2 emissions' to the atmosphere [8]. Decarbonization of biogas is a much more attractive option than CO_2 separation from fossil fuel power generation-derived flue gases. It arises from the fact that biogas contains less impurities such as particulate matter, NO_X, and SO_2. Further, biogas and biogas-derived post-processing gases are more enriched in CO_2 compared with air-fuel combustion-derived flue gases. This CO_2-enrichment provides an opportunity for low cost CO_2 capture due to the increased driving forces in CO_2 separation processes. Furthermore, by applying a WGS reactor the syngas is shifted to a CO_2-H_2 mixture from which the separation of CO_2 is technologically simpler than that from e.g. CO_2-N_2 or CO_2-CH_4 mixtures.

The aim of the presented decarbonized Oxy-Reforming Fuel Cell (OR-FC) power cycle is to generate fully decarbonized power via syngas route from renewable biogas. The examined idea relies upon oxy-fuel CH_4 splitting into CO and H_2. CO is then oxidized to CO_2 providing energy for the endothermic CH_4 as well as H_2O splitting. H_2 is further separated and used as a fuel for electrically efficient fuel cells. Since steam methane reforming (SMR) (equation (9.2)) is thermodynamically an equilibrium reaction a resulting unreacted CO/H_2O mixture can be either recycled back into an oxy-reforming reactor [3] or shifted to CO_2/H_2 by a selective catalytic water-gas shift (WGS) reaction. The present study analyzes only the latter option, however, a combination of those two approaches is also feasible as well as WGS equilibria can be shifted by separating one of the two products of the WGS reaction, i.e. H_2 or CO_2. Below, the OR-FC power cycle is described, simulated and discussed. The suitability of the OR-FC power cycle for fully decarbonized power production from renewable biogas, i.e. with 'negative net CO_2 emission' is addressed.

SHORTCOMINGS OF EXISTING BIOGAS-TO-ENERGY TECHNOLOGIES IN THE CCS CONTEXT

Existing energy technologies utilizing biogas, e.g. air-fuel combustion with electricity production release all CO_2 back to the atmosphere [9] Besides, combustion flue gases include CO_2 highly diluted by nitrogen which necessitates the costly separation of CO_2 from diluted CO_2-N_2 mixtures [10]. Another option is biogas upgrading to biomethane by separation of CO_2 and injection into natural gas grids. However, this technique involves the separation of CO_2 directly from biogas, i.e. from a CO_2-CH_4 mixture which is still relatively costly and the process is not fully decarbonized due to carbon remaining in the form of CH_4.

Further, there are two practiced methods for biogas to H_2 conversion. In accordance with the first method, CH_4 is separated from CO_2 (a costly process) and then pure CH_4 undergoes steam methane reforming (SMR). However, the endothermic nature of a SMR process necessitates supply of large quantities of heat which is usually delivered from combustion of fossil fuels in air, which, in turn, generates large amounts of CO_2 highly diluted by nitrogen. According to the second biogas to H_2 method, biogas is processed to syngas via a dry methane reforming (DMR) route, followed by water-gas shift (WGS) and the final H_2-CO_2 separation stage by e.g. pressure swing adsorption (PSA). A main disadvantage of this process is linked with its large endothermicity which necessitates external heat supply from e.g. air-fuel combustion which again produces CO_2 being highly diluted by nitrogen. Finally, biogas can also be converted to transportation fuels by catalytic dry reforming, to syngas and further to bio-methane, bio-methanol or other bio-hydrocarbons. Those attractive options have been experimentally validated [11] but still need some further refinements and demonstrations. For instance, the inclusion of low cost decarbonization of those processes can offer attractive "negative net CO_2 emissions" and hence additional economic advantages can arise from the utilization of renewable biogas.

Characterization of Sewage Sludge-Derived Biogas

Biogas is produced in Wrocław Sewage-Treatment Plant "Janówek" (WSTP) by treating of around 70,000 m^3/day of municipal sewage. Current WSTP's biogas production amounts to around 10,000 m^3/day. The production facility includes two dark anaerobic digesters having total volume of 13,000 m^3 which utilize a part of heat generated from downstream biogas combustion in order to maintain the temperature of 307 - 309 K relevant for mesophilic micro-organisms. Biogas is further desulfurized in a reactor utilizing activated bog ore operating according to the reaction $2Fe(OH)_3 + 3H_2S \rightarrow Fe_2S_3 + 6H_2O$. Further, biogas is purified mainly from organic sulfur compounds in an activated carbon adsorber.

Table 9.1. The composition of purified biogas from the WSTP as measured over the period of 2004 – 2010

biogas component	content	
	range	average
CH_4	60.2 - 67.7 %v	63.3 % (winter); 65.2 %v (summer)
CO_2	32.2 - 39.5 %v	35.7 %v
N_2	0.003 - 0.525 %v	-
O_2	0 - 0.083 %v	-
H_2O	0.1 - 19 $g \cdot m^{-3}$	3.5 g/m^3 (winter); 12.2 $g \cdot m^{-3}$ (summer)
H_2S	0 - 194 $mg \cdot m^{-3}$	39.8 $mg \cdot m^{-3}$
organic S (as H_2S)	0 - 140 $mg \cdot m^{-3}$	35.8 $mg \cdot m^{-3}$
total Cl	0 - 40 $mg \cdot m^{-3}$	10.6 $mg \cdot m^{-3}$
total F	0 - 5 $mg \cdot m^{-3}$	0.7 $mg \cdot m^{-3}$
NH_3	0 - 0.84 $mg \cdot m^{-3}$	0.13 $mg \cdot m^{-3}$
Traces of: methanethiol, ethanethiol, acetone, benzene, toluene, xylene, acetaldehyde, chloroethane and others.		

Note: Concentrations recalculated to conditions: 273 K, 101.3 kPa, dry basis [12].

Finally, purified biogas is subjected to air-fuel combustion and is converted into heat and electricity by means of two Jenbacher's gas engines having a nominal power of 601 MW_e and 738 MW_t each. Combustion flue gases from gas engines comprising original CO_2 and additional CO_2 from CH_4 oxidation undergo final deep oxidation by means of a catalytic converter and are released into the atmosphere. The flue gases are diluted with nitrogen due to the involvement of biogas combustion in the air-fuel mode.

The composition of municipal sewage sludge-derived biogas after its purification in the WSTP, i.e. in the form ready for further fuel to energy processing, is summarized in Table 9.1. Raw biogas composition strongly depends on its origin. As it has been indicated by Rasi et al [13], raw landfill biogas can include more N_2 (up to 17%v) than farm or municipal sewage sludge-derived biogases. High N_2 content limits the usefulness of landfill biogas in oxy-decarbonized H_2 production. Further, the amount of volatile organic compounds can range from 5 to 270 mg/m^3, and are lowest in raw biogas from a farm biogas plant. H_2S and other sulfur compounds occur in the lowest amounts in raw sewage sludge-derived biogas. From Table 9.1, it is also seen that municipal sewage sludge-derived biogas can be contaminated by Cl and F derivatives. In general, trace compounds present in biogas can affect fuel-to-energy processing of biogas, especially when it involves some more sophisticated techniques, such as membranes or catalytic reactors. Therefore, the selection of biogas processing technology should carefully take into account the origin of biogas fuel and the availability of adequate biogas purification technologies [14].

Oxy-Reforming Fuel Cell (OR-FC) Power Cycle

Biogas includes completely oxidized carbon in the form of CO_2 and unoxidized carbon in the form of CH_4. The objective of the current OR-FC process is to split CH_4 by means of oxidative reforming into free H_2 and simultaneously completely oxidize the remaining CO to CO_2. In order to avoid any dilution of reaction products with nitrogen pure oxygen is supplied as an oxidizer. Therefore, the fully decarbonized OR-FC process seeks for improved conditions for separating a CO_2-H_2 mixture at the expense of using oxygen as an oxidizer. Oxygen generation can be beneficially integrated in a closed mass and energy recirculation loop with a fuel cell unit and three reactors of the OR-FC cycle. Further, the addition of H_2O is beneficially another source of H_2 as well as oxygen thus reducing the consumption of pure oxygen and increasing the yield of H_2 in product gases. The investigated OR-FC process differs from a well-known SMR-WGS process by the utilization of pure oxygen as an oxidizer and by the beneficial integration of endothermic reforming and exothermic combustion reactions into one reaction unit. The product gases from the OR-FC process contain no nitrogen and hence the process can be fully decarbonized by cheap separation of a CO_2-H_2 mixture. Besides, the OR-FC process is autothermal while the SMR-WGS requires external heat supply which is frequently generated by air-fuel combustion with associated highly diluted CO_2 emissions. According to the OR-FC process main biogas components, i.e. CH_4 and CO_2 undergo the following chemical elementary reactions:

CH_4 combustion (irreversible):

$$CH_4 + 2\,O_2 \rightarrow CO_2 + 2\,H_2O \quad \Delta H^{298} = -802\ \mathrm{kJ \cdot mol^{-1}} \tag{9.1}$$

steam methane reforming (SMR) / methanation or reverse SMR (reversible):

$$CH_4 + H_2O = CO + 3\ H_2 \qquad \Delta H^{298} = 206\ \text{kJ·mol}^{-1} \tag{9.2}$$

water-gas shift (WGS) / reverse WGS (reversible):

$$CO + H_2O = CO_2 + H_2 \qquad \Delta H^{298} = -41\ \text{kJ·mol}^{-1} \tag{9.3}$$

H_2 combustion (irreversible):

$$H_2 + ½\ O_2 \rightarrow H_2O \quad \Delta H^{298} = -242\ \text{kJ·mol}^{-1} \tag{9.4}$$

CO combustion (irreversible):

$$CO + ½\ O_2 \rightarrow CO_2 \quad \Delta H^{298} = -283\ \text{kJ·mol}^{-1} \tag{9.5}$$

The reaction scheme given in equations (9.1) - (9.5) includes such elementary reactions as dry methane reforming (DMR) and methane partial combustion (MPC) which, however, do not reflect true chemical mechanisms and are the linear combinations of equations (9.1) - (9.5) [15-16]. By ensuring autothermal operating conditions and strict molar C/O ratio in substrates equal to 0.5 the overall reaction of the OR-FC process can proceed according to the following overall reaction:

$$CH_4 + \frac{1}{1.8} CO_2 + nO_2 + 2(1-n)H_2O \rightarrow \frac{2.8}{1.8} CO_2 + 2(2-n)H_2 \tag{9.6}$$

Further, from equation (9.6) it is seen that C/O = 0.5 and that C/H is:

$$C/H = \frac{2.8}{1.8(8-4n)} \tag{9.7}$$

Autothermal operation of the OR-FC process can be achieved when n = 0.3412 [3] and hence equation (9.6) becomes:

$$CH_4 + 0.5556\ CO_2 + 0.3412\ O_2 + 1.3176\ H_2O = 1.5556\ CO_2 + 3.3176\ H_2$$

$$\Delta H^{298} = 0 kJ/mol \tag{9.8}$$

From equation (9.8) it can be calculated that from biogas comprising e.g. 64.29%v CH_4 and 35.71%v CO_2 (i.e. CH_4:CO_2 = 1.8) with the addition of oxygen and H_2O in quantities satisfying C/O = 0.5 and C/H = 0.2344 one can obtain by autothermal conversion a mixture comprising 68.1%v H_2 and 31.9%v CO_2. Consequently, by utilizing the OR-FC process H_2 can be autothermally produced from renewable biogas with the yield of 3.3176 H_2:CH_4 (molar basis) and the remaining gas beneficially includes almost solely CO_2, i.e. is ready for compression and sequestration. The separation of this CO_2-H_2 mixture is technologically simpler than the separation of CO_2-N_2 or CO_2-CH_4 mixtures. However, the overall reaction given in equation (9.8) is reversible and hence the product gases from the OR-FC process include some CO and H_2O, since process equilibria are dominated by WGS chemistry. In order to force shifting of CO/H_2O to CO_2/H_2 two selective WGS reactors are proposed. Accordingly, a high temperature WGS (HT-WGS) reactor and a low temperature WGS (LT-WGS) reactor are sequentially set up after the oxy-reforming reactor. The flowsheet of the OR-FC cycle is presented in Figure 9.2. Accordingly, after the oxy-reforming reactor the reaction mixture still includes some quantities of CO and H_2O which needs to be shifted to CO_2 and H_2. This is accomplished by the inclusion of two WGS reactors. The OR-FC process involves two elementary reversible chemical reactions, i.e. SMR and WGS. The SMR reaction is strongly endothermic and produces moles thus, according to Le Chatelier's principle it is favored at high temperatures and low pressures. High temperatures are thus practiced in industrial conditions.

However, as far as pressure is involved it must be noted that the SMR reaction requires catalysts, which are better utilized under high-pressure conditions. Therefore, the oxy-reforming reactions are usually operated under high-pressure [17]. In contrast, the WGS reaction is thermodynamically favored by low temperatures.

However, in order to prevent reduction of reaction rate, temperatures must decrease gradually, i.e. by passing process gases from the oxy-reforming reactor through the HT-WGS reactor to the LT-WGS reactor. Further, according to the OR-FC process, the separated H_2 is sent to a fuel cell unit. Oxy-reforming and WGS reactors can operate under high-pressure which is beneficial for H_2/CO_2 separations by increasing their separation driving forces. Besides, high-pressure operation enables to improve the utilization of catalysts by increasing the amount of fuel gas flow per unit amount of catalysts. Finally, the fuel cell unit transforms H_2 into electricity with high fuel-to-electricity efficiency. An O_2 generation unit (OGU) can utilize either membrane air separation or H_2O electrolysis operated on electricity generated in the fuel cell stack. In the latter case, H_2 generated by electrolysis is fed to the fuel cell. The whole OR-FC process can thus be beneficially integrated by means of energy and mass recirculation loops. Oxy-reforming and WGS reactions proceed via catalytic routes. For CH_4 oxy-reforming attractive catalysts are those based on Rh, e.g. Ni-Rh/γ-Al_2O_3. For HT-WGS, the Fe_3O_4-Cr_2O_3 catalyst is better whereas for the LT-WGS a Cu-ZnO/Al_2O_3 catalyst is practiced.

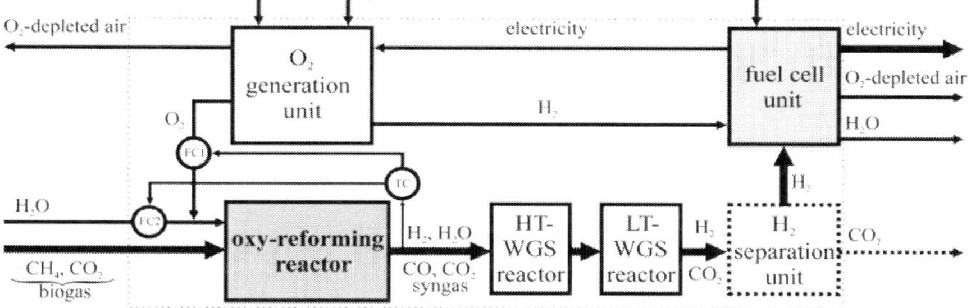

Note: FC - flow controller, TC - temperature controller.

Figure 9.2. The flowsheet of the OR-FC power cycle.

MATHEMATICAL MODELING

Thermodynamics of WGS reaction is favored under low temperatures. However, alleviated temperatures degrade the rate of reactions. Therefore, commercially, the temperature is reduced sequentially, e.g. by utilizing high temperature and low temperature WGS reactors. High temperature WGS enables to convert a part of CO and H_2O while low temperature WGS is responsible for full shifting of remaining minor quantities of CO/H_2O to CO_2/H_2. LT-WGS is a slow process at its low temperatures and it can require longer times on stream in order to thermodynamically equilibrate the WGS reaction. The catalytic reactors are modeled here with the use of differential equations. The governing equations arise from the material balance [18] of:
the gaseous phase:

$$\nabla \cdot \left(\rho^G y_i \sum_{j=1}^{j=NC} \left(D_{ij}^G \nabla x_j \right) \right) - \rho^G w \cdot \nabla y_i = 0 \tag{9.9}$$

the catalytic phase:

$$\nabla \cdot \left(\rho^G y_i \sum_{j=1}^{j=NC} \left(D_{ij}^{EFF} \nabla x_j \right) \right) - \rho^G w \cdot \nabla y_i + M_i R_i = 0 \tag{9.10}$$

The energy balance is provided for the gaseous phase:

$$\nabla \cdot \left(\lambda^G \nabla T \right) - \rho^G c_p^G w \nabla T = 0 \tag{9.11}$$

and the catalytic phase:

$$\nabla \cdot \left(\lambda^{EFF} \nabla T \right) + \sum_{r=1}^{r=NR} R_r \left(-\Delta H_r \right) = 0 \tag{9.12}$$

The momentum balance involves the gaseous phase:

$$-\rho^G (w \cdot \nabla) w + \nabla \cdot \left(-P\mathbf{I} + \mu^G \left(\nabla w + (\nabla w)^T \right) - \frac{2\mu^G}{3} (\nabla \cdot w)\mathbf{I} \right) = 0 \tag{9.13}$$

with the mass conservation equation:

$$\nabla \cdot \left(-\rho^G w \right) = 0 \tag{9.14}$$

the catalytic phase:

$$\nabla \cdot \left(\rho^G \left(\frac{\kappa \nabla P^C}{\mu^G} \right) \right) = 0 \tag{9.15}$$

the superficial gas velocity within the porous catalytic phase:

$$w^C = \frac{-\kappa \nabla P^C}{\mu^G} \tag{9.16}$$

The analyzed system includes 6 components that have different diffusivities. Hence, in order to account for cross-effects the Maxwell-Stefan (MS) multicomponent diffusion formulation is needed. The MS diffusivities are calculated according to the formula:

$$D_{ij}^G = k^{MS} \frac{\sqrt{\frac{1}{M_i} + \frac{1}{M_j}}}{\left(v_i^{D1/3} + v_j^{D1/3}\right)^2} \frac{T^{1.75}}{P} \tag{9.17}$$

The MS multicomponent effective diffusivities for the porous phase are then given by:

$$D_{ij}^{EFF} = D_{ij}^G \varepsilon^{1.5} \tag{9.18}$$

This formulation neglects Knudsen contribution to MS diffusivities which is justified when the pore radius of the catalyst supporting material is larger. Such reactor design is often preferred for millisecond contact time processes utilizing fast reactions such as oxy-reforming and expensive catalysts such as Rh [19]. The present 2D modeling approach offers improved insights into the nature of the reaction process compared with less rigorous 1D modeling. The governing equations are solved by using a finite element method. In the grid system, the grid distribution is denser adjacent to the wall, the centerline, the entrance and exit of the reactor as well as the interfaces between the porous region and the nonporous region. The boundary conditions used for the mass, energy and momentum balances define the flow, concentration and temperature at the inlet; assume the axial symmetry in the symmetry axis, the non-permeable external walls for mass, adiabatic external walls for energy, no-slip conditions for the flow at the walls, the pressure and the domination of convection at the outlet.

In reactor simulations main assumptions include steady state operation, neglecting of catalyst deactivation, idealized adiabatic conditions and neglecting of biogas impurities.

Reaction Kinetics

For the selected reactions one-step power-law kinetics exhibit adequate level of complexity for the engineering study of the oxy-reforming and WGS reactors. Kinetics for reactions in equations (9.1)-(9.5) are adapted from literature data [20-25].

In order to enable to model high-pressure reactors those kinetics are corrected for the effect of pressure. The correction allows for restraining the rate of the increase of catalytic reactivity with rising pressure due to the reduction of free active sites [26]. Accordingly, the molar production rate for a reaction r equals to:

$$R_r = A_r\,exp\left(\frac{-E_r}{R^G T}\right)\left(\prod_{i=1}^{i=NC} P_i^{\nu_{i,r}^{f'}} - \prod_{i=1}^{i=NC} \frac{P_i^{\nu_{i,r}^{b'}}}{K_r^{EQ}}\right)\left(\frac{P}{P^{REF}}\right)^{-1.06} \quad (9.19)$$

and the molar production rate for a component i is:

$$R_i = \sum_{r=1}^{r=NR} \nu_{i,r} R_r \quad (9.20)$$

The parameters for all 5 reactions, 6 components and 3 catalytic reactors are summarized in Tables 9.2 and 9.3.

Table 9.2. Parameters used in reaction kinetics, equations (9.19) - (9.20)

Reactant	Reaction eq. (9.1)			Reaction eq. (9.2) in oxy-reforming			Reaction eq. (9.4) in oxy-reforming			Reaction eq. (9.5) in oxy-reforming		
	$\nu^{f'}$	$\nu^{b'}$	ν	$\nu^{f'}$	$\nu^{b'}$	ν	$\nu^{f'}$	$\nu^{b'}$	ν	$\nu^{f'}$	$\nu^{b'}$	ν
CH_4	1	0	-1	1	0	-1	0	0	0	0	0	0
H_2O	0	0	2	1	0	-1	0	0	1	0	0	0
O_2	1	0	-2	0	0	0	1	0	-0.5	1	0	-0.5
CO_2	0	0	1	0	0	0	0	0	0	0	0	1
H_2	0	0	0	0	3	3	1	0	-1	0	0	0
CO	0	0	0	0	1	1	0	0	0	1	0	-1
A	2 ·- 10^6			1 -· 10^{-3}			9 -· 10^1			7 -· 10^0		
E	166 000			29 000			62 000			76 000		

Table 9.3. Parameters used in reaction kinetics, equations (9.19) - (9.20)

Reactant	Reaction eq. (9.3) in oxy-reforming			Reaction eq. (9.3) in HT-WGS			Reaction eq. (9.3) in LT-WGS		
	$\nu^{f'}$	$\nu^{b'}$	ν	$\nu^{f'}$	$\nu^{b'}$	ν	$\nu^{f'}$	$\nu^{b'}$	ν
CH_4	0	0	0	0	0	0	0	0	0
H_2O	1	0	-1	1	0	-1	1	0	-1
O_2	0	0	0	0	0	0	0	0	0
CO_2	0	1	1	0	1	1	0	1	1
H_2	0	1	1	0	1	1	0	1	1
CO	1	0	-1	1	0	-1	1	0	-1
A	4 ·- 10^{-1}			2 ·- 10^{-1}			3 -· 10^{-2}		
E	67 000			70 000			47 400		

THERMODYNAMIC EQUILIBRIA

The reactions given in equations (9.1) and (9.4)-(9.5) are irreversible since their equilibrium constants are very large. In contrast, the reactions given in equations (9.2)-(9.3) are reversible. Their thermodynamic equilibrium constants are adapted from [16]:

$$K_{SMR}^{EQ} = 10^{10} \cdot exp(30.114 - 26830 / T) \tag{9.21}$$

$$K_{WGS}^{EQ} = exp(-4.036 + 4400 / T) \tag{9.22}$$

CHARACTERIZATION OF SIMULATED REACTORS

The OR-FC process includes three catalytic reactors utilizing different catalysts and operating conditions. Figure 9.3 illustrates a reactor geometry as well as reaction substrates and products.

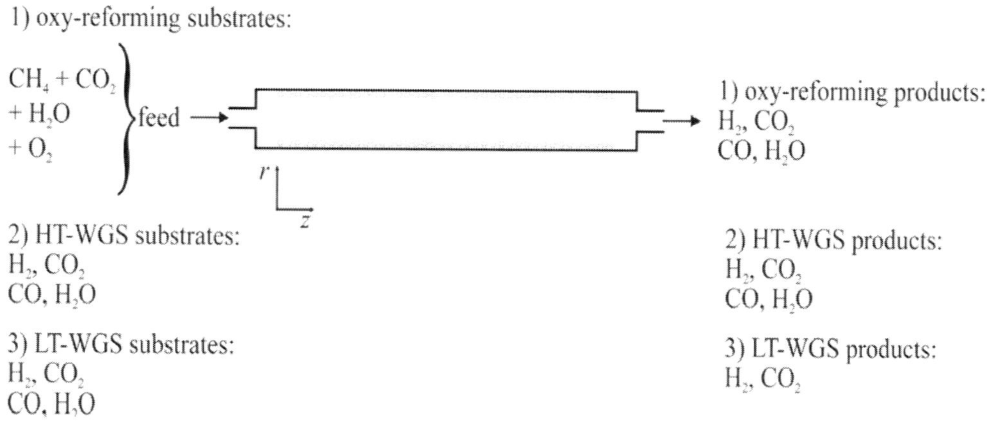

Note: Geometry used: length L^{CAT} = 2, 20 mm, ID = 0.4 mm, porous catalyst's layer thickness H^{CAT} = 30 μm.

Figure 9.3. The characterization of the simulated reactors.

Table 9.4. Characteristics of reactors and operating conditions used in simulations

Reactor type	Parameters					
	L^{CAT} [mm]	ID [mm]	H^{CAT} [μm]	$T^{SUB, IN}$ [K]	P [MPa]	$m^{FEED, IN}$ [kg s^{-1}]
oxy-reforming	2	0.4	30	900	2.0	10^{-7}
HT-WGS	2	0.4	30	700	2.0	10^{-8}
LT-WGS	20	0.4	30	400	2.0	10^{-8}

The current study uses structured reactors which allow for reduced pressure drops by more than an order of magnitude compared with standard fixed-bed reactors and thus higher reactor throughputs can be attained. All investigated reactors are simulated under elevated pressure of 2.0 MPa since reactors are better utilized under high-pressure operating conditions, i.e. they enable to convert more substrates per unit volume of a reactor per unit time. It is an advantageous feature since catalytic reactors are usually relatively expensive and thus their utilization must be maximized.

Below, a simulation case study is provided in order to validate the OR-FC process. A main discussion is limited to three catalytic reactors. Table 9.4 summarizes geometry and operating conditions utilized in the numerical simulations of the analyzed reactors.

PERFORMANCE OF THE OXY-REFORMING REACTOR

The objective of the oxy-reforming reactor is to convert all CH_4 into CO, H_2, CO_2 and H_2O products. High temperature oxy-reforming reactor operation favors the completion of SMR and combustion reactions and hence both CH_4 and O_2 are fully consumed. Figure 9.4 displays molar fractions of species and temperatures attained along the oxy-reforming reactor.

From Figure 9.4 it can be observed that the process is exothermic with increasing temperature along the reactor since here $n > 0.3412$ is applied. Further, CH_4 and O_2 are beneficially fully consumed. Relatively large quantities of CO and H_2 are produced. H_2O is partially reacted while the amount of CO_2 remains nearly unchanged. CO_2 is produced by combustion and consumed by reverse WGS with H_2 derived from SMR.

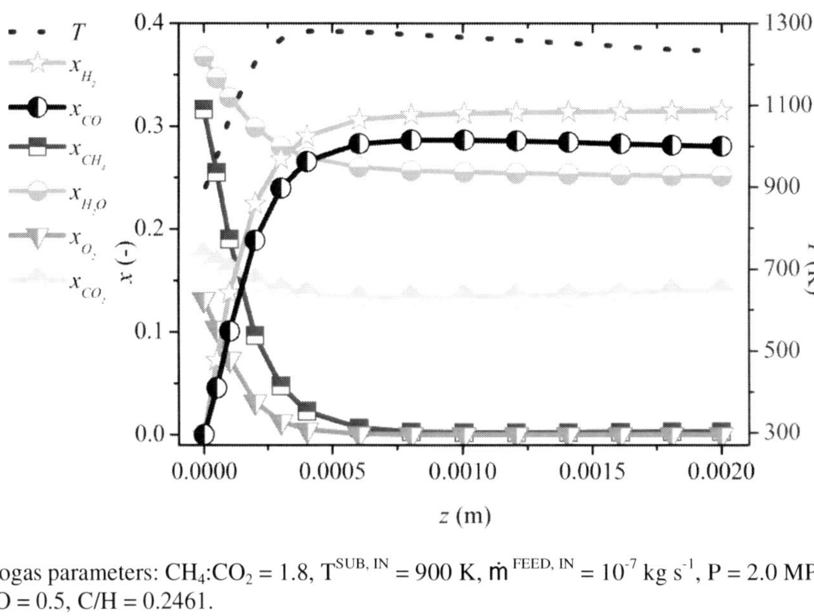

Note: Biogas parameters: CH_4:CO_2 = 1.8, $T^{SUB, IN}$ = 900 K, $\dot{m}^{FEED, IN} = 10^{-7}$ kg s^{-1}, P = 2.0 MPa, n = 0.42, C/O = 0.5, C/H = 0.2461.

Figure 9.4. Molar fractions of species x and temperature T along the oxy-reforming reactor.

Depending on operating conditions used the oxy-reforming process can be operated under reaction control, under mass transport control with O_2 as the limiting reactant or in an intermediate regime. For instance, high temperature favors mass transport control while radial miniaturization of channels favors reaction control. The rate control mechanisms of the oxy-reforming reactor have been recently investigated in details elsewhere [3] by 2D simulation.

In design of the oxy-reforming reactor the thermal conditions must be selected such so as to satisfy high temperatures required for WGS equilibria shifting and to assure catalysts thermal durability.

Further, fuel and O_2 injection techniques must address safety reasons. For instance, fresh reactants can be diluted by applying a large recycle [3]. Alternatively, fresh substrates can be pre-mixed with reactants withdrawn from the oxy-reforming reactor.

PERFORMANCE OF THE HT-WGS REACTOR

The main objective of the HT-WGS reactor is to selectively convert CO and H2O into CO2 and H2. Relatively high temperature enables to use relatively short rector characterized by moderate times on stream since reaction rates are not substantially degraded by still moderate operating temperatures. Figure 9.5 shows that under the thermal conditions studied CO molar fraction is partly reduced. Namely, CO content is reduced to around $x = 0.15$ while H_2 content reaches $x = 0.38$ and CO_2 reaches $x = 0.31$. Hence, due to equilibria limitations of the WGS reaction, a subsequent LT-WGS reactor located downstream is needed.

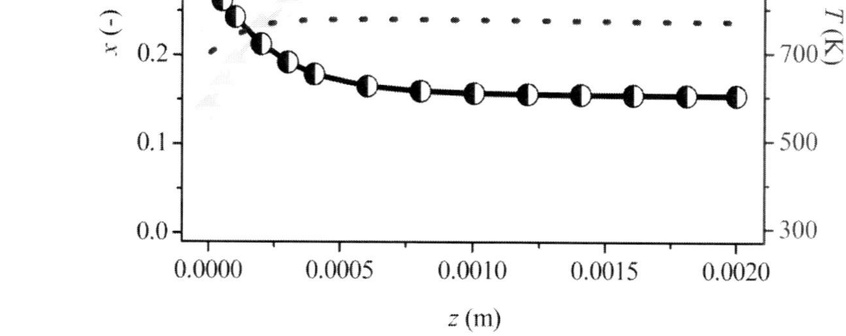

Note: Inlet composition is that from the oxy-reforming reactor output, $T^{SUB, IN} = 700$ K, $\dot{m}^{FEED, IN} = 10^{-8}$ kg s^{-1}, P = 2.0 MPa.

Figure 9.5. Molar fractions x and temperature T along the HT-WGS reactor.

PERFORMANCE OF THE LT-WGS REACTOR

By lowering reaction temperature the thermodynamic equlibria of the WGS reaction are shifted towards substrate formation. However, at the same time a substantial drop in reaction rate is attained. The decreased reaction rate leads to a situation that the reactor can not assure the completion of the WGS reaction and hence the WGS reaction can depart from its thermodynamic equilibrium. To investigate the departure of the WGS reaction from its thermodynamic equilibrium conditions a reaction quotient RQ_{WGS} is defined as follows:

$$RQ_{WGS} = \frac{\dfrac{P_{H_2} P_{CO_2}}{P_{CO} P_{H_2O}}}{K_{WGS}^{EQ}} \tag{9.23}$$

The interpretation of RQ_{WGS} is following. When RQ_{WGS} equals to 1 it means that the WGS reaction approaches its thermodynamic equilibrium.

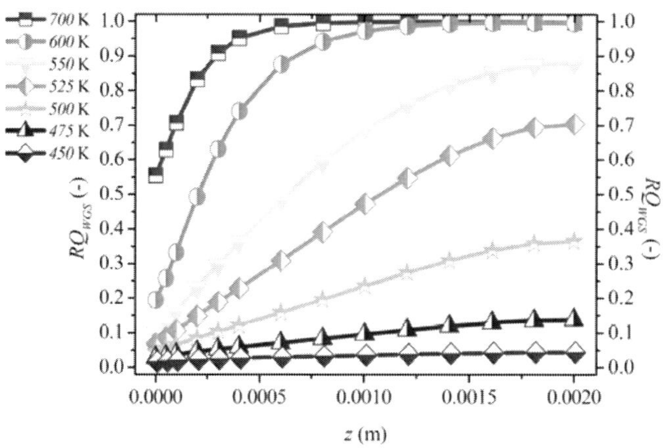

Note: Inlet composition is that from the HT-WGS output, $T^{SUB, IN}$ decreases from 700 to 450 K, $\dot{m}^{FEED, IN} = 10^{-8}$ kg s^{-1}, P = 2.0 MPa.

Figure 9.6. Effect of temperature on departure from thermodynamic equilibria of the WGS reaction in the LT-WGS reactor.

When $RQ_{WGS} > 1$ it means that product species (i.e. H_2 and CO_2) are in excess and thus the reverse WGS reaction proceeds faster than the WGS reaction. Finally, when $RQ_{WGS} < 1$ it means that substrates (i.e. CO and H_2O) dominate in a reaction mixture and hence the WGS reaction dominates over the reverse WGS reaction. When passing reactants from a HT-WGS reactor to a LT-WGS reactor the operating temperature is substantially reduced. Therefore, it is important to see how reduced temperature affects the thermodynamic equilibria in a reactor with specified dimensions. For this purpose Figure 9.6 presents the effect of reactor inlet T on RQ_{WGS}. From Figure 9.6 it is seen that when $T^{SUB,IN}$ drops the WGS reaction departs from its thermodynamic equilibrium. The observed effect is associated with both the degraded reaction rate and with the increase of WGS equilibrium constant (K^{EQ}_{WGS}). In the LT-WGS reactor studied with flow rate of 10^{-8} kg·s^{-1} it is seen that under $T^{SUB,IN}$ around 450 K the departure from equilibrium is very large.

Consequently, LT-WGS must assure longer time on stream for reactants in order to achieve conversion approaching the thermodynamic equilibrium, i.e. the reactor must be longer or must apply lower throughputs. Both those features are disadvantageous since they require either a larger LT-WGS reactor with expensive catalysts or the reduction of reactants throughputs which reduces catalysts utilization. From Figure 9.7 it is seen that in the present case study, under the operating temperature of around 500 K, the CO molar fraction can be reduced to around 0.04 while H_2 content reaches around 0.44 and CO_2 around 0.47. As it can be calculated from equation (9.22) the equilibrium constant of the WGS reaction (K^{EQ}_{WGS}) is moderate and it drops with decreasing T. For $T = 500$ K, K^{EQ}_{WGS} equals 117. This means that for a mixture comprising originally only four species, i.e. CO_2, H_2, CO and H_2O and having C/O = 0.5 in equilibrium conditions one obtains:

$$x_{CO} \cong \frac{\sqrt{K^{EQ}_{WGS} + 2} - 1}{2\left(K^{EQ}_{WGS} + 1\right)} \tag{9.24}$$

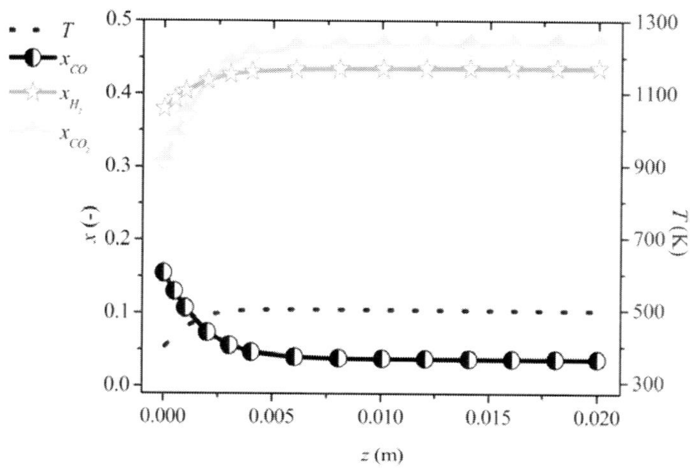

Note: Inlet composition is that from the HT-WGS reactor outlet, $T^{SUB,IN} = 400$ K, $\dot{m}^{FEED,IN} = 10^{-8}$ kg s^{-1}, P = 2.0 MPa.

Figure 9.7. Molar fractions x and temperature T along the LT-WGS reactor.

By substitution $K^{EQ}_{WGS} = 117$ one obtains $x \approx 0.04$ which is an equilibrium concentration of CO similar to that obtained in the current simulations. Thus in the presented case study the reaction mixture approaches equilibrium conditions.

In order to achieve further reduction of CO several techniques can be involved. For instance, a mass-recirculation process which provides additional enhancement of WGS equilibria shifting can be coupled to the OR-FC process [3]. Alternatively, one can use further reduction of operating temperatures, increase H_2O:CO ratio or utilize partial removal of CO_2 or H_2 from the reaction mixture by e.g. sorption enhancement (for CO_2) or membrane rectors (for H_2). From Figure 9.7 it is seen that the mixture exiting the LT-WGS includes mainly H_2

and CO_2. Thus by separating H_2 from this mixture a relatively highly enriched CO_2 stream is obtained which is ready for compression and sequestration.

Minor impurities of CO_2 such as CO can be recycled back to the oxy-reforming process [3]. The thinking underlying the proposed concept of biogas conversion via the OR-FC cycle with negative net CO_2 emissions can also be useful in other similar biogas conversion cycles.

Power Cycle II: Solid Biomass Fuelled Multi-Step Gasification Gas Turbine (MSG-GT)

Biomass is abundant, untapped and environmentally friendly renewable energy carrier. Sustainable solid biomass resource available for energy use includes: (i) agricultural residues (straw, grasses), (ii) forest residues (from e.g. sawmills), (iii) biomass production on surplus degraded land, (iv) organic wastes and (v) energy crops. Currently, most massive biomass-based power technologies are direct combustion (via e.g. CFB boilers) and co-combustion with coal.

However, direct combustion suffers from low energy conversion efficiencies (around 20%) due to high moisture content and thus low flame temperatures achieved. This problem is overcome in co-combustion but unfortunately biomass must be transported to centralized coal-fired power plants what substantially increases overall costs. Therefore, novel efficient biomass-based power cycles are needed.

Promising Technologies for Solid Biomass Fuelled Power Plants

Solid biomass can be converted to gaseous fuels by heating in a gasification agent such as oxygen/steam, air or CO_2. In a gasifier solid biomass is pyrolysed to volatile tars and to char. The char is then simultaneously oxidized and reformed to produce syngas. Biomass gasification is well suited for biomass IGCC (BIGCC) with combustion of syngas in a gas turbine (as a topping cycle) while hot gases from the gas turbine are used for steam generation in a heat recovery steam generators (HRSG) to run a steam turbine (a bottoming cycle).

Torrefaction is a mild pyrolysis process that improves the fuel properties of solid biomass. Biomass torrefaction is realized at temperatures below 300°C in absence of oxygen. The main products formed are torrefied biomass and volatilities - torrefaction gas. The torrefied solid biomass is characterized by low moisture content, high energy density and small volume. Besides improvements in relation to the ease of grinding/pelletisation, long-term storage potential and reduced long-distance transportation costs are achieved [27]. The torrefied biomass is suitable for direct combustion, co-combustion and gasification. Pre-treatment of biomass via torrefaction prior to gasification enables the use of a pressurized entrained flow gasifier and thus higher gasification temperatures and higher gasification thermodynamic efficiency can be achieved.

THERMODYNAMIC CONSIDERATIONS FOR EFFICIENT DESIGN OF SOLID BIOMASS FUELLED POWER PLANTS

Practical design principles for efficient energy conversion in solid biomass fuelled power plants can be derived directly from the First and Second Law of Thermodynamics. Namely, the First Law of Thermodynamics (the conservation of energy) necessitates that a power system under design should minimize energy losses to the environment. This can be achieved by thermal integration measures such as (i) minimizing radiation losses (via e.g. heat recovery or insulation techniques) and (ii) minimizing losses in output steams (via e.g. heat recirculation [28] or flue gas recirculation techniques [29]). More interesting power plant design insights arise from the Second Law of Thermodynamics (the quality of energy). Namely, in order to minimize exergy losses and thus entropy generation rates a solid biomass power plant should involve biomass conversion operations characterized by decreased irreversibility. For combustion the majority of the exergy destruction occurs during thermal energy exchange between hot reaction product molecules and cold reactant molecules [30]. The fuel oxidation itself is relatively efficient. This effect arise from the fact that in combustion processes fresh cold substrates are in direct contact with hot products, which is very inefficient, since the maximal temperature is reduced and thus less work can be delivered. Further, if an energy conversion process can take place with sufficient degrees of freedom, the optimal process follows a trajectory of optimal thermodynamic states, a so-called highway in state space, which is characterized by constant local entropy production [31]. This power plant design principle known as the equipartition of entropy production is satisfied in many biological systems such as human lungs (see [31] in relation to fuel cell design), and thus it can inspire highly efficient design of power plants.

MULTI-STEP GASIFICATION GAS TURBINE (MSG-GT) POWER CYCLE

In light of the aforementioned first- and second-law thermodynamic considerations one-step solid biomass combustion seems to be less efficient solution than multi-step solid biomass gasification. Namely, the multi-step solid biomass gasification technology can assure more reversible gradual fuel oxidation in multiple sub-units thus limiting thermal energy exchange between hot reaction products and cold reactants as well as enabling advanced energy integration. Figure 9.8 presents a flowsheet of a power plant utilizing multi-step solid biomass gasification technology. Accordingly, the proposed multiple biomass conversion steps involve biomass pre-treatment processes such as drying and torrefaction followed by gasification and final syngas combustion in a combined cycle gas turbine. Gasification itself can be realized in a high-temperature air-blown entrained flow gasifier involving chemical quench via the injection of torrefaction gases thus offering improved char and tar conversion even at low oxygen consumption as well as improved ash to slag fusion.

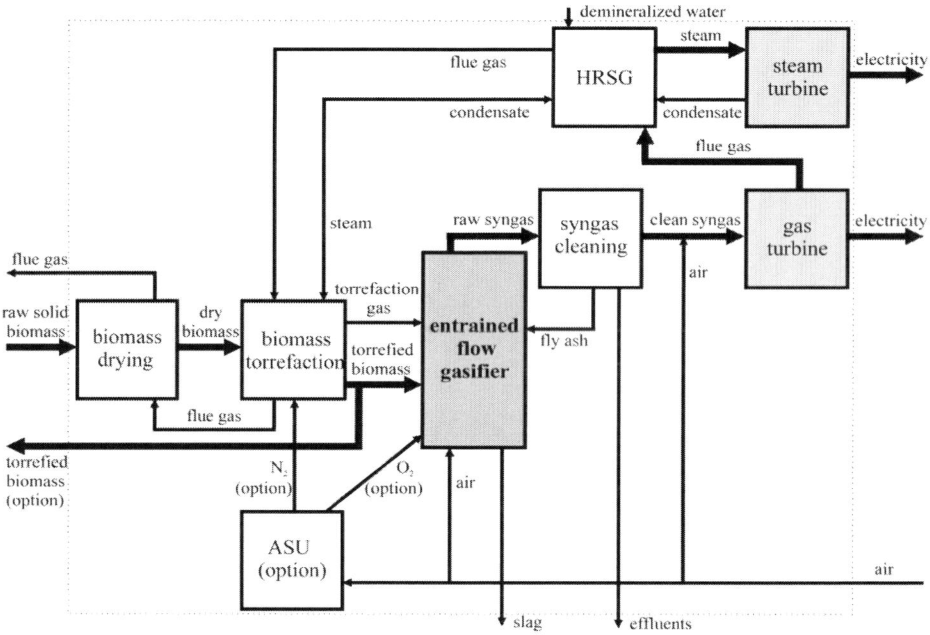

Figure 9.8. The flowsheet of the solid biomass fuelled power plant utilizing multi-step gasification gas turbine (MSG-GT) power cycle.

Moreover, the proposed flowsheet is in accordance with the principle of equipartition of entropy production, i.e. entropy is produced more uniformly in all sub-units of the proposed power plant. The biomass conversion route according to the flowsheet from Figure 9.2 can be more energy efficient, cleaner and optionally it can enable to manufacture a new valuable renewable product - the torrefied biomass.

Power Cycle III: Coal Fuelled Oxy-gasification Steam Chemical Looping Combustion Gas Turbine (OG-SCLC-GT)

Chemical looping combustion can potentially offer increased energy conversion efficiency. The gain in efficiency is possible due to the enhanced reversibility of the two redox reactions. In conventional direct combustion, the release of a fuel's chemical energy occurs in a highly irreversible manner - departing considerably from equilibrium [30].

In contrast, in CLC, both redox reactions can occur almost reversibly and at relatively low temperatures. This can allow a power plant using CLC to approach the ideal work output for an internal combustion engine without exposing its components to excessive operating temperatures. Moreover, CLC facilitates CO_2 capture because the two redox reactions generate two intrinsically separated flue gas streams with one consisting entirely of CO_2 and H_2O. This gives CLC clear benefits when compared with competing carbon capture technologies, as the latter generally involve a significant energy penalty associated with either post combustion scrubbing systems or the work input required for air separation plants.

Therefore, CLC-based technologies can achieve high efficiency power generation with low energy penalty carbon capture.

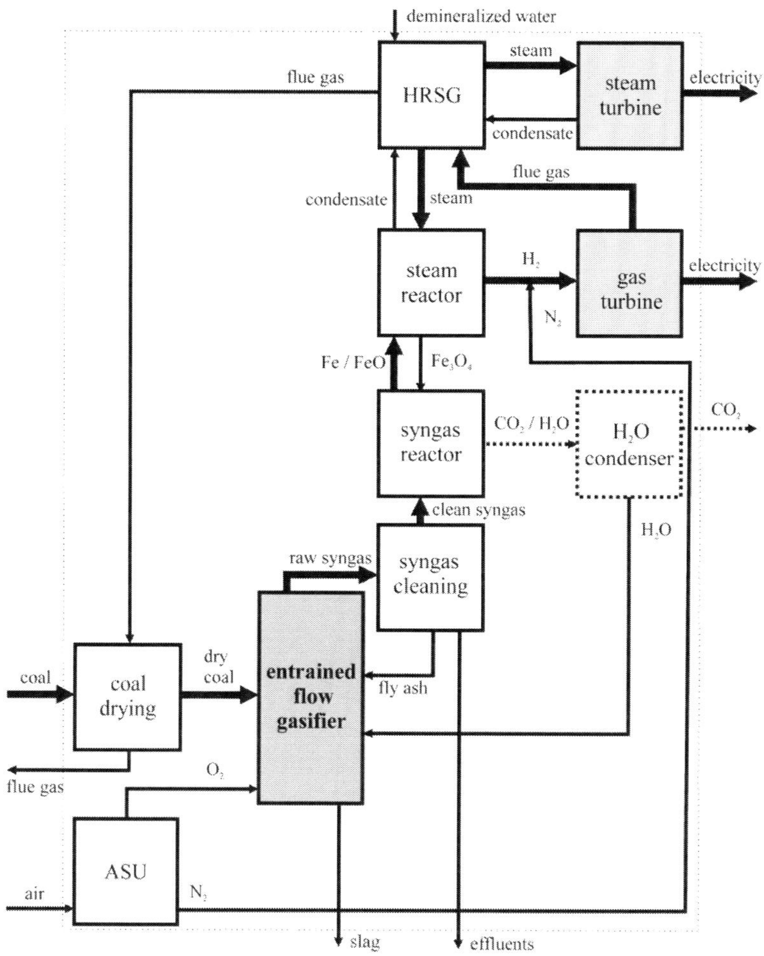

Figure 9.9. The flowsheet of the coal fuelled oxy-gasification steam chemical looping combustion gas turbine (OG-SCLC-GT) power cycle.

OXY-GASIFICATION STEAM CHEMICAL LOOPING COMBUSTION GAS TURBINE (OG-SCLC-GT) POWER CYCLE

A steam chemical looping combustion (SCLC) method consists of oxidation and reduction processes, which are undertaken in two separate reactors. In the reduction step, the syngas is reacted with an oxygen carrier such as iron oxide to form carbon dioxide and water. After condensing the water vapor, the captured carbon dioxide stream can be sent to the sequestration sites. The reduced form of iron is re-oxidized in an oxidation reactor to its original form using steam and recycled back to the reduction reactor.

In the fuel reactor the syngas is oxidized with iron oxide (magnetite) according to the following chemical reactions:

$$Fe_3O_4 + 4CO \rightarrow 3Fe + 4CO_2 \quad (9.25)$$

$$Fe_3O_4 + 4H_2 \rightarrow 3Fe + 4H_2O \quad (9.26)$$

In the steam reactor the reduced form of the oxygen carrier (iron) is oxidized back by using steam to regenerate the iron oxide and to produce hydrogen according to the reaction:

$$Fe + 4H_2O \rightarrow Fe_3O_4 + 4H_2 \quad (9.27)$$

Hydrogen is then sent to a combined cycle gas turbine. Figure 9.9 presents the flowsheet of the oxy-gasification steam chemical looping combustion gas turbine (OG-SCLC-GT) power system.

A main advantage of the OG-SCLC-GT is that it can offer CO_2 capture rate exceeding 99% while maintaining comparable to other CCS technologies power generation efficiency [5]. The consumption of oxygen can be minimized in optimized gasifier designs [4]. Besides, water recirculation loop (i.e. gasifier/CLC/condenser) minimizes water consumption thus mass and energy integration is ensured.

Other Emerging Decarbonized Power Cycles Involving Syngas Routes

Another interesting emerging concept is a catalytic membrane reactor [32]. According to this concept hydrogen can be produced from coal gasification-derived syngas without any syngas conditioning or desulfurization due to the use of sulfur resistant catalysts. The concept requires however further demonstration in regard to CO_2 capture potential. One another interesting process has been investigated by Ortiz et al. [33]. It relates to H_2 production with CSS by coupling steam methane reforming and CLC. Main benefits arise from the recycle of PSA tail gas into the CLC fuel reactor, in which they are oxidized to CO_2 and H_2O. Methane is simultaneously reformed in the system thermally integrated with the fuel reactor and methane-derived syngas is sent to the PSA unit.

Conclusion

The paper provided technological analyzes of decarbonized power generation via syngas routes. The role of syngas in low-carbon power cycles was presented. Three promising low-carbon power cycles were examined with emphasis put on their thermodynamics, integration opportunities and the CO_2 capture feasibility.

The first decarbonized power cycle, the OR-FC cycle, offered to produce hydrogen from renewable biogas with the maximal yield of ca. 3.3 H_2:CH_4 on molar basis. Through capturing CO_2 the OR-FC power cycle achieved negative net CO_2 emissions. According to the proposed method, the oxidative reforming of biogas to syngas was followed by the high-temperature shift reaction and the low-temperature shift reaction was performed to produce H_2. The OR-FC process was integrated with CO_2 capture, the fuel cell and the oxygen generator. The main advantage of the OR-FC process over other existing decarbonized processes included H_2 separation from highly concentrated CO_2-H_2 mixtures while remaining CO_2 could be compressed and sequestered. Also energy and mass integration was assured by involving relevant recirculation loops. The OR-FC process was illustrated by modeling and simulation of three reactors, i.e. the oxy-reforming reactor, the HT-WGS reactor and the LT-WGS reactor. In the oxy-reforming reactor nearly complete methane to syngas conversion was achieved. CO was shifted in the HT-WGS and further in the LT-WGS to its low equilibrium concentration. The OR-FC process produced the mixture containing mostly H_2 and CO_2 characterized by cost-effective separation. The results also revealed that the design of the LT-WGS reactor was crucial for WGS reaction completion due to its likely departure from the equilibrium conversion. Pure hydrogen was converted into electricity in the efficient fuel cell.

The second low-carbon power cycle, the MSG-GT cycle, offered to avoid problems encountered in two currently most massive biomass-based power technologies, i.e. in one-step direct combustion (it had low efficiency) and in co-combustion with coal in usually centralized power plants (it necessitated biomass transport from large distances). The MSG-GT cycle involved biomass pre-treatment via drying and torrefaction followed by biomass gasification to syngas and high efficiency syngas combustion. It was emphasized that such a multi-step cycle led to more reversible fuel processing than achieved in one-step combustion. Therefore, in accordance with the principle of entropy equipartition entropy was generated more uniformly what meant less exergy destruction and thus increased cycle efficiency could be achieved.

The third investigated low-carbon cycle, the OG-SCLC-GT cycle, benefited from both the improved cycle efficiency and the extremely high CO_2 capture rate. In the OG-SCLC-GT cycle coal was dried, gasified in th oxygen-blown gasifier and the resulting syngas was fed to the syngas CLC reactor that produced CO_2/H_2O mixture. Further, the reduced oxygen carrier was used in the steam reactor to produce hydrogen which was sent to the combined cycle gas turbine. The OG-SCLC-GT cycle was superior due to advanced mass and energy integration via several closed recirculation loops and due to the largest CO_2 capture rate.

AUTHOR'S BIOGRAPHY

Wojciech M. Budzianowski is an assistant professor at the Wrocław University of Technology, Wrocław, Poland. He received his Ph.D. from the Wrocław University of Technology in 1998. He has authored about 60 journal articles, 3 patent applications and 1 book. He is a member of the editorial advisory board of Recent Patents on Chemical Engineering, Section Editor of Journal of Power Technologies and an international reviewer. His current research interests include energy and fuels and energy engineering. Personal web site: http://works.bepress.com/wojciech_budzianowski

References

[1] K. M. K. Yu, I. Curcic, J. Gabriel and S. C. E. Tsang, *ChemSusChem* 1, 893 (2008).

[2] Stamatiou, P. G. Loutzenhiser, and A. Steinfeld, *AIChE J.* 57, n/a (2011) doi: 10.1002/aic.12580.

[3] W. M. Budzianowski, *Int. J. Hydrogen Energ.* 35, 7454 (2010).

[4] N. Nagasaki, Y. Takeda, T. Akiyama and T. Kumagai, *Hitachi Review* 59, 77 (2010).

[5] C. C. Cormos, *Int. J. Hydrogen Energ.* 36, 3726 (2011).

[6] Mansourizadeh and I. F. Ismail, *Desalination* 273, 386 (2011).

[7] F. N. Cayan, M. Zhi, S. R. Pakalapati, I. Celik, N. Wu and R. Gemmen, *J. Power Sources* 185, 595 (2008).

[8] W. M. Budzianowski, *Int J. Chem. Reactor Eng.* 8, A156 (2010).

[9] E. Ten Brummeler, *Water Sci. Technol.* 41, 849 (2000).

[10] W. M. Budzianowski, *Rynek Energii* 83(4), 21 (2009).

[11] N. Muradov and F. Smith, *Energ. Fuel* 22, 2053 (2008).

[12] K. Gaj, H. Cybulska-Szulc and F. Knop, in *Oczyszczanie Ścieków i Przeróbka Osadów Ściekowych* ed. by Sadecka Z. Zielona Góra 113 (2010).

[13] S. Rasi, A. Veijanen and J. Rintala, *Energy* 32, 1375 (2007).

[14] N. Abatzoglou and S. Boivin, *Biofuel. Bioprod. Bior.* 3, 42 (2009).

[15] Donazzi, A. Beretta, G. Groppi and P. Forzatti, *J. Catal.* 225, 259 (2008).

[16] Mahecha-Botero, Z. Chen, J. R. Grace E. E. S. H. Elnashaie, C. J. Lim, M. Rakib, I. Yasuda and Y. Shirasaki, *Chem. Eng. Sci.* 64, 3598 (2009).

[17] W. M. Budzianowski, *Int J. Chem. Reactor Eng.* 7, A20 (2009).

[18] R. B. Bird, W. E. Stewart and E. N. Lightfoot, *Transport Phenomena*, Wiley, New York, (2002).

[19] Bitsch-Larsen, R. Horn and L. D. Schmidt, *Appl. Catal. A-Gen.* 348, 165 (2008).

[20] W. Jin, X. Gu, S. Li, P. Huang, N. Xu and J. Shi, *Chem. Eng. Sci.* 55, 2617 (2000).

[21] R. H. de Smet, M. H. J. M. de Croon, R. J. Berger, G. B. Marin and J. C. Schouten, *Chem. Eng. Sci.* 56, 4849 (2001).

[22] V. A. Kirillov, A. S. Bobrin, N. A. Kuzin, V. A. Kuzmin, A. B. Shigarov, V. B. Skomorokhov, E. I. Smirnov and V. A. Sobyanin, *Ind. Eng. Chem, Res.* 43, 4721 (2004).

[23] W. H. Chen, M. R. Lin, T. L. Jiang and M. H. Chen, *Int. J. Hydrogen Energ.* 33, 6644 (2008).

[24] Y. Choi and H. G. Stenger, *J. Power Sources* 124, 432 (2003).

[25] Beretta, G. Groppi, M. Lualdi, I. Tavazzi and P. Forzatti, *Ind. Eng. Chem. Res.* 48, 3825 (2009).

[26] M. Reinke, J. Mantzaras, R. Schaeren, R. Bombach, R. Inauen and S. Schenker, *Combust. Flame* 136, 217 (2004).

[27] M. J. Prins, K. J. Ptasinski and F. J. J. G. Janssen, *Energy* 31, 3458 (2006).

[28] W. M. Budzianowski, *Rynek Energii* 91(6), 108 (2010).

[29] W. M. Budzianowski and A. Miller, *Recent Patents on Mechanical Engineering* 2, 228 (2009).

[30] W. R. Dunbar and N. Lior, *Combust. Sci. Technol.* 103, 41 (1994).

[31] S. Kjelstrup, M. -O. Coppens, J. G. Pharoah and P. Pfeifer, *Energ Fuel* 24, 5097 (2010).

[32] M. Abdollahi, J. Yu, P. K. T. Liu, R. Ciora, M. Sahimi and T. T. Tsotsis, *J. Membr. Sci.* 363, 160 (2010).

[33] M. Ortiz, P. Gayán, L. F. de Diego, F. García-Labiano, A. Abad, M. A. Pans and J. Adánez, *J. Power Sources* 196, 4370 (2011).

In: Syngas
Editors: Antonius Indarto and Jelliarko Palguandi

ISBN: 978-1-62100-870-5

Chapter 10

SYNGAS PRODUCTION AND APPLICATIONS FOR GREENHOUSE GAS MITIGATION

NorAishah Saidina Amin* and Muhammad Tahir
Universiti Teknologi Malaysia, Malaysia

ABSTRACT

Production of synthesis gas for various applications in response to greenhouse gas mitigation is discussed in this chapter. Carbon dioxide and methane are major contributors among greenhouse gases. Carbon dioxide reforming of methane to synthesis gas attracts larger interests amid continuous depletion of petroleum reservoirs and threats of global warming. At commercial level, syngas is being produced from energy intensified steam reforming of methane in catalytic reactors. It is essential to explore alternative, effective technologies for synthesis gas production. Thus, this chapter elaborates the potential applications of plasma and photo technologies for carbon dioxide reforming of methane. The plasma technology allows higher conversion and selectivity with lower input energy while photocatalysis consumes less energy than conventional methods by utilizing solar energy. With the aid of these technologies, new and innovative processes for sustainable development by mitigating greenhouse gases for synthesis gas production are possible.

INTRODUCTION

Energy crises and global warming has resulted in an accelerating push for finding new renewable energy resources for sustainable development in order to maintain quality life style that humans aspire to [1, 2]. Petroleum-derived fuels remained the primary source of energy in the last century and this trend is still continuing. In current circumstances, more than 80% of primary energy consumption is fulfilled by fossil fuels and of which 58% is directly consumed in transportation sector [3-5]. The sustainable development depends on the availability of sufficient energy resources, consumption over restoration ratio and their

* E-mail address: noraishah@cheme.utm.my.

environmental effects. However, regular consumption of fossil fuels has created imbalance between supply and demand of energy while combustion has resulted in unstable environment and global warming [6-9]. At present combustion of fossil fuels is increasing extremely along with looking for better quality of life, industrialization, and rapid increase of world population. It has been agreed that this increase not only deplete fossil reservoirs but also has significant effects on environment and anthropogenic activities has further added in increasing the concentrations of greenhouse gases, resulting in health risk and threats of global warming [10-15]. The greenhouse gases actively involve to environmental pollutions and climate damaging are carbon dioxide and methane as the concentrations of these gases continue to increase rapidly [16-21].The concentration of carbon dioxide was increasing at the rate of $< 1\%$, but this trend has shifted to $> 2\%$ in the last 25 years [22]. The energy sector, accounts for 90% generation of carbon dioxide and in developed countries, 75% CO_2 is emitted from combustion of fossil fuels [23]. In addition, the quantity of carbon dioxide in atmosphere was 1823 MMT (million metric tonnes) in 1990, which is expected to increase 2446 MMT till 2012 [24]. On the other hand, methane contributes significantly to global warming. The emission of methane from anthropogenic sources is responsible for 70% of the total methane being generated worldwide. The inclusive contribution of methane to global warming is substantial because in 100 year time-frame, methane arrested 21 times more heat than carbon dioxide. Hence, carbon dioxide and methane are the most harmful greenhouse gases for polluting our environment. If such trends of increasing CO_2 and CH_4 concentrations continue, then global temperature will be more than $4.5^{o}C$ by the middle of this century [25, 26]. Efforts for replacing fossil fuels with renewable energy resources are essential to overcome the challenges of global warming and energy crises. Currently renewable sources contribute around 14% of total demand for world energy. Renewable energy resources include biomass, hydropower, wind, solar, geothermal, tides, and ocean currents. All these energy resources are renewable, inexhaustible and except for biomass offers clean path to energy sources. Utilization of biomass, however, offers ways for waste to wealth concept by converting forest, agriculture or municipal solid wastes to bioenergy or biofuels. Sustainable development requires measuring the maximum energy supplies by these alternative sources and comparing them with expected increment of energy consumption in future [27, 28]. The leading source of renewable energy source seems to be biomass from which 7 to 10 TW (Terawatt) energy would be possible. For obtaining such amount of energy, entire agricultural lands would be necessary to be planted with fastest growing crops after which the plants will be harvested. The wastes or biomass are then available to be utilized and processed for energy production. But there are many concerns to follow such as energy production plans like input energy for crop growth, necessity of fertile land and efficient harvesting. Wind on land can produce 2.1 TW by saturating entire land with wind mills. On the other hand hydroelectric can supply 1.5 TW by installing dams on all available rivers. Cumulative energy from all tides, oceans currents in world can amount a total of less than 2 TW. The solar energy is considered freely available energy source but it is difficult to utilize direct solar energy due to the requirement of liquid fuels. However, solar energy can be saved through different renewable energy carrying chemical reactions. Thus a total energy expected from these resources would be 12 TW by efficient utilization of these renewable energy resources but energy demands is expected to reach 50 TW till 2050 [29]. The recycling of greenhouse gases to renewable synthesis gas is one of the more feasible solutions for sustainable development. Syngas is an important intermediate for synthesizing various types of chemicals and environment friendly clean fuels. Among the renewable chemicals and green fuels that can be produced from syngas are ammonia, methanol, acetic acid, methyl formate, dimethyl ether, synthetic gasoline and diesel. In addition, syngas is an intermediate in heavy hydrocarbons production through Fischer-Tropsch reactions including synthetic gasoline and diesel which are immediate replacements of fossil fuels for transportation sectors[30]. Furthermore, ammonia obtained from synthesis gas is consumed as raw material for industrial scale ammonium nitrate production while methanol is a source of renewable fuel and also utilized as intermediate for the production of various products. However, reforming of greenhouse gases to synthesis gas possess a challenging task amid current energy crises and global warming [31]. Hence, innovative methods will be necessary for efficient and economical conversions of both gases while utilizing minimum input energy for achieving the goal of sustainable development [32]. The reforming of CO_2 and CH_4 also called dry reforming, has been researched extensively. In dry reforming, synthesis gas is produced directly from greenhouse gas which ultimately mitigates its effects on environment. The CO_2 reforming of CH_4 not only reduce the atmospheric CO_2 and consumes CH_4 but also converts both greenhouse gases to syngas with suitable ratio H_2/CO, the most suitable as feedstock for the synthesis of liquid hydrocarbons [33, 34]. According to available technologies, methane reacts with carbon dioxide to produce CO and H_2 and reaction between CH_4 and CO_2 is highly endothermic requiring 247 KJ/mol of input energy at temperature of 800 – 1000°C with versatile catalysts [35-37]. The dry reforming of carbon dioxide of methane produces syngas with H_2/CO ratio of 1, as explained in equation (10.1).

$$CH_4 + CO_2 \xrightarrow{\text{catalyst, high temp}} 2CO + 2H_2 \quad \Delta H = +247\ \frac{KJ}{mol} \tag{10.1}$$

Although dry reforming is helpful for mitigating greenhouse gases, endothermic reaction requires higher input energy for synthesis gas production. Besides, catalyst deactivates due to coke deposition and sometimes blockages of reactor tubes reduces the conversion rate and ultimately decreases the overall production efficiency [38]. Synthesis gas ratio (H_2/CO) also depends on temperature, pressure and feed ratio of CH_4/CO_2. It has been reported that yield of syngas production increases with temperature but pressure has adverse effects. Higher temperature facilitates toward coke deposition on catalyst and inside reactor tubes resulting in deactivation of catalyst and clogging the reactor [39]. The possible reactions for carbon production at inclement temperature are illustrated by equation (10.2) and (10.3).

$$CH_4 \xrightarrow{\text{Catalyst, temp}} C + 2H_2 \quad \Delta H = +75.8\ \frac{KJ}{mol} \tag{10.2}$$

$$2CO \xrightarrow{\text{catalyst, temp}} C + CO_2 \quad \Delta H = -170.82\ \frac{KJ}{mol} \tag{10.3}$$

On commercial level catalytic type reactors are practiced for CO_2 reforming of CH_4 to synthesis gas and reforming process is carried out by passing the gas mixture through catalyst bed inside the reactor. The input energy required for endothermic reaction is supplied by combusting methane outside the reactor. The fixed-bed type reactors are practiced from long periods for producing syngas, but there are various drawbacks in this technology which triggers this route inappropriate for syngas production. The main obstacle is severe coke deposition which is intractable and results in deactivation of catalyst. The other disadvantages are lower heat and mass transfer rates, requiring higher input energy due to endothermic reactions, lesser reaction surface area with lower conversion and selectivity [40-43].

CO_2-CH_4 reforming to syngas by plasma reactors is considered a new technology and offers a better alternative comparative to catalytic reactors as the conversions are higher. In plasma reactors, reaction is conducted by electron induced chemistry and thermochemical reactions attain high conversion rate with acceptable selectivity. Plasma reactors can be used with or without catalyst and there are less possibilities of carbon deposition on the catalyst surface. In present years, plasma reactors have drawn increasing attentions due to the stability of catalysts. However, higher input energy is still a problem which can be overcome by applying engineering techniques [34, 44, 45]. Photocatalysis technology for reforming of CO_2-CH_4 by applying semiconductor materials has attracted researchers due to opportunity of utilizing sunlight for endothermic reaction. The renewable sunlight source is able to alleviate problems of higher input energy for converting greenhouse gas to renewable fuels and is considered a promising method for syngas production. However, researchers are mostly working in environment and solar energy related application and few studies discuss about renewable fuels [46]. The cost of higher input energy, catalyst life and reactor tubes plugging during synthesis gas production can be reduced by using this technology [47, 48].

In this chapter, CO_2 reforming of CH_4 to syngas by utilizing plasma and photo technologies has been discussed. The aim of this review is to elaborate the potential applications of plasma and photo technologies for syngas production through the recycling of greenhouse gases route. The basic principles of both technologies, mechanism of CO_2-CH_4 reforming and potential applications have been presented.

Plasma Technology for Reforming of CO_2-CH_4 to Syngas

Carbon dioxide and methane are stable molecules. Carbon dioxide reforming of methane is a highly endothermic process with 247 KJ/mole of input energy and reaction occurs at higher temperature and pressure. On commercial processes, this amount of energy is supplied by combustion of natural gas, which results in emission of carbon dioxide to atmosphere [49].The plasma system provides the solutions of most of these drawbacks. The plasma technology is considered more potential for reforming of CO_2-CH_4 to syngas but yet it is not practiced on commercial levels. In plasma, overall reforming reactions are the same as that of conventional reforming. However the energy and free radicals utilized for reforming reactions are provided by the plasma itself [39]. Syngas production through carbon dioxide reforming of methane by plasma is considered efficient method due to its electron induced chemistry. The plasma plays catalytic rule because of higher energetic species such as electrons, ions and radicals which significantly enhance reaction rate. Compared to catalytic processes, reforming of CO_2-CH_4 by plasma appears more efficient to produce syngas with higher conversion and selectivity due to its thermochemical reaction mechanisms. In addition, there are lesser chances of instability or deactivation of catalyst due to lower coke production, as compared to conventional processes. Plasma process provides unique ways to induce gas phase reactions which are utilized in several chemical reactions. The plasma processes showed higher conversion and thus in the last ten years or so the technology has drawn increasing attentions [31, 34]. The plasma systems can be applied for large number of hydrocarbons such as natural gas, gasoline, heavy oil and biomass. On the other aspect, the higher input energy probability for plasma excitation and feed flow rates need to be lowered for commercial applications. However, these problems can be overcome by applying efficient engineering approaches [33, 37].

Table 10.1. Basic type of reactions and their mechanisms which take place during plasma process

No.	Type of process	Reaction mechanism	Characteristics
1.	Excitation of atoms or ions	$e + A_2 \rightarrow A_2^{\bullet} + e$ $e + A \rightarrow A^{*} + e$	Leads to electronically excited state of atoms and molecules by energetic electron impact
2.	De-excitation	$e + A_2^{\bullet} \rightarrow A_2^{*} + e + hv$	Electronically excited state emits electromagnetic radiations on returning to the ground state.
3.	Ionization	$e + A_2 \rightarrow A_2^{+} + e$	Energetic electrons ionize neutral species through electron detachment and positively charged particles are formed.
4.	Dissociation	$e + A_2 \rightarrow 2A + e$	Inelastic electron impact with a molecule causes its dissociation without ions.
5.	Dissociative attachment	$e + A_2 \rightarrow A^{+} + A + e$	Negative ions are formed when free electrons attach themselves to neutral species.
6.	Dissociative ionization	$e + A_2 \rightarrow A + e$	Negative ions can also be produced by dissociative ionization reactions.
7.	Volume recombination	$e + A + B \rightarrow A + B$	Loss of charged particles from the plasma by recombination of opposite charges.
8.	Charge exchange	$A^{+} + B \rightarrow B^{+} + A$	Transfer of charge from incident ion to the target neutral between two identical or dissimilar partners.
9.	Recombination of ions	$A^{-} + B^{+} \rightarrow AB$	Two colliding ions recombine to form a molecule.
10.	Electron–Ion recombination	$e + A_2^{+} + M \rightarrow A_2 + M$	Charge particles are lost from the plasma by recombination of opposite charges.
11.	Ion-ion recombination	$A^{+} + B^{-} + M \rightarrow AB + M$	Ion-ion recombination can take place through three body collisions.

PLASMA PRINCIPLES

The plasma term, consists of ionized gas particles, was first used by the Irving Langmuir back in 1928. The ionized gas consists of mixtures of electrons, ions, and neutral particles. In plasma gas, there are basically two types of particles namely heavy and light charged particles. The light charged particles consist of electrons while the mass of ions and neutral particles is much larger than electrons called heavy particles.

On the basis of energy density levels, electrons density and temperature, plasma technology applied for reforming of CO_2 of methane to produce syngas can be classified as thermodynamic equilibrium and non-equilibrium plasma.

The thermodynamic equilibrium plasma is simply called thermal plasma which correlated directly with temperature and temperature of heavy particles is very close to electrons usually in the range of thousands of kelvin (K). The electrical power injected is higher than 1 kW and neutral species and electrons have the same temperature (around 5000 to 10000 K). The electron density in thermal plasma is much more than cold plasma and there is large amount of impact between electrons and heavy particles.

As a result electron energy from electric field moves towards heavy particles for achieving thermodynamic equilibrium between electrons and heavy particles. The chemical reactions takes place in thermodynamic equilibrium plasma are both thermo chemical and electron induced reactions [50].

On the other hand, thermodynamic non-equilibrium plasma is called cold plasma in which thermal kinetic energy of electrons is much larger than that of heavy particles. The non-equilibrium plasma technology has solved the disadvantages of higher temperature due to the reasons that majority of energy goes into the production of energetic electrons which proceed reforming reactions [51].The chemical reactions which take place in cold plasma are basically induced by energetic electrons and with the help of electric field these electrons are accelerated and energized.

During the process of electron-impact dissociation, ionization and excitation of gas molecules, the energetic electrons transfer their energies to reacting gas molecules by inelastic collisions. During these process free radicals, ions, as well as additional electrons are produced which participate for carrying desired chemical reactions. The atmospheric pressure non-thermal discharge methods creates highly reactive environment at normal temperature with highly flexible and environment friendly process. However, there is affinity to become unstable in reasons of rapid transition of arcs and filamentation [31, 52].

In plasma, reaction chemistry is quite complex and involves a large number of elementary reactions. The two main types of reactions take place in plasma namely homogenous and heterogeneous depending upon the use of catalyst. The homogenous reactions occurs between the species in gaseous phase due to the inelastic collisions of freely moving electrons while heterogeneous reactions takes place between the plasma species and solid surface immersed or in contact with plasma. In heterogeneous reactions, the same phenomenon as semiconductor photocatalysis happens. Different types of possible reactions in plasma process are explained in Table 10.1.

MECHANISM OF CO_2-CH_4 REFORMING TO SYNGAS BY PLASMA

Reforming of carbon dioxide of methane is highly endothermic due to the stability of molecules making the breaking of these molecules very difficult. However, in the presence of charge particle, ions and radicles they are reduced easily even under normal temperature and pressure. In general, the reforming of carbon dioxide with methane takes place under complex mechanism. However, the simple understanding of such reactions is explained by equations (10.4) - (10.9) [31, 32, 37, 53].

$$CH_4 \xrightarrow{e^* \text{ or pyrolysis}} C + 2H_2 \tag{10.4}$$

$$CH_4 \xrightarrow{e^*} CH_3 + H \tag{10.5}$$

$$CO_2 \xrightarrow{e^* \text{ or pyrolysis}} CO + O \tag{10.6}$$

$$2CH_3 \rightarrow C_2H_6 \tag{10.7}$$

The dehydrogenation of methyl radical can continue resulting in the production of coke from which CO can be produced while O produced reacts with hydrogen to generate water molecule as explained in equations (10.8) and (10.9).

$$C + O \rightarrow CO \tag{10.8}$$

$$H_2 + O \rightarrow H_2O \tag{10.9}$$

Equations (10.4) – (10.6) are highly endothermic reactions induced by energetic electrons impact or through thermochemical pyrolysis. However, equations (10.7) – (10.9) are radical reactions and occur without any energy barriers. From the reaction mechanism, it can be stated that for the conversions of CO_2-CH_4 to syngas by plasma, basically two reaction mechanisms are involved during energetic electrons impact. One is direct dissociation in which the molecules of CO_2 or CH_4 are collided by the energetic electrons with kinetic energy higher than that of dissociation energy threshold of ground state molecules, as explained by equation (10.10).

$$AB \xrightarrow{e^*} A + B \tag{10.10}$$

The other mechanism is vibrational excitation, which is mostly useful for non-electromagnetic molecules like CO_2 and CH_4. In vibrational excitation electron impact takes place through the formation of intermediate state, which excites chemical conversion in the reaction.

The input energy requires for electrons is less as compared to direct dissociation process and is most effective process for the reforming of CO_2 of CH_4 for economical syngas production. Such types of reactions can be explained by equation (10.11).

$$A + BC_{(v)} \rightarrow AB + C \tag{10.11}$$

Different researchers have evaluated that the ionization degree of thermal plasma is much higher than cold plasma and is proportional to input power. In addition, beside electrons, free radicals and ions, there are still many molecules present in plasma. By increasing input power, higher ionization energy is achieved which results in the formation of more radicals and fewer molecules.

As a result, the conversion of carbon dioxide and methane to produce syngas will be higher for increasing the overall efficiency of the process [31, 32].

Cold Plasma Technologies for CH_4-CO_2 Reforming

The reforming of carbon dioxide with methane by using cold plasma can be tested by using several types of plasma such as microwave discharge, corona discharge, dielectric barrier discharge, glide arc discharge and atmospheric pressure glow discharge [15, 33]. In cold plasma, the temperature of electrons varies from 1 to 10 eV while the temperature of heavy particles is in the range of hundreds of Kelvin. On the other hand local electron density varies from 10^{15} to 10^{20} m^{-3}. The cold plasma is considered inhomogeneous in discharge space resulting in lower reaction space with lower conversions. For increasing the conversion of reforming gases, catalyst is placed in discharge place or after the discharge place [54].

Corona Plasma

Figure 10.1 exhibits the schematic representation of corona discharge plasma. Corona discharge is an inhomogeneous discharge with low current density. It may be initiated at atmospheric pressure and has higher electric field intensity for ionizing or decomposing feed gas particles. Corona discharge uses inhomogeneous electrode geometries, such as pointed wire electrode with plate geometry.

It has small radius of curvature at top of the wire electrode that results in high electric field require to ionize neutral molecules. Corona reactor consists of an inner steel electrode and an outer electrode of aluminum foil wrapped around a dielectric tube such as glass. The shape of the glass tube depends on the type of corona electrode used. The glass allows a silent (glow corona) discharge, rather than a spark, to be generated between the electrodes.

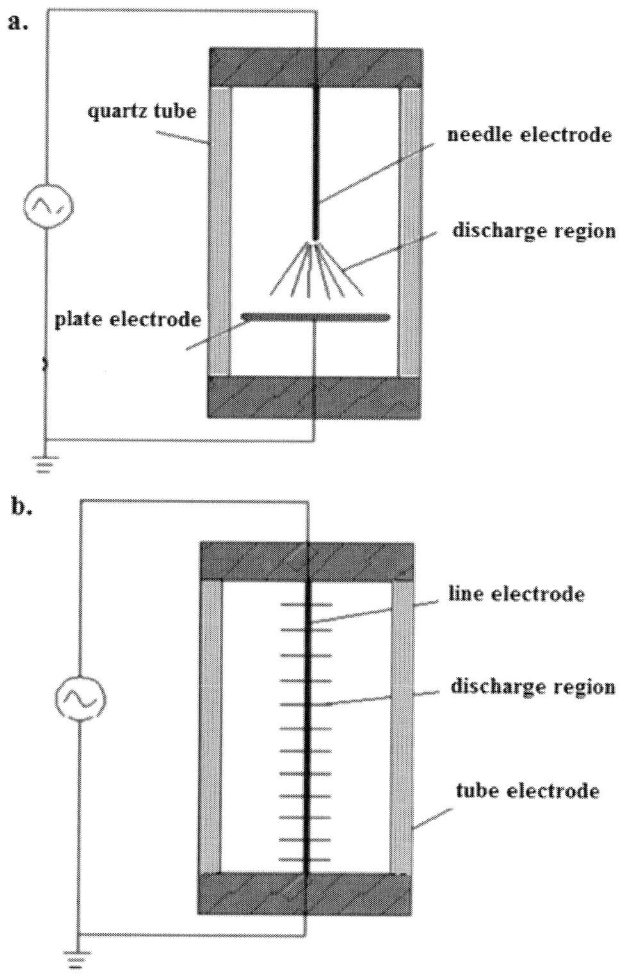

Figure 10.1. Schematic diagram of corona discharge reactor: a. Needle electrode; b. Line electrode [31].

The electron temperature of corona plasma is in range of 3.5 – 5 eV while gas temperature exists less than 400K with electron density of $10^{15} – 10^{19}$ m^{-3}. Higher electron density usually occupies in region around the high curvature electrodes. The CO_2-CH_4 reforming by corona discharge has been evaluated under different operating condition. In one experiment, discharge power of 46.3 was utilized for reforming of CO_2 and CH_4 and conversion of CH_4 was 62.4% while 47.8% of CO_2 and selectivity of H_2 and CO was 70% and 66.8%, respectively. However, it is reported that corona discharge is operated due to localized breakdown which restricts the available electron volume and it is difficult to handle large volume of CO_2 and CH_4 by corona discharge [31, 55].

Dielectric Barrier Discharge Plasma

In dielectric barrier discharge (DBD), the dielectric barrier distributes discharges throughout the volume at normal temperature. The reaction in dielectric barrier discharge takes place due to reactive species like ions, radicals and activated molecules [51, 56, 57]. The schematic representation of dielectric barrier discharge is depicted in Figure 10.2. DBD combines the large volume excitation of the glow discharge with high pressure characteristics of the corona discharge and it is also called silent discharge. In silent discharge two electrodes are placed in parallel arrangement with a gap of several millimeters and a dielectric layer covers at least one electrode with thickness of one or two mm, such that the entire electrode area will be effective for discharge reactions [37].

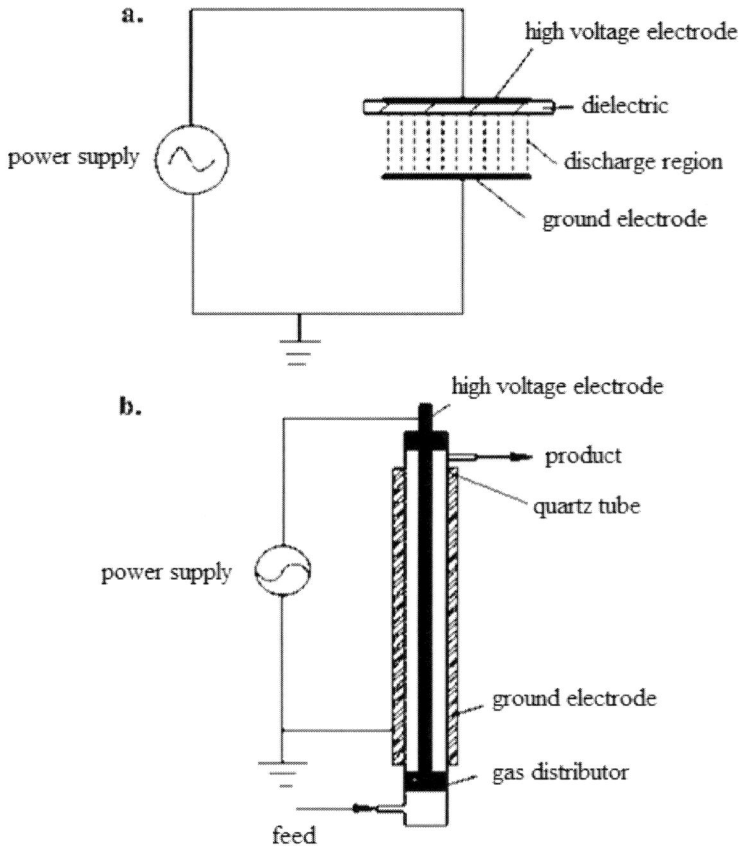

Figure 10.2. Schematic representation of dielectric barrier discharge plasma (DBD) [31].

The dielectric can be made from different materials such as glass, quartz, ceramics, polymers or other types of materials with low electric loss and higher breakdown strength. When the silent discharge is initiated at any location within the gap between electrodes, charge accumulates on the dielectric to form an opposite electric field and interrupt the current flow in a few nanoseconds to generate micro-discharges. The duration of the current pulse relates to the pressure, the properties of gases and the dielectric material applied. Therefore, dielectric barrier discharge can be referred to as barrier discharge or silent discharge with a specific type of AC discharge, which provides strong thermodynamic, non-equilibrium plasma at atmospheric pressure and at moderate gas temperature. In DBDs electrodes and discharges are separated by a dielectric barrier, which eliminates electrode etching and corrosion. Another fundamental difference is that the DBDs cannot be operated with DC voltage because the capacitive coupling of dielectric requires an alternating voltage to drive a displacement current. An AC voltage with amplitude of 1-100 kV and a frequency from line frequency to several megahertz is applied to DBD configurations. DBD cold plasma can be produced in various working mediums through ionization by high frequency and high voltage electric discharge [37, 51].

Recently, dielectric barrier discharge plasma for reforming of $CO_2$and CH_4 at low temperature and ambient pressure is reported by many researchers [58, 59]. The results indicate that syngas gas and higher hydrocarbons can be produced through this technology. However, they have lower conversion and selectivity which can be increased by increasing input power [60].

Microwave Plasma

The microwave plasma operating under atmospheric pressure has become the subject of increased attention in the last few years. The high density microwave plasma provides suitable conditions for efficient dissociation of molecules. The schematic representation of microwave plasma is shown in Figure 10.3.

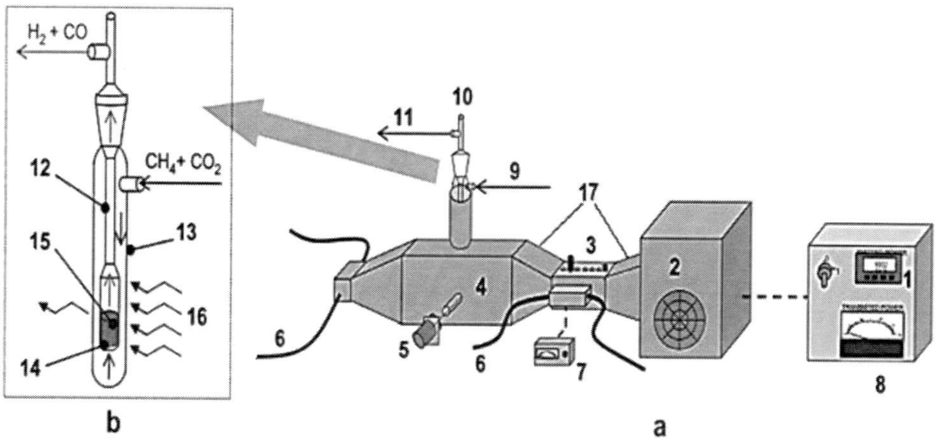

Note: 1. Input power control; 2. Microwave generator; 3. Manual 2-stub unit; 4. Resonance cavity; 5. Optical pyrometer; 6. Water sink; 7. Reflected power control; 8. Transmitted power control; 9. Inlet flow gas line; 10. Set quartz reactor-jacket catalyst and catalyst/microwave receptor; 11. Inflow gas line; 12. Quartz reactor; 13. Quartz jacket; 14. Porous plate; 15. Catalyst and microwave receptor bed; 16. Microwave radiation; 17. Wave guide.

Figure 10.3. a. Schematic diagram of plasma reactor (microwave discharge) setup; b. Close up view of quartz reactor jacket [31].

Microwave plasma is commonly used in microwave ovens and operated at frequency range of GHz and under wide pressure ranges. It consists of waveguide, microwave generator and resonance cavity where reactants are led into and plasma is generated. Microwave plasma has the advantages of easy operation, a chemical reactor without electrodes, contains high plasma density and operated under high electron mean energy. In microwave plasma heavy particle temperature is about 2000 K while the electron temperature is in the range of 0.4 – 0.6 eV [39, 54]. According to available literature, carbon dioxide reforming of methane cane be carried out by using low pressure microwave plasma and at atmospheric pressure microwave discharge has larger discharge space and better discharge uniformity than corona and dielectric barrier discharge. The results of atmospheric pressure microwave plasma reveals to be practiced for CO_2 reforming of CH_4 to syngas. The reforming of CO_2-CH_4 by microwave plasma indicates higher conversion and selectivity with larger treatment capacity and higher energy efficiency. The chemical reactions inside microwave plasma occur due to energetic electrons and high gas temperature. However, there is drawback in microwave plasma that this equipment is complicated, bulky and difficult to implement for industrial applications [31].

a.

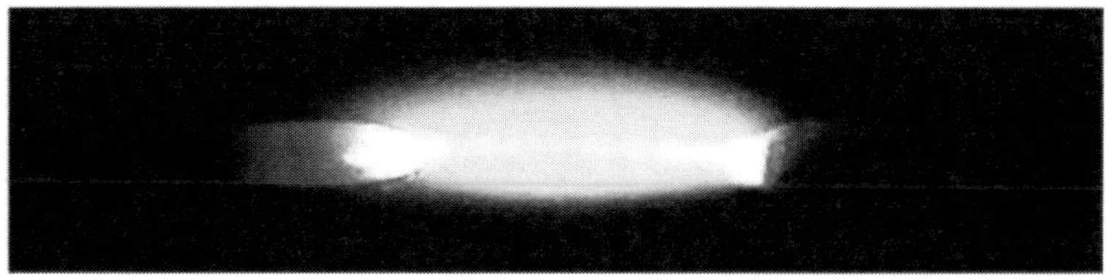

b.

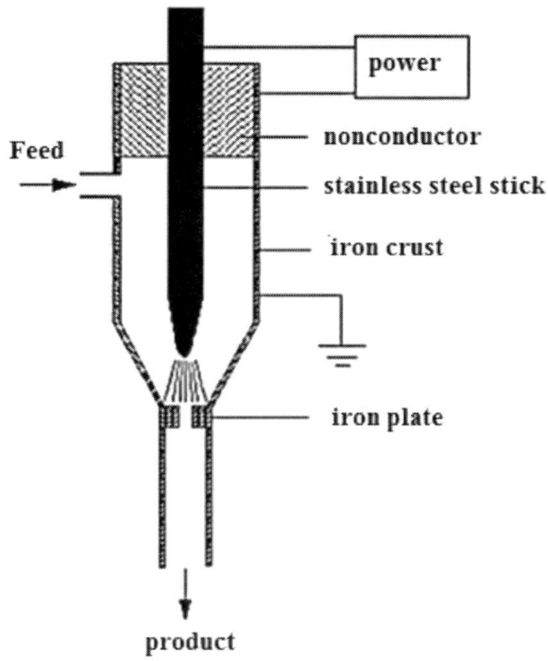

Note: a. Atmospheric pressure glow discharge phenomenon; b. Atmospheric pressure glow discharge plasma jet.

Figure 10.4. Schematic representation of glow discharge plasma [31, 34].

Atmospheric Pressure Discharge Plasma

Figure 10.4. shows the schematic diagram of glow discharge plasma. The glow terms indicates that plasma of discharge is luminous as compared to other relatively low power dark discharge. The glow discharge plasma can be operated under different operating pressures. The atmospheric pressure glow discharge is like arc luminous discharge as explained in Figure 10.4(a) which is obtained by applying volts of few kilovolts between two electrodes [31]. The atmospheric pressure glow discharge jet is shown in Figure 10.4 (b) [34]. The electrons in the glow discharge are highly energetic and the excited neutral atoms and molecules generate a typical glow. The homogenous plasma of the positive column with charge carrier generation by electron impact ionization of the molecules/atoms is applied as model plasma for fundamental investigations of plasma chemical processes. The generation of a positive column with higher current is possible by changing the cathode to a hollow cathode (using a hollow cylinder) or externally heated cathode. The generation of charge carriers in hollow cathodes is supported by pendulum electrons as well as by additional electron emission by the impact of photons and metal stables on the inner cathode surface. Such type of discharge operates with lower sustaining voltage and at higher current rates.

The atmospheric pressure glow discharge plasma has specific characteristic of electron density, electron energies and plasma temperature than non- thermal plasma such as corona discharge and dielectric barrier discharge plasma. The atmospheric pressure discharge plasma brings higher conversion which is several times higher than other plasmas [34]. The electron temperature in atmospheric pressure glow discharge plasma is in range of 1 – 2 eV and gas temperature is around 2000K with electron density in range of $10^{18} – 10^{19}/m^3$.

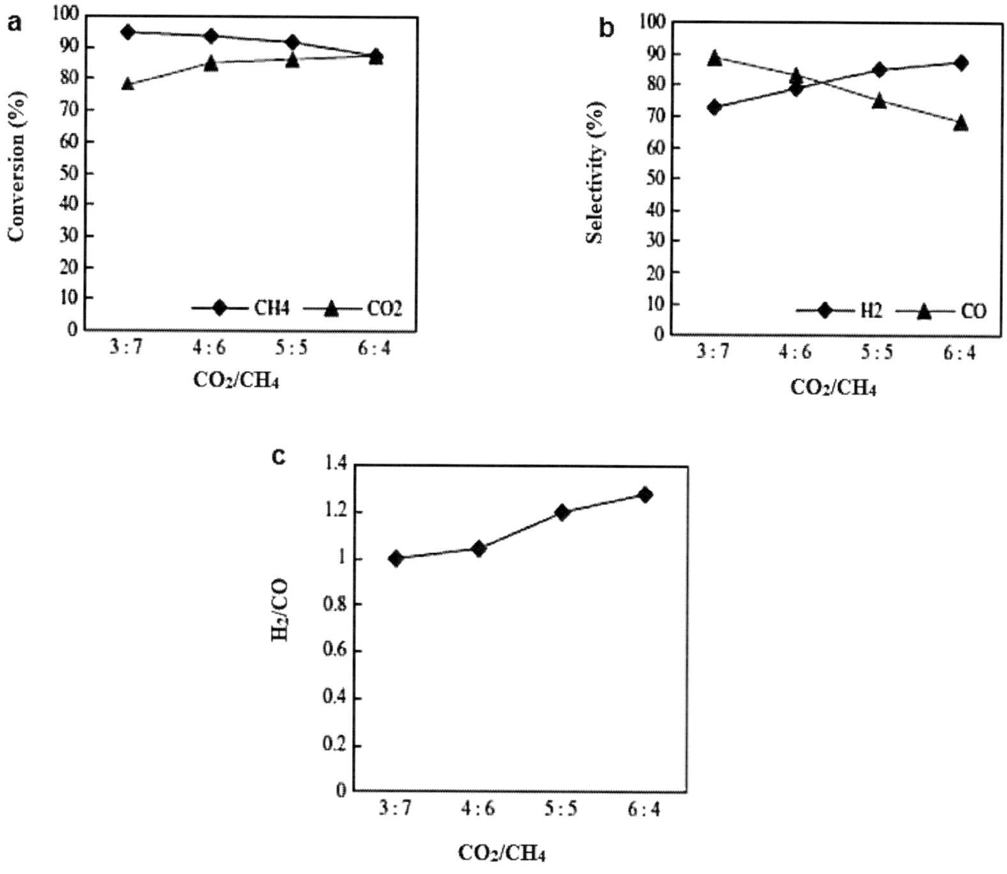

Note: The experiment was conducted at input power of 68.95W and total feed flux of 1000 mL/min.

Figure 10.5. Effects of CH_4/CO_2 rate on reaction [34]: a. Conversion of CH_4 and CO_2; b. Selectivity of H_2 and CO; c. H_2/CO mole rate.

The reforming of CO_2 with CH_4 by atmospheric pressure glow discharge plasma has been investigated by many researchers and found that product mainly consists of syngas gas and C_2 hydrocarbons. Under the operating conditions of discharge power of 23W, frequency of 4 KHz, feed gas flow rates of 120mL/min and CH_4/CO_2 ratio of 1:1, the maximum conversion of CO_2 and CH_4 obtained was 50% and 61% respectively. The selectivity of CO obtained was 63% while 77.5% of H_2 selectivity was attained [31]. CO_2 reforming of CH_4 was also reported in another study by applying atmospheric pressure discharge with the following operating conditions: CH_4/CO_2 rates of 3/7, input power of 69.95 W, the total feed flux at 1000 mL/min. The result indicates that highest conversion of CH_4 and CO_2 was 98.5% and 90.3% respectively with selectivity of H_2 obtained was 89.3% while CO selectivity obtained was 72.6%. The graphical representation of conversion and selectivity are presented in Figure 10.5 [34]. The atmospheric pressure glow discharge plasma is considered most promising for CO_2 reforming of CH_4 due to its higher electron density and better plasma temperature stability. The process is efficient for conversions of CH_4 and CO_2 into syngas. The higher selectivity is obtained by using this process with small amount of water. Compared with other plasmas, the glow discharge plasma has advantages for higher feed flow rate with higher conversion and selectivity [33].

THERMAL PLASMA TECHNOLOGIES FOR CH_4-CO_2 REFORMING

The thermal plasma has features of relatively higher temperature as well as higher active species density and has both thermal and chemical effects. By its thermal effects, it is usually used for plasma cut, plasma spray coating and for plasma metallurgy. On the other hand due to its chemical effects, it is usually practiced for the preparation of ultrafine powders. The thermal plasma usually has two types including continuous and uniform plasma. Thermal plasma usually operates at higher enthalpy contents as well as at higher temperatures. In thermal plasma, the temperature of electrons and heavy particles varies from 0.5 to 1 eV and higher electron density usually in the range of 10^{19}-10^{20}/m^3 [27]. Thermal plasma for production of syngas through CO_2 and CH_4 can be used with higher conversion and selectivity.

The CO_2 reforming of CH_4 has been studied for understanding the effects of feed flow rates and in one study results showed that conversion of CH_4 obtained was 96.33% while 84.63% of CO_2 was converted. On the other side, the selectivity of CO and H_2 obtained was 91.99% and 74.23% respectively, as shown in Figures 10.6 and 10.7. The comparison of different plasma technologies are also illustrated in Table 10.2 [32, 61].

With advancement in technologies, innovations in different research fields are aplenty. Photo technology is one of the innovations and many researchers are attracted towards it since it is a sustainable process utilizing abundantly available solar energy. Photocatalysis processes have been applied in different research areas and has gained much attractions due to its mild operating conditions [62].

The group of researchers working in physical chemistry, catalysis fields and particularly material science and reactor design has made it as their domain of research with such applications in air cleaning and wastewater treatments [63, 64]. Currently, syngas production by this technology has gain significant importance for replacing fossil fuels to overcome energy crises and environment pollutions.

Table 10.2. Comparison of conversion and selectivity for different types of plasma [49]

Plasma type	Feed flux (mL/min)	P (W)	Conversion (%)		Selectivity (%)		E (mmol/KJ)
			CH_4	CO_2	H_2	CO	
Pulsed corona plasma	25	42	63.7	60.2		62.6	0.26
DC corona discharge	60	50	36.0	26.6		80.9	0.75
Dielectric barrier discharge	60	100	64.3	43.1		32.2	0.26
Dielectric barrier discharge (AC)	150	500	64.3	55.1		33.3	0.13
Pulsed microwave plasma	200	65	70.8	68.8		75.0	1.56
Radio frequency plasma 3050PPS	100	29.5	26.7	17.2		19.8	0.58
Radio Frequency plasma 2920PPS	100	30.6	31.8	23.9		22.1	0.68
Abnormal glow discharge	350	85	75.4	64.1	76.20	82.1	2.03
Atmospheric pressure glow discharge	3.64×10^4	9.6×10^3	88.28	76.05	72.48	89.06	2.3

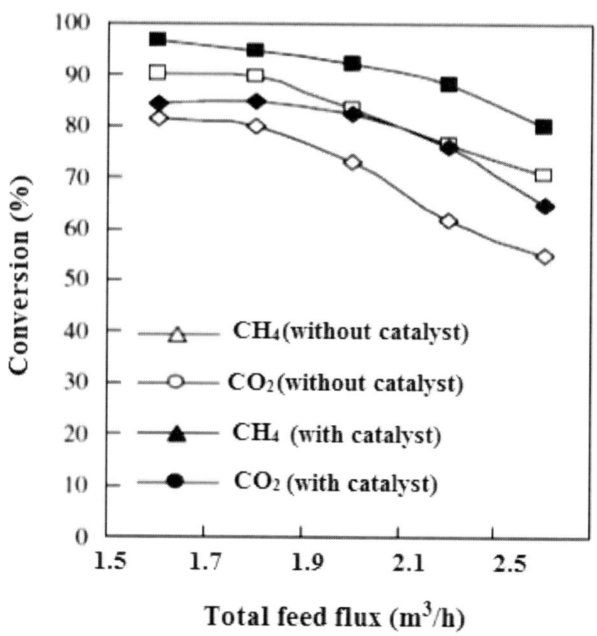

Figure 10.6. Effects of total flux on conversion of CO_2 and CH_4 in thermal plasma [61].

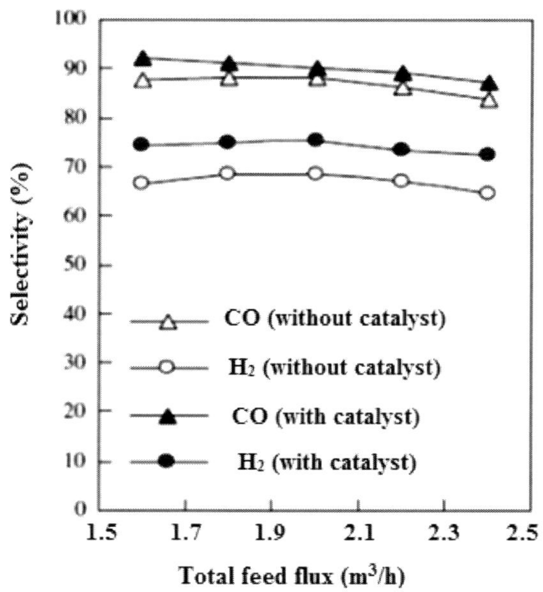

Figure 10.7. Effects of total flux on selectivity of H_2 and CO [61].

Photo Technologies for Syngas Production

Syngas production by reforming of carbon dioxide with methane is very challenging as the reactant molecules are stable and the endothermic reactions require higher input energy. The sustainable and economical syngas production has become the challenge because of higher input energy. On the other hand, emanation of greenhouse gas is serious problem because of its adverse effects on atmosphere, particularly in terms of global warming [65-67]. The conventional available conversion methods rely on higher input energy at higher temperature and pressures conditions. The current developments in phototechnology has introduced new pathway for syngas production through CO_2 reduction using different reducing agent under mild conditions practicing lower input energy by hunting solar energy [68]. The utilization of sunlight has many advantages, as it is renewable, readily available without any cost, relies on continuous and available power supply [69]. Moreover, during reduction of CO_2 emission in atmosphere, many valuable chemicals are produced that makes such approaches very attractive and economical [70, 71].

Photocatalytic Principles

The heterogeneous photocatalysis occurs with the help of three components:

1. Needs of photon with appropriate wavelength,
2. Semiconductor material,
3. Oxidizing agent for carrying reaction.

Semiconductors have characteristics like metals which consist of electronic state and exhibits higher energy region for reactions to take place. Semiconductors have higher band gap energies extends from filled valance band at top to vacant conduction band at bottom. The difference between top valance bands to bottom conductance band is responsible for higher rate of reaction and is called band gap or energy band gap (E_{bg}) and it is actually difference between two levels of band named filled valance band and empty conduction and measured in units of eV (electron volts) [72, 73].

The mechanism of photo excitation of semiconductor materials is depicted in Figure 10.8, which reveals that when light strikes on semiconductor surface, the photon energy is absorbed by surface molecules which become excited, resulting in the transfer of electrons to higher energy band gap, namely conductance band while positive charges are produced at lower energy band gap call valance band. The difference in energy levels between conductance band and valance band is call band gap energy which actually depends on the types of semiconductor materials. In addition, during striking of photons on semiconductor surface, molecules can excite but charge particles produce only when photon energy is greater than or equal to semiconductor band gap energy. The generated electrons are present at higher energy i.e. conduction band and positive holes retained at lower energy valance band in semiconductor catalyst.

During photoreduction, different types of processes involve including catalyst related, interface related, donor related and acceptor related processes, which contribute significantly for enhancing the efficiency of the process (Figure 10.8). In catalyst related process, there is production of electrons and holes by absorbing photon radiation and their instant recombination because charge particles are produced only for short time. The electrons and holes productions and recombination phenomena is illustrated by using equations (10.12) and (10.13) [26].

$$\text{photocatalyst} \xrightarrow{E_{ph} \geq E_{bg}} e^-_{cb} + h^+_{vb} \tag{10.12}$$

$$e^-_{cb} + h^+_{vb} \rightarrow \text{heat} \tag{10.13}$$

In equations (10.12) and (10.13), hv is called photon energy, represents a conduction band electron while called hole in valance band. During photocatalysis of semiconductor, excited electrons-holes pair lifetime is few nanoseconds and they recombine immediately but this time is sufficient for carrying various spontaneous and non-spontaneous reactions. On the other hand, interface elated process consists of transfer of electrons, adsorption on catalyst surface and mass transfer process. The electrons and holes which escape from the excitation regions are trapped by adsorbed species, which become active and contribute for various types of reduction and oxidation process as explained in Figure 10.8. The further explanation about surface phenomenon of photocatalytic reduction process is described in Figure 10.9. Owing to slow spontaneous reaction kinetics occurs between donor and acceptors, the reaction between adsorbed species A and D promoted by interaction of light and light absorber semiconductor is termed photocatalyzed.

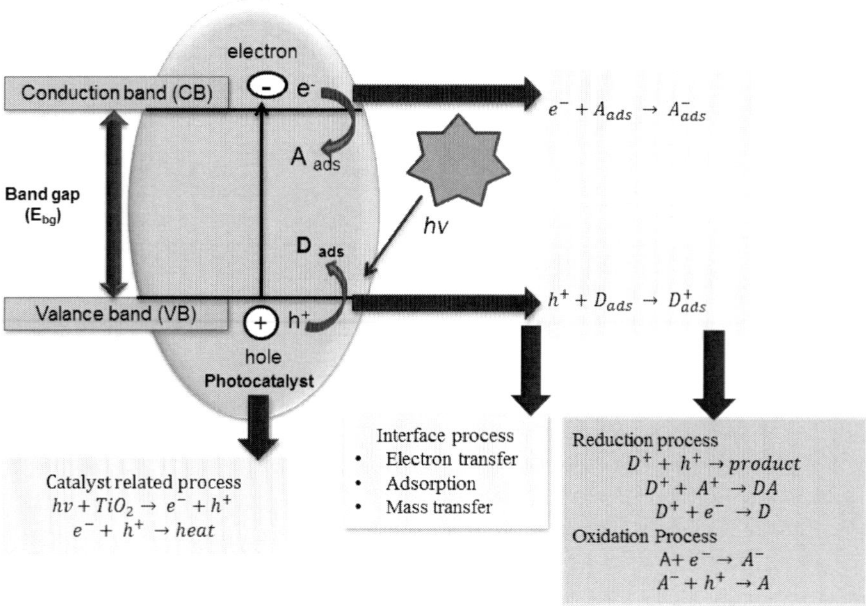

Figure 10.8. Schematic representations of band gap formation and process involved in photocatalytic reactions.

When light radiations with same band gap as semiconductor falls on its surface, then electron (e^-) excitation from valance band at bottom to conduction band at top are initiated. The photoreduction mechanism is very complex and there are many possibilities or paths which are expected after the production of charge particles. Because, life time of charge particles is in nanoseconds and if the band gap energy is very less then they recombine immediately after the production. On the other hand, if they have enough band gap energies to separate, then there are many possibilities or paths which can be adopted after separation. The separated charge particles can move toward the volume of the catalyst, where they can recombine named as volume recombination (path A), while the electrons and holes which succeeded to reach at semiconductor surface can recombine again with the generation of heat (path B). However, the other probabilities are that separated electron-hole couple migrates toward surface, where the carriers are eventually trapped and transfer electrons to absorbed molecules and this process becomes more efficient if species are already adsorbed on the semiconductor surface. At the surface, semiconductor donate electron to acceptors (pathway C) and holes can migrate their charges to donors (path D). All this mechanism of charge transfer depends on the bandage location of band gap and redox characteristics of absorbing species, respectively and reaction occurs irreversibly with acceptor and donor couples on the surface [26, 29]. The donor and acceptor related processes are explained by equations (10.14) – (10.19) [29].

$$\mathbf{D^+ + h^+ \rightarrow D^+} \quad \text{(oxidized radical)} \tag{10.14}$$

$$\mathbf{D^+ + h^+ \rightarrow P} \quad \text{(product)} \tag{10.15}$$

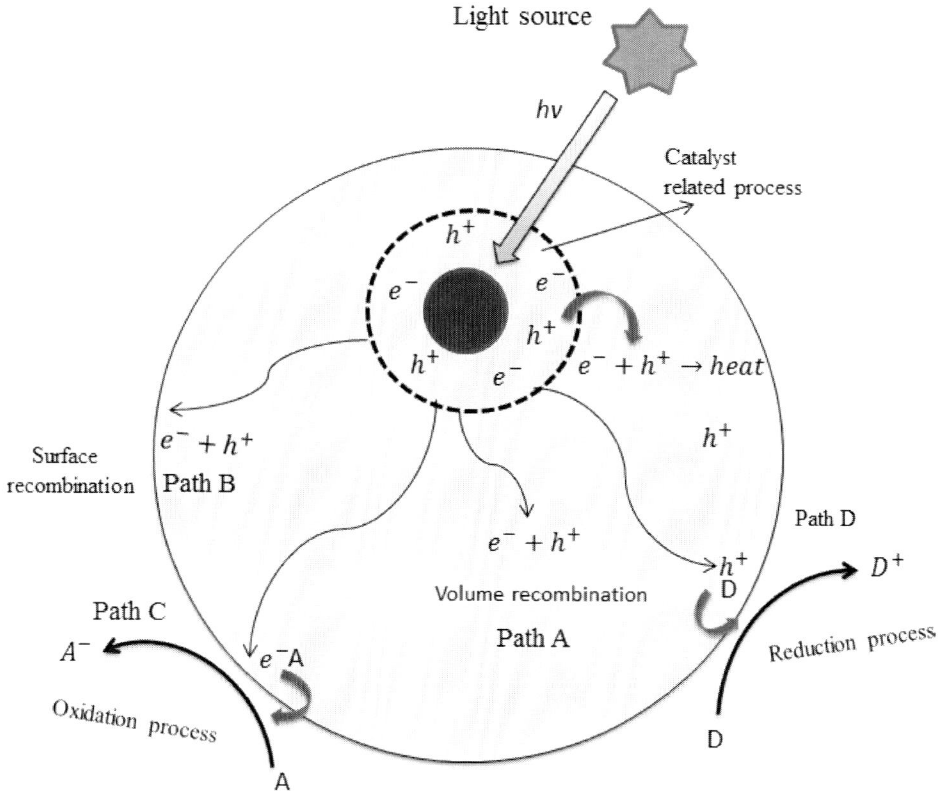

Semiconductor catalyst

Figure 10.9. Mechanism of photocatalytic reactions on semiconductor catalyst surface including surface and volume recombination, oxidation and reduction processes.

$$\mathbf{D^+ + e^- \rightarrow D} \quad \text{(back reaction)} \tag{10.16}$$

$$\mathbf{A + e^- \rightarrow A^-} \quad \text{(reduced radical)} \tag{10.17}$$

$$\mathbf{A^+ + h^+ \rightarrow A} \quad \text{(back reaction)} \tag{10.18}$$

$$\mathbf{D^+ + A^- \rightarrow DA} \quad \text{(product)} \tag{10.19}$$

Once excitation occurs across the band gap in semiconductor, then there is sufficient time that oxidation and reduction are possible. The recombination of electrons and holes is major problem in semiconductor and it can be prevented if surface defects or scavenger are present inside the surface of semiconductor to trap free elections and holes. The photocatalytic processes can be divided into different elementary processes that take place in different space –time position on the surface of semiconductor. The detail mechanism of different types of photoreduction processes are illustrated in Table 10.3.

Table 10.3. Photocatalytic processes steps involve during photoreduction [29]

Process type	Explanation	Reaction mechanism
1.Catalyst related processes	a. Light absorption	$TiO_2 + hv \xrightarrow{E_{ph} \geq E_{bg}} e_b^- + h_b^+$
	b. Thermalization of hot carriers	$e_b^- \rightarrow e_{cb}^-$, $h_b^+ \rightarrow h_{vb}^+$
	c. Bulk recombination of charges	$e_{cb}^- + h_{vb}^+ \rightarrow heat, \quad hv = E_{bg}$
	d. Separation of charges or migration of charges toward the surface	$e_{cb}^- \rightarrow e_s^-$, $h_{vb}^+ \rightarrow h_s^+$
2. Interface related processes	a. Surface trapping	$e_s^- + Ti-OH \rightarrow Ti + OH^-$
	b. Surface recombination	$h_s^+ + Ti + H_2O \rightarrow Ti - OH + H^+$
3. Donor/acceptor related processes	a. Donor reaction with h^+ and trapped h^+	$D + h^+ \rightarrow D^{\cdot+}$
	b. Donor back reaction	
	c. Acceptor reaction with e^+ and trapped e^+	$A + e^- \rightarrow A^{\cdot-}$
	d. Donor back reaction	$A^{\cdot-} + h^+ \rightarrow A$
	e. Donor-acceptor radical –ion reactions	
	f. Donor-donor, acceptor-acceptor reactions	$A^{\cdot-} + A^{\cdot-} \rightarrow A_2$

PHOTOREDUCTION OF CO_2 AND CH_4 TO SYNGAS

The photoreduction of carbon dioxide can be carried out by using different reductant including water (CO_2-H_2O) and methane (CO_2-CH_4). The use of photo energy through photocatalyst would break the thermodynamic barrier of endothermic reactions to occur at room temperature. In both processes, different types of products can be obtained depending upon semiconductor catalyst and design of photoreactor. The effective utilization of these processes is one of the important topics to be developed in current circumstances of energy crises and global warming. The syngas gas through these processes can be obtained efficiently because of CO_2 reduction to CO and that of CH_4 and H_2 to hydrogen [74]. According to available literature, most research is devoted to photoreduction of CO_2 by using water as reductant. Thus, greenhouse gas methane needs to be explored for various applications through photoreduction. On the other hand, photoreduction of CO_2 and CH_4 has attained significant importance in connection of energy crises and global warming.

The photoreduction of carbon dioxide with water as reductant is the most appropriate approach and helps for the production of renewable fuels and other chemicals. The photoreduction of CO_2 with water involves the initial excitation of solid surface resulting in the production of electrons and holes which transfer their energies to adsorbed species. In this process, two important species i.e. hydrogen atoms ($H^{\bullet}$) and carbon dioxide anion radicle ($\bullet CO_2^-$) are produced. Such reactions are explained by equations (10.20) and (10.21) [75, 76].

$$H^+ + e^- \rightarrow H^{\prime} \quad (10.20)$$

$$CO_2 + e^- \rightarrow \cdot CO_2^- \quad (10.21)$$

The radicals and cahrege paerrticles producted during initial photoexcitation reactions goes on further reduction process which results in the production of different types of products, equations (10.22) – (10.27) [77].

$$CO_2 + 2H^+ + 2e^- \rightarrow CO + H_2O \quad (10.22)$$

$$CO_2 + 4H^+ + 4e^- \rightarrow C + 2H_2O \quad (10.23)$$

$$CO_2 + 6H^+ + 6e^- \rightarrow CH_3OH + H_2O \quad (10.24)$$

$$CO_2 + 8H^+ + 8e^- \rightarrow CH_4 + 2H_2O \quad (10.25)$$

$$2H_2O + 4h^+ \rightarrow O_2 + 4H^+ \quad (10.26)$$

$$2H^+ + 2e^- \rightarrow H_2 \quad (10.27)$$

The photoreduction of carbon dioxide with methane to produce syngas is very helpful for mitigating greenhouse gases as well as for monitoring energy crises because during this process CO_2 and CH_4 greenhouse gases are involved. The possible reactions for CO_2 reforming of methane by photocatalysis are explained in equations (10.28) – (10.33) [74, 78].

$$2CH_4 \rightarrow C_2H_6 + H_2 \quad (10.28)$$

$$CO_2 + H_2 \rightarrow CO + H_2O \quad (10.29)$$

$$C + CO_2 \rightarrow 2CO \quad (10.30)$$

$$CO_2 + CH_4 \rightarrow CO + CH_2O + H_2 \quad (10.31)$$

$$CO_2 + CH_4 \rightarrow CO + CH_3OH \quad (10.32)$$

$$CH_4 + CO_2 \rightarrow 2CO + 2H_2 \quad (10.33)$$

During syngas production, there is possibility of production of several products like methanol which can be considered as green fuel and favors this process. However, other products can be controlled by using efficient catalyst as well as with appropriate design of photocatalytic reactor. CO_2 reforming by using different reductant through phototechnology is still facing many challenges due to lower conversion and selectivity. Researchers are working on photocatalytic reduction of CO_2 with H_2O but photoreduction of CO_2- CH_4 is still a demanding research area. The photoreduction of CO_2 - CH_4 by using semiconductor Cu/CdS–TiO2/SiO2 catalyst was performed at temperature of 373K and reported unsatisfactory results with very low conversion [79]. The photoreduction of CO_2-CH_4 to produce syngas CO and H_2 by using ZrO_2 is also studied with lower conversion and selectivity [35].

Table 10.4. Summary of photoreduction of CO_2-CH_4 literature for syngas production

Reactants	Catalyst	Light source	T, P	Products	Reactor type	Comments	Ref.
CO_2-CH_4	Ga_2O_3	300 W, Xe-lamp (220-300 nm)	473 K, 1 atm,	CO, H_2, C_2H_6	Quartz fixed bed reactor	Other product like C_2H_4, and C_3H_8 also observed. The conversion is near 1.22%.	[82]
CO_2-CH_4	Cu/CdS – TiO_2/SiO_2	125 W, Ultrahigh pressure, Hg lamp	393K, 1atm	CH_3COCH_3, CH_3CH_3, CO	Quartz fixed bed reactor	Conversion of methane 1.47% and CO_2 0.74% achieved	[79]
CO_2/CH_4, H_2	MgO	500 W, Ultrahigh pressure, Hg lamp	293 K, 1 atm	CO, H_2	Fixed bed, flat bottom reactor	Formaldehyde and acetaldehyde also used as reaction substrate. CO 3.6 μmol was observed.	[83]
CO_2-CH_4	ZrO_2	500 W, Ultrahigh pressure, Hg lamp	293 K, 1 atm	CO, H_2	Fixed bed, flat bottom reactor	The other products like HCHO and CH_3CHO was also observed	[26, 35]

The summary of CO_2 reforming of CH_4 for syngas production through photocatalysis is shown in Table 10.4, which describes that syngas, is the primary product. The results reveal that syngas can be produced definitely through photoreduction of CO_2-CH_4 as compared to CO_2-H_2O processes but the lower conversion and selectivity is still a challenge which needs to be focused. The conversion of CO_2 and CH_4 can easily be improved with the help of suitable selection of semiconductor catalyst and appropriate design of photocatalytic reactors. Successful design of photoreactor must have the ability to increasing the number of photon absorbed per unit area per unit volume. The design should also be efficient for utilizing the electron and holes towards carrying photoreactions with control selectivity [80].The addition of noble metals to semiconductor can change the photocatalytic properties due to change in semiconductor surface characteristics and increases the conversion, yield and selectivity of particular product. Loading of metals can also change the reaction products[81].

The photocatalytic reaction rate can be controlled by monitoring different steps including [65]; light absorption, temperature of the medium and photo generated charges transfer onto the surface. In addition, recombination of electron (e^-) and holes (h^+), reaction of electron and holes on semiconductor photocatalysis surface activated during photon radiation and quick removal of heat during photocatalysis need to be controlled too. Besides, efficient mass transfer at the semiconductor surface, morphology and characteristics of catalyst surface, particle size, and presence of metals ions are also important.

CONCLUSION

The syngas gas production through CO_2 reforming of CH_4 by using plasma and photo technologies is reviewed in this chapter. Both technologies have potentials for replacing conventional catalytic processes.

However, there are some disadvantages to practicing these technologies, which can be improved by efficient engineering approaches. By evaluating different processes of plasma, it seems that for achieving higher conversion and selectivity, three key factors are very important which needs to be focused. These factors are plasma temperature, electron density and plasma reactor configuration. The corona discharge and dielectric barrier discharge plasma are non-uniform plasma with lower electron density and have limited reaction volume for efficient conversion.

The microwave discharge plasma reveals uniform plasma with high plasma temperature and more discharge space for efficient conversion but equipment is more complicated and difficult for launching on industrial scales. Atmospheric pressure glow discharge plasma has many advantages as high electron energy, proper plasma temperature and higher electron density but this reactor is difficult to enlarge. On the other hand thermal plasma has higher specific enthalpy content, high temperature, more electron density, larger treatment volume and easy to enlarge with higher energy conversion efficiency.

The lower conversion and selectivity of this process is still a problem with higher input energy. Photocatalytic reduction of CO_2 reforming of CH_4 to syngas seems an innovative alternative technology for monitoring energy crises and mitigating global warming. Although there are fewer works relating to conversion of greenhouse gases to renewable fuels, photocatalysis is a potentially economical and environment friendly technology for sustainable development.

Lower conversions of CO_2 and CH_4 as well as selectivity of synthesis gas are still a problem for practicing this technology. The understandings of parameters that influence photocatalytic process are essential for the development of this technology for practical applications. Several key factors are necessary to be taken into consideration for achieving higher conversion and for improving the efficiency of photocatalytic process. The parameters include selection of semiconductor catalyst, design of photo reactor, suitable reductant and optimization of operating conditions.

ACKNOWLEDGMENTS

The authors would like to express their sincere gratitude to Universiti Teknologi Malaysia (UTM) for sponsoring this research under the Research University Grant (RUG) Project no Q.J130000.7125.00H49.

AUTHORS' BIOGRAPHIES

Professor Dr. NorAishah Saidina Amin (Prof NASA) is the head of Chemical Reaction Engineering Group (CREG) at the Faculty of Chemical Engineering, Universiti Teknologi Malaysia. Her fields of expertise are in catalytic reaction engineering, innovative reactor design and modeling of chemical reaction systems. Her research activities involve research development and consultancy (RDandC) on conversions of methane (GTL), palmoil, glycerol, and lignocellulossic biomass (BTL) to important chemicals such as hydrogen, syngas, C2+ and liquid fuels.

Muhammad Tahir is currently doing his PhD in Chemical Engineering under the supervision of Prof. Dr. NorAishah Saidina Amin at Universiti Teknologi Malaysia. He obtained his B.Sc (Chem. Eng) in 2004, M.Sc (Applied Environmental Science) in 2005 and M.Sc (Chem. Eng) in 2007 from University of Punjab, Lahore. He then worked as a research scholar at ICET, Lahore. He also has industrial experience working as the deputy manager of production at chemical and food processing industry (2006-2008). He also served as a lecturer at COMSATS Institute of Information and Technology, Lahore (2008-2010). His main research interests are in modeling and design of microreactors for chemical processes, synthesis of nanocatalyst for photocatalysis applications, and study on chemical and photochemical processes for conversion of biomass and greenhouse gases to renewable and sustainable fuels.

REFERENCES

[1] G. Veser, *Catalysis Today*, 157 (2010) 24.
[2] A. Singh, P.S. Nigam and J.D. Murphy, *Bioresource Technology*, 102 (2011) 10.
[3] P.S. Nigam and A. Singh, *Progress in Energy and Combustion Science*, 37 (2011) 52.
[4] H. Balat and E. Kırtay, *International Journal of Hydrogen Energy*, 35 (2010) 7416.
[5] L.T.H. Nam, T.Q. Vinh, N.T.T. Loan, V.D.S. Tho, X.-Y. Yang and B.-L. Su, Fuel, 90 (2011) 1069.
[6] R. Stanger and T. Wall, *Progress in Energy and Combustion Science*, 37 (2011) 69.
[7] G. Knothe, *Progress in Energy and Combustion Science*, 36 (2010) 364.
[8] T. Liu, G. Xu, P. Cai, L. Tian and Q. Huang, *Renewable Energy*, 36 (2011) 1284.

[9] A. Tolón-Becerra, X. Lastra-Bravo and F. Bienvenido-Bárcena, *Renewable Energy*, 36 (2011) 2067.
[10] M. Gallo, *Transport Policy*, 18 (2011) 413.
[11] O. Berglund and K. Berglund, *Soil Biology and Biochemistry*, 43 (2011) 923.
[12] H.F. Abbas and W.M.A. Wan Daud, *International Journal of Hydrogen Energy*, 35 (2010) 1160.
[13] G. Chen, C.-Y. Guo, X. Zhang, Z. Huang and G. Yuan, *Fuel Processing Technology*, 92 (2011) 456.
[14] E.-C. Jeon, S. Myeong, J.-W. Sa, J. Kim and J.-H. Jeong, *Applied Energy*, 87 (2010) 205.
[15] A. Indarto, D. Yang, J. Choi, H. Lee and H. Song, *Journal of Hazardous Materials*, 146 (2007) 309.
[16] M.B. Toftegaard, J. Brix, P.A. Jensen, P. Glarborg and A.D. Jensen, *Progress in Energy and Combustion Science*, 36 (2010) 581.
[17] K. Zeng and D. Zhang, *Progress in Energy and Combustion Science*, 36 (2010) 307.
[18] M. Arguelles, C. Benavides and B. Junquera, *Journal of Environmental Management*, 81 (2006) 249.
[19] J.L. Drouet, N. Capian, J.L. Fiorelli, V. Blanfort, M. Capitaine, S. Duretz, B. Gabrielle, R. Martin, R. Lardy and P. Cellier, *Environmental Pollution*, (2011) 1.
[20] S.S. Clair, J. Hillier and P. Smith, *Biomass and Bioenergy*, 32 (2008) 442.
[21] M. MacLeod, D. Moran, V. Eory, R.M. Rees, A. Barnes, C.F.E. Topp, B. Ball, S. Hoad, E. Wall and A. McVittie, *Agricultural Systems*, 103 (2010) 198.
[22] X. Xiaoding and J.A. Moulijn, *Energy and Fuels*, 10 (1996) 305.
[23] M. Ruth, S.A. Gabriel, K.L. Palmer, D. Burtraw, A. Paul, Y. Chen, B.F. Hobbs, D. Irani, J. Michael and K.M. Ross, *Energy Policy*, 36 (2008) 2279.
[24] H. Arakawa, M. Aresta, J.N. Armor, M.A. Barteau, E.J. Beckman, A.T. Bell, J.E. Bercaw, C. Creutz, E. Dinjus, D.A. Dixon, K. Domen, D.L. DuBois, J. Eckert, E. Fujita, D.H. Gibson, W.A. Goddard, D.W. Goodman, J. Keller, G.J. Kubas, H.H. Kung, J.E. Lyons, L.E. Manzer, T.J. Marks, K. Morokuma, K.M. Nicholas, R. Periana, L. Que, J. Rostrup-Nielson, W.M.H. Sachtler, L.D. Schmidt, A. Sen, G.A. Somorjai, P.C. Stair, B.R. Stults and W. Tumas, *Chemical Reviews*, 101 (2001) 953.
[25] M. Ritzkowski and R. Stegmann, *International Journal of Greenhouse Gas Control*, 1 (2007) 281.
[26] P. Usubharatana, D. McMartin, A. Veawab and P. Tontiwachwuthikul, *Ind. Eng. Chem. Res*, 45 (2006) 2558.
[27] N.L. Panwar, S.C. Kaushik and S. Kothari, *Renewable and Sustainable Energy Reviews*, 15 (2011) 1513.
[28] S. Mekhilef, S. Siga and R. Saidur, *Renewable and Sustainable Energy Reviews*, 15 (2011) 1937.
[29] G. Centi and R.A.V. Santen, Catalysis for Renewables, WILEY-VCH, Weimheim, 2007.
[30] E.H. Ale and E. Jamshidi, *Energy Conversion and Management*, 45 (2004) 345.
[31] X. Tao, M. Bai, X. Li, H. Long, S. Shang, Y. Yin and X. Dai, *Progress in Energy and Combustion Science*, 37 (2011) 113.
[32] X. Tao, M. Bai, Q. Wu, Z. Huang, Y. Yin and X. Dai, *International Journal of Hydrogen Energy*, 34 (2009) 9373.

[33] H. Long, S. Shang, X. Tao, Y. Yin and X. Dai, *International Journal of Hydrogen Energy*, 33 (2008) 5510.

[34] D. Li, X. Li, M. Bai, X. Tao, S. Shang, X. Dai and Y. Yin, International Journal of Hydrogen Energy, 34 (2009) 308.

[35] K. Teramura, T. Tanaka, H. Ishikawa, Y. Kohno and T. Funabiki, *J. Phys. Chem. B*, 108 (2004) 346.

[36] N. Burke and D. Trimm, *Catalysis Today*, 117 (2006) 248.

[37] V. Goujard, J. Tatibouet and C. Batiotdupeyrat, *Applied Catalysis A: General*, 353 (2009) 228.

[38] S. Futamura, H. Kabashima and G. *Annadurai, Catalysis Today*, 115 (2006) 211.

[39] J. Henriques, N. Bundaleska, E. Tatarova, F.M. Dias and C.M. Ferreira, *International Journal of Hydrogen Energy*, 36 (2011) 345.

[40] S. Ozkara-Aydınoglu, international Journal of Hydrogen Energy, 35 (2010) 1 2 8 2 1

[41] B. Michael, A. Donazzi and S.L. D., *Journal of Catalysis*, 265 (2009) 117.

[42] J.G. Jakobsen, T.L. Jørgensen, Chorkendorff and S. Jens, *Applied Catalysis A: General*, 377 (2010) 158.

[43] M. Minutillo and A. Perna, *International Journal of Hydrogen Energy*, 35 (2010) 7012.

[44] X. Tao, M. Bai, X. Li, H. Long, S. Shang, Y. Yin and X. Dai, *Progress in Energy and Combustion Science*, 37 (2011) 113.

[45] H. Long, S. Shang, X. Tao, Y. Yin and X. Dai, *International Journal of Hydrogen Energy*, 33 (2008) 5510.

[46] J.T. Carneiro, C.-C. Yang, J.A. Moulijn and G. Mul, *Journal of Catalysis*, 277 (2011) 129.

[47] B. Naik, S. Martha and K.M. Parida, *International Journal of Hydrogen Energy*, 36 (2011) 2794.

[48] Z.-Y. Wang, H.-C. Chou, J.C.S. Wu, D.P. Tsai and G. Mul, *Applied Catalysis A: General*, 380 (2010) 172.

[49] Q. Jiang, S. Faraji, K.J. Nordheden and S.M. Stagg-Williams, *Journal of Membrane Science*, 368 (2011) 69.

[50] Y.N. Chun, Y.C. Yang and K. Yoshikawa, *Catalysis Today*, 148 (2009) 283.

[51] L. Chen, X.-W. Zhang, L. Huang and L.-C. Lei, *Chemical Engineering and Processing: Process Intensification*, 48 (2009) 1333.

[52] A. Ağıral, T. Nozaki, M. Nakase, S. Yuzawa, K. Okazaki and J.G.E. Gardeniers, *Chemical Engineering Journal*, 167 (2011) 560.

[53] T. Nozaki, A. Ağıral, S. Yuzawa, J.G.E. Han Gardeniers and K. Okazaki, *Chemical Engineering Journal*, 166 (2011) 288.

[54] Y.-F. Wang, C.-H. Tsai, W.-Y. Chang and Y.-M. Kuo, *International Journal of Hydrogen Energy*, 35 (2010) 135.

[55] G. Zhao, S. John, J. Zhang, L. Wang, S. Muknahallipatna, J. Hamann, J. Ackerman, M. Argyle and O. Plumb, *Chemical Engineering Journal*, 125 (2006) 67.

[56] A. Indarto, J. Cho, H. Lee, H. Song and J. Palgunadi, *Journal of Rare Earths*, 24 (2006) 513.

[57] A. Indarto, D. Yang, J. Palgunadi, J. Choi, H. Lee and H. Song, Chemical Engineering and Processing: *Process Intensification*, 47 (2008) 780.

[58] I. Istadi and N. Amin, *Chemical Engineering Science*, 62 (2007) 6568.

[59] A. Indarto, *Journal of the Chinese Institute of Chemical Engineers*, (2008).

[60] Istadi and N. Amin, *Fuel*, 85 (2006) 577.
[61] X. Tao, *International Journal of Hydrogen Energy*, 33 (2008) 1262.
[62] Y.-Q. Zhang, W. Zhou, S. Liu and A. Navrotsky, *Chemistry of Materials*, 23 (2011) 1166.
[63] P. Du, J.T. Carneiro, Jacob A. Moulijn and Guido Mul, *Applied Catalysis A: General*, (2008) 119.
[64] R. Leary and A. Westwood, *Carbon*, 49 (2011) 741.
[65] K. Kočí, L. Obalová, L. Matějová, D. Plachá, Z. Lacný, J. Jirkovský and O. Šolcová, *Applied Catalysis B: Environmental*, 89 (2009) 494.
[66] S.C. Roy, O.K. Varghese, M. Paulose and C.A. Grimes, *American Chemical Society: Nano*, 4 (2010) 1259.
[67] Y. Ku, *Journal of Molecular Catalysis A: Chemical*, 212 (2004) 191.
[68] S.C. Dantas, J.C. Escritori, R.R. Soares and C.E. Hori, *Chemical Engineering Journal*, 156 (2010) 380.
[69] M.A. Malati, *Energy Conversion and Management*, 37 1345.
[70] M. Anpo, H. Yamashita, Y. Ichihashi and S. Ehara, *Journal of Electroanalytical Chemistry*, 396 (1995) 21.
[71] K. Ikeue, S. Nozaki, M. Ogawa and M. Anpo, *Catalysis Today*, 74 (2002) 241.
[72] d.H. Lasa, B. Serrano and M. Salaices, *Photocatalytic Reaction Engineering*, Springer Science, New York, 2005.
[73] S. Zhou and A.K. Ray, *Industrial and Engineering Chemistry Research*, 42 (2003) 6020.
[74] L. Yuliati, H. Itoh and H. Yoshid, *Chemical Physics Letters*, 452 (2008) 178.
[75] V.P. Indrakanti, H.H. Schobert and J.D. Kubicki, *Energy Fuels*, 23 (2009) 5247.
[76] R.J. Braham and A.T. Harris, *Ind. Eng. Chem. Res*, 48 (2009) 8890.
[77] H. Petek and J. Zhao, *Chemical Reviews*, 110 (2010) 7082.
[78] T. Kentaro, T. Tsunehiro, I. Haruka, K. Yoshiumi and F. Takuzo, *J. Phys. Chem. B*, 108 (2004) 346.
[79] D. Shi, Y. Feng and S. Zhong, *Catalysis Today*, 98 (2004) 505.
[80] H.d. Lasa, B. Serrano and M. Salaices, *Photocatalytic Reaction Engineering*, Springer, New York, 2005.
[81] A.L. Linsebigler, G. Lu, J.T. Yates and Jr, *Chemical Reviews*, 95 (1995) 735.
[82] L. Yuliati, H. Itoh and H. Yoshida, *Chemical Physics Letters*, 452 (2008) 178.
[83] P. Usubharatana, D. McMartin, A. Veawab and P. Tontiwachwuthikul, *Ind. Eng. Chem. Res*, 45 (2006) 2558.

In: Syngas
Editors: Antonius Indarto and Jelliarko Palguandi
ISBN: 978-1-62100-870-5

Chapter 11

LEAN PREMIXED SYNGAS COMBUSTION FOR GAS TURBINES APPLICATIONS

Salvatore Daniele*
Combustion Research Laboratory, Paul Scherrer Institute, Switzerland

ABSTRACT

Modifications of conventional natural-gas-fired burners for operation with syngas fuels using lean premixed combustion are challenging due to the different physicochemical properties of the two fuels. In this chapter, fuels ranging from pure methane to syngas (CO-H_2 blends) and pure hydrogen have been investigated, which were relevant to integrated gasification combined cycle (IGCC) power generation applications. A key differentiating parameter is the turbulent flame velocity, S_T, commonly expressed as its ratio to the laminar flame speed, S_L. This chapter reports an experimental investigation of lean premixed syngas combustion at gas-turbine like conditions, with emphasis on the determination of S_T. Experiments at pressures up to 2.0 MPa, inlet temperatures and velocities up to 773 K and 150 m/s respectively, and turbulence intensity to laminar flame speed ratios, u'/S_L, exceeding 100 are presented for the first time. The flame front corrugation was measured with Planar Laser Induced Fluorescence (PLIF) of the OH radical, while turbulence quantities were determined with Particle Image Velocimetry (PIV). Two different analyses based on mass balance were performed on the acquired flame images. The first one assessed absolute values of turbulent flame speeds and the second one, by means of an improved fractal methodology, provided normalized turbulent flame speeds (S_T/S_L). Comparisons between different syngas mixtures and methane clearly show much higher S_T/S_L values for the former fuel. It is shown that S_T/S_L is strongly dependent on preferential diffusive thermal (PDT) effects, co acting with hydrodynamic effects, even for very high u'/S_L. S_T/S_L increases with rising hydrogen content in the fuel mixture and with pressure. A correlation for S_T/S_L valid for all investigated fuel mixtures, including methane, is proposed in terms of turbulence properties (turbulence intensity and integral length scale), combustion properties (laminar flame speed and laminar flame thickness) and operating conditions (pressure and inlet temperature). The correlation captures effects of preferential diffusive thermal and hydrodynamic instabilities. Interaction between flame

* E-mail: salvatore.daniele@psi.ch.

front and turbulent eddies measured by the fractal dimension is shown to depend on their temporal activity. Along with the results on turbulent flame speed the operational window in terms of lean blow out and flashback limits is presented and its dependence on the combustion parameter is analyzed.

INTRODUCTION

Aim of this chapter is to characterize the combustion properties of high hydrogen containing fuels (syngas) relevant to turbulent lean premixed operations in gas turbine engines, with particular emphasis on the turbulent flame speed. Nowadays the popularity of power plants integrating gasification processes and gas turbine based power generation (IGCC, as briefly described in the next paragraph) rises many issues for combustor design engineering. In this scenario, where fuel-flexibility represents the key word, novel engines are required to operate with a wide range of fuel mixtures ranging from natural gas (mainly composed of methane) to pure hydrogen, including a variety of hydrogen/carbon-monoxide mixtures (syngas). This chapter contributes in answering the simple but key question coming from the related industry: "can we fire with syngas, in lean premixed mode, our natural gas burners?". Taking a look to the very recent literature, the answer seems to be not yet provided. For example, Robert Cheng summarized in 2010 [1]: "*At present, the basic properties of premixed turbulent flames at heated and pressurized conditions remain largely unexplored. There is an urgent need for data on the turbulent flame speeds and other scalar measurements to quantify the topology of the flamelets, the internal flamelets structures, and the heat release rate.*" The lack of data at high pressure and high temperature is seen as a major concern within the combustion community as also stated by Bradley who states in 2009 [2]:"*There is a dearth of necessary data, mostly at higher temperatures and pressures, necessitating many more measurements and computations, particularly for the new fuels.*". The need of a *"modern database"* containing quantities such as turbulent flame speeds, Markstein numbers and lengths etc., was also highlighted by Driscoll in 2008 [3].

The scarce knowledge acknowledged by the cited authors (and by many others), reflects to an industrial scenario wherein IGCC power plants are not yet able to safely fire syngas in premixed mode. This chapter takes up this challenge describing the first turbulent flame speed data at gas turbine like conditions available in the literature, published in 2009 [4] and 2011 [5]. These data, their analysis and the drawn conclusions represent the main outcome of the chapter. Along with the flame speed data, the operational window in terms of lean blow out and flashback limits is described.

INTEGRATED GASIFICATION COMBINED CYCLE AND SYNGAS

Within the gas turbine based power generation technologies, there is nowadays increased interest in Integrated Gasification Combined Cycle (IGCC) power plants, for reasons of security in fuel supply and optimization in energy use. These plants offer a big potential for clean electricity production with reduced CO_2 emissions when operated in the lean premixed combustion mode and when carbon capture technology is applied [6]. Via gasification, IGCC approaches convert solid and liquid hydrocarbons (biomasses, oil, coal, tars, etc.) to a gaseous fuel for gas turbine based power generation. The fuel obtained from the gasification process is conventionally termed syngas. It can be considered as a gaseous mixture containing mostly hydrogen (H_2) and carbon monoxide (CO), whose chemical composition varies according to the specific feedstock (e.g. biomass, coal, tar, etc.). It is commonly diluted with inert gases such as nitrogen (N_2), steam (H_2O) or carbon dioxide (CO_2), depending on the gasification process. A schematic of an IGCC plant is presented in Figure 11.1. Lean premixed combustion is considered as the state of the art technology in stationary gas turbines for highly efficient, low emissions power generation using natural gas fuel. It is therefore the prominent technology for application in IGCC power plants. To this purpose, fundamental properties of syngases need to be investigated for premixed flames at gas turbine like operating conditions. The combustion properties of syngases are mainly determined by their specific H_2 and CO content and are generally characterized by very high laminar flame speeds, a wide range of flammability limits and short ignition delay times [5,8,9], which all contribute to a high risk of autoignition and flashback phenomena. Beside the combustion characteristics, the physical properties of H_2 are also very different from those of natural gas. For example, the density of H_2 is an order of magnitude smaller than that of natural gas, resulting in increased volumetric flow rate requirements for the turbine, and at the same time the molecular diffusivity of H_2 is substantially higher, leading to thermal/diffusive imbalance effects during combustion.

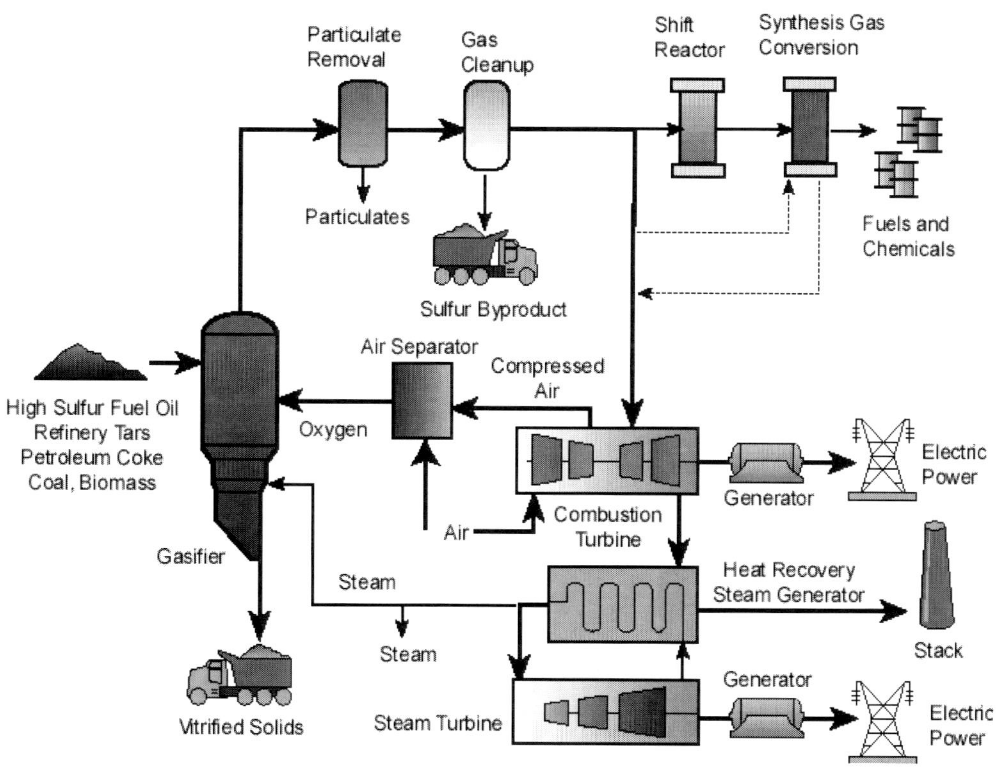

Figure 11.1. Integrated Gasification Combined Cycle and Syngas plant schematic [7].

OPERATIONAL WINDOW

As for any type of fuel, also for syngases the knowledge of the operational window is fundamental to the combustor design process with the purpose of safe engine operation. High H_2-containing fuels are characterized by higher burning rates compared to natural gas, leading to less restrictive lean blow out (LBO) limits and higher flashback (FB) propensity.

These two phenomena, LBO and FB, define the limits of the operational window of an engine (when thermoacoustic instabilities are not considered). Images of the flame front (acquired with planar laser induced fluorescence technique as it will be detailed in experimental section) are presented in Figure 11.2 for a stable flame (a), a LBO flame (b) and a flashback flame (c). The experimental facility in which these flames have been operated as described in the forthcoming section. LBO happens under very lean operative conditions (e.g. relevant to part load) and manifests itself with a detachment of the flame from its anchoring point leading to extinction. This phenomenon is usually due to the combination of reduced reactivity (hence lower flame speed) and fluctuations of parameters like heat release, pressure velocity etc., due to instabilities.

The term flashback defines the upstream propagation of the flame front into the fuel/air mixing section from its previous stabilization point in the combustion chamber. As it represents one of the major concerns in lean premixed syngas combustion, it deserved a more detailed description in this chapter.

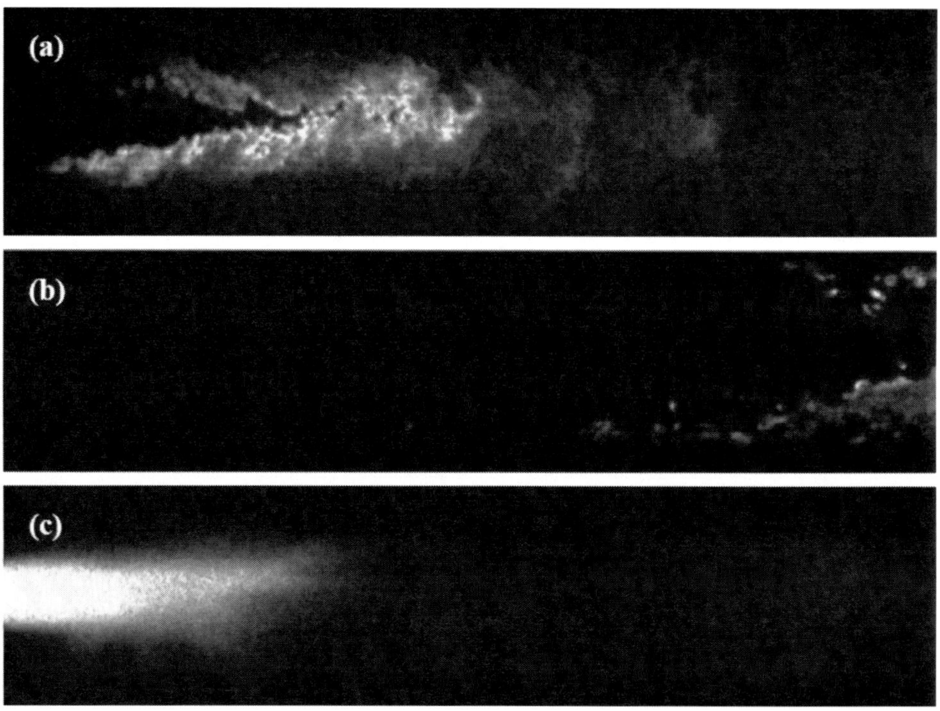

Figure 11.2. Stable (a), LBO (b) and FB (c) flame images. OH-PLIF (a) and (b), and OH-chemiluminescence (c). $P = 0.50$ MPa.

FB can happen via different mechanisms, a general classification is given in [10]. In this work four categories have been identified:

- propagation in the core flow
- propagation in the boundary layer
- propagation due to “combustion induced vortex breakdown” (CIVB)
- propagation due to combustion instabilities.

The first two mechanisms are prompted when the flame speed locally overtakes the velocity of the incoming flow [11-13]. CIVB derive from the coupling between turbulence and chemistry in swirled flows [14-16]. The last mechanism in the list is attributed to all flashback events derived from various instabilities (e.g. thermoacoustic, entropy waves, etc.).

In a scenario where fuel flexibility is the key of success for a modern, premixed, low emission gas turbine engine, flashback phenomena associated with highly reactive fuels represent a major issue. The backward propagation of the flame (flashback), from the combustion chamber through the mixing section toward the fuel injection point, induces serious damages to burner components and forces long-term shut downs. Consequently, experiments at gas turbine relevant conditions with the goal of describing and understanding the flashback propensity are of great value for combustor design. Performing flashback experiments is usually expensive due to the damages this phenomenon can incite. In addition, these experiments necessitate higher attention to safety issues due to their dangerous nature, especially when they are performed at high pressure and high temperature. For these reasons, facilities capable to operate such experiments are rare and there exist no comprehensive data sets available in the literature.

FLAME SPEED

When referring to stable and safe lean premixed gas turbine combustion, one of the most important parameters is the turbulent burning velocity, S_T. This property plays an ever increasing role in the design of new combustors operated with syngas fuels, as it assumes much higher values for hydrogen-rich gases compared to natural gas. When talking about turbulent flame speed, it is fundamental to specify that this parameter does not have a unique definition: in fact, four different definitions have been identified, as summarized in [1]:

- Global Consumption $\dfrac{\dot{m}}{\rho A}$

- Local Consumption $S_L \int_{-\infty}^{+\infty} \Sigma \partial \eta$

(11.1)

- Global Displacement $\dfrac{\partial r_{fl}}{\partial t} - U_r$

- Local Displacement $U_{fl} - U_r = U \sin(\theta)$

where: $\dot{m}$ is the mass flow rate, ρ is the density of the incoming fuel/air mixture, A is the flame front surface, S_L is the laminar flame speed, Σ is the flame surface density, $\partial r_{fl}/\partial t$ is the local flame front displacement velocity, U_r is the velocity of the reactants, U_{fl} is the velocity of the flame brush, θ is the angle between the normal to the flame front and the flow velocity. A more detailed description of the four definitions and their applications is provided in [1].

The data presented in this chapter have to be regarded as "consumption" data belonging to the first category, where the mass balance is applied at the combustor inlet and at the average flame front surface.

It has been shown, as summarized in [1,3], that these different definitions lead to different results generating a big spread in the data. This has led to the generally accepted conclusion that turbulent flame speed data measured in a particular facility and with a specific methodology (according to one of the four definitions) cannot be directly compared to each other. It may be worth stating beforehand that the author does not fully agree with this conclusion. This topic is addressed in [5] and summarized in the results section.

In order to describe S_T for syngas mixtures and compare it with that of natural gas, insights on the effects of molecular transport have to be advanced. The significance of such effects on S_T has been highlighted by Lipatnikov and Chomiak in their review [17], concluding that in fuel lean syngas/air turbulent combustion, molecular transport effects are still important and dominated by the H_2 component.

Turbulent flame speed is determined by the extent of the flame front surface area wrinkling and by the effect the flame wrinkles have on the local propagation speed of the flamelets. In any premixed flame (laminar or turbulent), the growth of an initial wrinkle is affected by the hydrodynamic Darrieus Landau (DL) instability [18]. DL instabilities enhance the flame front wrinkling and therefore S_T with the following mechanism: due to the flame induced convergence/divergence of the incoming flow field ahead of the concave/convex flame front, the flow velocity increases/decreases whereas the flame speed stays constant. Therefore both concave and convex bulges grow and the perturbation is increased [17]. Effects of pressure on DL instabilities in turbulent combustion have been firstly reported for CH_4 fuels in Kobayashi et al. [19] and more recently for syngas mixtures in Daniele et al. [5] and Ichikawa et al. [20]. Nonetheless, there is still controversy regarding the precise impact of the DL instability at elevated pressures, as discussed recently in Lipatnikov and Chomiak [21].

Another important instability affecting the flame front propagation characteristics occurs for thermo-diffusively imbalanced systems (i.e. for limiting reactants with Lewis numbers $Le < 1$). In case of $Le < 1$ and lean combustion, a bulge pointing towards the fresh mixture will act as a local attractor for the limiting reactant (fuel), and additionally due to the relative lower heat diffusivity the local temperature and therefore the local flame speed are enhanced resulting in a further growth of the bulge. Hydrogen fuels, in particular, are strongly affected by preferential diffusive thermal (PDT) effects in lean premixed combustion, as summarized in [17].

Understanding the DL and PDT effects, as well as their coupling and their interaction with the local flame stretch, is of fundamental importance for turbulent combustion. This understanding leads to a description of the flame front propagation at widely disparate scales: at large scales with relevant velocity S_T and at small scales with corresponding velocity the stretched laminar flame speed $S_{L,k}$.

Despite the increasing research activities in premixed syngas combustion, as discussed in [1-3], there are still many open questions that call for pressing answers. In contrast to the extensive knowledge gained in the last decades for hydrocarbons (especially CH_4), very few experimental data are available in the literature for syngas fuels at elevated pressures and preheats [4,5,20]. Moreover, literature turbulent flame speed data are generally not directly comparable due to the different adopted definitions for S_T and the different experimental methodologies used. The lack of syngas data becomes very evident especially when considering gas turbine relevant conditions, involving elevated pressures (1 - 3 MPa), high preheat temperatures (600 - 700 K) and large turbulence intensities ($u'/S_L > 50$).

FLAME FRONT/TURBULENCE INTERACTION

The propagation characteristic of a flame front in a turbulence flow field (represented by the turbulent flame speed, S_T) depend (apart from reactivity, exothermicity and diffusivity of the fuel/air mixture, represented by the laminar flame speed, S_L) on the interaction between the combustion wave itself and the turbulent eddies. This interaction has been addressed in early studies by Damköhler [22] who introduced the wrinkling concept, which was subsequently adopted and expanded in the classification of turbulent combustion in various regimes by Borghi [23], Peters [24], Abdel-Gayed et al. [25] and Poinsot et al. [26].

The most frequently used diagram to indentify and explain the different combustion regimes is the so called Borghi-Peters diagram [27]. The diagram (represented in Figure 11.3) depicts several combustion regimes in the two-dimensional space defined by L_T/δ_L and u'/S_L.

L_T/δ_L is the ratio between the integral length scale of turbulence and the unstretched laminar flame thickness: this parameter spatially quantifies the interaction between eddies and flame front.

u'/S_L is the ratio between the turbulence intensity and the unstretched laminar flame speed; this parameter quantifies the interaction between eddies and flame front from a temporal point of view in terms of the competition between the local microscopic flow velocity and the local flame speed.

The flame front structure is understood as composed by a preheating zone O(l) and a reaction zone O(l); these two zones are separated by the so called inner layer, having size of about one order of magnitude smaller than the other two O(0.1l). The preheating zone is chemically inert; the fuel gets attacked by radicals only in the thin inner layer and is oxidized via chain reaction to CO and H_2. These two compounds react then in the last layer: the reaction zone [27].

Different combustion regimes are identified by non-dimensional numbers as:

$$Re_T = \frac{u' L_T}{\nu} \approx \frac{u'}{S_L}\frac{L_T}{\delta_L},$$

$$Da = \frac{S_L}{\delta_L}\frac{L_T}{u'}, \qquad (11.2)$$

$$Ka = \left(\frac{u'}{S_L}\right)^{3/2}\left(\frac{L_T}{\delta_L}\right)^{-1/2},$$

$$Ka_\delta = \delta^2 Ka,$$

and by physical boundaries indicating $u' = S_L$ and $l = l_m$. Where the last expression identifies the size of eddies equal to the maximum distance l_m at which preheated fluid can be transported ahead of the flame. The parameter l_m is commonly identified as Zimont scale [28].

The flames investigated in this chapter are mostly contained within the thin reaction zones regime. In this regime, small eddies of the size of the Kolmogorov scale (η) can enter the preheating zone and thus they can modify the (turbulent) heat transport ahead of the flame front. Evidences of the effects of turbulence transport on the temperature fields are provided again by the group of N. Peters in [29], and further analyzed by Gulder in [30]. The eddies of size η are, however, larger than the inner layer and therefore they cannot penetrate that layer and thus they cannot proceed further and interact with the chemical processes happening within the reaction zone. More detailed information about the combustion regimes can be found in [31].

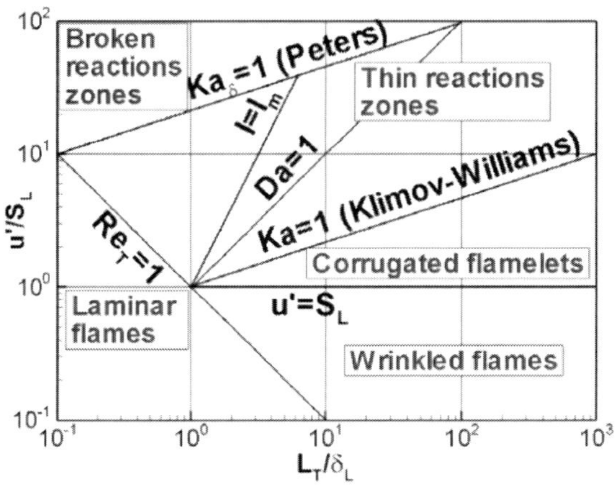

Figure 11.3. Borghi-Peters diagram [27].

EXPERIMENTAL

Test Rig

Experiments have been performed in a high pressure, optically accessible combustion chamber, delivering a maximum thermal power of 0.5 MW depicted by the in Figure 11.4 and 11.5. In Figure 11.4 it is possible to distinguish the preheater and the mixing section upstream of the combustion chamber. The figure also offers in the bottom part a detailed view of the burner in operation.

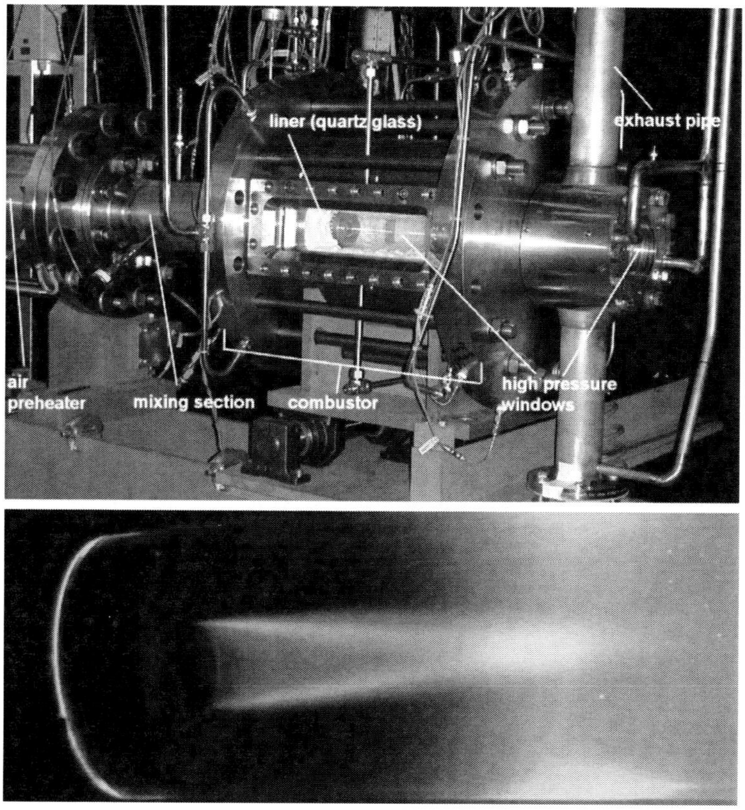

Figure 11.4. Test rig with detail of the burner in operation.

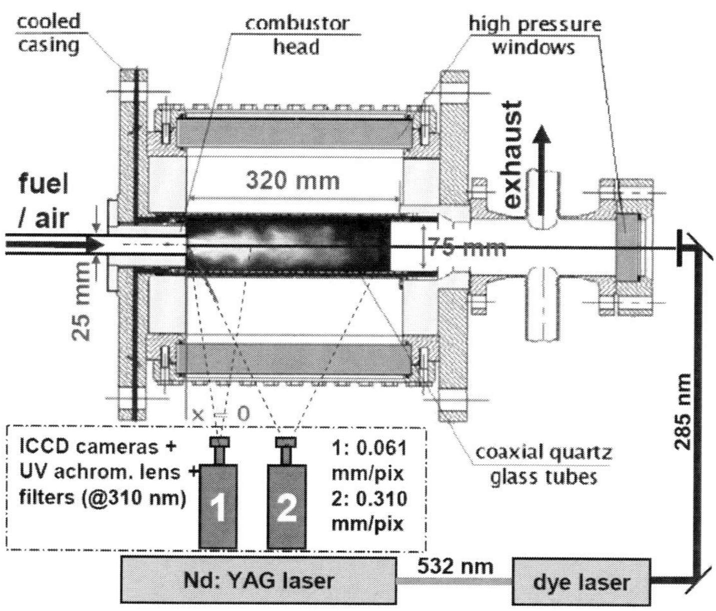

Figure 11.5. Experimental setup: high pressure turbulent combustor and OH-LIF optical arrangement.

The combustion chamber is cylindrical, with a length of 320 mm and an inside diameter of 75 mm. It is made of a double wall quartz tube with cooling air flowing in the annular section. A 25 mm diameter tube, coaxial to the reactor delivers the fuel/air mixture. Within this tube, fuel is injected and mixed with electrically preheated air 400 mm upstream of the combustion chamber. Fuel injection is accomplished in a co-flow configuration with the preheated air, through an array of five jets (1 mm in diameter) uniformly distributed within the delivery tube. For the above arrangement, mixing of the two streams is expected to be perfect and this is supported by analysis of the resulting NO_x emissions at the reactor exit.

A sudden radial expansion by a factor of three between the delivery tube and the combustion chamber induces an outer toroidal recirculation zone, which stabilizes the flame. There is no swirl or piloting flame applied. Pressurization is achieved by a cylindrical steel tank enclosing the reactor and delivery tube. High pressure air is supplied by an oil-free compressor. Optical access to the reaction zone is provided by three quartz windows positioned on the tank, two side ones and one along the reactor axis (see Figure 11.5).

Flashback Passive and Active Control System

Due to the frequent occurrence of flashback connected with the operations of H_2-containing fuels, a passive and active automatic control system was designed and implemented. The control system is described in this section.

Passive Control System

The passive control consists in the injection of N_2 into the boundary layer of the fuel/air delivery pipe through an annular slot positioned 150 mm upstream of the combustion chamber inlet section, as sketched in Figure 11.6. The injected N_2 dilutes the reactive mixture in the boundary layer with the purpose of reducing flashback propensity.

The N_2 flow rate was kept 1% of the main air flow rate for all operative conditions. It was verified by analyzing OH-chemiluminescence 2-D pictures and temperature measurements of the exhaust gas temperature that the amount of N_2 injected does not modify in an appreciable way either the flame position as shown in Figure 11.7 nor the flame temperature (differences on the order of 1 K were measured). Figure 11.7 presents OH-chemiluminescence images of the flame:

- on the left hand side the effect for different injected N_2 levels is depicted, the upper image is taken in absence of injected N_2 and therefore it represents the reference case.
- on the right hand side the normalized difference between the actual injected N_2 level image and the image corresponding reference case (upper left) is shown. The upper right image corresponds to the subtraction operated between two reference cases images in order to evaluate the “noise” associated with this operation.

The figure shows a very mild effect for the 1% case and a gradual increment of the flame length up to 4%. When 5% N_2 is applied the flame stabilization mechanism changes and the flame front detaches from the burner lips.

Active Control System

In order to avoid major damages and time consuming shut downs due to flashback occurrence the facility was equipped with an active control system. The sensor triggering the control action is given by the signal of a photomultiplier (sampled at 100 kHz) who capture the chemiluminescence signal coming from the mixing section. The optical access to the mixing section is provided by a high temperature resistant optical fiber. The Control System Is Equipped With Two Solenoid Valves as Actuators. the Control Logic Is Explained with the Help of Figure 11.8 which shows the hardware components in the top and the control block diagram below.

In a nominal condition the valve V1 is open and V2 is closed. As soon as FB happens and the flame is detected from the PM, the valve V2 is switched to open. This action forces part of the fuel to be injected through piloting holes surrounding the burner directly into the combustion chamber.

If no flame is detected by the PM the normal burning conditions can be recovered by redirecting all the fuel through the main injector slowly and readjusting parameters like Φ to safe operation. With this procedure, flashback can be controlled without blowing out the flame.

If after opening valve V2 combustion is still detected within the air/fuel mixing section, the next action is to close valve V1. With this configuration all the fuel is injected through the piloting holes and no flame can be present in the mixing section anymore. At this point the normal conditions can be slowly reestablished.

The control system has shown very good efficiency and fast time response (~0.5 s). The re-injection of all the fuel through the main injector is not done automatic but manually (it can easily be automated by employing different hardware). This procedure was successful for pressures below 1.0 MPa. At higher pressures auto-ignition phenomena has been observed at the main fuel injector as soon as the fuel was re-injected.

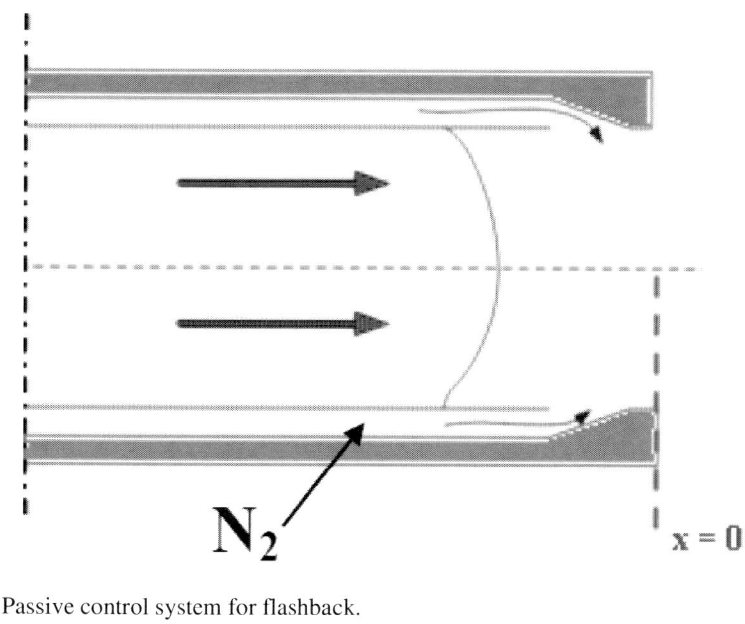

Figure 11.6. Passive control system for flashback.

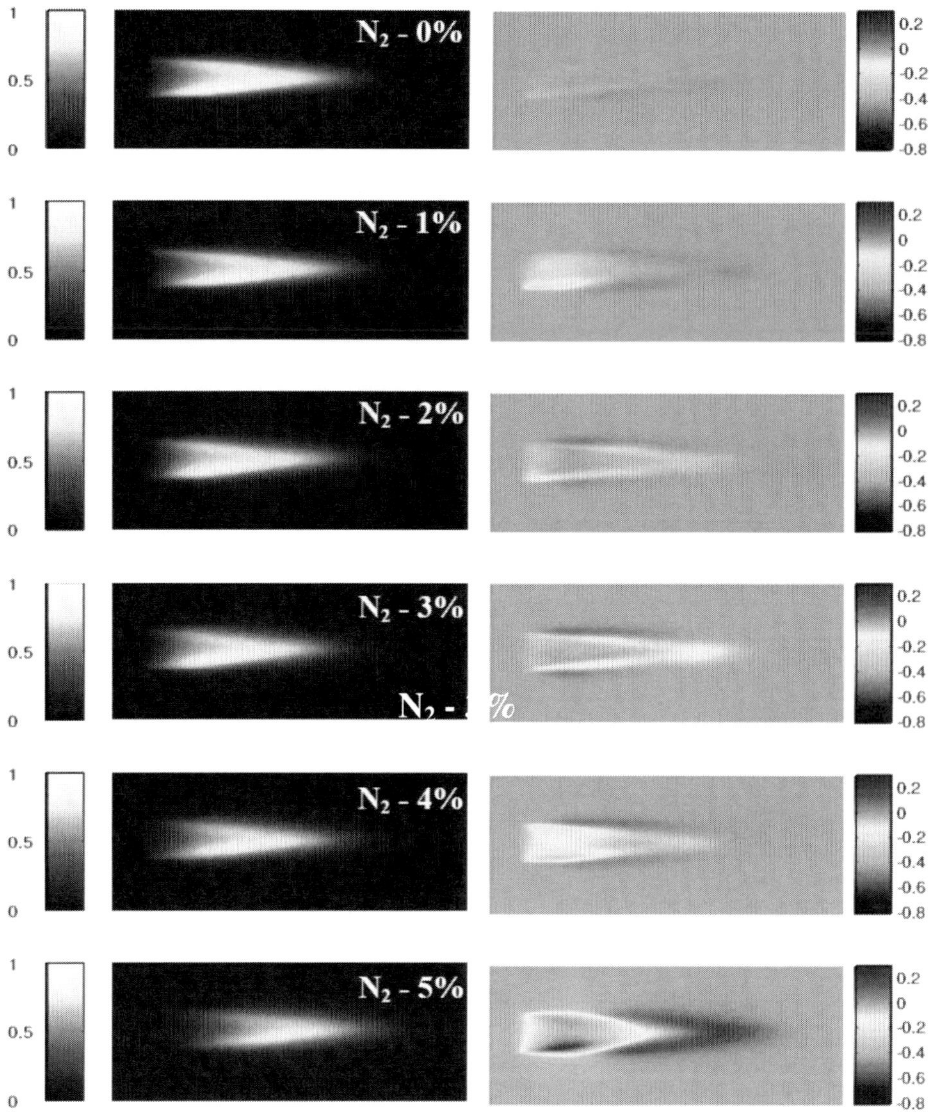

Figure 11.7. Effect of N_2 injection on the flame. OH-chemiluminescence images.

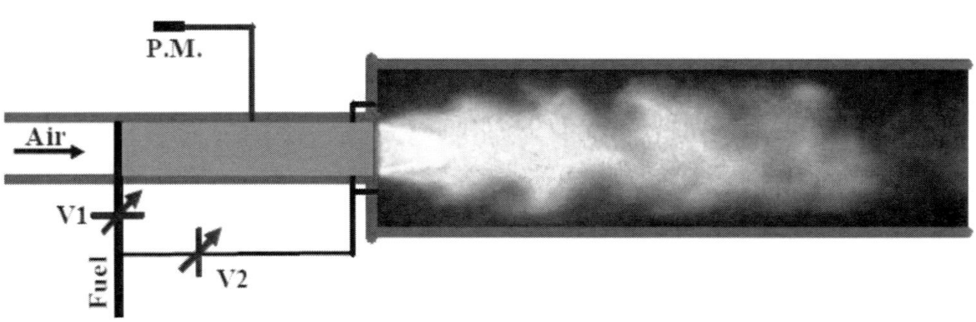

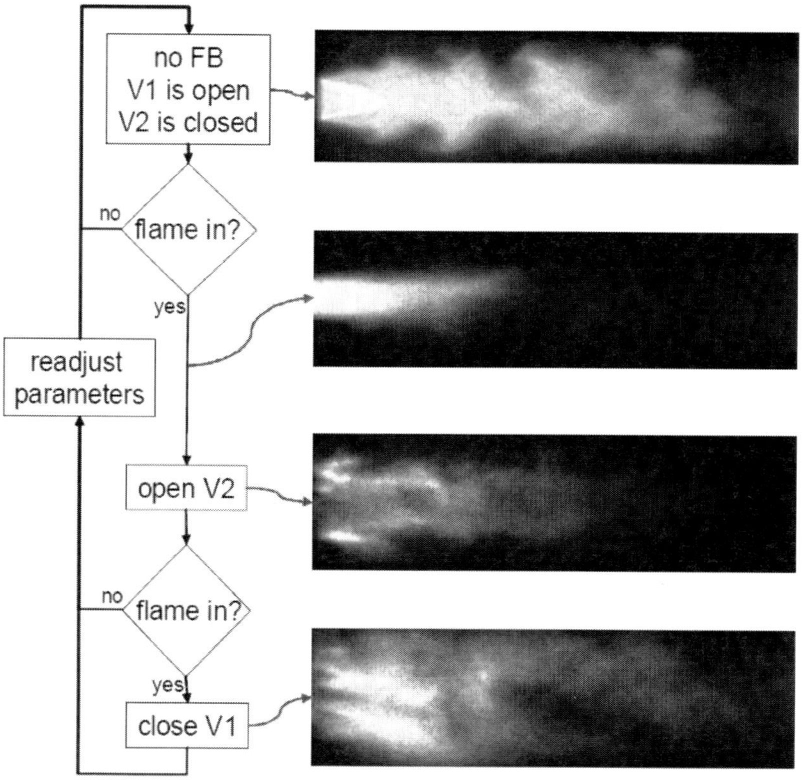

Figure 11.8. Active control system for flashback.

MEASURING TECHNIQUES

The main measurement campaigns operated within the work here summarized were dedicated to the flame front imaging, achieved by means of planar laser induced fluorescence of the OH radical (OH-PLIF). The data set also includes instantaneous turbulent flow field data, acquired with particle image velocimetry (PIV). These last data were collected in a former work [32].

Following the approach of earlier works [4,5], flame speeds derived from the OH-PLIF images are correlated with local turbulence properties extracted from the PIV experiments. Such properties are the turbulence intensity (u') and the integral length scale (L_T) at the centerline position of the flame front. Centerline values of the turbulent intensity are actually representative of all flame front positions, since measurements have shown that the average flame front surface closely corresponds to a u' isosurface [32].

The choice of local flame front rather than of inlet values for reference turbulence flow properties has the purpose of deriving parameters that are generally representative of the local flame/turbulence interactions, as discussed in [5] and described in the results section. The PIV database consists of non-reacting flow measurements, described in detail elsewhere [32].

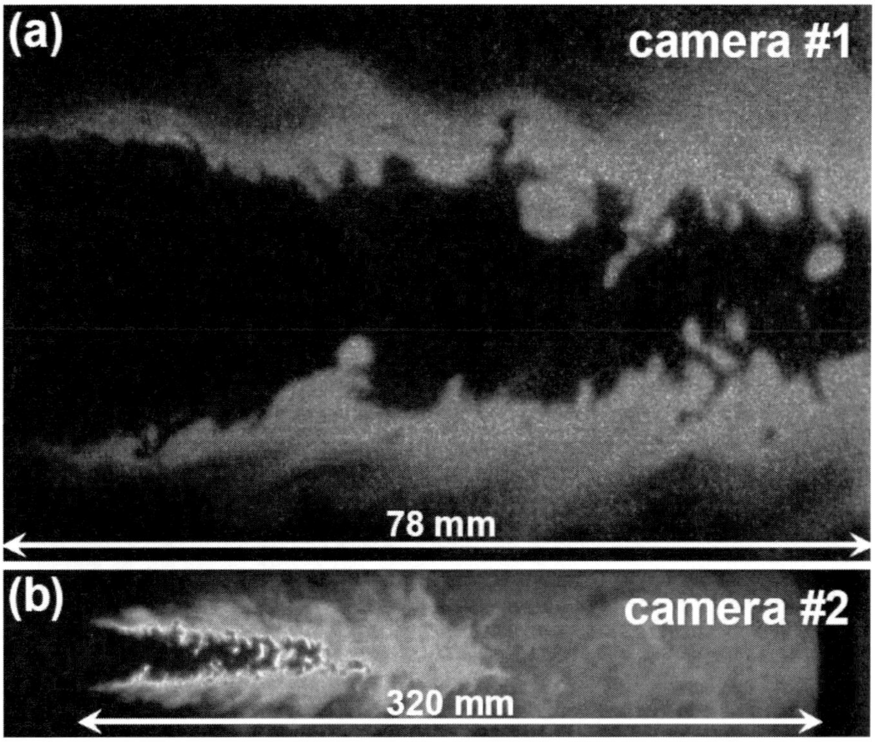

Figure 11.9. Flame imaging of the two cameras. (a) Camera #1, resolution: 0.061 mm/pixel; (b) Camera #2, resolution: 0.310 mm/pixel.

The OH-PLIF setup is depicted in Figure 11.5. An Nd:YAG / dye laser system produced a beam at 285 nm, which was converted to a light sheet by a cylindrical lens. The light sheet entered the combustor through the quartz window at the rear flange of the tank (see Figure 11.5). The fluorescence of both OH (1-1) and (0-0) transitions at 308 and 314 nm, respectively, was collected through a dielectric band pass filter centered at 310 nm. The laser sheet thickness was estimated to be ~0.1 mm, at least over the region imaged with Camera #1 (Figure 11.9.a).

The emitted fluorescence light was collected at 90° angle using two intensified CCD cameras equipped with a UV achromatic lens system and appropriate dielectric filters. The two intensified cameras were herein used for imaging different viewing areas. The first camera acquired high resolution images for detailed analysis of the flame front corrugation at small scales, while the second one recorded the entire flame front area for the ensuing global mass consumption analyses. The resolutions of the two cameras were 0.061 and 0.310 mm/pixel, respectively. The difference in the image sizes of the two cameras is depicted in Figure 11.9 for the same flame sample.

Test Matrix

The investigated fuel mixtures vary from pure CH_4 to pure H_2 (the latter diluted with N_2), including mixtures with different H_2-CO compositions, and two cases whereby CH_4 is co-fired with syngas. The fuel mixtures are presented in Table 11.1, according to the color coding used for the figures throughout the entire manuscript. These mixtures are representative of different scenarios relevant to IGCC power plants, accounting for different feedstocks and various gasification processes. The strategy of the fuel mixtures selection is justified by [33-36].

The produced experimental database comprises over 300 different flames. The data were acquired within the following ranges for the different combustion parameters:

-inlet velocity (U_o) : 40-150 m/s
-inlet temperature (T_o) : 423-773 K
-pressure (P) : 0.10-2.00 MPa

The resulting test matrix is relevant to gas turbine applications. All investigated conditions are positioned in Figure 11.10 on the Borghi diagram as modified by Peters [27]. The bulk of the data fall in the thin reaction zones regime, a major part characterized by Damkohler numbers (Da) < 1, with only a few data points in the broken reaction zones, characterized by Karlovitz number (Ka_δ) > 1. In Figure 11.10, l_m denotes the Zimont mixing scale [27, 31].

Figure 11.11. provides OH-PLIF single-shot images at different equivalence ratios, Φ. An elongation of the flame length is observed with decreasing Φ, while no differences in the flame morphology are evident when crossing $Da = 1$. The probabilities of a broken flame surface and the presence of burning isolated pockets increase with decreasing Φ. For $Ka_\delta > 1$ the reaction zone is broken; however, analysis of CO emissions at the exhaust confirms that combustion is still complete. Flame images for $Ka_\delta > 1$ have been processed with the same methodology, as it was always possible to define an average continuous reaction zone.

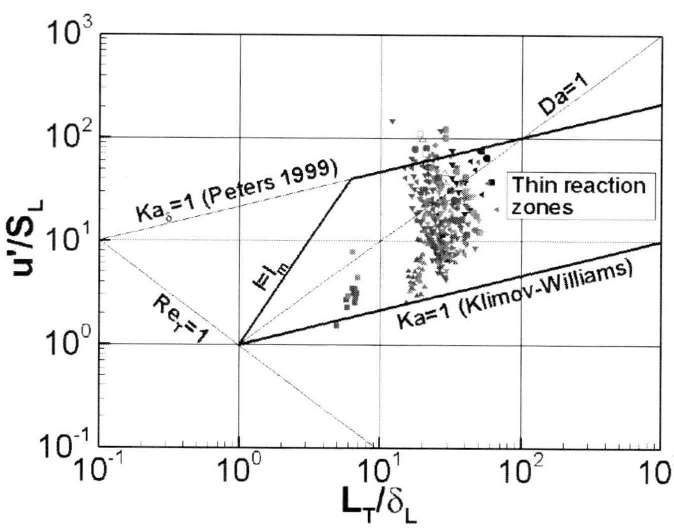

Note: Color coding according to Table 11.1.

Figure 11.10. Placement of measuring points on the Borghi-Peters diagram.

Table 11.1. Fuel mixtures used in the study

#	Mixture	Vol %	Process	Color coding
1	H_2-CO	33-67	coal gasification	green
2	H_2-CO	50-50	oil gasification	red
3	H_2-CO	67-33	partial CO_2 removal	blue
4	H_2-CO-N_2	40-40-20	biomass / air blown gasification	violet
5	H_2-N_2	70-30	CO_2 removal	light blue
6	H_2-CO-CH_4	20-20-60	natural gas / cofiring	black
7	H_2-CO-CH_4-N_2	10-10-60-20	natural gas / cofiring	gray
8	CH_4	100	natural gas	orange

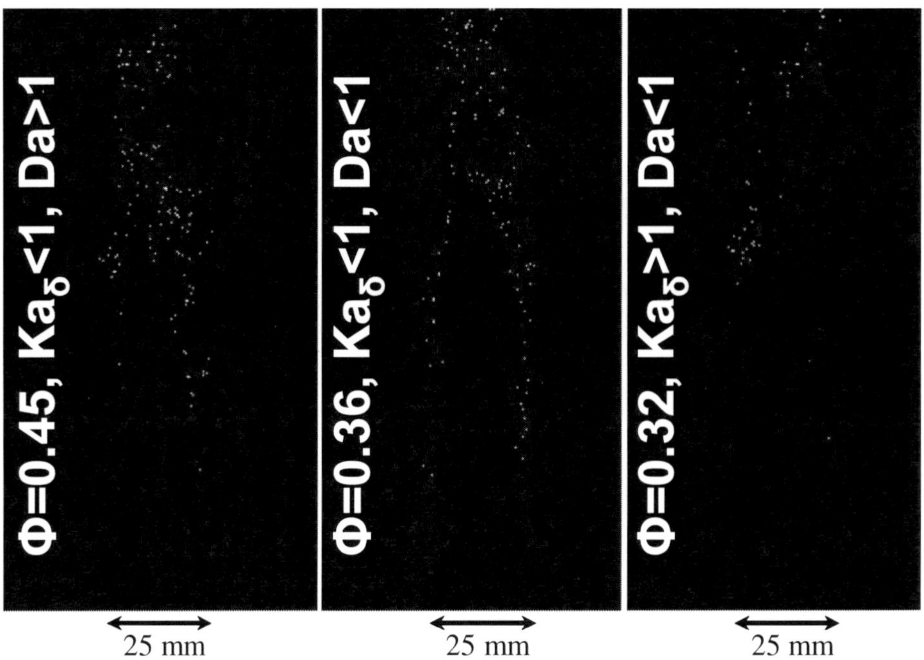

Figure 11.11. OH-PLIF images corresponding to three different regimes at P = 1.5 MPa.

RESULTS: OPERATIONAL WINDOW

Within this paragraph, the operational window in terms of lean blow out (LBO) and flashback (FB) limits is presented for different syngas mixtures and in comparison with CH_4. Along with the exhaust CO measurements, needed to analyze LBO phenomena, also NO_x emissions were collected and included in this chapter. A detailed analysis of NO_x emissions is however not the scope of this chapter.

LEAN BLOW OUT

Methodology

Detection of LBO limits was based on the OH-chemiluminescence signal recorded by a photomultiplier sampled at 1.2 kHz. The sensor was mounted aligned with the burner symmetry axis detecting the OH-chemiluminescence from the rear tank window shown in Figure 11.5 (sensor is not shown). The resulting signal is therefore proportional to the OH-chemiluminescence integrated over the whole flame.

Tests to determine LBO limits are carried out, starting from a stable flame at a fixed equivalence ratio and then slowly decreasing the fuel flow rate in steps, keeping the other parameters constant, until LBO occurs. The exhaust gas concentrations and the OH-chemiluminescence signal are simultaneously recorded. However, the temporal resolution of the exhaust gas concentration measurements is much smaller (1 Hz) than for the chemiluminescence signal. Additionally there is a certain time delay in the response of the exhaust gas analysis system to a concentration change because of the (high) gas volume in pipes, pump, filters etc. needs to be purged. At each step a certain period of time is allowed to elapse to verify whether the flame is stable or not (criteria: exhaust gas temperature and species concentrations are steady in time). Then, towards LBO, the flame sensing signal begins to fluctuate heavily and regions of local extinctions are observed (periods of zero intensity; see signal trace shown in Figure 11.12.a) (was also observed in [37]). Proximity to lean extinction corresponds to a rapid increase in the CO concentration indicating incomplete combustion due to (local) extinction events. The flame then enters a self exciting mode (increased amplitude) in which its position, shape and length change rapidly until LBO occurs. An example of the OH-chemiluminescence time series of a lean blowout event is shown in Figure 11.12.

The OH-chemiluminescence signal has been acquired with a photomultiplier using an interference filter. The sampling rate was 1200 Hz which led to a resolution of 600 Hz in the Fourier spectrum, a "time window" of one second was selected to analyze the signal (Figure 11.12.b).

The obtained signal was analyzed in both time and frequency domains. This analysis led to a definition of the lean blow out limit as the starting point of low frequency oscillations occurring as the flame approaches extinction. These oscillations are attributed to local extinction/re-ignition phenomena which can trigger pressure waves feeding into feedback loops. In some cases LBO could be detected not as a direct consequence of incomplete combustion (high CO emissions), but as a result of these coupling phenomena.

The LBO limit has been defined as the point in which the combustion process starts to show oscillations characterized by high amplitude and low frequency. These characteristics were already observed also in a previous work [38]. Approaching lean extinction, the local mechanism of extinction/re-ignition modulates the overall pressure field, resulting in a heat release fluctuation.

In Figure 11.12, a typical example of the OH-chemiluminescence signal acquired approaching LBO, is presented (a). In the bottom part of the same figure (b) a detailed time window of one-second is shown (b). The flame instability results in a self-excited oscillation with growing amplitude until the lean extinction is reached. Referring to the typical flame behavior shown in Figure 11.12, after about $t = 60$ s it would not be possible anymore to "recuperate" the flame via addition of fuel. Looking at the increasing amplitude in the signal, onset of LBO pulsations can be detected about 30 seconds before extinction (in these specific conditions). Based on this an automatic control system for LBO instability was designed and presented in a past work [38]. Around $t = 60$ s, as it can be seen in the detailed time-window in Figure 11.12.b the oscillations are characterized by a well defined frequency. This frequency can be analyzed with a Fourier transformation of the time signal. Examples of frequency spectra for several equivalence ratios are shown in Figure 11.13. The dominant frequency of 5 Hz, which can be identified in the peak of the black colored spectrum representative of the LBO condition, has been found independent of fuel composition; the same frequency can be seen in the previous Figure 11.12b counting the peaks (time window set to 1 s).

In Figure 11.13 the OH-chemiluminescence spectra of the coal gasification simulating mixture (H_2-CO / 33-67) at different equivalence ratio values are reported. These spectra were obtained as an average over 10 "single shot" spectra. The figure shows that by reducing the fuel, energy is stored in the low frequency range of the spectrum (below 100 Hz), as it has been observed also by Muruganandam et al. for natural gas fuel in a swirl-stabilized dump-combustor [39] and in past works in a bluff-body stabilized combustor [38,40]. In a previous work [40] chemiluminescence spectra and axial velocity fluctuation spectra (recorded by means of laser doppler anemometry technique) were found to be strongly correlated.

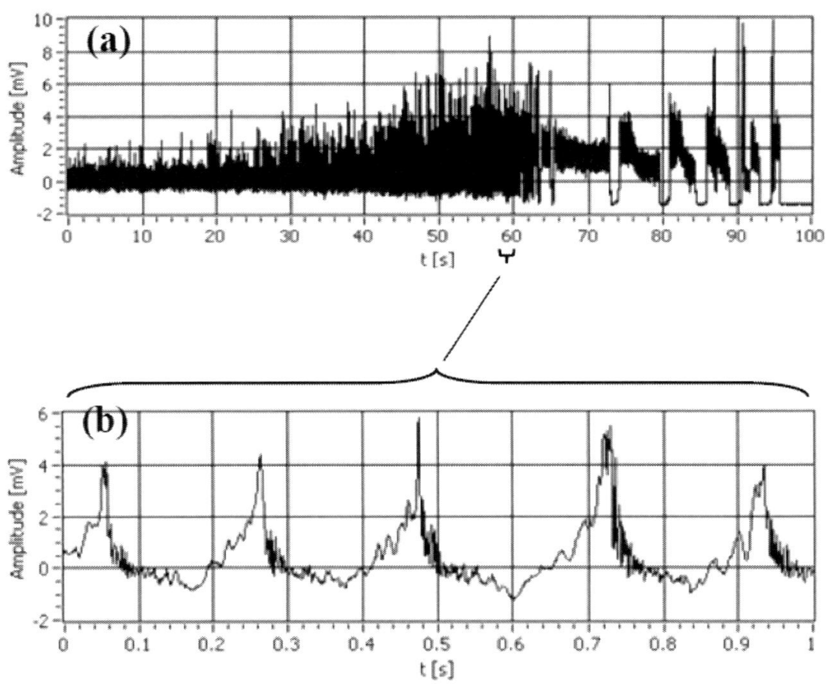

Note: $P = 0.50$ MPa, $U_o = 40$ m/s, $T_o = 673$ K.

Figure 11.12. Time series of the photomultiplier signal approaching LBO H_2-CO-CH_4 / 20-20-60 (a), and detailed time window (b).

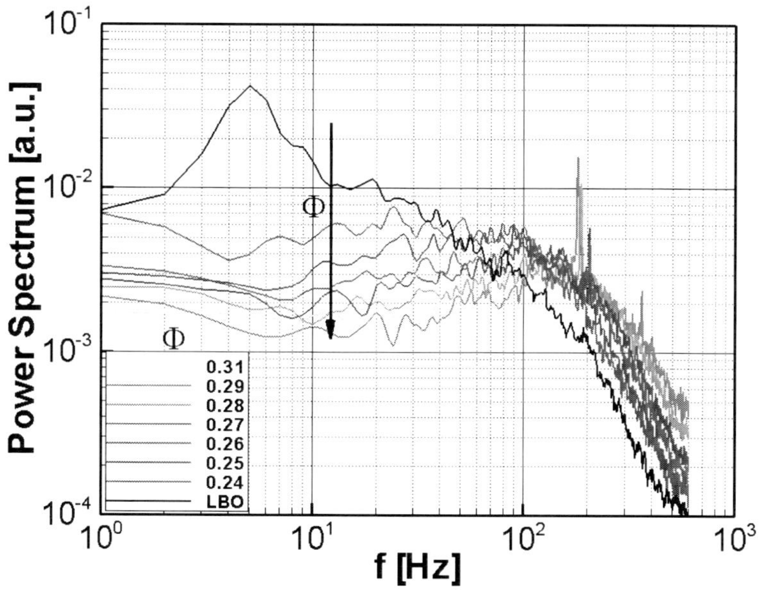

Note: $P = 0.50$ MPa, $U_o = 40$ m/s, $T_o = 673$ K. Φ in labels.

Figure 11.13. OH-chemiluminescence power spectra for H_2-CO / 33-67.

The mentioned work concluded that chemiluminescence radiation can be assumed as a passive scalar and used as a "natural" seeding to trace turbulent combustion dynamics. Taking into account this work, the energy increase at low frequency in the OH-chemiluminescence spectra can be seen as indicative of a corresponding energy increase in the "large scales" (low frequency range) of the turbulence field, occurring in the overall flame (considering that the sensor integrate the signal over the whole flame). The oscillations result in lean extinction (LBO) when the fluctuations of heat release (acquired with the photomultiplier), pressure field and turbulent flow field are coupled and exceed a certain amplitude.

LBO Limits and NOx Emissions

LBO Limits

In Figure 11.14, lean stability limits (LBO) are summarized in terms of equivalence ratio and pressure. As it is shown in Figure 11.14 the LBO limits were found to be almost pressure independent for the CH_4 containing mixture (black symbols). All other mixtures show slightly increased pressure dependence: the flame extinguishes at higher equivalence ratio with increasing the pressure. No significant dependence of the LBO limits on the inlet velocity was found for the three "pure" syngas mixtures and therefore diagrams for different velocities are not reported here (more detailed descriptions can be found in [41]).

In the Figure 11.15 and Figure 11.16, NO_x and CO emissions are presented. All the acquired points are representative of a steady-state conditions unless for the last and highest CO points, in which the flame is close to extinction.

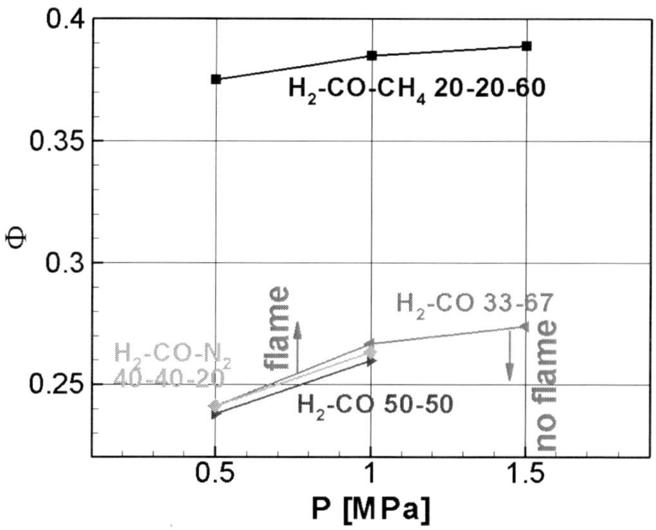

Figure 11.14. LBO limits at $U_o = 40$ m/s, $T_o = 673$ K.

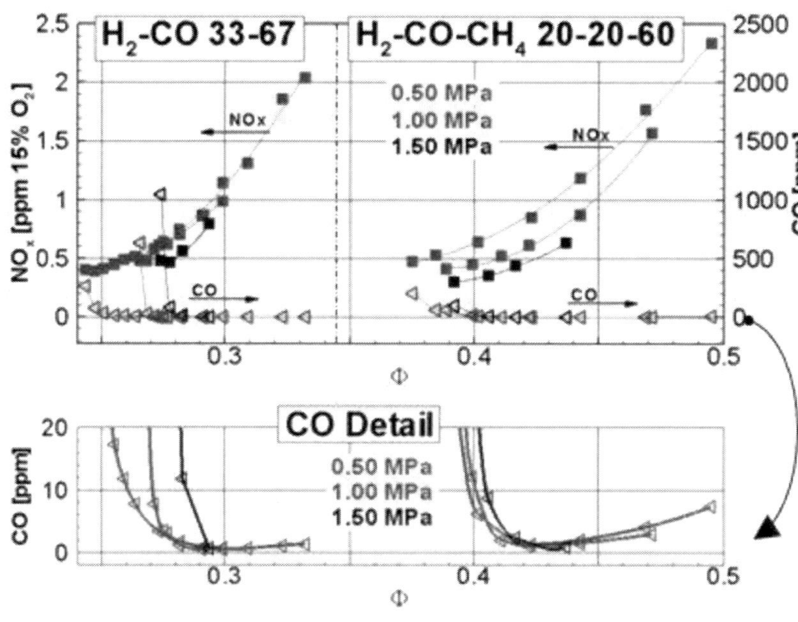

Figure 11.15. NO_x and CO emissions: effect of P on H_2-CO / 33-67 (left hand side) and H_2-CO-CH_4 (right hand side) at $U_o = 40$ m/s, $T_o = 673$ K.

Apart from extremely lean conditions (close to LBO), the measured CO emissions are always close to zero. Because of the low flame temperature at equivalence ratios close to LBO and a higher probability of local extinction events at these conditions the CO concentration increases while approaching LBO. For these low temperature conditions the residence time is then not long enough for complete oxidation of CO.

Figure 11.15 shows the dependence of the LBO limits on the pressure for the fuel mixtures H_2-CO / 33-67 (left hand side) and H_2-CO-CH_4 / 20-20-60 (right hand side). The presented data were taken at 0.50, 1.0 and 1.5 MPa, with preheating the mixture up to 673 K. For all "pure" syngas mixtures, the blow out takes place at higher equivalence ratios with increasing pressure and the CO emissions show a steeper gradient approaching LBO at high pressures. A different behavior is observed regarding the methane containing mixture (right hand side), where the LBO limit is almost at the same equivalence ratios for each pressure and the CO gradient approaching LBO changes only very little with increasing pressure. In the following an explanation concerning the different pressure behavior, based on the difference in CO gradients, is presented.

In Figure 11.15, a detail of the increasing CO emissions is shown. It is noted that for the syngas mixture (left hand side), incomplete combustion, characterized by non-zero CO emissions, starts for all the investigated pressures (0.5, 1.0 and 1.5 MPa) around almost the same equivalence ratio of 0.285. For leaner conditions, the steep increase in CO for higher pressures indicates extinction at less lean conditions. The phenomenology behind the CO gradient can be investigated with the help of the chemiluminescence spectra. As it has been shown in Figure 11.13, more energy is accumulated in the low frequency range approaching LBO which results in an oscillation characterized by a dominant frequency of 5 Hz. The energy increase can be interpreted as an evidence of the coupling between heat release oscillations (detected with the photomultiplier) and the fluctuations in the pressure field. The feedback between these two parameters is given by means of the acoustic characteristic (i.e. geometry) of the combustion chamber and the gain of this closed loop is proportional to the output power of the combustor.

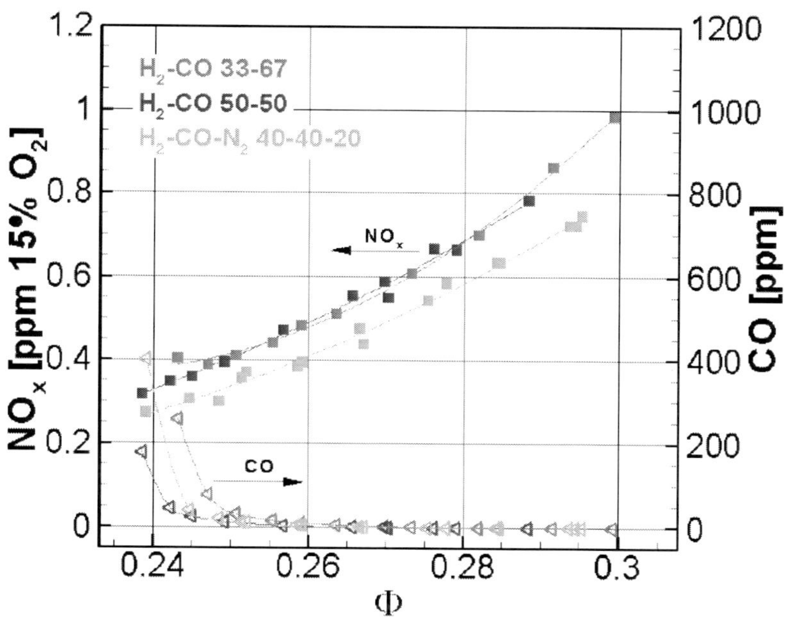

Note: $P = 0.50$ MPa, $U_o = 40$ m/s, $T_o = 673$ K.

Figure 11.16. LBO limits and NO_x mixtures: H_2-CO / 50-50, H_2-CO / 33-67, H_2-CO-N_2 / 40-40-20.

The increase of power with pressure thus results in a stronger amplification of the oscillations leading to quick extinction. This energy increase in the large scales has been measured by comparing the OH-chemiluminescence spectra at different pressures (not reported here).

The described behavior is also valid for the CH_4 containing mixture (Figure 11.15, right hand side). Though in this case, due to slower hydrocarbon oxidation chemistry even at low pressure (0.50 MPa) enough time for amplification to the thermoacoustic feedback loop is given. As can be seen in Figure 11.15 (right hand side) the CO increases in a similar fashion for all pressure levels investigated.

In Figure 11.16, a comparison between the three "pure" syngas mixtures is presented. Looking at the CO emissions (as usual the last and highest acquired value is representative of the LBO limit), the oil-syngas mixture (#2, Table 11.1) shows the most extended LBO limits, although the difference with the other syngas mixtures is not very pronounced. This might be attributed to the higher H_2 content in this fuel mixture which provides higher concentrations of OH radicals and a higher laminar flame speed leading to an enhancement of the flame stability [42].

NOx Emissions

NO_x emissions for a wide, representative set of measuring points are presented in Figure 11.17; to compare different mixtures, the results are plotted over the adiabatic flame temperature instead of the equivalence ratio. For sake of consistency with the previous diagrams, the latter is also displayed in the upper part of the figure. In the middle part the O_2 percentage measured in the exhaust is also displayed. The measured points belonging to the "pure" syngas mixtures are contained in a narrow band which shows an exponential trend (note the logarithmic scale on the NO_x axis), according to the characteristics of thermal NO_x production. The small difference between the three mixtures could be seen more clearly in Figure 11.16. The N_2-diluted mixture (simulating biomass gasification), characterized by lower adiabatic flame temperatures (the difference is on the order of 30 degree), exhibits lower NO_x, while undiluted coal and oil gasification syngas mixtures with slightly higher adiabatic flame temperature (for a given equivalence ratio) lead to the higher NO_x emissions.

Compared to the "pure" syngas mixtures, Figure 11.17 shows that the data belonging to the CH_4 containing fuel mixture (in black) do not fall in the same range but are characterized by lower NO_x emissions. This can be attributed to two different factors. First, the difference in the residence time between the flame front and the exhaust probe has to be taken in account. Figure 11.18 presents two OH-LIF (Laser Induced Fluorescence) images which help to visualize the distances (d_1, d_2) leading to the different residence times in the hot exhaust flow up to the sampling point.

In Figure 11.18, two flames (H_2-CO-CH_4 / 20-20-60 and H_2-CO / 50-50) are compared at the same adiabatic flame temperature of about 1700 K. It can be estimated, looking at the different flame lengths, that the mentioned residence time is more than doubled in the case of the H_2-CO / 50-50 mixture, considering that the exhaust probe is placed at the very end of the chamber (e.g. right edge of the pictures). The difference in the residence time is estimated to be on the order of 10 ms for the given flames (corresponding to a distance d_2-d_1 of about 100 - 120 mm; Figure 11.18.).

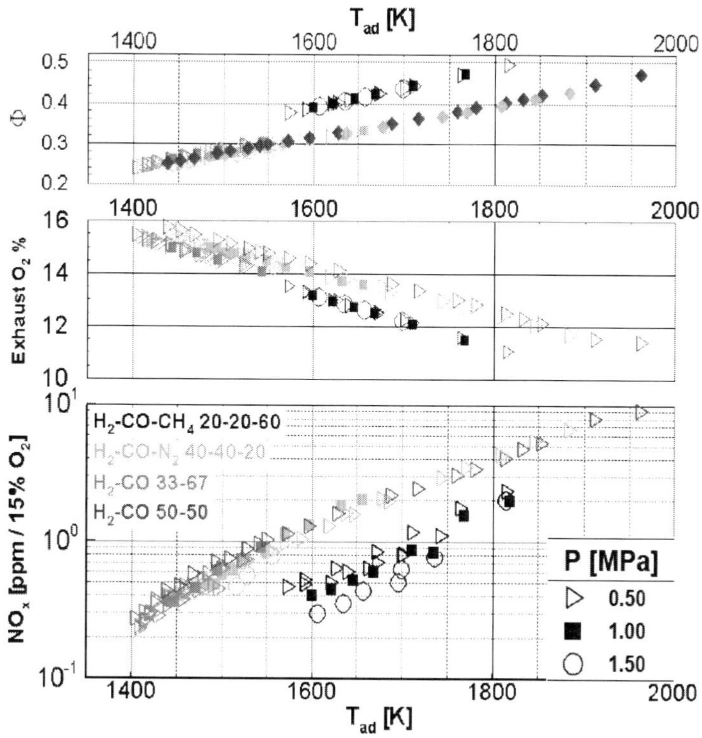

Figure 11.17. NO_x emissions.

According to chemical kinetic calculations of thermal NO_x formation this leads to a difference in the NO_x values on the order of 0.1 ppm (at 1700 K), which partially justifies the gap in the emission data in Figure 11.17.

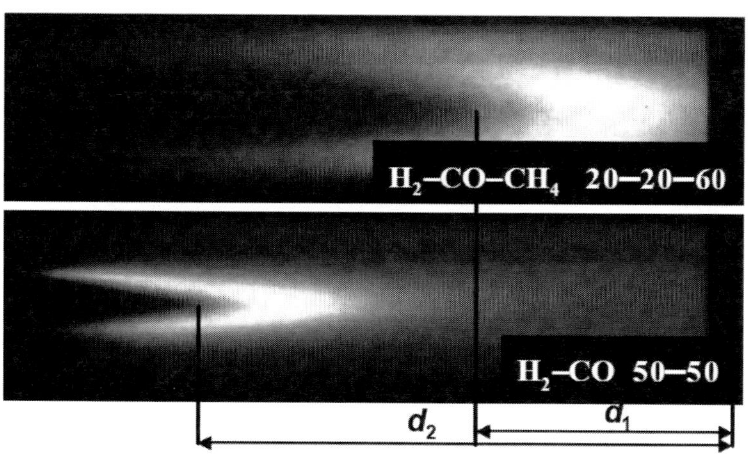

Note: T_o = 1700 K, P = 0.50 MPa, U_o = 40 m/s, T_o = 673 K.

Figure 11.18. OH-PLIF images: H_2-CO-CH_4 / 20-20-60 and H_2-CO / 50-50.

The other effect to consider is the dependence of the "thermal NO" formation rate on the oxygen (O_2) concentration in the exhaust gas as described by [43]:

$$\frac{d(NO)}{dt} \propto (O_2)_{eq}^{0.5} \cdot (N_2)_{eq} \qquad (11.3)$$

Comparing the CH_4 containing mixture with pure syngas mixtures, it has to be taken into account that for a fixed adiabatic flame temperature, the percentage of oxygen available after the flame front is considerably different (Figure 11.17.b). E.g. at an adiabatic flame temperature of 1700 K, the oxygen percentage is between 12 Vol.% and 13.5 Vol.% for co-firing syngas and for pure syngas, respectively. This difference results in a difference in the NO formation rate of about 6%, according to equation (11.3), which justifies a NO_x increase again on the order of 0.1 ppm.

Only for the methane containing fuel gas mixture an effect of operating pressure on the NO_x emissions can be discerned. Higher operating pressure leads to lower NO_x emissions indicating a reduced peak concentration of radicals promoting NO_x formation (this can be due to faster recombination reactions at elevated pressure conditions). Apparently due to the different chemical kinetic pathways for hydrocarbon oxidation, only with the presence of hydrocarbons, pressure sensitive elemental reaction steps become important which influence NO_x formation via concentrations of O/H and CH_i radicals. This effect is obviously mostly limited to the main reaction zone (flame front) as the slope of NO_x emission data with respect to adiabatic flame temperature is largely maintained for all pressure conditions (indicating the general importance of thermal NO_x formation in the hot exhaust gas at all operating conditions and for all fuel mixtures).

FLASHBACK

Methodology

The measurements presented in this section had the purpose of checking the feasibility of firing syngas in the described experimental facility (Figure 11.5), which was conventionally fired with CH_4. A removal of all the potentially flame holders from the mixing section (perforated plates, static mixers) was needed. Then measurements were performed to obtain the operational window.

Several flashbacks were provoked by keeping constant inlet temperature and pressure and increasing in little steps the equivalence ratio until flashback happened. The occurrence of flashback was determined detecting the chemiluminescence signal captured in the mixing section by a photomultiplier equipped with an optical fiber (as afore detailed when talking about the control system in the previous section).

For some conditions the same experiment has been performed many times in order to check its repeatability. The results are shown in Figure 11.19. This particular case corresponds to a pressure of 0.50 MPa and preheating temperature of 673 K; the Φ_{FB} is 0.56±0.02. This result shows a reasonable repeatability of the flashback event.

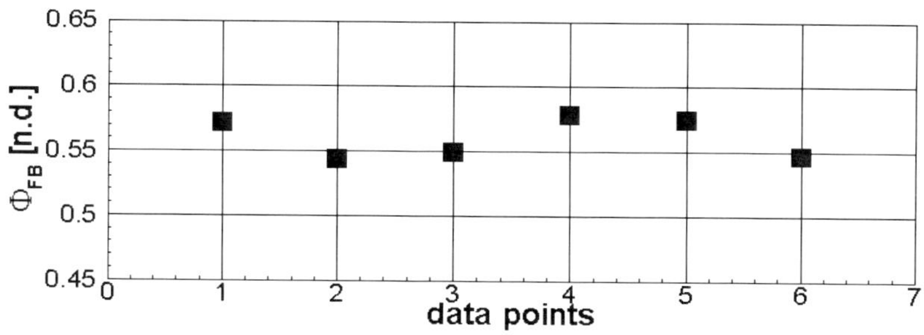

Note: $P = 0.50$ MPa, $T_o = 673$ K, $\underline{U}_o = 45$ m/s.

Figure 11.19. Variation of the measured FB propensity.

It should be mentioned, for the sake of completeness, that while operating the rig some flashbacks happened as consequence of various instabilities (deriving from coupling between mass flow rates and pressure throttle valve regulator) these data points are not repeatable, in a consistent way, and therefore not reported.

Based on the observation of the flame the authors believe that for the experimental facility used in this study the mechanism for flashback is the propagation of the flame front in the boundary layer of the mixing section.

The non-swirled flow conditions do not produce instabilities in the bulk flow and consequently no back propagation of the flame due to CIVB should be possible. Furthermore propagation through the core flow region is also improbable; the flow velocity in this region is higher then the bulk velocity [32], and the S_T (global consumption), measured near FB conditions (as reported in [4] and shown here following), is only on the order of 18% of the bulk velocity.

Evidence of the boundary layer flashback is also given by Figure 11.20 which shows a flame close to FB (left hand side) and a flame which after flashback (right hand side) is anchored at the slot used for the N_2 injection depicted in Figure 11.6. The images were acquired by means of OH planar laser induced fluorescence technique. The absence of any OH signal in the core of the flame for both cases demonstrates that the flame does not advance into the mixing section in the central part of the flow. Moreover if the propagation had happened through the core, most likely the flame would not stabilize on the slot but proceed further anchoring at the fuel injector (placed 400 mm upstream). In this case no flame front could be detected in the combustion chamber.

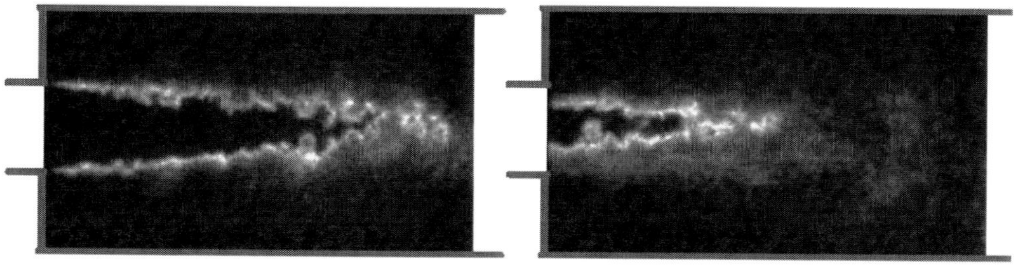

Figure 11.20. OH-PLIF: Flame fronts. Flame close to FB (left) and FB (right).

FB Limits and Dependence on the Combustion Parameters

FB Limits

Figure 11.21 reports the operational window in terms of Φ_{FB} versus pressure. Lean blow out limits are also reported for completeness.

Results are shown for different inlet temperatures and different inlet velocities. The figure depicts how the operational window drastically narrows with increasing the pressure and/or the inlet temperature (no margin left for $P > 1.1$ MPa).

As the hypothesis of flashback through the core region was rejected, no analysis by means of the Peclet number correlations was performed like it was done by other authors [11]. As it is explained in [11], this kind of analysis provides good insights for core flow flashbacks, because it doesn't account for boundary layer conditions. In the same study, which also deals with boundary layer flashback, this analysis was applied demonstrating its inefficiency in this case.

The propagation of the flame front through the boundary layer is dependent on the possibility of the flame to propagate along a surface (quenching distance) and on a local balance between the flow velocity of the incoming fresh air/fuel mixture in the boundary layer and the local flame speed. Both velocities vary upon changes of pressure, inlet velocity and inlet temperature.

The dependence of Φ_{FB} on these parameters (P, T_o, U_o) is analyzed in the following sections.

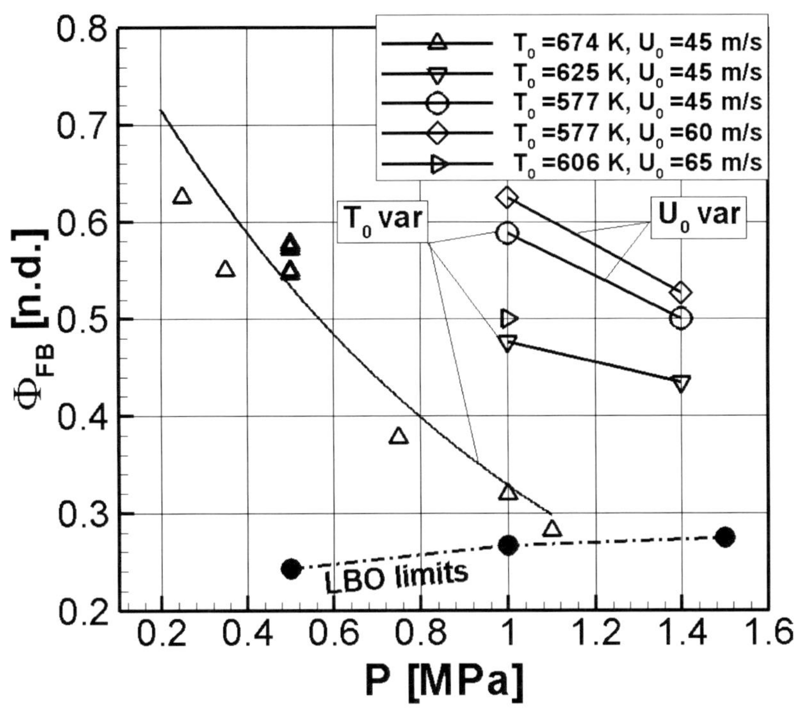

Note: LBO (filled symbols) and FB (non-filled symbols) limits.

Figure 11.21. Operational window.

Inlet Velocity Dependence

In Figure 11.21 data for different inlet velocities (circles, and diamonds; taken for 1.0 MPa and 1.4 MPa) highlight that variations of this parameter do not have a strong impact on the flashback propensity. On first sight increasing of the inlet velocity could be seen as an approach to avoid FB, i.e. trying to blow the flame further away from the burner lips. This strategy shows to have little effect. The minor difference in the FB limits is accompanied by a low sensibility of the flame front position on the inlet velocity (as highlighted in [4] and depicted in the following Figure 11.30). This is explained by the increase of the Reynolds number at higher velocities, i.e. different turbulent characteristics, which enhance the overall flame surface by adding wrinkling. This enhancement leads to a higher fuel consumption rate, S_T. As result of this effect, the flashback propensity stays almost unchanged. The balance between the turbulent flame speed and U_o, expressed as $S_T^{0.05}/U_o$, undergoes a very little variation as depicted by Figure 11.22, where the ratio $S_T^{0.05}/U_o$ is plotted versus U_o for inlet velocities up to 150 m/s ($S_T^{0.05}$ is the turbulent flame speed evaluated at a progress variable of $\langle c \rangle = 0.05$; the procedure will be detailed in the forthcoming section). Increase of U_o of almost a factor of 4 yields to a decrease of $S_T^{0.05}/U_o$ of about 5% only. This decrease is probably due to stretching effects and consequent local extinctions increasing with the increase in the *Re* number.

The magnitude of variation in Φ_{FB} in Figure 11.21 of the curves representative for the U_o variation (circles for 45 m/s and diamonds for 60 m/s) is in very good agreement with the magnitude of the variation of $S_T^{0.05}/U_o$ for the same inlet velocities which can be estimated from Figure 11.22. The good matching of these two values leads to the conclusion that the slight reduced flashback propensity for higher inlet velocities must be totally attributed to the growth of the importance of stretching effects and their influence on the turbulent flame speed.

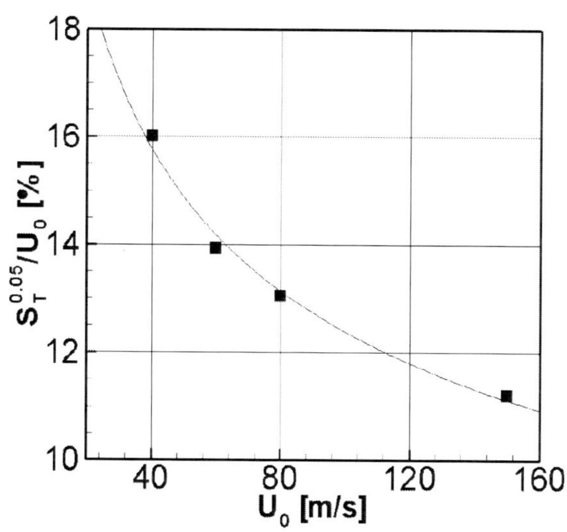

Figure 11.22. Variation of for different U_o at $P = 0.50$ MPa, $\Phi = 0.42$.

Inlet Temperature Dependence

It was found that FB propensity is significantly mitigated by reducing the inlet temperature. The variation of the inlet temperature at constant pressure has an impact on the reactivity of the mixture, summarized by the laminar flame speed, ($S_{L,o}$, in this work computed using the 1-D freely propagating flame package from Cantera [44] with the GRI 3.0 chemical reaction mechanism [45]) and on the turbulence conditions due to the correspondent variation in the Reynolds number. For the temperature variation described in this section (573 - 673 K), changes in the turbulence conditions have been neglected (*Re* varies of about 20%).

The left hand side of Figure 11.23 reports the same data points already presented in Figure 11.21 for 1.0 MPa and different inlet temperatures. In the current diagram data are plotted in the form of $1/\Phi_{FB}$ versus T_o. As highlighted in this figure flashback propensity increases by rising the inlet temperature with a power exponent around 4 (solid line is the power-fit).

In order to correlate the flashback propensity with the $S_{L,o}$, the temperature exponents, x, of the law $S_{L,o}$~T_o^x have been calculated for a pressure of 1.0 MPa and for different values of Φ. The results are reported on the right hand side of Figure 11.23 with filled symbols; the linear interpolation of those results is represented by the solid line. As shown here the dependence of $S_{L,o}$ on T_o varies upon the equivalence ratio showing that lean flames are more sensible to inlet temperature variations. This observation suggests that using an inlet temperature reduction strategy against flashback is a favorable option for syngases. For these gas mixtures the design operative condition is leaner then for hydrocarbon flames; this is clear when comparing the adiabatic temperature for CH_4 and syngas (see figure Figure 11.17).

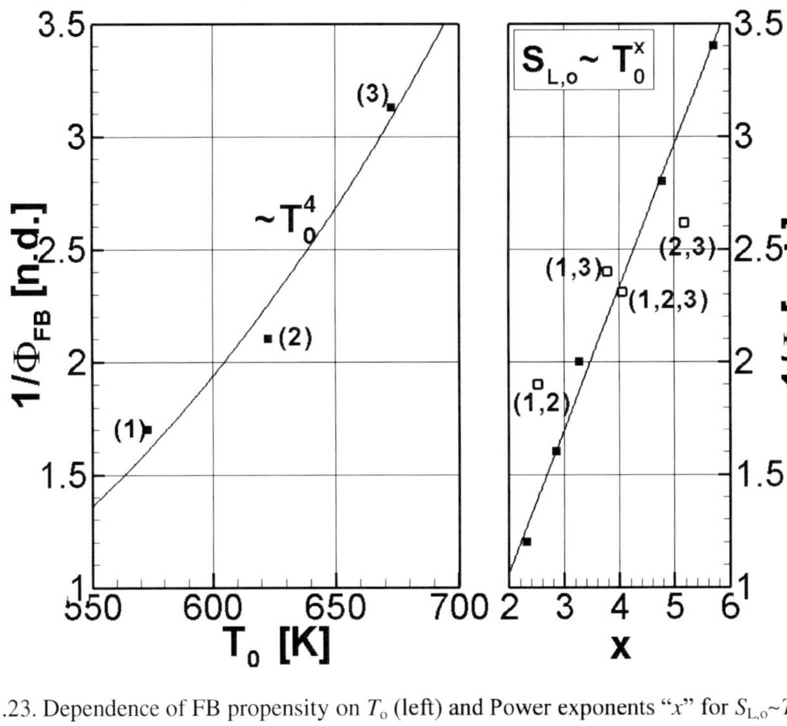

Figure 11.23. Dependence of FB propensity on T_o (left) and Power exponents "x" for $S_{L,o}$~T_o^x at different Φ (right).

The non filled symbols in the same figure represent the power exponent of the law $1/\Phi_{FB}{\sim}T_o^x$ calculated from the points on the left hand side diagram. For example "(1,2,3)" represents the power exponent calculated by using the points "(1)","(2)" and "(3)".

The average exponent "(1,2,3)" of $1/\Phi_{FB}{\sim}T_o^x$ is in very good agreement with the trend of the exponents of $S_{L,o}{\sim}T_o^x$ (solid line on right hand side), the other exponents show a small deviation. Unfortunately there are no more data points available for a more accurate analysis.

The agreement leads to the conclusion that $S_{L,o}$ is the main parameter influencing FB propensity when varying the T_o.

Pressure Dependence

As it was observed for the lean blow out limits in also the flashback limits show a dependence on the pressure. An analysis of flashback propensity in relation with pressure variations includes considerations on several parameters and effects. The major effects of an increase in pressure can be summarized as follows, and then analyzed separately:

- the velocity profile in the fuel/air mixing pipe changes due to the increase of *Re* number. This change yields to a steeper velocity gradient at the wall [46] which contributes in a reduction of FB propensity.
- the quenching distance decreases opening to the flame flashback pathways characterized by lower flow velocities, as they are closer to the walls [43].
- the increase in pressure has different effect on the S_T depending on the Φ. For equivalence ratios close to flashback S_T has been observed to rise.
- the fluctuations in the combustion chamber increase as well as they can be understood as proportional to the output power of the combustor

Due to the many unknown and unmeasured variables and due to the scarcity of data collected, herewith only a phenomenological analysis is proposed.

Power exponents of the law $\Phi_{FB}{\sim}P^n$ have been derived for the data presented in Figure 11.21 in general this exponent can be approximated with $n = -0.5$ (exact exponents are: $n = -0.51$, -0.48, -0.44, -0.27; due to the big scattering an exact average of these coefficients has no sense: due to scarcity of measurement points at certain conditions this analysis has no quantitative aims).

Concerning the velocity profile in the mixing pipe, no measurements or simulations are available yet for the experimental facility. For fully developed turbulent flows the distance from the wall "y" at which the local velocity of the flow matches the turbulent flame speed ($u_{local} = S_T$) scales like $y{\sim}P^{-0.2}$, as estimated from Kays (pp. 248-249) [46]. It should be remembered that the assumption of undisturbed boundary layer is in general not correct due to the peculiar geometry (see Figure 11.6).

Concerning the quenching distance, its dependence on the pressure obeys to $d_q{\sim}P^{-1}$ [43]. For different fuels this dependence varies, for example using the data of a recent work conducted on pure H_2 [11], the estimated d_q behaves like $P^{-1.6}$. Slight different values can be calculated, for other fuels, from other data available in the literature [47].

The evaluation of the effect of pressure on the turbulent flame speed reveals the combined effect of $S_{L,o}$ and flow conditions at the flame front. The influence of S_T was evaluated for an equivalence ratio close to flashback ($\Phi = 0.42$) for the given conditions. The dependence is weak; the relative power exponent calculated with a best power fit (not shown) is $n = 0.12$. Considering the dependence of S_T on Φ to be $S_T \sim \Phi^2$ (as well not shown), the final contribution is $\Phi_{FB} \sim P^{0.06}$, which can be neglected.

Summarizing FB propensity is enhanced by the quenching distance behavior $d_q \sim P^{-1}$. This is counteracted by the increase of the velocity at the wall, which leads the streamline for which $u = S_T$ to move away from the wall with law $P^{-0.2}$. The combination of these effects gives a global dependence of $P^{0.8}$. This global dependence alone does not balance the behavior in Figure 11.21 $\Phi_{FB} \sim P^{-0.5}$.

It is therefore understood that this simple analysis is not sufficient to explain this dependence. For this reason, and also with other purposes, it was decided to investigate the phenomenon by means of large eddy simulations. The numerical analysis is conducted in collaboration with ENEA-Casaccia (in Italy) by Giacomazzi et al. Preliminary results are available in [48].

RESULTS: FLAME SPEED

In this section flame speed data are collected and described. After having detailed the data analysis, turbulent flame speed values will be first introduced as absolute values. Following, behaviors of the normalized turbulent flame speed and of the (mean) flamelets speed (laminar stretched flame speed) will be discussed. The database used for these analyses includes over 300 different flame conditions.

DATA REDUCTION

The following paragraphs illustrate the procedures employed to analyze the flame speed. Within this chapter, the turbulent flame speed assumes the meaning of consumption per unit time and it was evaluated with two different approaches both based on the conservation of the mass between the combustor inlet section surface and a flame front surface.

In order to evaluate the mass balance two sets of information are needed: the knowledge of the flame front surface area and of the flame front topology. The evaluation of these two sets of information is based on analysis applied on the OH-PLIF raw images as described in the following paragraphs.

Mass Balance Equations

Turbulent flame speed values have been evaluated by applying global consumption analyses. The methodology is clarified with the schematic in Figure 11.24, depicting the inlet section and the flame front (due to axisymmetry, only half of the domain is shown).

Flame speeds were evaluated by applying continuity analysis and considering that mass throughput is conserved through the following two surfaces in Figure 11.24: inlet section (A_o) and average flame front surface (A_{av}).

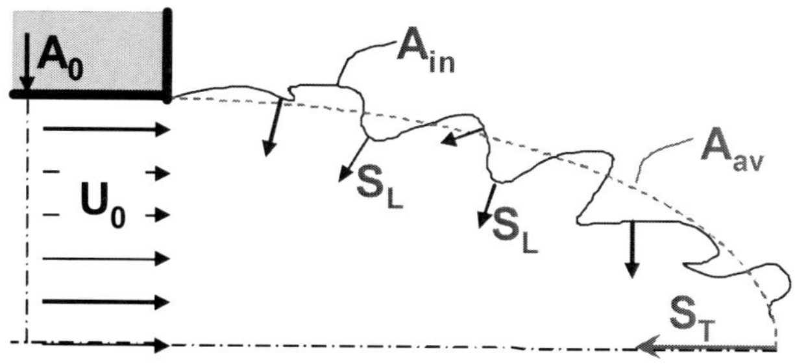

Figure 11.24. Schematic of mass balance across the flame: A_{in} is the instantaneous flame front area and A_{av} the average flame front area.

The flow velocities normal to these two surfaces are the inlet bulk velocity (U_o) and the negative of the turbulent flame speed ($-S_T$), respectively. The mass balance equation 11.4 can be recast to equation 11.5 (assuming the density of the fresh gas mixture at the inlet of the combustor and just upstream of the flame front is maintained [31]) so as to derive absolute values for S_T and the normalized flame speed S_T/S_L, respectively.

$$\rho_o \cdot A_o \cdot U_o = \rho_o \cdot A_{av} \cdot S_T \quad (11.4)$$

$$S_T = \frac{A_o}{A_{av}} \cdot U_o \quad (11.5)$$

Flame Front Detection

In order to measure and derive parameters such as the turbulent burning velocity, flame front position, flame brush thickness etc., experimental campaigns using planar laser induced fluorescence of OH radicals (OH-PLIF) were conducted (setup as described in experimental section). To ensure a good statistic, for each measured operative condition, 400 images (instantaneous shots) were acquired (at about 20 Hz) with the camera #2 as depicted in

Figure 11.5. An example of an instantaneous shot is depicted in Figure 11.9.b.

The data post processing was done applying an already developed methodology [49] which has been further improved and tuned on ultra-lean syngas flames.

Key issue is the determination of the flame front from the raw OH-PLIF image. The value of each single pixel of the picture is directly proportional to the fluorescence emission described by the following relation:

$$I_{OUT} = I_{IN} \cdot c_{dye} \cdot \varepsilon(\lambda) \cdot \phi(\lambda, T) \quad (11.6)$$

It is well known and clear from this formula [50] that LIF measures either concentration distribution or temperature fields assuming that the other of these two parameters is constant. In the case of a flame both temperature and dye concentration are varying in the entire physical domain. The seeding dyes (OH radicals) are produced in the flame front region, heated up, cooled down downstream and then at the end destroyed. It has not been attempted to derive quantitative information on OH radical concentration and/or local flame temperature. Despite of this deficiency, indeed, the position of the flame front in the plane of the laser sheet can be identified considering that the two effects of interest happening in the flame front, temperature raise and dye concentration increase, give both a positive contribution to the detected fluorescence intensity. Considering this it is now clear that the flame front should be placed where the steepest gradient of the acquired OH signal is obtained; this assumption bears the flame front directly to a local property (gradient) instead of relating it to global quantities such as for example the maximum intensity of the picture. The modifications applied to the previous flame front detection procedure are exactly based on this last observation. Due to the peculiarity of the pictures it was impossible to apply standard edge detection algorithm based on gradient-based threshold methodologies successfully. For this reason, a threshold was applied to the raw data. With defining a threshold the issue of selecting an appropriate normalization factor arises. The first idea, the simplest, is to normalize by the maximum pixel value of the average picture built up from the 400 instantaneous shot ones. This methodology gives good results when the image has good quality (high signal intensity) and a not too much difference in distribution of intensity between root, tip of the flame and post flame front zone; ideal conditions are non ultra-lean flames and low pressure (Figure 11.25 a_1 - a_2). In the case of syngas flames the condition of interest for gas turbine combustion are often ultra-lean, considering also that for a typical preheating temperature of 673 K, an equivalence ratio of 0.4 yields an adiabatic flame temperature of about 1800 K (see Figure 11.17) which is already within the upper limits for a stationary engine. With raising the pressure the image intensity decreases. This is partially attributed to the drop in concentration of OH radicals for fixed Φ according to Le Chatelier principle (the system counteracts the increase in pressure reducing the radical concentration; e.g. promoting third body reactions). Another reason for having weaker signal at high pressure has spectroscopic reasons: the OH peak decreases and broadens with the increase of this parameter. The problem now arising with a threshold proportional to the maximum intensity is the risk to lose trace of the combustion wave in the flame root region, characterized especially by weak signal levels in the case of ultra-lean and/or high pressure conditions as shown in Figure 11.25 b_1 - b_2.

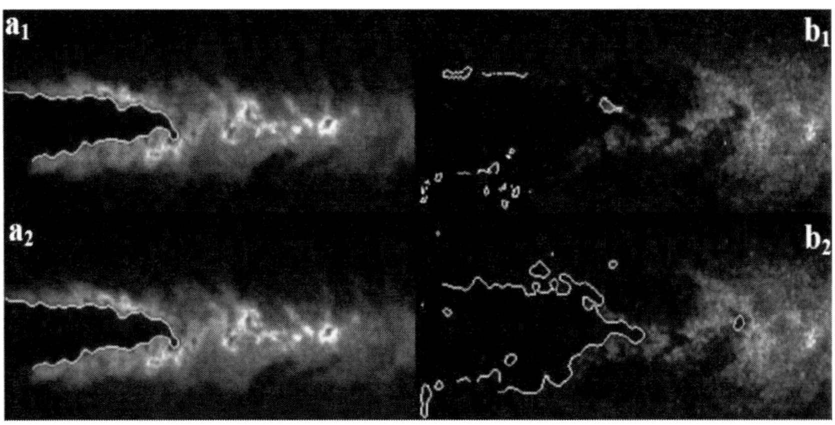

Figure 11.25. Good quality picture "a" (P = 0.50 MPa, Φ = 0.36) and bad quality picture "b" (P = 1.0 MPa, Φ = 0.29).

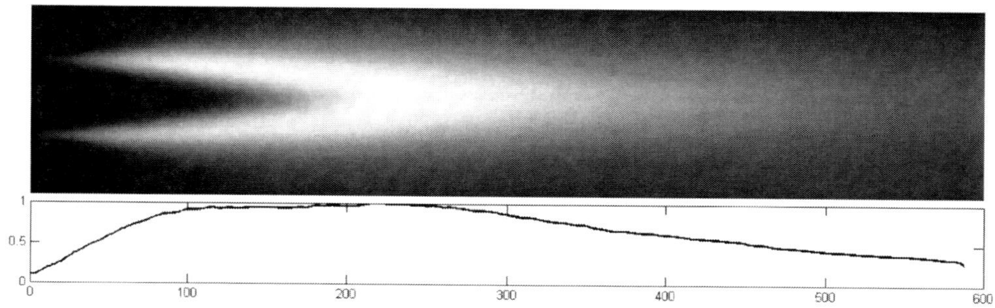

Figure 11.26. Geometrically non uniform threshold concept.

In order to avoid/reduce this risk a geometrically non uniform threshold has been defined as it is depicted in Figure 11.26. The threshold value, for every single pixel, is proportional to the maximum intensity of its respective column (axial distance from combustor inlet) in the average image (Figure 11.26).

To clarify the effect of this threshold method in Figure 11.25 are reported two flames representative of mixture #2 (Table 11.1): a) $P = 0.50$ MPa, $\Phi = 0.36$, b) $P = 1.0$ MPa, $\Phi = 0.29$. The white lines in the pictures are representative of the flame front. The pictures a_1, b_1 have been processed assuming a uniform threshold based on the absolute maximum. The pictures a_2 and b_2 were obtained with the geometrically non uniform threshold concept. As it can be seen, this method shows good detection characteristics of the flame front even at high pressure, ultra-lean mixture conditions Figure 11.25 b_2, while the first method fails (b_1). No relevant differences are present for the high quality pictures, as shown in Figure 11.25 a_1-a_2. The good detection is achieved because the flame front is assigned to the steepest gradient of the signal, which has not strong relation with the absolute maximum intensity, but is related to local characteristics. As it can be seen in Figure 11.25, the maximum signal intensity is located somewhere further downstream of the flame front.

The average flame front surface is worked out starting from the 400 instantaneous flame front planar contours using the progress variable ($<c>$) approach. Iso-contours of $<c> = 0.05$ were selected to represent the combustion wave, where $<c> = 0$ and $<c> = 1$ stand for the fresh mixture and completely burned gas respectively. The three-dimensional flame front surface is obtained by rotating the iso-contour line about the x-axis assuming axis-symmetric flames which is the case here.

Absolute Values of Turbulent Flame Speed

Absolute values of turbulent flame speed are presented in this paragraph in a first section where their dependence on the combustion parameters will be highlighted, and in a second section where the relation between S_T and the flame front position will be demonstrated.

The first section targets mainly engineering concerns providing direct comparisons of S_T for different fuels and hinting to the importance of the coupling between stretch and PDT effects for H_2-containing lean fuel mixtures which is already detectable in the absolute values of S_T and will be fully addressed in the forthcoming paragraphs.

The second section, describing the dependence of S_T on the combustor geometry (and hence on the turbulence characteristics), provides a link between the absolute values and the normalized values of S_T and shall be also understood as a preamble for the forthcoming paragraph. In this section, it is in fact explained how S_T values are correlated with "local" values of turbulence intensity. It is moreover highlighted what are the consequences of picking "local" values and its advantages compared with what is commonly done in the literature, where conventionally burner-exit center-section values are used.

Dependencies on the Combustion Parameters

The flame stabilizes in the combustor via the sudden expansion geometry (see Figure 11.5). It can be stated in general that the position of the flame front shows appreciable dependence on the mixture composition, equivalence ratio and air-fuel preheating temperature; pressure and inlet velocity variations affect this parameter in a weaker way. It is universally valid that the flame moves closer to the combustor inlet for: higher equivalence ratio, higher preheating temperature and higher H_2 percentage in the fuel mixture as it was already observed in a previous work conducted in this Lab. [37] and also by other authors [51, 52].

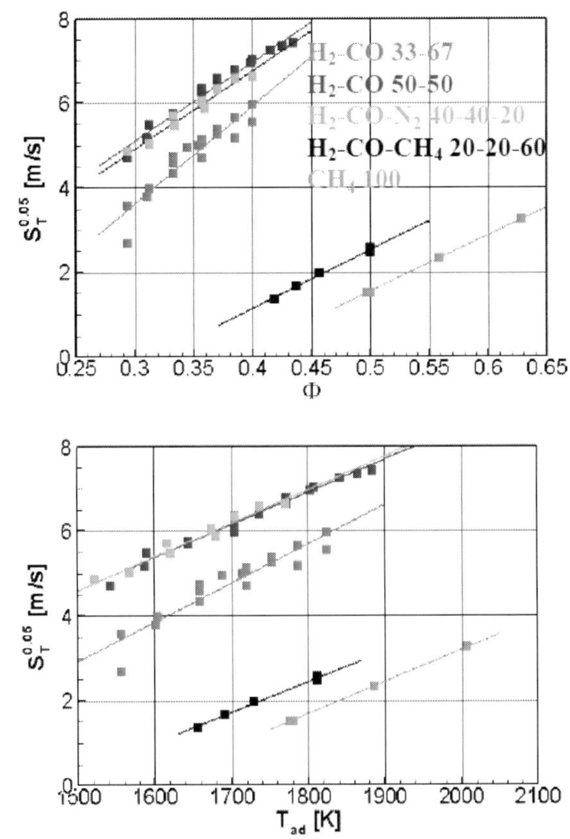

Note: $P = 0.50$ MPa, $T_o = 673$ K. Color coding according to Table 11.1.

Figure 11.27. versus Φ (a) and T_{ad} (b) for all investigated mixtures.

Turbulent burning velocity data versus the equivalence ratio for all the investigated mixtures are given in Figure 11.27a. As it was expected the three pure syngas mixtures presented in the figure show higher values of than the CH_4 co-fired syngas (black symbols). Pure methane has been included as reference (in orange). Increase of the H_2 content in the mixture results in higher burning velocity. This was already observed in a previous work by Boschek et al. [51], where, analyzing mixtures of CH_4 doped with H_2 it was also highlighted that the increase in turbulent flame speed due to the H_2 addition cannot be totally attributed to the higher $S_{L,o}$, but must be credited also to other factors like preferential diffusion and stretch effects. Hawkes and Chen [53] also looked at stretch effects on H_2 enriched methane flames concluding that H_2 fuelled flames release heat at much higher rate than pure CH_4 flames due to a different laminar heat release rate, differences in flame surface and effects of flame curvature. Comparison between mixtures #2 (H_2-CO / 50-50) and #4 (H_2-CO-N_2 / 40-40-20) (Table 11.1) shows that the effect of nitrogen dilution is very small and can be totally attributed to the slightly lower adiabatic flame temperature characterizing the diluted mixture (about 30 K on average for a given equivalence ratio); in fact when plotting versus the adiabatic flame temperature, data for mixtures #2 and #4 collapse as it is presented in Figure 11.27b. Figure 11.27.b could also indicate preliminary combustor design rules. It highlights the difference between turbulent burning velocities of syngas mixtures and pure CH_4, for fixed adiabatic temperature. The analyzed range goes from ultra-lean conditions ($T_{ad} < 1600$ K) relevant for engine part load to equivalence ratios prompting flame temperatures well above the conventional turbine inlet temperature ($T_{ad} = 1900$ K). These high temperature conditions are also of interest for gas turbines applications because commonly the reactants access the combustion chamber in not perfectly mixed conditions. This will be the case especially for syngas fuelled systems which have to pay specific attention to avoid auto ignition risks. This leads to the necessity of having data at higher temperatures in order to predict the impact on the global consumption rate of the hotter pockets characterized by locally higher values of equivalence ratio. Chemical properties of the mixtures remain relevant when analyzing the differences between mixtures #2 (H_2-CO / 50-50) and #4 (H_2-CO-N_2 / 40-40-20) compared to mixture #1 (H_2-CO / 33-67) in Figure 11.27b. They play an even stronger role when comparing mixture #6 (H_2-CO-CH_4 / 20-20-60) with the pure syngas mixtures.

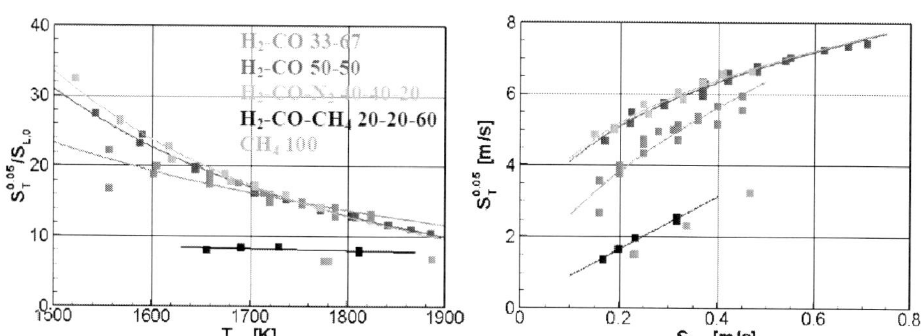

Note: $P = 0.50$ MPa, $T_o = 673$ K. Color coding according to Table 11.1.

Figure 11.28. (a) $S_T^{0.05}/S_{L,o}$ versus T_{ad}. (b) versus $S_{L,o}$.

In order to account for both effects (temperature and chemistry), the data are presented in Figure 11.28a as $S_T^{0.05}/S_{L,o}$ versus the T_{ad}. If all phenomena could be described by effects related only to chemical kinetic properties (comprised by the laminar un-stretched flame speed) all data should collapse into a single correlation function (i.e. a horizontal line). Remaining differences between fuel mixtures must thus be attributed to interaction effects of turbulent flow field conditions at the flame front with specific physical properties of fuel mixtures components e.g. such as hydrogen. As it can be seen there are still certain differences in $S_T^{0.05}/S_{L,o}$ for the pure syngas data; in particular mixture #1 (H_2-CO / 33-67) shows lower values of $S_T^{0.05}/S_{L,o}$ for ultra-lean conditions (low adiabatic temperature). In the same figure it is possible to see that mixture #1 (H_2-CO-CH_4 / 20-20-60) exhibits an almost flat slope while the other data present an exponential decay. Understanding of the different behavior of the pure syngas mixtures involves considerations about the molecular diffusion and stretch effect that are typical for H_2-rich mixtures. As it is known from literature an increasing H_2 content in the mixture correspond to a decrease of Lewis number (defined as the ratio of thermal diffusivity to mass diffusivity). Decreasing the Lewis number of the deficient reactant enhances the local burning rate; the effect has been observed to be very strong especially for lean mixtures [17]. This is exactly what is observed in Figure 11.28, where the ratio $S_T^{0.05}/S_{L,o}$ increases with lower adiabatic flame temperature (i.e. leaner fuel/air mixtures). The effect is more pronounced for the H_2-richer mixtures (#2 and #4), data belonging to mixture #1 (lower H_2 content: 33%) do not follow the exactly same trend especially for very lean conditions (T_{ad} < 1650 K). The phenomenology can be further analyzed with the help of Figure 11.28.b. Starting from less lean conditions (i.e. high laminar flame speed values) going towards the ultra-lean conditions, data of mixture #6 (co-firing) correlate on a straight line indicating a pure chemistry based dependence.

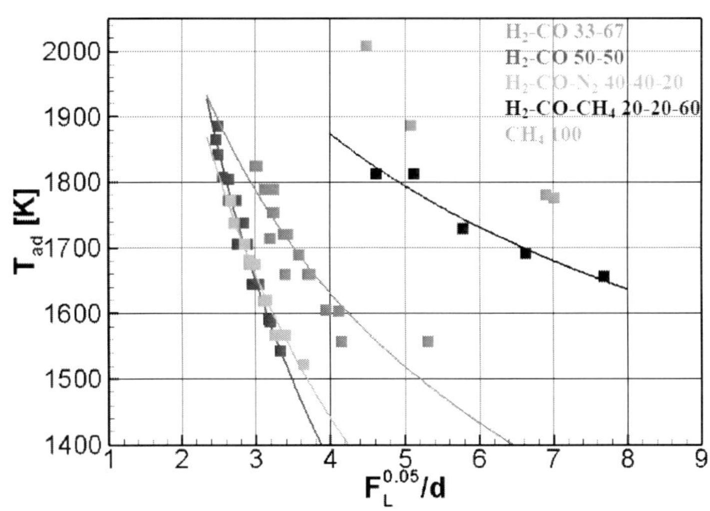

Note: Color coding according to Table 11.1.

Figure 11.29. Flame front position ($F_L^{0.05}/d$) on the combustor centerline.

For the pure syngas mixture a high level of is kept before the curves bend. The conclusion that for high CH_4 content (60% for mixture #6) the increase in turbulent flame speed achieved by moving to richer conditions depends purely on chemical effects and is only valid for natural gas in general and for mixtures of CH_4-H_2 with H_2 percentage up to 20% [51]. For the sake of completeness it has to be mentioned that turbulence effects influences also the different behavior of mixture #6 (H_2-CO-CH_4 / 20-20-60) and the pure syngases. These two different classes of flames stabilize in significantly different zones of the combustor thus they "see" different local turbulence levels, as it is described in Figure 11.29. The figure depicts the position of the flame front (on the combustor centerline) corresponding to the progress variable $\langle c \rangle = 0.05$, $F_L^{0.05}$, normalized by the injector diameter, $d = 25$ mm. A detailed evaluation of the influence of local turbulence properties on the local flame propagation is presented in the following paragraphs.

The general trends described for an inlet velocity of 40 m/s are also valid for different values of this parameter (60, 80, 150 m/s). The effect of inlet velocity on the turbulent burning rate can be further analyzed with the help of Figure 11.30.

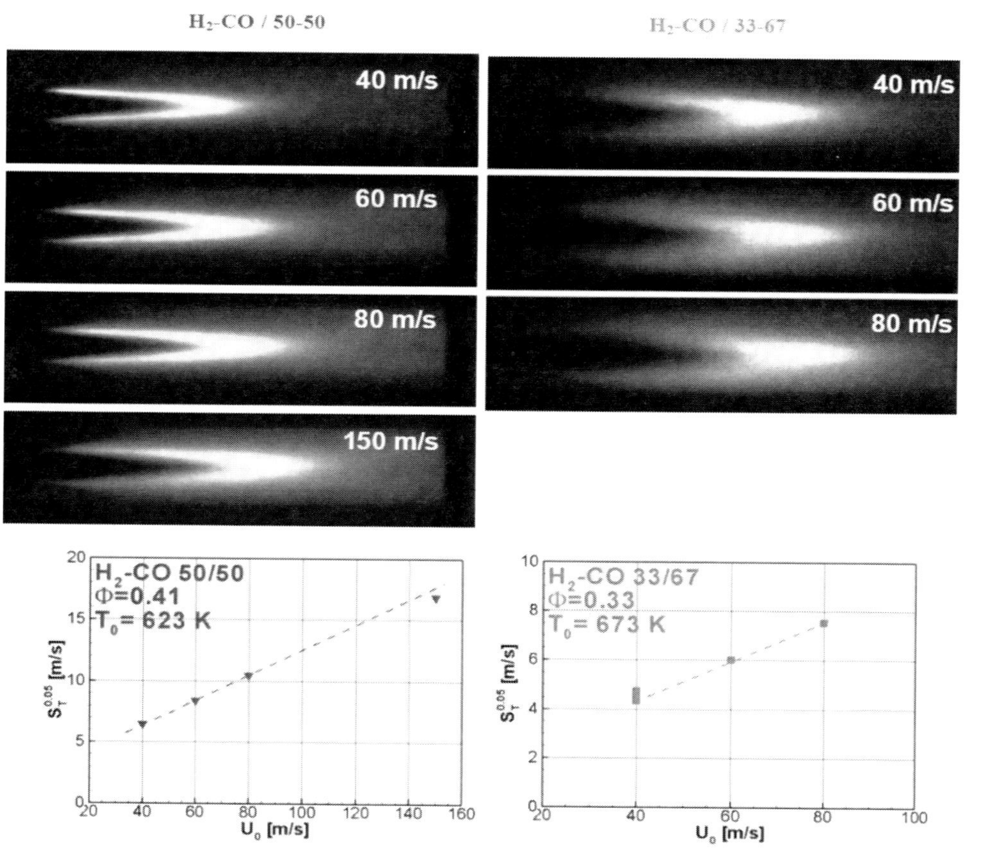

Note: Color coding according to Table 11.1. $P = 0.50$ MPa.

Figure 11.30. Influence of U_o on flame front position and on $S_T^{0.05}$. (above) OH-PLIF averaged images and (below) data.

In Figure 11.30 turbulent flame speed data are reported versus the inlet velocity for mixtures #2 (H_2-CO / 50-50) and #1 (H_2-CO / 33-67). On the upper part of Figure 11.30 averaged PLIF images corresponding respectively to different inlet velocities are reported. Within the presented range a fairly linear dependence of S_T on the inlet velocity is observed for all the analyzed conditions. Figure 11.30 shows that there is just a negligible elongation of the flame varying the inlet velocity and, in addition, taking a closer look to the PLIF images it is also clear that the flame front mean shape is not appreciably affected by this parameter neither. From those observations it is obvious that with increasing the inlet velocity the instantaneous flame front is able to adapt/to consume the higher incoming mass flow rate ($\rho_o U_o A_o$) without changing too much the average flame surface (A_{av},). In order to compensate for the higher mass flow rate the instantaneous flame front surface (A_{in}) must be increased e.g. due to enhanced wrinkling/corrugation. The enhanced corrugation of the flame front at elevated inlet velocities is prompted by higher turbulence intensity levels justified by the different Reynolds numbers ($Re_d = U_o\, d/\upsilon$) corresponding to the various inlet conditions (which for $P = 0.50$ MPa: $Re = 78000$, 118000, 157000 and 300000 for 40, 60, 80 and 150 m/s respectively). Looking at Figure 11.30 an observation can also be made regarding the sensitivity flashback propensity on the inlet velocity. Given the increase of with U_o and the corresponding poor effect on the flame front position, it is concluded that fighting FB by increasing U_o is not a good strategy.

Relation between Turbulent Flame Speed and Flame Front Position

Recently [1,3,54], it has been stressed that turbulent flame speeds may depend on the geometry of the particular experimental rig, since local flame front wrinkles bear the memory of the ones generated upstream. In this respect, flames generated with similar geometries should preferably be compared. To eliminate most of the geometry dependence, turbulent flame speeds values, S_T, are correlated with turbulence characteristics such as integral length scale, L_T, and turbulence intensity, u', measured under non-reactive flow conditions at the flame location. The turbulent flow field in the present configuration has been studied in the past [32]. Several experiments were conducted with different perforated plates mounted in the fuel/air mixing section. In the current study no perforated plates were used: it was impossible to position any element in the mixing section (which could act as potential flame holder) when firing syngas because of its high flashback propensity. Previous studies have shown that the effect of different types of perforated plates on the turbulent flow field is only restricted to core flow regions close to the combustor inlet ($x < 2d$, $d = 25$ mm: diameter of fuel/air delivery pipe), revealing that the turbulence field at the location of the flame front is essentially dominated by the shear generated turbulence due to the backward facing step geometry [32]. Therefore, turbulence intensity and integral length scales were inferred in the present study by averaging over four different flow conditions corresponding to four perforated plate geometries. The resulting average values of u' and L_T on the combustor symmetry axis are presented in Figure 11.31 versus the flame front position, $F_L^{0.05}$ normalized by the injector diameter, d. The variation of the centerline u' in each of the four cases with respect to the average value, was at most 1 m/s. Centerline values are considered representative of the turbulence characteristics along the entire flame front: it has been shown that the flames stabilize along contours of nearly constant u' and L_T [32]. Figure 11.31 presents a wide selection of measurements comprising inlet temperatures from 423 to 773 K, equivalence ratios between 0.25 and 0.75, pressures from 0.1 to 2.0 MPa, and various fuel mixtures (Table 11.1). In the same figure, f_{BT} values are also plotted. The dashed and the dashed dotted lines represent values corresponding to a flame with an ideal conical shape and to a rotationally symmetric ellipsoid, respectively. The correlation between and $F_L^{0.05}/d$ is nearly independent on the gas mixture composition and operating conditions. "Short" flames assume conical shapes, while "longer" flames rather resemble axisymmetric ellipsoidal shapes as depicted in Figure 11.31 by the average OH-PLIF images (the arrows connect the data points to the corresponding flame locations, $F_L^{0.05}$). Moreover, Figure 11.31, shows that the angle method used by Kobayashi [19] and other authors [55] for the determination of S_T, can lead for the actual combustor geometry to errors not only in the absolute value of S_T but also in the observed trends (see non constant difference between the dashed and the solid line). Venkateswaran et al. [55], using the same configuration as Kobayashi, reported an error by the angle method in estimating absolute values of S_T, stating however that trends were not affected. In direct contrast, Kobayashi showed no difference between values calculated with the angle method and the progress variable method (<*c*> = 0.1) [56].

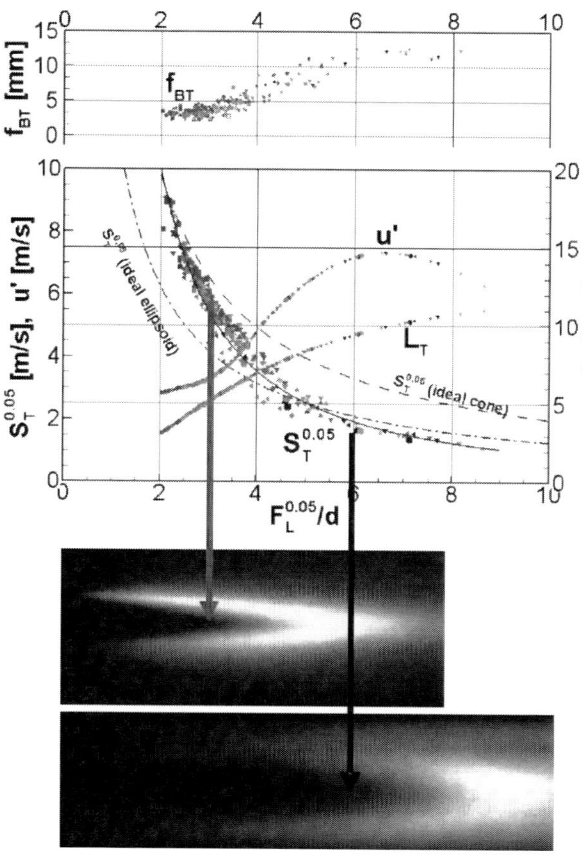

Note: Color coding according to Table 11.1.

Figure 11.31. $S_T^{0.05}$, u', L_T, f_{BT} versus $F_L^{0.05}/d$, and averaged OH-PLIF images for two flames.

With the approach herein presented, it is possible to generate data at different u' (at the same inlet velocity) without varying the overall combustor geometry. The flame stabilizes at a certain location, where it encounters specific values of u' and L_T, according to its propagation capability.

Until now there are no values of global fuel consumption rate, S_T, reported in the literature for syngas compositions at high preheating temperatures, high pressures and high turbulence intensity. In [55], S_T (global consumption) values for only atmospheric pressure and temperature are reported. A one to one comparison with this work is not possible as the conditions are different and the methodology used to obtain and post process the data is also different.

Normalized Values of Turbulent Flame Speed

Mixture Composition Effect on Turbulent Flame Speed

Figure 11.32. reports measurements for a constant pressure of 0.5 MPa, $T_o = 623$ K and $U_o = 40$ m/s for four different fuel mixtures. An additional dataset for CH_4 at 0.25 MPa is provided. Data points characterized by the same Φ are connected by additional solid lines.

It is well known that $S_T/S_{L,o}$ values exhibit a first zone, for low values $u'/S_{L,o}$, wherein $S_T/S_{L,o}$ is almost linearly proportional to $u'/S_{L,o}$, and a second zone where the $S_T/S_{L,o}$ versus $u'/S_{L,o}$ curve bends and appears to level off as reviewed in [57]. The bending occurs for a particular $u'/S_{L,o}$ depending on fuel mixture composition. Data for CH_4 were apparently acquired within this second zone only. The effect of pressure on S_T for CH_4 is depicted by the difference between the two data sets indicated with open and filled triangles. Increase in

$/S_{L,o}$ with rising pressure has already been reported in [19,56]. This effect was attributed [19] to the growth of the hydrodynamic instability, DL, according to Darrieus and Landau [18] who described this instability for laminar flamelets.

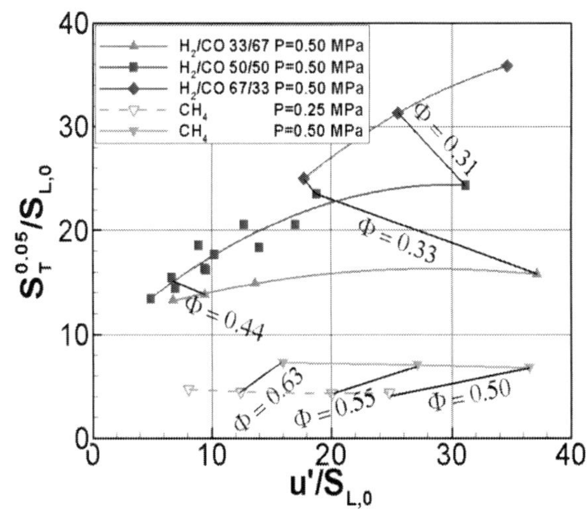

Figure 11.32. $S_T^{0.05}/S_{L,o}$ versus $u'/S_{L,o}$: H_2-content effect for syngas and pressure effect for CH_4.

It is noted that Kobayashi's findings [19] were recently questioned by Lipatnikov and Chomiak [21], arguing that there is no evidence for the DL instability being alone responsible for the increase in $S_T/S_{L,o}$ with rising pressure. Moreover, they concluded [17] that DL instability can be significant only for weak turbulence. The decrease of importance of DL effects at high $u'/S_{L,o}$ is further supported by Peters [58], although he provided no insights on the impact of pressure. Therefore, the effect of DL instabilities at elevated pressures is still an open question. Data at constant pressure confirm trends presented elsewhere at significantly lower u'/S_L and reviewed in [17]: the higher the H_2 content in the mixture, the larger the S_T/S_L. This behavior can be attributed to preferential diffusive thermal instability, PDT, as described in [17]. As u' is taken at the flame front position along the centerline, and since u' varies upon Φ, it is not possible to acquire data at constant Φ and $u'/S_{L,o}$ simultaneously. Therefore, it was verified numerically that the Lewis number, Le_{H2}, for every fuel mixture, does not change appreciably within the range of equivalence ratios presented, such that the difference in $S_T^{0.05}/S_{L,o}$ for $P = 0.50$ MPa (Figure 11.32) can be predominantly attributed to PDT effects due to different absolute H_2 levels in the mixture.

Pressure Effect on Turbulent Flame Speed

Figure 11.33. depicts data at different pressures (0.1 to 2.0 MPa) and various Φ (0.30 to 0.50), for the H_2- CO 50-50 mixture. In addition, linear regression lines for constant Φ (dashed lines) and constant $S_{L,o}$ (dashed dotted lines) are shown. Regression lines for constant u' roughly follow the $\Phi = const.$ lines, and are thus not shown. This behavior is based on the experimental observation that the flame front position for constant Φ is only weakly affected by pressure, thus the reaction zone encounters roughly the same u' (see Figure 11.31). An increase in pressure leads to an increase in $S_T^{0.05}/S_{L,o}$, irrespective of the way the pressure increase is attained (at constant $u'/S_{L,o}$, constant u', constant $S_{L,o}$, or constant Φ).

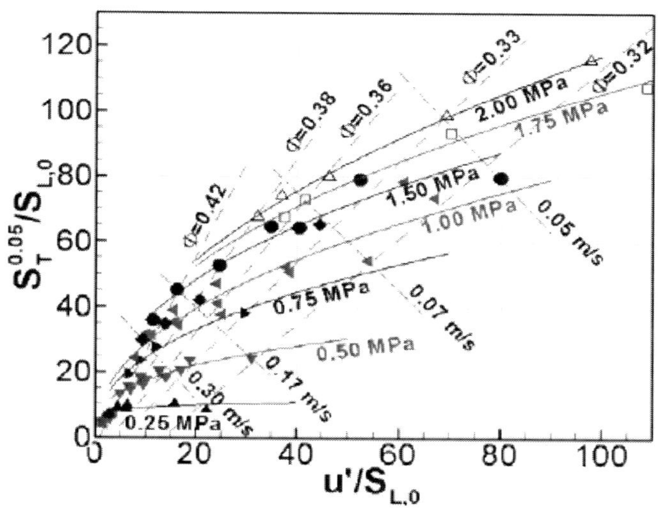

Symbol	P [MPa]
■	0.10
▲	0.25
▼	0.50
►	0.75
◄	1.00
♦	1.25
●	1.50
□	1.75
Δ	2.00

Figure 11.33. $S_T^{0.05}/S_{L,o}$ versus $u'/S_{L,o}$: pressure effect for syngas (H_2-CO / 50-50).

This trend is similar to that reported for CH_4 [56] and also observed in the present data (Figure 11.32). As a consequence, the phenomenological explanation for CH_4 [56], invoking the role of DL instability for the increase of $S_T^{0.05}/S_{L,o}$ with pressure, could be considered valid for syngas as well, however, at higher $u'/S_{L,o}$. The increase in $S_T^{0.05}/S_{L,o}$ with rising pressure is stronger when moving along a Φ isoline. In this case u' is also kept roughly constant, such that the increase in $u'/S_{L,o}$ can be studied solely due to the decrease in $S_{L,o}$. Due to the decrease in $S_{L,o}$, for constant u' the flame front becomes more sensitive to hydrodynamic wrinkling, leading to the accelerated increase in $S_T^{0.05}/S_{L,o}$.

In case of syngas ($Le_{H2} < 1$), PDT effects are expected to be important and to further discriminate data acquired at different pressures. To evaluate the role of PDT effects, the $S_T^{0.05}/S_{L,o}$ ratios at two pressures, 0.25 and 0.50 MPa for the syngas mixture presented in Figure 11.33. and for the CH_4 in Figure 11.32 is compared. Comparisons are performed within a range of $u'/S_{L,o}$ (20-40) whereby $S_T^{0.05}/S_{L,o}$ is weakly dependent on $u'/S_{L,o}$ for syngas, and fairly independent for CH_4. These ratios are roughly 2.5 for syngas and 1.6 for CH_4. Hence, it appears that the syngas mixture undergoes an additional enhancement of $S_T^{0.05}/S_{L,o}$, most probably attributable to PDT and to the coupling between DL and PDT instabilities. The difference in these two factors cannot be attributed to the difference in laminar flame thickness; in fact, within the range of $u'/S_{L,o}$, considered for this comparison, the syngas mixture is characterized by a larger thickness, due to the smaller Φ (larger flame thickness reduces the possibility of flame wrinkling at small scales and does not contribute to an increase in $S_T^{0.05}/S_{L,o}$). An account for different flame thicknesses is addressed in the next section.

Correlation for Normalized Turbulent Flame Speed

Figure 11.34. reports a database of over 300 measured operating conditions. On the abscissa, the terms P/P_R and T_R/T_o normalize for pressure and inlet temperature variation as in [59] ($P_R = 0.1$ MPa, $T_R = 1$ K). The term L_T/δ_L accounts for the capability of eddies having size between δ_L and L_T to wrinkle the flame front at a given u'; this term appears to also capture Lewis number effects via the dependence of δ_L on the thermal diffusivity.

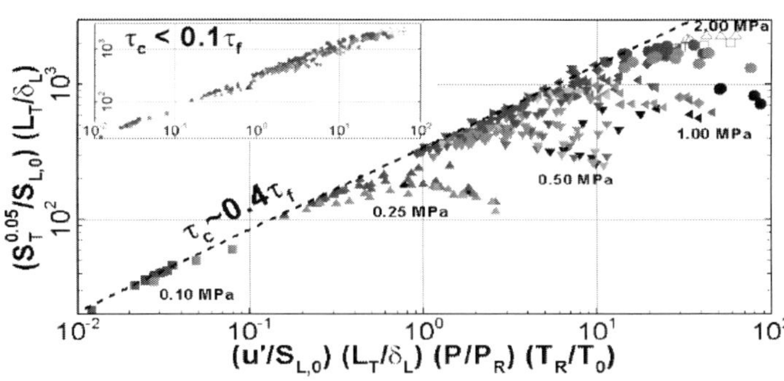

Note: Color coding according to Table 11.1, pressure legend as in Figure 11.33.

Figure 11.34. Normalized turbulent flame speed data.

A first data set, where $S_T^{0.05}/S_{L,o}$ increases with a well defined power law (proportional range, see insert), and a second one where $S_T^{0.05}/S_{L,o}$ values level off (bending range) can be distinguished. The bending occurs at $u'/S_{L,o}$ values dependent on pressure, and not on mixture composition. The onset of bending is found to depend on the ratio of the following time scales:

$$\tau_f = \frac{L_T}{u'} \text{ and } \tau_c = \frac{f_{BT}}{S_T}, \tag{11.7}$$

where τ_f is the turnover time of an eddy having the size of the integral length scale, thus representing the longest turbulent timescale interacting with the flame front; τ_c is the time necessary for the flame front to travel through the flame brush thickness, f_{BT}, thus denoting the timescale at which the flame can respond to flow perturbations.

Eddies characterized by τ_f much shorter than τ_c, have too high frequencies ($1/\tau_f$) to interact and contort the flame front, thus their contribution to the wrinkling process and consequently to the increase of $S_T^{0.05}/S_{L,o}$ is meager or negligible. This concept was demonstrated in [60,61]. The criterion $\tau_c = 0.1\tau_f$ appears to be appropriate to separate the dataset in a "linear log-log" behavior (see insert in

Figure 11.34. and a "bending" one. As a matter of fact, this criterion ($\tau_c = 0.1\tau_f$) roughly corresponds to = u' (as it was found that f_{BT}~$10L_T$).

The dashed line in Figure 11.34 describes the trend of those data in the "linear" range with comparable τ_f and τ_c ($\tau_c / \tau_f \approx 0.4$). Above this limiting line it is not possible to find any data point: at these conditions, eddies can wrinkle the flame front apparently with maximum efficiency and no further wrinkling (i.e. acceleration of fuel consumption) is possible. A correlation for these data can be written as:

$$\frac{S_T^{0.05}}{S_{L,0}} = a\left(\frac{u'}{S_{L,0}}\right)^{0.63}\left(\frac{L_T}{\delta_L}\right)^{-0.37}\left(\frac{P}{P_R}\right)^{0.63}\left(\frac{T_0}{T_R}\right)^{-0.63} ;a = 337.45 \tag{11.8}$$

The comparison of this correlation with other ones available in the literature leads to the following considerations:

1. A dependence of $S_T/S_{L,o}$ on $u'/S_{L,o}$ with a similar exponent (n = 0.7) was also reported by Klimov [62].
2. The negative exponent n = -0.37 of L_T/δ_L is not in contrast with other correlations [13; 26]; for the data set supporting the above correlation this term is related to P/P_R as:

$$\left(\frac{L_T}{\delta_L}\right) = b\left(\frac{P}{P_R}\right)^{0.66} ;b = 8.3 \tag{11.9}$$

   By substituting equation (11.9) in (11.8), the exponent of L_T/δ_L appears to be positive (0.58), in agreement with the previous mentioned correlations [13,26].
3. Kobayashi [56,63] has proposed similar correlations, accounting just for the pressure ratio. Simpler correlations were also attempted by plotting $S_T^{0.05}/S_{L,o}$ versus

$(u'/S_{L,o})(P/P_R)(T_R/T_o)$, not reported here for brevity. This correlation produced also good results; however data points characterized by large differences in L_T/δ_L were not correlating on a single straight line as in Figure 11.34. but on parallel lines having roughly the same exponent, with proportionality coefficients being higher for lower pressures. These differences were particularly obvious for data at 0.1 MPa, which are characterized by much smaller L_T/δ_L. In [56,63], Kobayashi obtained a correlation for $S_T/S_{L,o}$ versus $(u'/S_{L,o})(P/P_R)$ with power laws having exponents $n = 0.38$ and 0.40. These measurements were performed within the corrugated flamelets regime. Working out the global exponent, n, in P/P_R by substituting again equation (11.9) in (11.8), it is obtained $n = 0.385$. The agreement with [56,63] is seen as a confirmation that there is no experimental evidence that the two regimes (corrugated flamelets and thin reaction zones) differ when comparing the S_T, as observed in [17].

CONCLUSION

Lean premixed turbulent flames were investigated at gas turbine relevant conditions. Extensive measurements were performed for a variety of syngas mixtures (including pure hydrogen fuel) and for methane. Measurements were mainly conducted in the thin reaction zones regime. The experimental campaigns mainly involved the recording of OH-chemiluminescence and planar laser induced fluorescence (PLIF) of the OH radical for assessing the flame front corrugation, and the use of a previously established database consisting of particle image velocimetry (PIV) data for characterizing the turbulent flow field. Flame images were simultaneously acquired at both low and high resolutions in order to investigate flame propagation characteristics at the large and small scales, respectively. Pollutant emissions were also measured with a standard gas analysis system. In the following paragraphs, the main conclusions for every topic addressed in this chapter are drawn.

OPERATIONAL WINDOW

Lean Blow Out

1. A methodology for the detection of LBO limits based on the time history of flame OH-chemiluminescence signal and its frequency analysis has been presented. With this approach, LBO can be detected before actual "flame out" occurs. Thus, this non-intrusive method based on the flame chemiluminescence signal -captured by means of an optical fiber arrangement- could potentially be applied in real gas turbine combustors and be used to avoid tripping of the machine with all its related negative consequences.
2. Not surprisingly, the results show that the higher the H_2 content in the fuel mixture the leaner the equivalence ratios for which LBO occurs. The difference is significant for fuel mixtures with/without methane and can be attributed to the higher propagation speeds of hydrogen containing syngas mixtures.

3. Lean blow out limits were found to have a weak dependence on pressure for the methane containing mixtures, whereas for pure syngas mixtures this dependence was stronger: lean blow out limits move toward less lean operation with increasing pressure. No significant dependence on the inlet velocity was found.
4. LBO was assessed not as a direct consequence of incomplete combustion, but as a result of strong coupling between heat release waves and pressure field. This interpretation also explained the different pressure behavior of the various fuels.

Flashback

Dependence of flashback on pressure, inlet temperature and inlet velocity was described. Interpretations of the described behaviors were proposed.

1. The results show strong pressure dependence; the observed trend narrows down dramatically the operational window for high pressures and highlights the need of experiments at these conditions. Based on a phenomenological analysis, the pressure dependence was attributed mainly to the decrease of the quenching distance; however, further investigations on the flow field are required to validate the proposed explanations and to describe the full picture.
2. Flashback propensity was found almost independent on the inlet velocity, demonstrating that increasing U_0 when fighting (boundary layer) flashback is not a good strategy.
3. With the aim of mitigating flashback propensity, the most encouraging results were obtained by reducing the inlet temperature. In this case flashback propensity has been observed to scale according to the laminar flame speed. Due to the strong power exponent (Φ_{FB}~T_o^4) especially for ultra-lean conditions, a local reduction of the inlet temperature is particularly suggested for syngas applications as for these fuels the operative conditions must be leaner than the ones typically encountered in natural gas combustion.
4. The architecture and the logic of an automatic control system for flashback were outlined. The control system showed fast response and capability of avoiding damages. The system controls the back propagation of the flame, re-establishing it in the combustion chamber without the need of shutting down the rig.

Flame Speed

For the first time, experimental turbulent flame speed data were reported for syngas at pressures up to 2.0 MPa, preheat temperatures up to 773 K and $u'/S_L > 100$. In contrast to many other works, S_T in terms of global consumption has been related to the turbulence properties measured at the flame front and not at the inlet section, so as to minimize the dependence of this parameter on the geometrical characteristics of the burner.

1. The dependence of $S_T/S_{L,o}$ on H_2 content in the fuel mixture and on pressure was assessed. The increase of $S_T/S_{L,o}$ on H_2 content was attributed to preferential diffusive-thermal (PDT) effects, being active also at high turbulence intensities. These effects have been shown to also play a role in the increase of $S_T/S_{L,o}$ with rising pressure, attributed to the coupling of PDT with hydrodynamic instabilities.
2. For the first time in the literature, it has been shown that $S_T/S_{L,o}$ versus $u'/S_{L,o}$ data for syngas show linear and bending characteristics, similar to that known for CH_4. The onset of bending depends on the competition between the temporal activity of the turbulent eddies (turnover time of eddies of integral length-scale size) and the timescale at which the flame front can respond to flow perturbations. The latter was expressed as the ratio of the flame brush thickness to S_T. The nature of flame / eddies interaction based upon temporal arguments, theorized in the literature, has been demonstrated experimentally for the first time with this work.
3. A new correlation for $S_T/S_{L,o}$ independent on the fuel mixture composition, capable of capturing both hydrodynamic and preferential diffusive-thermal instabilities has been proposed. The correlation expresses the ratio $S_T/S_{L,o}$ in terms of turbulence properties (turbulence intensity and integral length scale), combustion properties (laminar flame speed and laminar flame thickness) and operating conditions (pressure and inlet temperature). The success of the correlation is due to the choice of measuring the turbulence parameters at the flame front position rather than at the combustor inlet as commonly done in the literature.

CONCLUSION

By providing the information summarized in the previous paragraphs, this chapter presents a contribution to the studies of lean premixed turbulent syngas combustion, addressing the lack of data and knowledge reported by several other authors as summarized in the introduction. This is especially true for turbulent flame speeds at gas turbine relevant conditions.

Important contributions to the field are the detailed illustrations of the flame / eddies interaction and of the preferential-diffusive-thermal effect of hydrogen and their consequence on the flame front propagation. The results are supported by an extensive experimental database (over 300 flames) which serves also as a valid benchmark for numerical simulation and combustion modeling.

The presented database and derived trends of flame speeds along with the information on the operational window, especially with regards to flashback propensity, are valuable also from an industrial point of view. The work summarize here eventually contributes in providing key answers to the question postulated in the introduction: "can we fire with syngas, in lean premixed mode, our natural gas burners?".

Acknowledgments

This work has been realized thanks to the support of the Swiss Federal Office of Energy (BFE) and Alstom Power (Switzerland).

The author would like to thank Prof. Konstantinos Boulouchos, Dr. Peter Jansohn, PD Dr. John Mantzaras, Dr. Rolf Bombach, Dr. Wolfgang Kreutner, Dr. Alexey Denisovand and Mr. Daniel Erne for their valuable contribution to this work.

Nomenclature

ν	Molecular kinematic viscosity
λ	Taylor scale
Φ	Fuel to air equivalence ratio
δ_L	Flame thickness of unstreched laminar flame
$<c>$	Progress variable
A_o, A_{in}, A_{av}	Inlet, instantaneous and average flame areas
CIVB	Combustion Induced Vortex Breakdown
Da	Damköhler number
DL	Darrieus-Landau (hydrodynamic instability)
dq	Quenching distance
FB	Flashback
f_{BT}	Turbulent flame brush thickness
$F_L^{<c>}$	Flame length at specified progress variable <c>
IGCC	Integrated Gasification Combined Cycle
LBO	Lean Blow Out
L_T	Integral length scale
P	Pressure
PDT	Preferential-Diffusive-Thermal (instability)
PIV	Particle Image Velocimetry
PLIF	Planar Laser Induced Fluorescence
Re_T	Turbulence Reynolds number, $u'LT/\nu$
S_L	Laminar flame speed
S_T	Turbulent flame speed
T_{ad}	Adiabatic flame temperature
T_o	Preheat temperature
u'	Turbulence intensity
U_o	Inlet velocity
η	Kolmogorov scale of turbulence

References

[1] Cheng, R.K., *Synthesis Gas Combustion - Fundamentals and Applications*. 2009, CRC Press, Boca Raton: Florida, USA. p. 129-168.

[2] Bradley, D., Combustion and the design of future engine fuels. *Proc. IME C J. Mech. Eng. Sci.*, 2009. 223(12): p. 2751-2765.
[3] Driscoll, J.F., Turbulent premixed combustion: Flamelet structure and its effect on turbulent burning velocities. *Prog. Energy Combust. Sci.*, 2008. 34(1): p. 91-134.
[4] Daniele, S., P. Jansohn, and K. Boulouchos. Flame front characteristic and turbulent flame speed of lean premixed syngas combustion at gas turbine relevant conditions. in *Proc. ASME Turbo Expo 2009*. 2009. Orlando, Florida.
[5] Daniele, S., et al., Turbulent flame speed for syngas at gas turbine relevant conditions. *Proc. Combust. Inst.*, 2011. 33(2): p. 2937-2944.
[6] Hoffmann, S., et al., Performance and Cost Analysis of Advanced Gas Turbine Cycles With Precombustion CO2 Capture. *J. Eng. Gas Turb. Power*, 2009. 131(2): p. 021701.
[7] *Gasification Technologies Council*. [cited; Available from: www.gasification.org.
[8] Sung, C.J. and C.K. Law, Fundamental combustion properties of H-2/CO mixtures: Ignition and flame propagation at elevated pressures. *Combust. Sci. Technol.*, 2008. 180(6): p. 1097-1116.
[9] Chaos, M. and F.L. Dryer, Syngas combustion kinetics and applications. *Combust. Sci. Technol.*, 2008. 180(6): p. 1053-1096.
[10] Lieuwen, T., et al., Burner development and operability issues associated with steady flowing syngas fired combustors. *Combust. Sci. Technol.*, 2008. 180(6): p. 1169-1192.
[11] Wang, Q., et al. Correlating Flashback Tendencies for Premixed Injection of Hydrogen and Methane at Elevated Temperature and Pressure. in *Proc. ASME Turbo Expo 2009*. 2009.
[12] Schafer, O., R. Koch, and S. Wittig, Flashback in lean prevaporized premixed combustion: Nonswirling turbulent pipe flow study. *J. Eng. Gas Turb. Power*, 2003. 125(3): p. 670-676.
[13] Kurdyumov, V.N., E. Fernandez, and A. Linan, Flame flashback and propagation of premixed flames near a wall. *Proc. Combust. Inst.*, 2000. 28: p. 1883-1889.
[14] Noble, D.R., et al. Syngas Mixture Composition Effects Upon Flashback and Blowout in *Proc. ASME Turbo Expo 2006*. 2006. Barcelona, Spain.
[15] Nauert, A., et al., Experimental analysis of flashback in lean premixed swirling flames: conditions close to flashback. *Exp. Fluids*, 2007. 43(1): p. 89-100.
[16] Konle, M., Sattelmayer, T. Time Scale Model for the Prediction of the Onset of Flame Flashback Driven by Combustion Induced Vortex Breakdovn *(CIVB)*. in *Proc. ASME Turbo Expo 2009*. 2009. Orlando, Florida.
[17] Lipatnikov, A. and J. Chomiak, Molecular transport effects on turbulent flame propagation and structure. *Prog. Energy Combust. Sci.*, 2005. 31(1): p. 1-73.
[18] Landau, L., On the theory of slow combustion. *Acta Physicochim. URSS*, 1944. 19(1): p. 77-85.
[19] Kobayashi, H., et al., Burning Velocity of Turbulent Premixed Flames in a High-Pressure Environment. *Proc. Combust. Inst.*, 1996. 26: p. 389-396.
[20] Ichikawa, Y., et al., Flame structure and radiation characteristics of CO/H2/CO2/air turbulent premixed flames at high pressure, *Proc. Combust. Inst.*, 2011. 33(1): p. 1543-1550.
[21] Lipatnikov, A. and J. Chomiak, Effects of Premixed Flames on Turbulence and Turbulent Scalar Transport. *Prog. Energy Combust. Sci.*, 2010. 36: p. 1-102.

[22] Damkohler, G., The effect of turbulence on the combustion rate in gas compounds. *Z. Elktrochem. Anegw*. P., 1940. 46: p. 601-626.

[23] Borghi, R., On the structure and morphology of turbulent premixed flames in Recent advances in the aerospace sciences. 1985, *New York, Plenum Press*. p. 117-138

[24] Peters, N., Laminar flamelet concepts in turbulent combustion. *Proc. Combust. Inst.*, 1988. 21(1): p. 1231-1250.

[25] Abdelgayed, R.G., D. Bradley, and F.K.K. Lung, Combustion Regimes and the Straining of Turbulent Premixed Flames. *Combust. Flame*, 1989. 76(2): p. 213-218.

[26] Poinsot, T., D. Veynante, and S. Candel, Diagrams of premixed turbulent combustion based on direct simulation. *Proc. Combust. Inst.*, 1991. 28(1): p. 613-619.

[27] Peters, N., The turbulent burning velocity for large-scale and small-scale turbulence. *J. Fluid Mech.*, 1999. 384: p. 107-132.

[28] Zimont, V.L., Theory of Turbulent Combustion of a Homogeneous Fuel Mixture at High Reynolds-Numbers. *Combust. Explos. Shock Waves*, 1979. 15(3): p. 305-311.

[29] Kortschik, C., T. Plessing, and N. Peters, Laser optical investigation of turbulent transport of temperature ahead of the preheat zone in a premixed flame. *Combust. Flame*, 2004. 136(1-2): p. 43-50.

[30] Gulder, O.L., Contribution of small scale turbulence to burning velocity of flamelets in the thin reaction zone regime. *Proc. Combust. Inst.*, 2007. 31: p. 1369-1375.

[31] Peters, N., Turbulent Combustion,*Cambridge University Press*. 2005.

[32] Siewert, P., Flame front characteristics of turbulent premixed lean methane / air flames at high-pressure and high-temperature, in Swiss Federal Institute of Technology ETH-Zurich. 2005: Zurich, Switzerland.

[33] Jansohn, P. *A Swiss Initiative for High Efficiency/Low CO2 Gas Turbine Plants*. in *POWER-GEN Europe*. 2007. Feria de Madrid.

[34] Reiss, F., K. Reyser, and T. Griffin. The ALSTOM GT13E2 Medium BTU Gas Turbine. in *Proc. ASME Turbo Expo 2002*. 2002. Amsterdam, The Netherlands.

[35] Jones, R.H., J.H. Jones, and G.J. Thomas, *Materials for the Hydrogen Economy*. CRC Press ed, ed. C. Press. 2007: CRC Press.

[36] Zheng, L. and E. Furinsky, Comparison of Shell, Texaco, BGL and KRW gasifiers as part of IGCC plant computer simulations. *Energy Convers. Manage.*, 2005. 46: p. 1767-1779.

[37] Griebel, P., E. Boschek, and P. Jansohn, Lean blowout limits and NOx emissions of turbulent, lean premixed, hydrogen-enriched methane/air flames at high pressure. *J. Eng. Gas Turb. Power*, 2007. 129(2): p. 404-410.

[38] Bruschi, R., et al., Optical Unsteady Characterization of Turbulent Combustion: Application in an Active Control System in AIAA-2005-4328 41st AIAA/ASME/SAE/ASEE Joint Propulsion Conference and Exhibit. 2005 Tucson, Arizona.

[39] Muruganandam, T.M., et al., Optical And Acoustic Sensing Of Lean Blowout Precursors, in AIAA 2002-3732 38th AIAA/ASME/SAE/ASEE Joint Propulsion Conference and Exhibit. 2002: Indianapolis, Indiana.

[40] Giacomazzi, E., et al., Experimental and Numerical Unsteady Characterization of Turbulent Combustion, in AIAA 2005-3579 41st AIAA/ASME/SAE/ASEE Joint Propulsion Conference and Exhibit. 2005: Tucson, Arizona.

[41] Daniele, S., P. Jansohn, and K. Boulouchos. Lean Premixed Combustion of Undiluted Syngas at Gas Turbine Relevant Conditions: NOx Emissions and Lean Operational Limits. in *Proc. ASME Turbo Expo 2008*. 2008. Berlin, Germany.

[42] Dagaut, P. and G. Dayma, Hydrogen-enriched natural gas blend oxidation under high-pressure conditions: Experimental and detailed chemical kinetic modeling. *Int. J. Hydrogen Energy*, 2006. 31(4): p. 505-515.

[43] Glassman, I., Combustion, ed. I. *Academic Press*. 1987, San Diego, Ca.

[44] Cantera. 1998 [cited; Available from: www.cantera.org.

[45] Smith, G.P., et al. *An optimized detailed chemical reaction mechanism for methane combustion*. Gas Research Institute, 2000 [cited; Available from: http://www.me.berkeley.edu/gri_mech.

[46] Kays, W., M., Crawford,M.,E., *Convective heat and mass transfer*. 1993: Mc Graw Hill.

[47] Friedman, R. and W.C. Johnston, Pressure Dependence of Quenching Distance of Normal Heptane, Iso-Octane, Benzene, and Ethyl Ether Flames. *J. Chem. Phys.*, 1952. 20(5): p. 919-920.

[48] E. Giacomazzi, et al. *LES Analysis of a Syngas Turbulent Premixed Dump-Combustor at 5 Bar*. in *Proc. PTSE2010*. 2010. Ischia (It).

[49] Griebel, P., P. Siewert, and P. Jansohn, Flame characteristics of turbulent lean premixed methane/air flames at high pressure: Turbulent flame speed and flame brush thickness. *Proc. Combust. Inst.*, 2007. 31: p. 3083-3090.

[50] Roesgen, T., *Quantitative Flow Visualization; Institut fuer Fluiddynamik ETH Zurich*, ed. I.f.F.E. Zurich. 2005.

[51] Boschek, E., P. Griebel, and P. Jansohn. *Fuel Variability Effects on Turbulent, Lean Premixed Flames at High Pressures*. in *GT2007-27496 ASME Turbo Expo 2007: Power for Land, Sea, and Air*. 2007. Montreal, Canada.

[52] Littlejohn, D., et al., Laboratory Investigations of Low-Swirl Injectors Operating With Syngases. *J. Eng. Gas Turb. Power*. 132(1).

[53] Hawkes, E., Chen, JH., *Turbulent Stretch Effects On Hydrogen Enriched Lean Premixed Methane-Air Flames*. Combustion Research Facility Sandia National Laboratories Livermore, CA, 94551, USA, Report, 2002.

[54] Shepherd, I.G. and R.K. Cheng, The burning rate of premixed flames in moderate and intense turbulence. *Combust. Flame*, 2001. 127(3): p. 2066-2075.

[55] Venkateswaran, P., et al. Global Turbulent Consumption Speeds ($S_{T,GC}$) of H_2/CO Blends. in *Proc. ASME Turbo Expo 2009*. 2009. Orlando, Florida.

[56] Kobayashi, H., et al., Burning velocity correlation of methane/air turbulent premixed flames at high pressure and high temperature. *Proc. Combust. Inst.*, 2005. 30: p. 827-834.

[57] Lipatnikov, A. and J. Chomiak, Turbulent flame speed and thickness: phenomenology, evaluation, and application in multi-dimensional simulations. *Prog. Energy Combust. Sci.*, 2002. 28(1): p. 1-74.

[58] Peters, N., H. Wenzel, and F.A. Williams, Modification of the turbulent burning velocity by gas expansion. *Proc. Combust. Inst.*, 2000. 28: p. 235-243.

[59] Metghalchi, M. and J.C. Keck, Laminar Burning Velocity of Propane-Air Mixtures at High-Temperature and Pressure. *Combust. Flame*, 1980. 38(2): p. 143-154.

[60] Ashurst, W.T., G.I. Sivashinsky, and V. Yakhot, Flame Front Propagation in Nonsteady Hydrodynamic Fields. *Combust. Sci. Technol.*, 1988. 62(4-6): p. 273-284.

[61] Mantzaras, J., Geometrical Properties of Turbulent Premixed Flames - Comparison between Computed and Measured Quantities. *Combust. Sci. Technol.*, 1992. 86(1-6): p. 135-162.

[62] Klimov, A.M., "Premixed Turbulent Flames-Interplay of Hydrodynamic and Chemical Phenomena", in Flames. *Laser and Reactive Systems*, 1983. 88: p. 133-146.

[63] Kobayashi, H., Y. Kawabata, and K. Maruta, Experimental Study on General Correlationof Turbulent Burning Velocity at High Pressure. *Proc. Combust. Inst.*, 1998. 27: p. 941-948.

CONCLUSION

In principle, synthesis gas (primarily consisting of CO and H_2) can be produced from any hydrocarbon feedstock, including: methane, coke oven gas, petroleum coke, coal, and biomass. Currently, the lowest cost routes for syngas production are based on natural gas, the cheapest option being remote or stranded reserves. Economic considerations and current fossil-based source limitation dictate that the syngas production should consider other energy sources, such as coal, organic fuels, coke oven gas, woods. The technology development for syngas production should be improved to exceed beyond the conventional one. In this case, creative idea and extensive research are necessary. The extensive research and development efforts devoted to syngas production are documented in this report. The purpose of this report is to review the many alternative and promising syngas productions and syngas conversion processes and summarize the advantage of technique based on experimental and simulation.

Syngas production in high pressure condition is proposed in the Chapter 1. The use of a Chemical Compression Reactor (CCR) shows quite promising technique for syngas and hydrogen production with high conversion of methane. As currently there are wide industrial bases for manufacturing cheap CCRs on the base of internal combustion engine, this technique shows a great possibility to be applied in near future.

Another technique of syngas production from methane is based on noncatalytic methane reaction with metal oxides. This method could avoid the coke deposition and catalyst deactivation. The method is ideal for natural gas carrier with '*in situ*' hydrogen/electricity production. The microwave-assisted CO_2 reforming of CH_4 over carbon-based catalysts could also be the answer as it combines the catalytic and dielectric properties of carbonaceous materials with the advantages of microwave heating. Therefore, the process promises to be competitive, both from the point of view of conversion and energy consumption.

One of the abundant methane-based sources of synthesis gas production is coke oven gas, contains of 25% methane and 58% hydrogen. With world's annual output of coke is more than 500 million tons and there are more than forty-five billion cubic meters coke oven gas without any recovery and utilization action for every year, this energy source could be very promising in near future. Coke oven gas, however, could be converted easily into syngas with current technologies, such as: catalytic steam reforming, catalytic partial oxidation and autothermal reforming.

The use of coal as a transition feedstock for a hydrogen and synthetic fuels economy is likely to be accomplished via gasification. Coal gasification is a well established mature technology to produce a syngas (CO and H_2) capable of being compatible with biomass and wastes as feedstock, reducing both fossil fuels dependency and CO_2 emissions. The proposed technique, called Integrated Gasification Combined Cycle (IGCC) technology, has clear environmental advantages over traditional coal gasification plants, being considered the cleanest and most efficient method for the production of electricity. ELCOGAS power plant, located in Puertollano (Spain), is an example of this technology, participating in two projects for generation, capture and storage of CO_2 and H_2 via WGSR and for obtaining "green diesel" via Fisher-Troops.

The use the wood as an energy feedstock for syngas production is also promising. In order to have better conversion and yields, plasma technique has been introduced. It shows that plasma gasification as compared with autothermal gasification and pyrolysis have many advantages, such as higher process efficiency. Waste processing by the newest methods will essentially decrease the ecological load on the environment. The potential applications of plasma and photo technologies for carbon dioxide reforming of methane are discussed in Chapter 10. The plasma technology allows higher conversion and selectivity with lower input energy while photocatalysis consumes less energy than conventional methods by utilizing solar energy. With the aid of these technologies, new and innovative processes for sustainable development by mitigating greenhouse gases for synthesis gas production are possible.

The application of syngas-inclusion power generation processes has also been reviewed with very promising result (Chapter 9 and 11). The technology development to transform our abundant biomass resources into clean, affordable, and domestically-produced biofuels, biopower, and high-value bioproducts could result in economic development, energy supply options, and energy security. In this context, commercially available and near-commercial syngas conversion processes were evaluated on technological, environmental, and economic bases.

This report serves as a first step. Additional, more detailed analyses are required to identify promising, cost-effective synthesis technologies where syngas synthesis and syngas conversion could make an impact.

GLOSSARY

Adiabatic Index: The ratio of specific heats.

Autothermal Process: Gasification processes differ on autothermal and allothermal after the manner of energy supply. In autothermal process, thermal energy comes from combustion of a part of raw material for achievement of required temperature level, and in allothermal it is brought from the outside.

Autothermal Reforming (ATR): Methane reforming to hydrogen and carbon monoxide using steam, oxygen, and/or carbon dioxide.

Bio Fuels: The biofuels are chemical made fuels by conversion of biomass. The biomass is renewable organic matter such as agricultural crops and waste, wood and wood waste, animal waste, municipal and other wastes.

Borghi-Peters Diagram: Classical diagram for turbulent premixed flames

Brayton and Steam Cycle: The Brayton cycle is a thermodynamic cycle that describes the workings of the gas turbine engine, basis of the airbreathing jet engine and others. Steam (Rankine) cycle - describes a model of steam-operated heat engine.

Brunauer-Emmet-Teller (BET): The term BET refers to the BET theory that describes the physical adsorption of gas molecules on a solid surface and serves as a basis for the measurement of the specific surface area of a material.

Carbon Dioxide Oxidative Coupling of Methane (CO_2 OCM): Methane conversion to higher hydrocarbons (especially C_2 hydrocarbons) and hydrogen and carbon monoxide using carbon dioxide. The process follows oligomerization and coupling reactions.

Carbon dioxide Reforming of Methane (CORM): Methane reforming to hydrogen and carbon monoxide using carbon dioxide.

Carbon of Catalyst/Microwave Receptor (CC/MR): Carbon of a carbonaceous material that acts as catalyst and receptor of microwaves.

Carbonaceous Deposit from the Decomposition of Methane (CCH_4): Carbon resulting from the decomposition of methane.

Catalyst and Microwave Receptor (C/MR): Carbonaceous material that acts as catalyst and receptor of microwaves.

Chemical Compression Reactor: A heat engine of cyclic operation that uses the operating principle of an internal combustion engine in which a piston compresses reactants up to the temperature necessary for rapid occurrence of chemical conversion.

Chemical Looping Reforming (CLR): A cyclic process for syngas production.

Coke Oven Gases (COG): Mixture of gases released during the coke-making process. It consists of H_2 (~55 - 60%), CH_4 (~23 - 27%), CO (~5 - 8%) and N_2 (~3 - 5%), along with other light hydrocarbons, H_2S and NH_3 in small proportions.

Combustion Induced Vortex Breakdown (CIVB): Recirculation bubble motion induced by interaction between flow field and flame

Conduction Band (CB): A partially filled band or a band of vacant energy levels just higher in energy than a filled band or the upper energy band in semiconductor separated by the energy from the valance band in which electron can conduct electricity.

Dielectric barrier discharge plasma (DBD): In Dielectric Barrier Discharge, the generation of plasma undertakes by means of two metallic electrodes, in which an AC voltage is applied and an insulating material is attached between them. As a result of the discharge at normal temperature and atmospheric pressure, a space between the electrode and the insulating material is created, so-called plasma.

Dry Reforming: Methane reforming with carbon dioxide to syngas.

Equilibrium Plasma: The equilibrium plasmas also call thermal plasmas are characterized by charged particles in which temperature of heavy particles is very close to electrons, usually in range of thousands of Kelvin.

Fischer-Tropsch (FT): The Fischer-Tropsch process is a set of chemical reactions that convert a mixture of carbon monoxide and hydrogen into liquid hydrocarbons. It is a key component of gas-to-liquids technology (GTL).

Flashback (FB): Back propagation of the flame in the mixing section

Fossil Fuel: The fossil fuels are the hydrocarbons which are consumed without any chemical process. These fuels were formed by the microbial decomposition of organic matters like plants remains millions year ago under high temperature and pressure.

Free-Piston Ballistic Compressor: A heat device of pulse operation in which a piston compresses reactants up to the temperature necessary for rapid occurrence of chemical conversion in the single compression stage.

Gas to Liquids (GTL): Chemical conversion of natural gas into liquid hydrocarbons.

Gasification Process: Gasification is a process that converts organic or fossil based carbonaceous materials into carbon monoxide, hydrogen, carbon dioxide and methane. This is achieved by reacting the material at high temperatures, without combustion, with a controlled amount of oxygen and/or steam. The resulting gas mixture is called syngas (from synthesis gas or synthetic gas) or producer gas and is itself a fuel. Gasification is a method for extracting energy from many different types of organic materials.

Infrared (IR): The part of the invisible spectrum that is contiguous to the red end of the visible spectrum and that comprises electromagnetic radiation of wavelengths from 800 nm to 1 mm.

Internal Combustion Engine: A heat engine that converts chemical energy in a fuel into mechanical energy, usually made available on a rotating output shaft.

International Energy Agency (IEA): The International Energy Agency is an autonomous intergovernmental organization established in the framework of the Organisation for Economic Co-operation and Development (OECD) in 1974, in the wake of the 1973 oil crisis. The International Energy Agency works to ensure reliable, affordable and clean energy for its 28 member countries and beyond.

Integrated Gasification Combined Cycle (IGCC): Power plants combining gasification process with gas turbine based power generation.

Ionization Degree: For plasma to exist, ionization is necessary. The degree of ionization of a plasma is the proportion of atoms that have lost (or gained) electrons, and is controlled mostly by the temperature.

Lean Blow Out (LBO): Detachment of the flame from the anchoring point in lean conditions

Liquified Natural Gas (LNG): Cryogenic liquification of natural gas.

Low Heating Value (LHV): Value characterising the energy of fuel combustion. The quantity is determined by subtracting the heat of vaporization of the water vapor from the higher heating value. The higher heating value (HHV) is determined by bringing all the products of combustion back to the original pre-combustion temperature, and in particular condensing any vapor produced.

Low-grade Fuel: Fuel with low amount of LHV.

Mass Spectrometer (MS): Instrument for gas analysis by electron bombardment and detecting based on mass to charge ratio.

Metallurgical Reactions: Chemical reactions in the metallurgical industries.

Methane to Gasoline (MTG): Methane oligomerization to gasoline.

Non-equilibrium Plasma: Also called low temperature or cold plasma are characterized by charged species with a much higher kinetic energy than neutral species. In this plasma, temperature of electrons, ions and neutral atoms is the same.

OH Chemiluminescence: Natural chemiluminescence of OH radical

Oxidative Coupling of Methane (OCM): Methane conversion to higher hydrocarbons (especially C_2 hydrocarbons) and hydrogen and carbon monoxide using oxygen. The process follows oligomerization and coupling reactions.

Particle Image Velocimetry (PIV): Imaging laser technique based on mie-scattering of seeding particles.

Partial Oxidation of Methane (POM): Methane conversion to hydrogen and carbon monoxide using oxygen as an oxidant. The oxidation is carried out partially using low oxygen content.

Photocatalytic Reduction: The photocatalytic reduction is the process of reducing stable molecules by using photocatalyst in which charged particles are produced during photo irradiation.

Planar Laser Induced Fluorescence (PLIF): Imaging laser technique based on induced fluorescence of molecules used as dyes.

Preferential Diffusive Thermal (PDT): Interaction between diffusion of species and thermal diffusion in flames

Proportional Integral Derivative (PID): A proportional integral derivative controller (PID controller) is a generic control loop feedback mechanism (controller) that calculates an error value as the difference between a measured process variable and a desired set point, and attempts to minimize the error by adjusting the process control inputs.

Refuse Derived Fuel (RDF): A fuel produced by shredding and dehydrating solid waste.

Reverse Water Gas Shift (RWGS): Reaction of hydrogen product and carbon dioxide to water and carbon monoxide that accompanies CO_2 reforming of methane reaction. RWGS is slightly endothermic chemical reaction.

Scanning Electron Microscopy (SEM): Surface characterization technique that images a sample by scanning it with a high-energy beam of electrons in a raster scan pattern.

Semiconductor Catalyst: An element, that is the intermediate in electrical conductivity between the conductors and the insulator is called semiconductor. The semiconductor material can absorb photon energy and become excited and if used for chemical reaction then called semiconductor catalyst.

Steam Reforming of Methane (SRM): Methane reforming to hydrogen and carbon monoxide using steam.

Sulphur Passivated Reforming Process (SPARG): Process based in the selective poisoning with sulphur of a nickel catalyst that produces a synthesis gas with low H_2/CO ratio and minimizes the carbon deposits.

Superadiabatic Compression in the CCR Cylinder: The compression of gas during its flow from one vessel to another (with an increase in entropy), for example, in a divided cylinder of an ICE, that makes it possible to attain higher temperatures in comparison to adiabatic compression to the same maximum pressure or compression ratio.

Thermal Activator: A heat recuperator with a developed surface, for example, a set of metal plates or a porous metallic or ceramic body located in the auxiliary chamber of the divided cylinder.

Thermogravimeter (TG): Instrument for measuring of weight change versus temperature or time.

Turbulent Flame Speed: Speed of the flame in a given turbulent environment.

Turbulent Lean Premixed: Flames where fuel and oxidizer are premixed with an excess of the second component (lean), burning in a turbulent flow field.

Valance Band (VB): The outermost electron shell of atoms in semiconductor, in which the electrons are too tightly bound to the atom to carry electric current. The highest occupied energy level in a solid crystal semiconductor or insulator that is occupied by electrons also called filled band.

Vibrational Excitation: The excitation is increase in the internal energy of an atomic or molecular system caused by a collision with another particle of greater energy. For atoms in a discharge, excitation usually refers to increasing the energy level of a bound electron. In plasma, vibrational excitation is process in which electron impact takes place through the formation of intermediate state, which excites chemical conversion in the reaction with less input energy.

Volumetric Hourly Space Velocity (VHSV): Variable used to described a catalytic reactor and defined as the ratio between the feed flow rate, given in $L \cdot h^{-1}$, and the mass of catalyst, given in g.

Water Gas Shift (WGS): A slightly exothermic chemical reaction in which carbon monoxide reacts with water vapor to form carbon dioxide and hydrogen. WGS accompanies steam reforming of methane.

INDEX

A

acetaldehyde, 33, 248, 290
acetic acid, 33, 67, 68, 226, 271
acetone, 248
acid, 33, 38, 55, 67, 68, 127, 159, 211, 226, 246, 271
activated carbon, 137, 138, 139, 140, 141, 142, 143, 144, 145, 169, 248
activation energy, 2, 6, 7, 242
active site, 141, 165, 167, 169, 253
actuality, 77
adjustment, 123
adsorption, 38, 53, 159, 165, 242, 246, 248, 285, 351
advancement, 282
advancements, 158
adverse effects, 271, 284
aerospace, 345
agriculture, 182, 270
Air Force, 200
alcohols, ix, 163, 164, 175, 211, 212, 228
aldehydes, 163
algae, 244
algorithm, 328
alkenes, 163, 228
alternative energy, ix
aluminium, 227
aluminum oxide, 73
ambient air, 214, 235
ammonia, vii, ix, x, 33, 34, 67, 68, 69, 71, 92, 95, 101, 111, 159, 160, 214, 218, 219, 226, 232, 271
ammonium, 33, 271
amplitude, 129, 278, 313, 315
anaerobic digesters, 248
anchoring, 300, 321, 353
argon, 4, 40, 41, 43, 44, 159
aromatics, 33
atmosphere, xii, 32, 39, 42, 127, 129, 142, 180, 213, 214, 243, 245, 247, 249, 270, 272, 284
atmospheric pressure, 52, 89, 184, 187, 274, 276, 278, 279, 280, 281, 282, 336, 352
atoms, 100, 105, 169, 187, 197, 273, 280, 289, 353, 354
attachment, 273
automation, 127
automobiles, 37
avoidance, 229
awareness, 158

B

band gap, 285, 286, 287
barium, 37
barriers, 275
base, x, xii, 1, 3, 6, 32, 36, 37, 39, 41, 42, 122, 131, 143, 147, 160, 164, 165, 173, 214, 234, 240, 243, 265, 349
basicity, 165, 170, 171, 173
batteries, 210
behaviors, 326, 341
Beijing, 94, 96
bending, 336, 338, 339, 342
benefits, 124, 146, 159, 234, 241, 262, 264
benzene, 248
biodiesel, 90, 153, 163, 210
bioenergy, 182, 247, 270
biofuel, 182, 210
biogas, 125, 145, 146, 241, 245, 247, 248, 249, 250, 253, 260, 264
biological systems, 261
biomass, vii, x, xiii, xiv, 102, 122, 139, 147, 151, 152, 153, 154, 155, 156, 158, 160, 161, 163, 173, 182, 183, 187, 199, 203, 210, 214, 241, 244, 245, 247, 260, 261, 262, 265, 270, 273, 292, 299, 312, 318, 349, 350, 351
biopower, 350
biosphere, 180, 213

blends, 154, 173, 297
boilers, 260
bone, 199
Brazil, 182, 210
breakdown, 229, 277, 278, 301
burn, 203
butyl ether, 33, 226
by-products, 158, 159, 164

C

calcium, 43, 44, 165, 169
calibration, 40, 44
campaigns, 309, 327, 340
capillary, 40, 169
capital intensive, 229
carbon atoms, 169, 197
carbon dioxide, xii, xiv, 7, 13, 16, 17, 21, 23, 33, 34, 35, 67, 69, 76, 99, 100, 101, 102, 103, 108, 113, 118, 119, 121, 125, 157, 159, 160, 180, 183, 189, 191, 192, 194, 195, 201, 202, 234, 235, 236, 237, 239, 263, 269, 270, 272, 275, 276, 279, 284, 288, 289, 299, 350, 351, 352, 353, 354
carbon materials, 138, 146
carbon monoxide, vii, 7, 13, 16, 17, 21, 22, 23, 24, 25, 26, 33, 34, 39, 45, 58, 67, 69, 87, 115, 122, 156, 157, 159, 163, 165, 169, 180, 187, 189, 196, 299, 351, 352, 353, 354
carbon nanotubes, 169, 174, 175
carbonization, 72
case study, 256, 259
catalysis, viii, 91, 174, 175, 282
catalyst deactivation, 29, 141, 227, 253, 349
catalytic activity, 121, 125, 133, 137, 138, 139, 140, 141, 142, 161, 170
catalytic effect, 37
catalytic hydrogenation, 228
catalytic system, xii, 127, 163
category b, vii
C-C, 90, 227
CCR, 1, 2, 3, 4, 5, 6, 13, 14, 16, 17, 19, 20, 26, 349, 354
cell surface, 229
ceramic, 5, 35, 52, 127, 169, 234, 240, 354
ceramic materials, 169
challenges, ix, xii, 146, 211, 224, 270, 290
chemical industry, 3, 67, 68, 90
chemical reactions, 69, 70, 78, 87, 106, 187, 245, 251, 263, 270, 273, 274, 280, 352
chemical stability, 164
chemicals, ix, xiii, xiv, 100, 101, 153, 154, 156, 163, 226, 228, 270, 284, 288, 292
chemiluminescence, 300, 306, 307, 308, 313, 314, 315, 317, 318, 320, 340, 353
chemisorption, 165
China, 67, 68, 72, 91, 92, 93, 94, 95, 96, 153, 234
chlorine, 156, 160
classification, 203, 301, 303
clean energy, 352
clean technology, 38
cleaning, 82, 160, 181, 202, 203, 223, 245, 246, 282
climate, 158, 243, 270
climate change, 158
closure, 17, 18, 32
coal, vii, ix, x, xi, xiii, xiv, 3, 32, 90, 93, 94, 96, 147, 151, 152, 153, 154, 155, 156, 158, 159, 160, 161, 163, 173, 180, 183, 187, 199, 205, 210, 214, 226, 228, 241, 243, 245, 260, 263, 264, 265, 298, 312, 314, 318, 349
cobalt, xii, 33, 71, 163, 164, 165, 167, 169, 171, 172, 173, 228
coding, 311, 312, 330, 331, 332, 333, 335, 338
cogeneration, viii
coke, vii, viii, x, xi, xii, xiii, xiv, 29, 32, 33, 35, 36, 37, 39, 58, 67, 68, 69, 71, 78, 82, 83, 85, 87, 89, 90, 91, 92, 94, 96, 99, 100, 108, 110, 112, 115, 117, 119, 125, 137, 138, 140, 155, 158, 160, 173, 199, 271, 272, 275, 349, 352
coke formation, xii, 58, 82, 99, 108, 110, 112, 115, 119, 137
collaboration, 326
collisions, 70, 273, 274
combined effect, 325
combustibility, 21
combustion processes, 261
commercial, ix, 20, 77, 121, 125, 130, 131, 146, 159, 160, 161, 163, 173, 218, 219, 223, 224, 227, 237, 269, 271, 272, 350
community, 298
comparative analysis, 179, 184
compatibility, 43, 209, 233
compensation, 187, 189, 191, 200, 214
competition, 303, 342
competitiveness, 153, 223, 232, 233
complexity, 3, 192, 224, 253
composition, vii, x, xii, 7, 9, 13, 17, 18, 26, 34, 52, 58, 60, 61, 74, 99, 102, 103, 106, 108, 117, 119, 133, 136, 143, 154, 156, 157, 160, 171, 173, 179, 181, 182, 183, 184, 186, 187, 188, 189, 191, 192, 193, 194, 196, 197, 199, 201, 202, 203, 224, 225, 227, 228, 248, 249, 257, 258, 259, 299, 314, 330, 335, 336, 339, 342
compounds, xiii, 32, 67, 68, 71, 82, 100, 123, 125, 156, 158, 160, 161, 169, 173, 228, 248, 249, 303, 345

compression, 1, 2, 3, 4, 5, 6, 7, 8, 9, 11, 13, 16, 17, 18, 19, 20, 25, 26, 211, 214, 251, 260, 352, 354
computation, 61
computer simulations, 345
condensation, 29, 55, 169, 197, 224, 226, 245, 246
conditioning, 160, 264
conductance, 285
conduction, 144, 218, 285, 286
conductivity, 223, 243, 353
conductors, 353
conference, 95, 96
configuration, 80, 159, 220, 291, 306, 307, 334
conservation, 56, 252, 261, 326
construction, 74, 77, 153, 204, 228
consumption, xiii, 2, 3, 6, 8, 12, 23, 24, 25, 32, 35, 36, 69, 75, 82, 83, 89, 90, 93, 121, 123, 131, 132, 143, 144, 145, 146, 152, 179, 180, 181, 182, 193, 211, 212, 213, 214, 220, 221, 223, 229, 233, 249, 261, 264, 269, 302, 310, 321, 323, 326, 331, 336, 339, 341, 349
contact time, 136, 253
contamination, 161, 181
convergence, 302
conversion rate, 89, 271, 272
conversion reaction, 71
cooling, 14, 25, 159, 187, 201, 203, 216, 221, 245, 246, 306
copper, 122, 161
coproduction, 159
corona discharge, 276, 277, 278, 281, 283, 291
correlation(s), 109, 110, 165, 170, 297, 332, 335, 339, 340, 342, 346
corrosion, 214, 222, 278
cost, xii, 4, 9, 32, 35, 69, 74, 82, 125, 137, 142, 159, 181, 182, 187, 193, 195, 214, 218, 219, 220, 221, 222, 229, 230, 231, 232, 233, 234, 241, 247, 248, 265, 272, 284, 349, 350
counterbalance, 224
CPC, 93
cracks, 41
crises, 269, 283, 288, 289, 291
critical value, 25
criticism, 210
crops, 182, 260, 270, 351
crude oil, 29, 32
crystalline, 164
crystallization, 35
cultivation, 244
cycles, vii, 15, 17, 192, 241, 243, 244, 245, 260, 264
cycling, 213, 223, 240
Czech Republic, 206

D

damages, 301, 307, 341
danger, 184
data analysis, 326
data set, 301, 309, 336, 338, 339
database, xiv, 298, 309, 311, 326, 338, 340, 342
decay, 332
decomposition, 40, 43, 44, 70, 79, 87, 88, 108, 110, 112, 115, 118, 121, 123, 124, 127, 131, 133, 135, 139, 141, 142, 163, 181, 182, 187, 200, 351, 352
decomposition reactions, 70, 87
defects, 128, 287
deficiency, 196, 328
deficit, 183
deforestation, 243
degradation, 222, 223, 233, 237
degradation rate, 222, 223
dehydration, 213, 227
Denmark, 223, 235
Department of Energy, 235
deposition, viii, xi, xiii, 29, 35, 58, 74, 92, 121, 124, 125, 126, 137, 146, 158, 170, 271, 272, 349
deposits, 121, 123, 126, 133, 134, 135, 136, 140, 141, 142, 181, 354
derivatives, 122, 123, 154, 198, 249
desorption, 246
destruction, 191, 261, 265
detachment, 273, 300
detectable, 161, 329
detection, 328, 329, 340
developed countries, 181, 270
deviation, 40, 325
dew, 245, 246
diamonds, 323
diaphragm, 20
dielectric constant, 127
dielectrics, 127
diesel fuel, x, xiv, 4, 32, 33, 153, 228
differential equations, 38, 60, 252
diffraction, 165
diffusion, 56, 60, 80, 81, 84, 88, 169, 218, 242, 253, 331, 332, 353
diffusivities, 253
diffusivity, 30, 242, 299, 302, 303, 332, 338
digestion, 125, 211
dipoles, 127
discharges, 39, 68, 204, 277, 278
discontinuity, 198, 216
dispersion, 30, 35, 164, 165, 167, 169, 171
displacement, 21, 25, 169, 184, 192, 278, 302
dissociation, 7, 8, 37, 215, 218, 236, 273, 274, 275, 276, 279

distillation, xiv, 35, 214
distribution, 3, 78, 83, 84, 85, 89, 105, 128, 129, 132, 163, 165, 167, 168, 169, 170, 173, 174, 211, 253, 327
divergence, 186, 302
DME, 76, 211, 213, 226, 227
donors, 286
doppler, 314
drying, 127, 192, 198, 261, 265
durability, 219, 220, 222, 223, 227, 233, 238, 257
dyes, 327, 353

E

economic competitiveness, 153, 223, 232, 233
economic development, 350
economic evaluation, 92, 93
economic integration, 155
economics, 156, 159, 205
education, 93
efficiency of use, 179, 180
effluent, 74, 75, 151, 154, 157, 160, 173
effluents, 125, 151
electric current, 131, 354
electric field, 127, 129, 274, 276, 278
electrical conductivity, 223, 353
electricity, vii, ix, xiv, 29, 32, 125, 145, 146, 151, 152, 153, 158, 159, 180, 183, 184, 185, 186, 187, 188, 192, 194, 195, 198, 199, 203, 209, 210, 211, 212, 213, 214, 215, 216, 217, 220, 222, 224, 227, 229, 230, 231, 232, 233, 234, 244, 245, 247, 249, 251, 265, 298, 349, 350, 352
electrochemistry, 219
electrodes, 193, 202, 203, 237, 238, 276, 277, 278, 279, 280, 352
electrolysis, 125, 209, 211, 212, 213, 215, 216, 217, 218, 219, 220, 221, 222, 223, 224, 225, 226, 227, 229, 232, 233, 234, 235, 236, 237, 238, 251
electrolyte, 215, 218, 219, 220, 222, 224, 229, 231, 236
electromagnetic, 126, 127, 128, 129, 139, 273, 275, 352
electromagnetic fields, 128
electron(s), 41, 139, 140, 272, 273, 274, 275, 276, 277, 279, 281, 282, 285, 286, 291, 352, 353, 354
elongation, 311, 334
emission, xiii, 32, 35, 36, 37, 39, 41, 52, 75, 91, 187, 200, 210, 214, 243, 247, 270, 272, 281, 284, 301, 319, 320, 327
endothermic, xi, xii, 34, 35, 36, 58, 60, 69, 77, 78, 79, 82, 91, 108, 112, 114, 122, 123, 124, 133, 145, 157, 158, 187, 189, 191, 215, 216, 222, 226, 247, 249, 251, 271, 272, 275, 284, 288, 353
energy consumption, 3, 69, 75, 121, 123, 131, 132, 143, 144, 145, 146, 152, 179, 180, 181, 193, 214, 269, 349
energy density, 151, 152, 184, 211, 260, 274
energy efficiency, 112, 114, 152, 158, 183, 184, 186, 192, 193, 218, 225, 241, 244, 280
energy expenditure, 5
energy input, 6, 129, 187, 197, 217
energy security, 158, 350
energy supply, ix, 153, 181, 191, 350, 351
energy transfer, 153
engineering, vii, viii, 3, 29, 38, 92, 95, 181, 182, 183, 184, 185, 187, 192, 199, 204, 224, 253, 265, 272, 273, 291, 292, 298, 329
entropy, 4, 5, 6, 215, 261, 262, 265, 301, 354
environment, 67, 68, 69, 82, 94, 184, 202, 204, 261, 270, 272, 274, 283, 291, 350, 354
environmental contamination, 181
environmental effects, 270
environmental impact, vii, 159
environmental protection, 153
equality, 9, 17
equilibrium, 34, 39, 61, 68, 79, 82, 99, 100, 101, 102, 103, 104, 105, 106, 108, 109, 110, 112, 114, 116, 117, 118, 119, 133, 184, 187, 215, 224, 225, 226, 227, 242, 245, 247, 255, 258, 259, 262, 265, 274, 278, 352, 353
erosion, 193
etching, 278
ethanol, 68, 102, 235
ethylene, 33, 99, 109, 110, 119, 154, 214
Europe, 181, 210, 218, 345
European Commission, 153
European Union, 210, 234
evaporation, 5, 42, 44, 181, 198, 216, 217, 221
evidence, 236, 317, 336, 340
evolution, 10, 72, 133, 161, 223
excitation, 273, 274, 275, 278, 285, 286, 287, 288, 354
expenditures, 5
expertise, 29, 292
exploitation, 151, 173
exporter, 153
extinction, 300, 313, 315, 316, 317, 318
extraction, 182, 214
extracts, 246

F

feedstock, viii, x, xi, xiii, xiv, 71, 76, 122, 123, 124, 125, 146, 151, 152, 179, 181, 182, 183, 184, 185, 186, 187, 189, 192, 193, 194, 195, 198, 200, 202,

204, 212, 213, 216, 217, 227, 229, 271, 299, 311, 349, 350
fermentation, 214
fertilizers, 218, 219
fiber, 129, 131, 307, 320, 340
filament, 41
filters, 310, 313
finite element method, 38, 57, 253
Fischer-Tropsch process, xi, 101, 158, 352
Fischer-Tropsch synthesis, ix, 71, 154, 157, 158, 162, 163, 165
fixed bed reactors, 35
flame, 70, 72, 78, 81, 83, 95, 260, 297, 298, 299, 300, 301, 302, 303, 304, 306, 307, 308, 309, 310, 311, 313, 315, 316, 318, 320, 321, 322, 323, 324, 325, 326, 327, 329, 330, 331, 332, 333, 334, 335, 337, 338, 339, 340, 341, 342, 343, 344, 345, 346, 352, 353, 354
flame propagation, 333, 340, 344
flammability limit, 299
flashbacks, 320, 321, 322
flexibility, viii, xv, 127, 223, 234, 298, 301
flow field, 78, 89, 302, 303, 309, 315, 332, 334, 340, 341, 352, 354
fluctuations, 300, 315, 317, 325
flue gas, 32, 75, 214, 247, 249, 261, 262
fluid, 38, 228, 304
fluidized bed, 33, 91, 200, 205, 245
fluorescence, 300, 309, 310, 321, 327, 328, 340, 353
food, 127, 210, 292
food processing industry, 292
formaldehyde, 33, 87, 226
formation, xi, xii, 7, 58, 70, 74, 77, 82, 99, 103, 108, 109, 110, 112, 115, 118, 119, 133, 136, 137, 142, 163, 164, 169, 181, 222, 227, 258, 275, 276, 286, 319, 320, 354
formula, 15, 163, 186, 188, 189, 253, 327
fractal dimension, 298
free energy, 61, 99, 100, 102, 103, 104, 105, 106, 107, 108, 119, 215
free radicals, 272, 274, 276
freedom, 73, 261
FTIR, 40, 52, 53, 54
fuel cell, 3, 37, 52, 60, 70, 90, 102, 111, 153, 211, 219, 222, 234, 236, 240, 241, 242, 246, 247, 249, 251, 261, 265
fuel consumption, 233, 323, 336, 339
fusion, 200, 261

G

geometry, 15, 129, 255, 256, 276, 317, 325, 330, 334, 335
Germany, 206, 209, 210, 218, 219, 223, 228, 232, 233, 234, 346
global warming, xii, 32, 37, 151, 152, 269, 284, 288, 291
glow discharge, 276, 278, 280, 281, 282, 283, 291
glycerol, 292
glycol, 33
grain boundaries, 222
granules, 203
graph, 85, 86, 109, 110, 111, 112, 113, 114, 115, 116, 117
graphite, 183, 203, 215
grasses, 260
greenhouse, x, xii, xiii, 1, 30, 32, 33, 35, 36, 37, 38, 39, 52, 124, 146, 153, 158, 180, 181, 210, 243, 269, 270, 271, 272, 284, 288, 289, 291, 292, 350
greenhouse gas(es), xii, xiii, 1, 30, 32, 33, 35, 36, 37, 38, 39, 52, 124, 146, 153, 180, 181, 210, 243, 269, 270, 271, 272, 284, 288, 289, 291, 292, 350
growth, x, 165, 169, 180, 183, 187, 189, 193, 195, 197, 198, 199, 204, 228, 247, 270, 302, 323, 336

H

harvesting, 270
hazardous substances, 181
health, 270
heat capacity, 17, 99
heat loss, 8, 194
heat release, 3, 8, 71, 73, 78, 84, 89, 114, 139, 228, 298, 300, 313, 315, 317, 331, 341
heat transfer, 2, 3, 5, 6, 8, 9, 15, 74, 127, 199
heavy oil, 273
heavy particle, 274, 276, 279, 282, 352
height, 30, 40, 41, 52, 53, 54, 78
history, 236, 340
homogeneity, 25
hot spots, 121, 128, 129, 130, 133
human activity, 32
humidity, 41, 53, 156, 187, 196, 214
hunting, 284
hybrid, 119
hydrocarbons, viii, ix, x, xii, 3, 32, 33, 35, 68, 94, 101, 102, 106, 110, 123, 131, 154, 157, 158, 159, 163, 164, 167, 168, 169, 170, 172, 175, 180, 183, 211, 226, 228, 235, 244, 245, 248, 271, 273, 279, 282, 298, 303, 320, 351, 352, 353
hydrocracking, 33
hydrogen atoms, 289
hydrogenation, 33, 161, 165, 170, 175, 226, 228, 239
hydrolysis, 127
hydroxide, 214
hypothesis, 322

I

ICE, 2, 3, 4, 5, 6, 13, 19, 27, 354
ideal, vii, 29, 81, 90, 104, 157, 158, 262, 328, 335, 349
ignition source, 4
images, 55, 135, 297, 300, 306, 308, 309, 310, 311, 312, 318, 319, 321, 326, 327, 333, 334, 335, 340, 353
impregnation, 164, 170, 173
impurities, 128, 181, 183, 188, 189, 191, 192, 193, 222, 247, 253, 260
incomplete combustion, 313, 317, 341
India, 206
Indonesia, 99, 119
induction period, 85, 88
industrial experience, 292
industrialization, 243, 270
industry, vii, xiv, 3, 67, 68, 71, 75, 90, 94, 125, 151, 161, 182, 206, 292, 298, 353
inefficiency, 322
infrastructure, xiii, 3, 211
insertion, 225
insulation, 217, 221, 261
insulators, 127, 139
Integrated Gasification Combined Cycle (IGCC), viii, 151, 155, 298, 350, 352
integration, xii, 145, 153, 155, 223, 234, 241, 244, 249, 261, 264, 265
interaction effect, 332
interface, 36, 37, 49, 85, 222, 285
interference, 127, 313
International Energy Agency (IEA), 147, 210, 352
inventions, 204
investment, 69, 72, 75, 155, 211, 220
ionization, 30, 41, 42, 129, 273, 274, 276, 279, 281, 353
ions, 272, 273, 274, 275, 276, 277, 291, 353
Iran, 29, 177
iron, ix, 32, 33, 38, 122, 161, 163, 164, 167, 228, 263, 264
irradiation, 353
isolation, 227
issues, 26, 92, 93, 94, 132, 153, 159, 219, 233, 298, 301, 344

K

kerosene, x, 153, 233
kinetic equations, 6, 8
kinetic model, 346
kinetic parameters, 39, 40, 48, 51
kinetic studies, 40
kinetics, 7, 34, 36, 37, 223, 253, 254, 285, 344
knots, 182
KOH, 219

L

lakes, 213
laminar, 15, 297, 299, 302, 303, 318, 324, 326, 331, 332, 336, 338, 341, 342, 343
lasers, 26
laws, 340
LCAT, 242, 255
lead, 15, 34, 36, 74, 79, 82, 121, 124, 142, 146, 184, 211, 223, 224, 302, 318, 335
leakage, 75, 218
leaks, 131
legend, 338
lens, 310
lifetime, xiii, 125, 220, 222, 229, 230, 233, 234, 285
light, 101, 125, 141, 169, 210, 214, 261, 274, 285, 286, 291, 310, 312, 352
linear dependence, 198, 334
liquefied natural gas, 32
liquid fuels, x, xiv, 90, 101, 153, 154, 162, 163, 187, 188, 189, 195, 204, 209, 210, 211, 212, 213, 232, 245, 270, 292
liquid phase, 67, 68
liquids, 29, 32, 154, 204, 210, 352
low temperatures, 34, 126, 170, 181, 251, 252, 262
LPG, 20, 211, 227
Luo, 240

M

magnesium, 165
magnetic field, 129, 203, 207
magnitude, 194, 197, 199, 256, 299, 303, 323
majority, 163, 169, 174, 261, 274
Malaysia, 119, 153, 228, 269, 292
man, 182
management, 127, 219, 220, 233
manufacturing, x, 1, 349
marketing, 229
mass, 2, 7, 8, 14, 15, 16, 29, 30, 36, 39, 40, 47, 48, 50, 54, 55, 56, 60, 90, 105, 145, 146, 179, 182, 184, 185, 187, 192, 194, 195, 197, 198, 199, 202, 219, 241, 242, 249, 251, 252, 253, 257, 259, 264, 265, 272, 274, 285, 291, 297, 302, 310, 321, 326, 327, 332, 334, 346, 353, 354
mass spectrometry, 40
materials, vii, xi, xiii, xiv, 3, 40, 67, 68, 72, 121, 127, 128, 129, 138, 139, 141, 142, 146, 153, 160, 161,

163, 169, 193, 202, 210, 221, 229, 233, 234, 240, 272, 278, 285, 349, 352
matrix, 242, 311
measurements, 20, 21, 129, 131, 170, 298, 306, 309, 312, 313, 320, 325, 334, 336, 340
mechanical properties, 181
media, 95, 131, 189, 201, 204
melt, 200, 203
membranes, 243, 246, 249
Mendeleev, 180
mercury, 40, 159
metal oxides, 29, 60, 62, 164, 349
metallurgy, 127, 282
metals, 37, 125, 137, 138, 139, 161, 164, 176, 200, 285, 291
meter, 20, 131, 214
methane reforming, vii, ix, 4, 53, 58, 95, 103, 123, 124, 125, 143, 145, 157, 242, 247, 250, 264
methanol, vii, ix, x, xi, xiv, 29, 32, 33, 35, 38, 55, 60, 62, 67, 68, 71, 81, 90, 91, 92, 93, 96, 101, 102, 122, 125, 153, 158, 189, 213, 225, 226, 227, 235, 239, 240, 248, 271, 290
methodology, 297, 302, 311, 326, 327, 328, 336, 340
methyl methacrylate, 226
methyl tertiary, 33
microwave heating, 121, 127, 128, 129, 133, 135, 139, 141, 143, 145, 146, 349
microwave radiation, 127, 139, 140, 143, 144
microwaves, 127, 128, 129, 131, 136, 139, 143, 144, 351
migration, 223, 288
military, 222
miniaturization, 257
Ministry of Education, 93
missions, 121, 151, 156, 241, 298
mixing, 42, 78, 79, 80, 83, 91, 154, 157, 300, 301, 304, 306, 307, 311, 320, 321, 325, 334, 352
MMA, 226
modelling, 229
models, 38, 48, 51, 143
modifications, xiii, 154, 328
modulus, 30, 57, 219, 227
moisture, 20, 179, 181, 184, 187, 190, 192, 194, 196, 197, 198, 199, 202, 214, 260
molar ratios, 55, 60, 74, 158, 161, 162, 169, 226
molar volume, 31
mole, xii, 6, 9, 12, 31, 35, 39, 42, 43, 45, 52, 53, 54, 55, 57, 58, 87, 104, 105, 106, 107, 109, 110, 111, 112, 113, 114, 115, 116, 117, 183, 272, 281
molecular mass, 242
molecular weight, 30, 158, 167, 169, 170
molecules, 54, 70, 83, 261, 272, 273, 274, 275, 276, 277, 279, 280, 284, 285, 286, 351, 353
molybdenum, 71
momentum, x, 252, 253
morphology, 291, 311, 345
Moscow, 26, 27, 28, 205, 207
MSW, 182, 186
MTBE, 226
municipal solid waste, xiv, 183, 270

N

nanofibers, 135, 174, 175
nanostructures, 174, 175
National Aeronautics and Space Administration, 218
natural gas, vii, viii, ix, x, xi, xiii, xiv, 1, 3, 29, 32, 33, 34, 35, 36, 37, 69, 70, 71, 77, 95, 100, 101, 102, 121, 122, 123, 124, 125, 145, 146, 153, 154, 156, 157, 158, 161, 163, 180, 210, 211, 214, 226, 228, 243, 245, 247, 272, 298, 299, 300, 301, 302, 312, 314, 333, 341, 342, 346, 349, 352, 353
Nd, 310
negative consequences, 340
Netherlands, viii, 90, 159, 174, 175, 205, 345
neutral, 214, 221, 235, 273, 274, 276, 280, 353
New Zealand, 227
NH2, 7
nickel, xi, xii, 29, 35, 36, 62, 70, 71, 73, 125, 140, 141, 163, 219, 354
Nigeria, 228
nitrogen, 7, 40, 52, 53, 71, 159, 183, 184, 185, 192, 193, 196, 247, 248, 249, 299, 331
noble metals, 125, 161, 164, 291
non-polar, 54
Norway, 218, 223
NPC, 206
NREL, xv, 235
nucleation, 38, 125
numerical analysis, 326

O

oceans, 270
octane, 69
OECD, 234, 352
OH, 7, 86, 87, 88, 219, 248, 297, 300, 305, 306, 308, 309, 310, 311, 312, 313, 314, 315, 318, 319, 321, 326, 327, 333, 335, 340, 353
oil, vii, xi, xiii, 3, 29, 32, 122, 153, 154, 158, 169, 180, 182, 210, 218, 228, 243, 245, 273, 298, 306, 312, 318, 352
olefins, vii, 33, 163, 228
oligomerization, 33, 351, 353
operating range, 156, 160
optical fiber, 307, 320, 340

optimization, 93, 152, 154, 161, 163, 173, 179, 183, 292, 298
organic chemicals, 226
organic compounds, 158, 249
organic matter, 351, 352
oscillation, 313, 317
osmosis, 214, 235
oven gas, vii, 67, 68, 69, 71, 78, 79, 80, 82, 83, 85, 87, 89, 90, 91, 92, 94, 96, 125, 349
ox, 67, 68, 123, 140, 163, 164, 170, 228
oxalate, 43, 44
oxidation, xi, xii, xiii, 1, 3, 4, 6, 9, 13, 14, 17, 18, 21, 26, 29, 30, 34, 35, 36, 38, 61, 67, 68, 70, 71, 72, 74, 76, 77, 78, 81, 82, 83, 84, 85, 87, 88, 89, 91, 94, 95, 96, 99, 101, 102, 103, 110, 117, 119, 123, 126, 131, 143, 157, 158, 175, 192, 229, 234, 241, 245, 246, 249, 261, 263, 285, 287, 316, 318, 320, 346, 349, 353
oxygen consumption, 82, 83, 89, 261

P

palladium, 34, 35
parallel, 73, 75, 80, 218, 278, 340
partial differential equations, 60
partition, 1, 4, 5, 6
patents, 146, 174, 204
pathways, 210, 228, 232, 320, 325
PCT, 239
peat, 182, 186, 205
peer review, 234
penalties, 159
permeability, 35, 243
permission, 126, 130, 135, 137, 139
permit, 226
petroleum, ix, x, xi, xiii, 125, 152, 153, 155, 157, 199, 210, 211, 232, 269, 349
Petroleum, xiii, 27, 63, 153, 209, 269
phenomenology, 317, 332, 346
photocatalysis, xiii, 269, 274, 284, 285, 289, 290, 291, 292, 350
photons, viii, 281, 285
photosynthesis, 214, 243, 247
physical chemistry, 282
physical properties, 171, 299, 332
physicochemical properties, 297
physics, 26, 204, 205, 234
plants, x, xi, xii, xiii, xiv, 29, 32, 35, 37, 38, 39, 51, 55, 60, 62, 68, 73, 75, 122, 125, 145, 151, 153, 158, 159, 181, 182, 210, 214, 218, 219, 228, 232, 236, 243, 245, 247, 260, 261, 262, 265, 270, 298, 311, 350, 352
poison, 71, 161
polar, 54
polarity, 203
polarization, 127, 128, 139, 236
policy, 94
pollutants, x, 37, 159
pollution, 1, 29, 67, 69, 82
polymers, 127, 215, 219, 278
polymerization, 163
population, 100, 105, 180, 182, 206, 270
porosity, 30, 40, 53, 56, 164, 243
positive correlation, 165, 170
potential benefits, 159
power generation, vii, 90, 158, 180, 184, 187, 204, 206, 210, 227, 236, 241, 243, 244, 247, 263, 264, 297, 298, 350, 352
power plants, x, 32, 125, 158, 159, 214, 245, 260, 261, 265, 298, 311
precipitation, 181
preparation, 38, 89, 172, 181, 282
principles, 115, 261, 272
probability, 152, 165, 169, 273, 316
process control, 195, 353
process gas, 251
production technology, 95, 96
project, 93, 174, 218, 219, 224, 226, 234
promoter, 165, 167, 169, 172
propagation, 300, 301, 302, 303, 321, 322, 333, 336, 340, 341, 342, 344, 352
propane, 70, 156
proportionality, 170, 340
propylene, 154
protection, 72, 153, 184
prototype, 52
pulp, 155
purification, ix, 82, 92, 144, 145, 154, 159, 204, 214, 227, 249
purity, 40, 53, 159, 213, 220, 241
pyrolysis, xiv, 82, 83, 90, 102, 179, 189, 191, 192, 193, 196, 200, 260, 275, 350

Q

quality of life, 270
quartz, 130, 131, 278, 279, 290, 306, 310

R

radar, 127
radial distance, 30, 31
radiation, 126, 127, 128, 130, 131, 136, 139, 140, 143, 144, 217, 261, 279, 285, 291, 315, 344, 352
radical reactions, 81, 275
radicals, 272, 274, 276, 277, 289, 303, 318, 320, 327

radicle, 289
radio, 127
radius, 31, 50, 51, 155, 253, 276
raw materials, vii, xi, 3, 40, 67, 68, 72, 153, 160, 163, 193, 202
reactants, vii, 1, 3, 6, 13, 14, 15, 16, 20, 26, 30, 80, 82, 83, 103, 106, 118, 257, 258, 259, 261, 279, 302, 331, 332, 351, 352
reaction mechanism, 87, 88, 227, 272, 275, 324, 346
reaction order, 6, 243
reaction rate, 22, 30, 49, 57, 60, 77, 78, 79, 82, 85, 103, 157, 164, 183, 224, 227, 242, 251, 257, 258, 272, 291
reaction rate constants, 49
reaction temperature, 4, 70, 82, 99, 109, 110, 119, 161, 162, 167, 168, 169, 173, 258
reaction time, 80
reaction zone, 52, 57, 79, 80, 303, 304, 306, 311, 320, 337, 340, 345
reactivity, 39, 121, 133, 140, 141, 253, 300, 303, 324
reagents, 158, 169, 204
reality, 153
recombination, 273, 285, 286, 287, 288, 291, 320
recommendations, 179
recovery, 1, 5, 6, 9, 10, 26, 31, 42, 67, 90, 229, 244, 260, 261, 349
recycling, 90, 214, 227, 229, 233, 236, 270, 272
reformers, 52, 58, 73, 95
reforms, 71, 83, 91
regenerate, 264
regeneration, 1, 17, 18, 38, 121, 133, 141, 211, 218, 236
regression, 337
regression line, 337
reliability, 159
renewable energy, 152, 153, 182, 204, 212, 223, 224, 237, 239, 244, 260, 269
renewable fuel, 245, 271, 272, 288, 291
reparation, 38
requirements, 71, 72, 84, 92, 146, 160, 299
researchers, ix, 272, 276, 279, 282
reserves, 32, 153, 158, 182, 349
residuals, 202
residues, xiii, 139, 260
resistance, 37, 38, 52, 125, 164, 199, 221, 231
resolution, 310, 313
resource utilization, 239
resources, vii, ix, 67, 68, 90, 91, 153, 180, 181, 182, 186, 203, 210, 214, 237, 269, 350
response, 223, 238, 269, 307, 313, 341
response time, 223
restoration, 269
reverse osmosis, 214, 235
reverse reactions, 7
risk, 270, 299, 328, 329, 331
room temperature, 53, 239, 288
root, 328
rotations, 12
routes, vii, 103, 209, 212, 213, 214, 234, 241, 244, 246, 251, 264, 349
Royal Society, 235, 236
rubber, 127, 182, 206
rubbers, 182
rules, 331
Russia, 1, 26, 27, 179, 182
ruthenium, 73, 163

S

safety, 20, 35, 72, 127, 131, 224, 257, 301
scaling, 144, 145, 146, 197, 214
scarcity, 325
scattering, 325, 353
science, 282
scientific papers, 174, 175
scientific publications, 146, 147
second generation, 76, 228
security, 153, 158, 298, 350
seeding, 315, 327, 353
segregation, 237
selectivity, xiii, 101, 123, 126, 128, 138, 146, 161, 164, 165, 167, 168, 169, 170, 173, 225, 269, 272, 277, 279, 280, 282, 283, 284, 290, 291, 350
semiconductor, xiii, 272, 274, 285, 286, 287, 288, 290, 291, 292, 352, 353, 354
sensing, 313
sensitivity, 30, 34, 40, 41, 42, 229, 230, 334
sensors, 129
sewage, 181, 248, 249
shape, 30, 78, 133, 276, 313, 334, 335
shear, 334
showing, 170, 324
signs, 12
silica, 169
silicon, 169, 172, 222
simulation, 24, 29, 35, 57, 60, 83, 95, 143, 229, 241, 256, 257, 265, 342, 345, 349
simulations, 253, 255, 256, 259, 325, 326, 345, 346
sintering, 141, 142
SiO2, 125, 138, 169, 290
slag, 183, 200, 201, 203, 261
sludge, 185, 249
sodium, 170, 211
software, 103
solid oxide fuel cells, 102, 222, 234, 240, 246
solid phase, 35, 57, 59, 245

solid waste, xiv, 158, 183, 270, 353
solidification, 55
solution, 38, 40, 57, 72, 99, 100, 104, 105, 127, 157, 219, 221, 226, 261
solvents, 214
sorption, 259
South Africa, 153, 158, 228
Soviet Union, 92
Spain, xiv, 95, 121, 146, 151, 158, 161, 174, 175, 344, 350
species, 40, 84, 85, 89, 99, 100, 103, 104, 105, 161, 164, 216, 242, 243, 245, 256, 258, 259, 272, 273, 274, 277, 282, 285, 286, 289, 313, 353
specific heat, 2, 8, 15, 242, 351
specific surface, 53, 351
specifications, 160
sponge, 33
spot market, 210, 232
stability, viii, xiii, 82, 125, 138, 142, 164, 210, 229, 234, 240, 272, 275, 282, 315, 318
stabilization, 21, 40, 192, 300, 306
starch, 210
steel, 38, 52, 68, 214, 276, 306
stoichiometry, 188
storage, xiv, 37, 62, 124, 147, 151, 152, 159, 175, 193, 210, 214, 224, 234, 235, 243, 260, 350
stress, 183
stroke, 3, 5, 8, 9, 13, 14, 15, 16, 17, 20
strontium, 37
structural changes, 38, 60
structure, x, 164, 169, 303, 344, 345
style, 269
substitution, 143, 223, 259
substrate, 258, 290
substrates, 243, 250, 255, 256, 257, 258, 261
subtraction, 306
sulfur, xiv, 32, 33, 71, 72, 82, 92, 159, 160, 161, 173, 211, 222, 245, 248, 249, 264
sulphur, xiii, xiv, 82, 123, 125, 153, 185, 354
summer program, 239
Sun, 64, 147, 177, 238
supervision, 292
surface area, 2, 8, 53, 80, 122, 169, 229, 272, 302, 326, 351
surface chemistry, 130
surface properties, 139
surplus, 211, 260
sustainability, 153, 210, 234
sustainable development, xiii, 269, 291, 350
sustainable energy, vii, ix, 236
Sweden, 175
Switzerland, 237, 297, 343, 345
symmetry, 253, 313, 334
synergistic effect, 227

T

tanks, 211
tar, 153, 261, 299
target, 3, 222, 243, 273
tariff, 232
taxes, 232
techniques, x, xiii, 1, 4, 38, 244, 246, 249, 257, 259, 261, 272
technologies, vii, viii, x, 1, 27, 77, 129, 153, 154, 158, 186, 187, 203, 204, 205, 209, 210, 212, 214, 215, 233, 236, 243, 247, 249, 260, 263, 264, 269, 272, 282, 291, 350
temperature dependence, 8, 10, 216
testing, 219, 222
thermal decomposition, 43
thermal energy, 6, 77, 127, 143, 180, 181, 183, 187, 188, 192, 194, 196, 197, 198, 215, 216, 217, 261, 351
thermodynamic analysis, vii, viii, 102
thermodynamic cycle, 9, 351
thermodynamic equilibrium, 82, 101, 102, 103, 133, 242, 255, 258, 259, 274
thermodynamics, vii, viii, 61, 241, 264
thermogravimetry, 30, 36, 37, 39
threats, 269, 270
tides, 270
time series, 313
time-frame, 270
titania, 169, 172
titanium, 222
toluene, 248
topology, 298, 326
total energy, 215, 216, 217, 270
total product, 214
trade, 220
trade-off, 220
trajectory, 261
transesterification, 163
transformation, 74, 82, 85, 88, 127, 158, 187, 314
transformations, 5, 8
transition temperature, 74, 82
transport, vii, 3, 32, 37, 151, 152, 153, 182, 205, 209, 210, 212, 214, 232, 233, 234, 235, 257, 265, 302, 304, 344, 345
transportation, ix, x, xiii, 29, 32, 33, 124, 156, 181, 211, 212, 224, 239, 248, 260, 269
treatment, 91, 184, 186, 187, 189, 204, 260, 261, 265, 280, 291
trial, 199
triggers, 272

tungsten, 40
turbulence, 205, 297, 301, 303, 304, 309, 315, 324, 330, 333, 334, 336, 337, 341, 342, 343, 345, 346
turbulent flows, 325
turnover, 339, 342
two-dimensional space, 303

U

uniform, 25, 83, 129, 132, 282, 291, 329
Union Carbide, 160
United Kingdom (UK), 95, 147, 206, 235
universal gas constant, 2, 7, 100, 105, 242
urea, 33
United States (US), 20, 95, 146, 182, 183, 205, 206, 210, 218, 239, 240, 343, 346
USSR, 27
UV, 310

V

vacuum, 20, 40
valuation, 346
valve, 5, 13, 19, 20, 307, 321
vanadium, 211
vapor, 7, 13, 17, 36, 42, 52, 55, 70, 71, 82, 83, 87, 263, 353, 354
variables, 21, 102, 325
variations, ix, 323, 324, 325, 330
vector, 153
vegetable oil, 210
vehicles, ix, 153, 159, 213
velocity, 2, 8, 31, 70, 83, 121, 133, 136, 145, 146, 155, 161, 163, 169, 242, 253, 297, 300, 301, 302, 303, 311, 314, 315, 321, 322, 323, 325, 326, 327, 330, 331, 333, 334, 335, 341, 343, 345, 346, 354
vessels, 4, 211
VHSV, 122, 136, 140, 141, 142, 354
viscosity, 243, 343
volatile organic compounds, 249
volatility, 42
vulcanization, 127

W

wall temperature, 8
waste, x, xiii, xiv, 20, 67, 68, 69, 122, 123, 125, 127, 158, 180, 181, 182, 184, 185, 186, 187, 189, 191, 192, 193, 194, 200, 201, 202, 203, 204, 205, 210, 222, 270, 351, 353
water evaporation, 216, 217, 221
water purification, 204, 214
water resources, 214
water vapor, 7, 13, 17, 42, 70, 82, 83, 87, 263, 353, 354
wavelengths, 352
wealth, 270
web, 240, 265
weight loss, 42
wind power, 210, 231, 232
wood, 127, 155, 179, 182, 184, 186, 187, 189, 190, 191, 192, 193, 194, 195, 196, 197, 198, 199, 201, 202, 203, 350, 351
worldwide, ix, 159, 210, 218, 219, 270

X

X-ray diffraction (XRD), 52, 54, 55, 152, 165, 170, 171

Z

zeolites, 125, 175
zinc, 29, 31, 36, 37, 38, 39, 40, 42, 44, 51, 52, 55, 60, 62, 71, 160, 211
zirconia, 215, 218, 219, 237
ZnO, ix, 29, 31, 36, 37, 38, 39, 41, 42, 43, 44, 45, 46, 47, 48, 49, 50, 51, 52, 53, 58, 60, 61, 62, 125, 160, 161, 226, 239, 251